CLIMATE VARIABILITY AND ECOSYSTEM RESPONSE AT LONG-TERM ECOLOGICAL RESEARCH SITES

CLIMATE VARIABILITY AND ECOSYSTEM RESPONSE AT LONG-TERM ECOLOGICAL RESEARCH SITES

Edited by

David Greenland

Douglas G. Goodin

Raymond C. Smith

OXFORD

UNIVERSITY PRESS

2003

OXFORD
UNIVERSITY PRESS

Oxford   New York
Auckland   Bangkok   Buenos Aires   Cape Town   Chennai
Dar es Salaam   Delhi   Hong Kong   Istanbul   Karachi   Kolkata
Kuala Lumpur   Madrid   Melbourne   Mexico City   Mumbai   Nairobi
São Paulo   Shanghai   Taipei   Tokyo   Toronto

Library of Congress Cataloging-in-Publication Data
Climate variability and ecosystem response at long-term ecological research sites /
edited by David Greenland, Douglas G. Goodin, Raymond C. Smith.
p.   cm. — (Long-Term Ecological Research Network series)
Includes bibliographical references.
ISBN 0-19-515059-7
1. Bioclimatology.   2. Ecology—Research.   3. Climatic changes.
I. Greenland, David, 1940–   II. Goodin, Douglas G. (Douglas Gard)
III. Smith, R. C. (Raymond Calvin), 1934–   IV. Series.
QH543 .C578 2003
577.2'2—dc21   2002033393

9 8 7 6 5 4 3 2 1

Printed in the United States of America
on recycled, acid-free paper

Preface

Global climate change is a central issue facing the world today. The topic has received intense national and international attention as exemplified by the continuing series of books produced by the Intergovernmental Panel on Climate Change. The issue of potential global warming is constantly addressed by the popular news media. Long-Term Ecological Research (LTER) sites can provide unique perspectives on this topic because of their large legacy of past ecosystem research and observations and their ability to act as a network and provide intersite comparisons. Furthermore, well over half the LTER sites have a climate investigation component as one of their main working hypotheses, and all the sites study the interaction of climate and ecosystems to some degree. The question-oriented organization adopted in this volume and some of the results described in it will provide an important stimulus to future research on the topic.

Climate variability and ecosystem response has been one of the ongoing areas of interest in the LTER since the inception of the program in 1980. The theme has been part of the fifth core area of research in the LTER program, namely, the study of patterns and frequency of disturbances, and the theme also has relevance to several of the other core areas of research. Climate Variability and Ecosystem Response was the title of a workshop held in 1988 that gave rise to a monograph written by the LTER Climate Committee in 1990 (Greenland and Swift 1990, 1991). A second monograph entitled "El Niño and Long-Term Ecological Research Sites" (Greenland 1994) arose out of another workshop held at the 1993 LTER All Scientists Meeting at Estes Park, Colorado. Climate Variability and Ecosystem Response was again the title of the Fall 1997 Long-Term Ecological Research (LTER) Coordinating Meeting in Santa Barbara. Some of the abstracts from the papers at this meeting will be posted at the LTER Climate Committee web site http://

intranet.lternet.edu/committees/climate/. Ongoing interest in this topic led to the planning of three workshops on the same topic for the August 2000 All Scientists Meeting LTER at Snowbird, Utah. The Snowbird workshops dealt with the overall theme as it applied to the quintennial, decadal, and century to millennial timescales. A series of questions were employed to focus the presentations and discussion on the topic. Material from these meetings and workshops together with selected additional material form the basis of the content of this book.

The compilation of work such as this is the result of a very large amount of time, energy, and resources provided by numerous people and organizations. We thank all those who helped run the various workshops over the years and all those who contributed papers and presentations to them. We especially thank the successive leaders of the LTER program, former leader Dr. Jerry Franklin, University of Washington, and current leader Dr. Jim Gosz (Chair) and Dr. Bob Waide (Executive Director), both of the University of New Mexico, for their ongoing support. Dr. David Coleman, University of Georgia, chair of the LTER Publications Committee, Mr. Kirk Jensen, Executive Editor, and Mr. Keith Faivre, Production Editor, Oxford University Press, have also been extremely supportive. The copyeditor at Oxford University Press made major improvements to the text. We also thank Dr. Andrew Fountain, Portland State University, Dr. Berry Lyons, Ohio State University, and Dr. Glen Juday, University of Alaska, who were earlier coeditors of the project. All of these people are noteworthy for extremely selfless action, and they continue the long tradition in the LTER program of putting the good of the program before the interests of individuals. We also thank the National Science Foundation Division of Environmental Biology, Long-Term Programs for funding to all parts of the LTER Program and especially the Network Office, which supported several of the meetings that culminated in this book. Any opinions, findings, conclusions, and recommendations expressed in the material are those of the author(s) and do not necessarily reflect the views of the National Science Foundation.

We would also like to acknowledge the following institutions and journals for permission to reproduce some previously published material: *BioScience, Physical Geography*, and *Geografiska Annaler Series A*.

Several of the chapters in this book went through multiple internal review processes—especially those chapters with multiple authors. At least one person has reviewed all of the chapters in this book. Five outside reviewers reviewed the overall outline of the book. The concluding materials were kindly and constructively reviewed by Dr. Bruce Hayden of the University of Virginia. Specifically, he is responsible for suggesting that future attention be given to the problem of detecting the climate signal in ecosystem dynamics. Other persons who provided help for reviews include Dr. John A. Harrington, Jr., Dr. Peter Lamb, Dr. Nathan Mantua, Dr. Charles W. Martin, Dr. Aaron Moody, Dr. Timothy R. Seastedt, and Dr. Marilyn Walker. We thank them for their help, and we apologize to anyone we may have inadvertently omitted.

References

Greenland, D., editor. 1994. *El Niño and Long-Term Ecological Research Sites*. Publication No. 18. LTER Network Office: University of Washington, Seattle, WA. 57 pp.

Greenland D., and L. W. Swift, Jr. 1990. *Climate Variability and Ecosystem Response*. USDA Forest Service. Southeastern Forest Experimental Station. General Technical Report SE-65. 90 pp.

Greenland D., and L. W. Swift, Jr. 1991. Climate Variability and Ecosystem Response: Opportunities for the LTER Network. *Bulletin of the Ecological Society of America* 72(2):118–126.

Contents

Contributors

Full contact information of LTER members may be obtained from the LTER Network web site personnel directory at http://lternet.edu/people/.

Valerie Barber
Bonanza Creek LTER
University of Alaska

Julio L. Betancourt
Sevilleta LTER
USGS Tucson

Neil Bettez
Arctic LTER Site
Woods Hole Oceanographic Institution

Fred Bierlmaier
H. J. Andrews LTER
USDA Forest Service

Emery R. Boose
Harvard Forest LTER
Harvard University

Anthony J. Brazel
Central Arizona Phoenix LTER
Arizona State University

Barton D. Clinton
Coweeta LTER
USDA Forest Service

Aaron R. Cooper
Coweeta LTER
North Carolina State University

Linda A. Deegan
Arctic LTER Site
Woods Hole Oceanographic Institution

Peter T. Doran
McMurdo LTER
University of Illinois at Chicago

Andrew W. Ellis
Central Arizona Phoenix LTER
Arizona State University

Scott Elias
Niwot Ridge LTER
University of Colorado

Andrew G. Fountain
McMurdo LTER
Portland State University

William "Bill" R. Fraser
Palmer LTER
Polar Oceans Research Group

Philip A. Fay
Konza Prairie LTER
Kansas State University

Stuart H. Gage
Kellogg Biological Station LTER
Michigan State University

David Greenland
Andrews and Niwot Ridge LTER sites
University of North Carolina at Chapel Hill

Douglas G. Goodin
Konza Prairie LTER
Kansas State University

Mark Harmon
Andrews LTER
Oregon State University

Bruce P. Hayden
Virginia Coast Reserve LTER
University of Virginia

Nils R. Hayden
Coweeta LTER
University of Georgia

John E. Hobbie
Arctic LTER
Woods Hole Oceanographic Institution

Clive Howard-Williams
McMurdo LTER
National Institute of Water and
   Atmosphere, New Zealand

Chris Jaros
McMurdo LTER
University of Colorado

Julia Jones
Andrews LTER
Oregon State University

Glenn Patrick Juday
Bonaza LTER
University of Alaska

Brian D. Kloeppel
Coweeta LTER
University of Georgia

James A. Laundre
Arctic LTER
Woods Hole Oceanographic Institution

Jiping Liu
Palmer LTER
NASA Goddard Space Flight Center

W. Berry Lyons
McMurdo LTER
Ohio State University

Sally MacIntyre
Arctic LTER
University of California at Santa Barbara

Arthur McKee
Andrews LTER
Oregon State University

Maurice J. McHugh
Louisiana State University

Diane M. McKnight
McMurdo LTER
University of Colorado

Douglas G. Martinson
Palmer LTER
Columbia University

Joseph Means
Andrews LTER
Oregon State University

Bruce T. Milne
Sevilleta LTER, New Mexico
University of New Mexico

H. Curtis Monger
Jornada LTER
New Mexico State University

Douglas I. Moore
Sevilleta LTER, New Mexico
University of New Mexico

Thomas Nylen
McMurdo LTER
Portland State University

W. John O'Brien
Arctic LTER
University of North Carolina at Greensboro

Steven Oberbauer
Arctic LTER
Florida International University

James A. Parks
Sevilleta LTER, New Mexico
University of Arizona

Robert R. Parmenter
Sevilleta LTER, New Mexico
University of New Mexico

William T. Pockman
Sevilleta LTER, New Mexico
University of New Mexico

Scott Rupp
Bonanza Creek LTER
University of Alaska, Fairbanks

Douglas Schaefer
Luquillo LTER
University of Puerto Rico

Gaius Shaver
Arctic LTER
Woods Hole Oceanographic Institution

Karie Slavik
Arctic LTER
Woods Hole Oceanographic Institution

Raymond C. Smith
Palmer LTER
University of California at Santa Barbara

Sharon E. Stammerjohn
Palmer LTER
University of California at Santa Barbara
and Columbia University

Frederick "Fred" J. Swanson
Andrews LTER
USDA Forest Service

Thomas W. Swetnam
Sevilleta LTER
University of Arizona

James M. Vose
Coweeta LTER
USDA Forest Service

Diana Wall
McMurdo LTER
Colorado State University

Katherine A. Welch
McMurdo LTER
Ohio State University

Cathy Whitlock
Andrews LTER
University of Oregon

Martin Wilmking
Bonanza Creek LTER
University of Alaska, Fairbanks

Xiaojun Yuan
Palmer LTER
Columbia University

John Zasada
Bonanza Creek LTER
USDA Forest Service

Abbreviations

Code letters commonly used for LTER sites.
Details may be found at http://lternet.edu/sites/

AND	H. J. Andrews Experimental Forest, Oregon
ARC	Arctic Tundra, Alaska
BES	Baltimore Ecosystem Study, Maryland
BNZ	Bonanza Creek, Alaska
CAP	Central Arizona–Phoenix, Arizona
CDR	Cedar Creek, Minnesota
CWT	Coweeta Hydrologic Laboratory, North Carolina
HFR	Harvard Forest, Massachusetts
HBR	Hubbard Brook, New Hampshire
JRN	Jornada Basin, New Mexico
KBS	Kellogg Biological Station, Michigan
KNZ	Konza Prairie, Kansas
LUQ	Luquillo Experimental Forest, Puerto Rico
MCM	McMurdo Dry Valleys, Antarctica
NWT	Niwot Ridge, Colorado
NTL	North Temperate Lakes, Wisconsin
PAL	Palmer Station, Antarctica
PIE	Plum Island Ecosystem, Massachusetts
SEV	Sevilleta, New Mexico
SGS	Shortgrass Steppe, Colorado
VCR	Virginia Coast Reserve, Virginia
FCE	Florida Coastal Everglades, Florida
GCE	Georgia Coastal Ecosystems, Georgia
SBC	Santa Barbara Coastal, California

CLIMATE VARIABILITY AND ECOSYSTEM RESPONSE AT LONG-TERM ECOLOGICAL RESEARCH SITES

1

An Introduction to Climate Variability and Ecosystem Response

David Greenland

Douglas G. Goodin

Raymond C. Smith

The regularities of our planet's climate determine a large part of the form and function of Earth's ecosystems. The frequently nonlinear operation of the atmosphere gives rise to a rich complexity of variability superimposed on the fundamental regularities. A traditional definition of climate is "the long-term state of the atmosphere encompassing the aggregate effect of weather phenomena—the extremes as well as the mean values" (Barry and Chorley 1987). Ecosystems share some of the same properties as the climate system. At one level their operation is fairly straightforward. Ecologists, to a certain extent, understand the flows of energy and matter through these systems. A good deal of ecosystem operation over time is characterized by some degree of homeostasis. On the other hand, nonlinear change and multiple variables have placed uncertainty and surprise at the forefront of much ecological research. In both the climate and the ecosystem the only certainty often appears to be change. The task of this book is to focus on some of this change at the interface between the climate and the ecosystem and by doing so gain insights into the operation of both systems.

The Theme of the Book

Millennial-scale (1000-year) climate variability has driven large changes of vegetation and fauna at almost all of the Long-Term Ecological Research (LTER) sites. Decadal climate variability at some sites has seen dramatic changes in fish catches and has altered tree species composition. During the first two decades of study, LTER sites have been affected by two super El Niño events and several more "normal" El Niños and La Niñas. Major droughts have affected species diversity and killed some trees. Severe storms and floods have damaged stream restoration structures. Coastal sites have measured a rise in sea level. Antarctic sites have documented the decrease of some penguin populations and a rise in other populations as a result of climatic warming over 50 or more years. Climate variability has constantly been on investigators' minds. It is little wonder that ecologists clearly recognize climate as a driver of biotic systems. Parmesan and her coworkers describe how climate affects individual fitness, population dynamics, and the distribution and abundance of species, as well as ecosystem structure and function (Parmesan et al. 2000). They relate how regional variation in climatic regimes creates selective pressures for the evolution of locally adapted physiologies, and morphological and behavioral adaptations. They quote the curious fact that climate even determines gender in some species. Map turtles (*Graptemys*) produce only males if the incubation temperatures are below 28°C and only females if the incubation temperatures are above 30°C (Bull and Vogt 1979). The implications of a steep warming trend for this species are dire! The role of climate as a driver of ecosystems has important practical implications for ecology. For example, Swetnam and Betancourt (1998) make clear that regional climate signals existing in ecosystems must be extracted before variations in ecosystem components can be attributed to other causes.

The theme of this book is how ecosystems respond to climate variability. This theme is examined at a variety of LTER sites and over a variety of timescales. The subject matter of the book is focused on a series of questions that are outlined here. The theme of climate variability and ecosystem response is inherently deterministic and implicitly carries with it the notion of climatic cause and ecosystem result. The analyses in this volume will amply demonstrate that this is a valid and fruitful working assumption. However, we acknowledge that this approach is limited in several senses. First, we recognize that, although in many instances climate may be recognized as the prime ecosystem driver, it is becoming increasingly clear that many ecosystem functions directly or indirectly affect the climate (e.g., Hayden 1998a). Second, there are many factors, both biotic and abiotic, that affect ecosystems besides climate. Third, many internal operations of ecosystems lead to ecosystem response and change. Fourth, many aspects of climate variability and ecosystem response have important implications for human systems. Human activities can sometimes overwhelm or strongly modify climatic influences. The change from grassland to shrubland over the last 150 years at the Jornada and Sevilleta LTER sites is an interesting example (chapters 17 and 15). It is impossible for us to deal with all these aspects, and so some degree of focus is necessary. That focus is provided by the more "simple" climate variability and ecosystem response approach. We also concentrate, for the most part, on results of research conducted at LTER sites. We are well aware that many other researchers and groups are addressing the issue of climate variability and ecosystem response within other contexts.

Despite these caveats, we think it is legitimate to treat climate variability as a prime driver of ecosystem responses. In this volume we also tend to approach climate in isolation from other factors. Climate differs from other ecosystem drivers: It has a certain regularity, expectedness, and predictability. Even in the areas of uncertainty, it is often possible to put outer bounds on the kinds and sizes of variability that might be expected. This cannot be said with so great a confidence for many biotic factors. Directional evolutionary trends in some cases, and complete extinction in other cases, make the biotic world a very surprising one. When one adds such anthropogenic factors as land-use change, genetic engineering, and the development of new technologies, the uncertainties mount ever higher. Our approach in dealing with what we know about climate variability and ecosystem response is simple, but it contains the possibility of developing new knowledge.

The LTER Program

The LTER program conducts and facilitates ecological research at 24 sites in the United States and the Antarctic. More sites are likely to be added to the LTER network in the future. There is also an important and growing International LTER (ILTER) program (LTER Network Office 1998). The U.S. LTER research sites operate as a network with a network office located at the University of New Mexico at Albuquerque. The network is a collaborative effort involving more than 1100 scientists and students. The current 24 LTER sites are located in various biomes throughout the United States and Antarctica (figure 1.1; Callahan 1984; Franklin et al. 1990; Van Cleve and Martin 1991; http://lternet.edu/). One of the missions of the LTER program is to conduct a cross-site synthesis. LTER research, like much Global Change research, focuses mostly on timescales of months to centuries. The operation as a network enables LTER to address large-scale questions concerning ecological phenomena such as the variations in stream organic matter budgets across the United States (Webster and Meyer 1997). The network also creates opportunities for comparisons between ecosystems across regional, continental, and global gradients such as organic matter decomposition (Long-Term Intersite Decomposition Experiment Team [LIDET] 1995). The network operation also allows scientists to distinguish system features controlled by absolute and relative scales. Neither the large-scale questions, such as what the decomposition rates are across the country, nor questions of absolute and relative scale, such as how decomposition rates vary along soil moisture gradients within LTER sites, can usually be answered without a detailed specification of the climate of LTER sites. The importance of cross-site synthesis has been expressed by an external review of the program as follows: "The power of the network approach of the LTER program rests in the ability to compare similar processes (e.g., primary production or decomposition of organic matter) under different ecological conditions. As a result, LTER scientists should be able to understand how fundamental ecological processes operate at different rates and in different ways under different environmental conditions" (Risser and Lubchenco 1993).

Two other features of the LTER program are important in the present context. First, the program prides itself on its interdisciplinary nature. The wide range of ecosystems studied demands that these studies be made in an interdisciplinary manner and that no single subdiscipline dominate. The LTER program also prides itself on its environmental information management system. This information man-

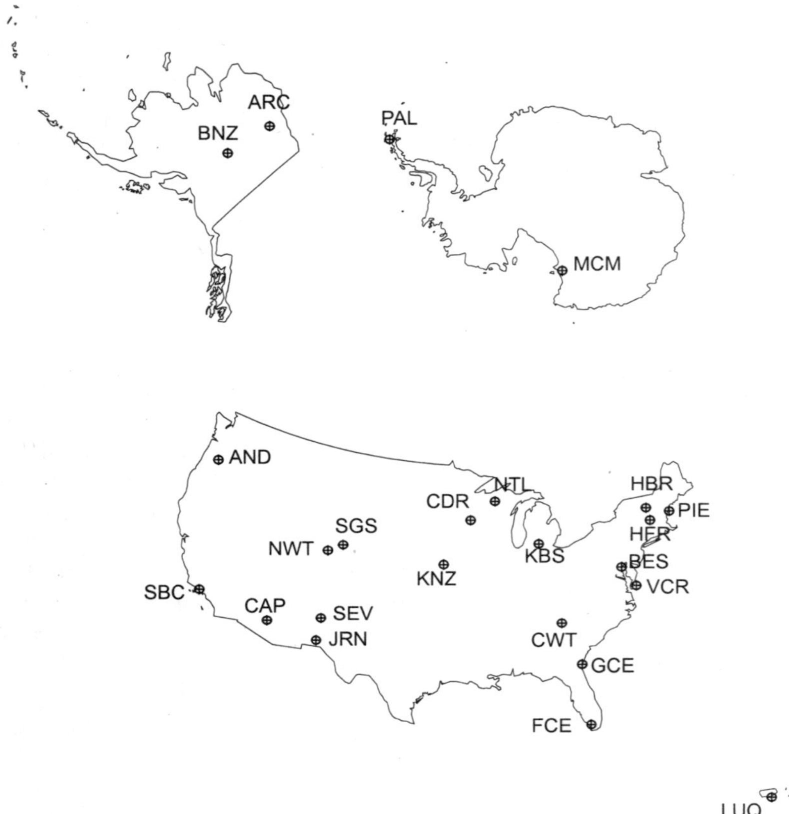

Figure 1.1 Location of the LTER sites. For an explanation of codes, see table 1.1 or the list of abbreviations in the frontmatter of the book.

agement system, and its climate data component, is regarded as a model for such systems worldwide (Michener et al. 1998; Baker et al. 2000).

The LTER program encourages coherence in ecological research over the long term to take advantage of the fact that many ecosystem processes operate at long time scales and show directionality and periodicity. Studies that have recognized this (e.g., at Hubbard Brook [Likens and Bormann 1995; Likens et al. 1996]) have made fundamental contributions to ecology. Within these sites it was found that human-derived as well as natural perturbations act over a long time period.

Studies at the LTER sites are organized around five core themes: (1) pattern and control of primary production, (2) spatial and temporal distribution of populations selected to represent trophic structure, (3) pattern and control of organic matter accumulation in surface layers and sediments, (4) patterns of inorganic input and movement through soils, groundwater, and surface waters, and (5) patterns and frequency of disturbance. Although climatic aspects affect all these themes, the role of climate is paramount in the last theme.

The LTER sites (table 1.1; figure 1.1) were not selected primarily to give good geographic coverage. They were selected first based on the quality of research proposed at the site. As a result, the sites together do not necessarily provide a systematic spatial coverage of the country or its climate and biomes. The network was not designed to replicate the spatial cover of meteorological observations given by the National Weather Service stations. The temporal rather than the spatial emphasis of the LTER network is one of the reasons why this book takes on a structure categorized by timescale.

The LTER Program and Climate

Both ecologists and climatologists recognize climate research as having a key role in long-term ecological research. Climate is one of the largest driving forces of ecological and hydrological processes at all of the LTER sites. Each LTER site is required to organize its 6-year research program around a central fundamental working hypothesis. A majority of the sites have climate as a central component of their research hypothesis. For example, one of the central questions of the H. J. Andrews Experimental Forest LTER research is, How do land use, natural disturbance, and climate change affect three key ecosystem properties: carbon dynamics, biodiversity, and hydrology? The goals of the Arctic LTER Project are to understand how tundra, streams, and lakes function in the Arctic and to predict how they respond to changes, including changes in climate. It is therefore essential to investigate the climate of the LTER sites in a systematic manner. Each LTER site maintains its own climate program and, at many sites, climate data represent the longest time sequence of data available. Increasing attention to possible ecological consequences of global climate change requires that we understand how climate varies and what the potential is for rapid directional climate change (LTER 1989; Greenland and Swift 1990 and 1991; IPCC 2001).

An example of the importance of long-term climate, or climate-related, information to ecosystem science may be taken from an aquatic LTER site. The number of days of ice cover on Lake Mendota, Wisconsin, which is part of the North Temperate Lakes (NTL) LTER site, illustrates the importance of long-term records and the need for benchmark climatic studies (Magnuson 1990; Robertson et al. 1992; Magnuson et al. 2000). If one started observing in 1998, one might conclude there are about 50 days of ice cover on the lake. However, the data for the decade 1989–1998 indicate that the average length of ice cover was about 100 days and that the 1998 value was "unusual." Fifty years of data (1949–1998) show a downward trend from about 110 to 90 days, with El Niño years having very short values of ice cover, as in 1998. The complete observed record starting in 1856 confirms the downward trend in the number of ice cover days as well as suggests interesting interdecadal variability. The duration of ice cover in this aquatic ecosystem determines the productivity and activity at all trophic levels during the ice-free summer period.

Although many of the analyses presented in this volume could be made with any subset of data from U.S. climate stations or climate divisions, there are specific rea-

Table 1.1 Long-Term Ecological Research (LTER) Sites

Site	Abbreviation	Ecosystem	Climate
H. J. Andrews Exp. Forest (Oregon)	AND	Coniferous Forest	Marine West Coast
Arctic Tundra (Alaska)	ARC	Arctic Tundra	Arctic Tundra
Baltimore Ecosystem Study (Maryland)	BES	Urban Ecosystem	Moist Subtropical (urban)
Bonanza Creek Exp. Forest (Alaska)	BNZ	Boreal Forest	Subarctic
Central Arizona, Phoenix (Arizona)	CAP	Urban Ecosystem	Desert (urban)
Cedar Creek Nat. History Area (Minnesota)	CDR	Hardwood Forest/ Tallgrass Prairie	Humid Continental
Coweeta Hydrol. Lab. (N. Carolina)	CWT	Deciduous Forest	Humid Continental
Hubbard Brook Exp. Forest (New Hampshire)	HBR	Northern Hardwood	Humid Continental
Harvard Forest (Massachusetts)	HFR	Hardwood/Whitepine/ Hemlock	Transition Humid Continental
Jornada (New Mexico)	JRN	Desert	Subtropical Desert
Kellogg Biological Station (Michigan)	KBS	Agricultural	Humid Continental
Luquillo Exp. Forest (Puerto Rico)	LUQ	Tropical Rainforest	Tropical Rainforest
Konza Prairie (Kansas)	KNZ	Tallgrass Prairie	Midlatitude Steppe
North Temperate Lakes (Wisconsin)	NTL	N Temperate Lake Mixed Forest	Humid Continental
Niwot Ridge/Green Lakes Valley (Colorado)	NWT	Alpine Tundra	Highland
Plum Island Ecosystem (Massachusetts)	PIE	Coastal Estuary	Moist continental
Sevilleta (New Mexico)	SEV	Desert/Grassland/ Forest Transition	Low-latitude Desert
Shortgrass Steppe Formerly Central Plains Exp. Range (Colorado)	SGS/CPR	High Plains Grassland	Midlatitude Steppe
Virginia Coast Reserve (Virginia)	VCR	Barrier Island	Humid Subtropical
Florida Coastal Everglades (Florida)	FCE	Freshwater Marsh, Coastal Estuary	Humid Subtropical
Georgia Coastal Ecosystems (Georgia)	GCE	Barrier Island	Humid Subtropical
Santa Barbara Coastal (California)	SBC	Semiarid Coastal and Marine	Mediterranean
McMurdo Dry Valleys	MCM	Desert Oases	Polar Ice Cap
Palmer Station Antarctica	PAL	Coastal and Ocean Pelagic	Polar Marine

sons for concentrating on LTER sites. First, the analyses are directly focused on the LTER sites that have a legacy of ecosystem research. Second, the sites have ongoing, coherent programs of ecosystem research. Third, several of the LTER sites have climate stations at places rarely sampled by national weather observing systems. The alpine tundra NWT D1 site at an elevation of 3749 m (12,300 ft.) is a case in point.

It is helpful to pause and reflect on exactly what the "climate" in climate variability and ecosystem response actually is. This question is raised by Goodin et al. (chapter 20) for the context of Net Primary Productivity (NPP) at the Konza Prairie. In this specific context the "climate" has been defined using values of air temperature, precipitation, and pan evaporation with various indexes derived from these variables, while bearing in mind subsets of time such as the "growing season." We use the term *climate* differently for almost every different ecosystem considered in this book. The climate that ecosystems experience is most truly represented by values of heat, moisture, gas, and momentum exchange at what the Russian scientist Alexander I. Voeikov called in 1884 the "outer effective surface" of the ecosystem components. Except in cases of the most detailed microclimatological studies, ecologists and climatologists usually deal with values of variables such as air temperature and precipitation that act only as surrogates of the variable that we ought to be measuring. Thus we see "through a glass darkly." This approach is forced on us partly by practical and economic considerations and partly because most meteorological observing networks are established with weather forecasting rather that climate/ecosystem interaction purposes in mind.

Climate Variability and Ecosystem Response in the LTER Program

The LTER community has provided insights into the area of climate variability and ecosystem response at several meetings over the last two decades. The insights may act as a point of departure for the present volume. In several cases the insights previously noted have become even more important as new discoveries have been made.

First, we are reminded that long-term studies are especially suited to exploring four major classes of long-term ecological phenomena (Strayer et al. 1986). Strayer and coworkers identify these phenomena as (1) slow processes, (2) rare events, (3) subtle changes in the systems, and (4) complex processes involving multivariate studies where the long-term context can add degrees of freedom to the solution of the problem. The first three of these classes of change may readily be identified in climate data and the fourth is also applicable to climate data in certain circumstances.

The 1988 LTER Climate Committee focused on four main areas of climate variability and ecosystem response (Greenland and Swift 1990, 1991): (1) the importance of terminology, (2) the ubiquitous importance of time and space scale, (3) a consideration of climatic indexes, other than temperature and precipitation, which may be useful in ecosystem studies, and (4) the similarities and the dissimilarities among the LTER sites. Scale is so important that we will consider it throughout this volume.

Regarding terminology, the consensus was that climate variability should be taken as a given and we should concentrate on "episodes" and "events" within the existing variability. An *event* is taken as a single occurrence such as an individual large rainstorm often embedded in the functioning of the synoptic climatic scale. An *episode* is taken as a string of items and is in some way related to the time constant of the system. Events or short-lived episodes often have the characteristic of resetting the time clock of the system. They are marked by a large change in the ecosystem at the time of the occurrence, followed by a long tail of less obvious adjustments. The operation of streams is a good example of this. Although not all the authors in this volume use this terminology, we find it very useful in the concluding section (chapter 21) of this book when comparing the climatic variability and ecosystem response among LTER sites.

There are at least three, often overlapping, kinds of climate episodes. Each of these must be distinguished to minimize confusion. First, there are climate episodes defined by the data of the climatic series themselves, their time series, and indications of changes of states. Second, there are climatic episodes as perceived by humans, which, though often described by means of climatic data, are importantly frequently related to the timescale of the human life span, somewhere between 40 and 80 years. An example would be the drought of the Dust Bowl years in the 1930s in the United States. Third, there is the type of climate episode as perceived, or defined, by the individual components, or groups of components, of the ecosystems themselves. The latter type is especially scale dependent and important to Long-Term Ecological Research. There is a tendency to impose human-oriented concepts of scale on our systems instead of letting the functions of the ecosystems themselves define the scale that is most important.

Similarities and dissimilarities across the LTER network were considered in 1988, and many of the issues remain the same today. Many LTER sites do not yet show clear or obvious ecosystem effects from slow trends or even from intermediate-scale events but do show a marked effect to a severe atmospheric event. As the LTER program has developed over the past two decades, the presence in the ecosystem of the legacy of a severe atmospheric event or episode has emerged as a signature finding at almost every LTER site. The Hubbard Brook ecosystem, for example, was not markedly affected by the droughts of the 1960s, but the ecosystem still shows the effect of a single hurricane that traversed its area in 1938 (Merrens and Peart 1992). Major ecosystem changes stem from catastrophic events at many LTER sites. Windthrow of trees is a repeated catastrophic event. However, many ecological events that owe their existence to atmospheric occurrences are mediated through the operation of geomorphic processes. The redistribution of sediment, for example, in the dry Jornada, New Mexico, site during an intense rainstorm may have marked consequences on the biota either by covering them or by providing new microhabitats.

Most LTER sites follow hemispheric, or at least regional, trends in temperature and precipitation (Greenland and Kittel 2002). This bodes well for the extrapolation of results from the LTER network to larger areas. Yet, occasionally, as in the case of the Niwot Ridge, Colorado, data, larger spatial and temporal trends are not displayed by an individual LTER site. Even more specifically, Pielke et al. (2000) conclude the spatial variation in climate variables indicate that the direction and magnitude of regional climate trends cannot necessarily be inferred from single-site records, even over relatively homogeneous terrain. They based their analysis on the other Colorado LTER site, the Short Grass Steppe site. When examined in 1988, several sites showed time coincidence for changes in the values of certain variables. The change in the lake freezing data of the North Temperate Lakes in 1880, 1940, and possibly 1980 was reflected in different series at other sites and is also reflected in general climate data. Since that time, a major LTER-related project at NTL has demonstrated the hemispherewide concurrence in the thawing dates of lake ice (Magnuson et al. 2000).

We should also note that the geography of the LTER network is such that particular spatial scales are emphasized. The individual LTER site is typically 50 km or less between boundaries. A few—PAL, NTL, CWT, and SEV—are rather larger. An exercise to investigate the spatial representativeness of individual LTER sites concluded that most sites generally represented a larger area than the size of the site itself, but that area was quite variable from site to site (http://lternet.edu/collaborations/syn_09.html). Consequently, the emphasis of many, although not all, LTER studies is at the local or regional scale. On the other hand, the distribution of sites (figure 1.1) does sample much of the North American continent and part of the Caribbean and Antarctica and a wide variety of climates (figure 1.2). Indeed, there is a significant latitudinal gradient between the Arctic Tundra (ARC, 68.6° N) and Palmer Station (PAL, 64.7° S) and the stations in between. This sampling is not systematic in terms of spatial distribution. The current network of sites is biased toward mid- and high latitudes. Results from cross-site studies therefore represent gradients of variables and processes rather that the systematic geographic distribution of the variables and processes. The LTER network of sites is oriented primarily to "long-term" rather than "large-area" studies.

The 1988 workshop suggested several fertile areas for further research related to the similarities and dissimilarities of climate variability and ecosystem response across LTER sites. These include an investigation of (1) the importance of catastrophic events in relation to slower trends and cycles, (2) the time coincidence of certain major climatic breakpoints that appear to exist at several sites and the effects on the ecosystems of the related changes from one episode to another, and (3) the relationship of climate and phenological studies across the LTER network. Some progress has been made on the first two, but LTER scientists have paid little attention to the third even though the topic is receiving considerable attention elsewhere (Schwartz 1999). Participants in the 1988 workshop also identified some exciting ways, such as airmass analysis, by which we can go beyond the use of simple temperature and precipitation values in defining breakpoints between climatic episodes. This technique has been explored effectively for the Konza Prairie LTER site by Hayden (1998b).

A 1997 LTER workshop on climate variability and ecosystem response was equally fruitful. The growth of the LTER network has led to a greater diversity of ecosystems studied and consequently a wider range of the types of interactions be-

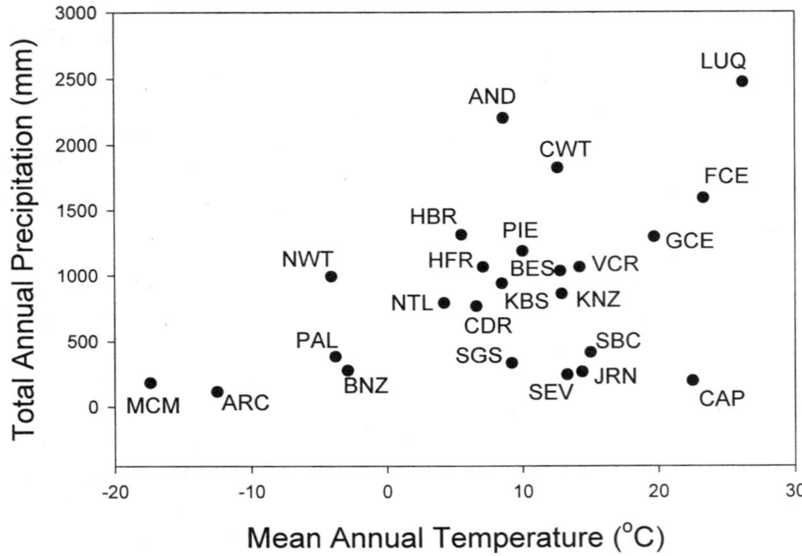

Figure 1.2 Distribution of LTER sites by annual mean temperature (°C) and total annual precipitation (mm). Data are for the period 1961–1990. For an explanation of codes, see table 1.1 or the list of abbreviations in the frontmatter of the book. Reprinted with permission from Greenland et al. 2003. Long-Term Research on Biosphere-Atmosphere Interactions. *BioScience* 53(1):33–45. Copyright American Institute of Biological Sciences.

tween climate and ecosystems. A sampling of the papers presented at the workshop demonstrates this, as discussed subsequently.

Some papers of the 1997 workshop were consistent with the suggestion of warming in the high latitudes of Earth. Dr. Fraser of the Palmer Marine Antarctic site examined, with some success, the hypothesis that changes in the population abundance of penguins occur when environmental frequencies no longer match the requirements of evolved life histories. The environment has seen a decrease in frequency of cold years with heavy ice over the last 50 years and a 4–5°C increase in temperature (chapter 9). Drs. Chapin and Juday of the Bonanza Creek Boreal Forest site in Alaska documented strong climate warming in last three decades, which has led to the melting of permafrost and the earlier breakup of ice from rivers. Furthermore, a higher snow amount tends to open the crowns of trees, providing more suitable conditions for the outbreak of spruce budworm infestations (chapter 12).

Workers from other LTER sites investigated the longer term paleoclimatic aspects of their environments. Caroline Yonker noted three periods of climatic instability in the Holocene paleosoils of the Shortgrass Steppe in Colorado. Dr. Laura Huenneke of the Jornada site is interested in separating the climatic and human influences on desertification processes. She uses evidence from C_3 and C_4 vegetation in buried soils to suggest that human modifications of the landscape are superimposed on natural long-term cycles of landscape stability and instability. Further evidence is found on terraces in the Rio Grande valley and nearby eolian deposits. By way of contrast, Janice Fuller, at the Harvard Forest site, provided pollen evidence to suggest that European settlement activities may have obscured the effect of natural spatial climate change in the New England area. Dr. Fountain reported that the lakes of the McMurdo Dry Valleys in Antarctica have a layer of saline water that could possibly be sea water left over from the past. Furthermore, organic carbon in the Taylor Valley may be associated with a paleolake in the valley (chapter 16).

Scale

Scale is an ever-present issue in many disciplines of science. Scale is so important that, in many ways, it determines the kinds of questions that may be asked about the operation of the ecosystem, and it often determines the answers to the questions as well. A specific recurrent issue is how to relate the scales at which climate systems operate to those scales at which the biotic parts of the ecosystems operate. The 30-year period over which "climatic normals" are taken is an artificial human construct and may have little bearing on ecosystem realities. Decadal averages of climate data might be more meaningful. At the very least, we should recognize that the averaging period will have a very large role in what we consider to be an "episode." The definition of climate as perceived by the individual component of the ecosystem is directly related to scale. A soil microorganism might regard an individual rainstorm as a significant climatic event, whereas a tree at the Andrews LTER site in Oregon would be acclimated to a "climate" far exceeding any 30-year climatic period. The ecosystem responder defines its own climatic scale.

The Framework Questions

In planning this volume we decided to focus on a set of questions that emphasize the dynamic nature of climate variability and ecosystem response. An important consideration was the need for generalization. Within the LTER program, modeling is a fertile method for generalization. Whereas the material we deal with does not lend itself to cross-site modeling per se, we decided to ask questions that will lead to a modeling framework. With this in mind, we next discuss the questions that were used at the outset.

The first framework question is, What kind of climate variability is being investigated? We must first recognize that there are several types of climate variability. The principal types according to Karl (1985) are as follows: (1) a trend is a smooth monotonic increase or decrease; (2) a fluctuation is two changes of mean whereby two maxima (minima) and one minimum (maximum) are evident; (3) a discontinuity is a single abrupt change in the mean; (4) a vacillation is a series of climate fluctuations but with mean values drifting about two or more average values; (5) an oscillation is a gradual transition between a maximum and minimum value that tends to repeat itself in the time series; (6) an oscillation in which the interval between the maximum and minimum values is approximately equal is called a periodicity, particularly where the maximum and minimum values are more or less equal over the period of interest. Even at the outset, we recognize that one or more of these types of climatic variability may operate simultaneously at any one LTER site. In addition, the distinction between the different types of climatic variability is not always clear, as is pointed out by McHugh and Goodin (chapter 11).

The next part of the framework consists of a series of questions.

1. Are there any preexisting conditions that will affect the impact of the climatic event or episode? For example, the effect of an intense rainstorm will be different, depending on whether the soil is already saturated.

2. Is the climate effect direct or does it go into a cascade? If a cascade is entered, how many levels does it have and is the interaction between each level linear or nonlinear? A cascade system is generally regarded as one that exhibits flow of material, energy, or information (Chorley and Kennedy 1971; Strahler 1980; Thomas and Huggett 1980). This is one of the more important questions. In introducing the framework questions, we note that the question about the existence of cascades, or a cascading set of events, lays the groundwork for systems analysis and modeling approaches. During this cascade identification, or modeling process, the parts of the cascade about which little is known are sometimes highlighted, thus establishing a potential agenda for further research needs.

3. Is the primary ecological effect completed by the time of the next climatic event or episode (or part thereof) or not? If the effect is complete, we may consider the next part of the cascade (if any). If the primary ecological effect is not complete (i.e., reaches a new constant level), is it still of sufficient magnitude to have an effect on the rest of the ecosystem? If so, we should pass the effect along the cascade.

4. Does the climatic event or episode or the ecological response have an identifiable upper or lower limit? If a limit exists, we can stop the consideration if necessary at the limit but keep the cascade going until it reaches limits that may exist in later parts of the cascade.

5. Does the climatic event or episode or ecosystem response reverse to some original state? If so, what timescales are involved? Does the climate state go back to the original position or beyond? Do cascades reverse? Can we identify the timing of these events?

6. After the climatic event or episode has occurred, do the values of the climatic or ecosystem variables return along their outward path or is there hysteresis or some other trajectory in operation? If the latter, how does this affect the cascade?

All of these questions relate to a deterministic, nonchaotic system. We may also ask whether the system is chaotic or random. If the system is random, no further explanation is possible, except that in some cases it may be possible to proceed using probability theory. If the system is chaotic, we must compute, or otherwise find, the parameters of the chaos such as its attractors and Lyapunov exponents.

This initial framework is summarized for convenience in a schematic in figure 1.3. A complete answer to these questions would place investigators in a good position to develop a model of the important climate variability and ecosystem re-

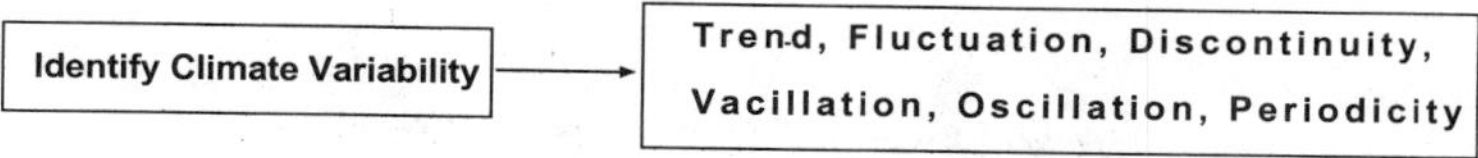

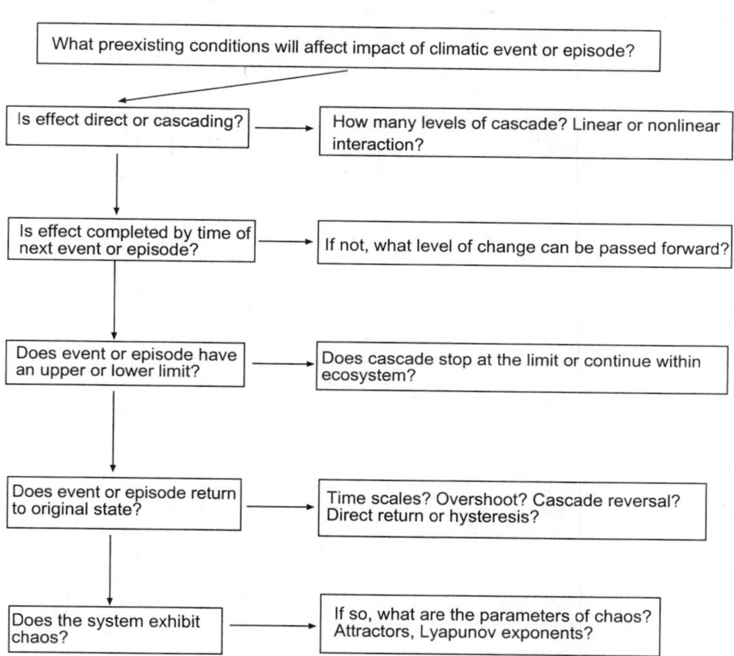

Figure 1.3 Schematic of the original framework questions used in the book.

sponse factors for the LTER site in question. Possibly more important, the aim of the questions is to ensure that the topic is treated in a systematic and thorough manner. The questions were "field tested" in conference presentation and in print (Greenland 1999) and found to be quite useful. Once more we recognize the limitations of this "one size fits all" approach, but we believe the need for focus and the quest for generality surpass the inherent limitations of any particular set of questions. The authors of the chapters in this book were presented with an early version of these questions and asked to address at least one or more of them in the preparation of their chapter. They were free to choose whether to deal with the question implicitly or explicitly.

After all the individual investigations that form the chapters of this book were complete, we reexamined the framework questions. We found that some changes in the ordering of the questions was necessary and that some questions are more fruitful than others. In retrospect, the framework questions fall into two categories (fig-

CLIMATE VARIABILITY & ECOSYSTEM RESPONSE (CVER)

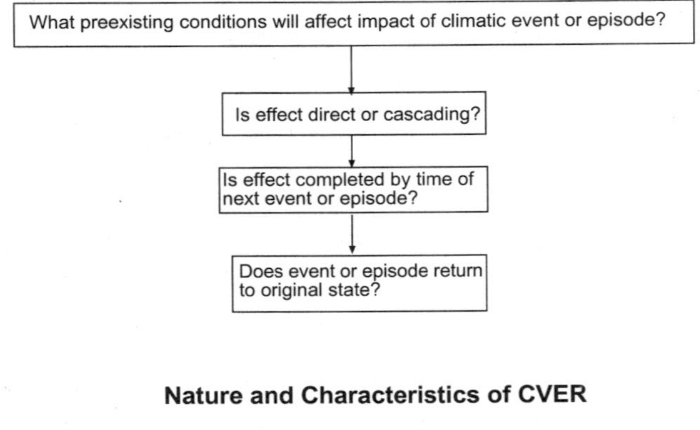

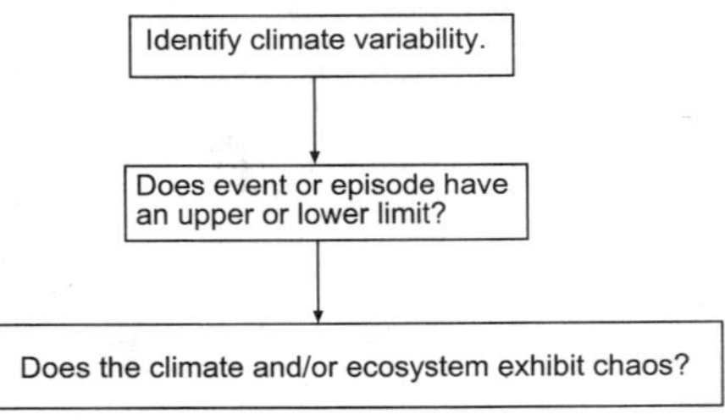

Figure 1.4 Schematic of the revised framework questions used in the book.

ure 1.4). The first category deals with the dynamics of climate variability and ecosystem response and assumes an underlying temporal sequence. The questions that fall most naturally into this category are those dealing with the preexisting conditions, the cascade of effects, whether the effects are completed by the time of the next climatic event or episode, and whether the event or episode and/or the ecosystem return to some original state. The second category of questions deals with the nature and characteristics of climate variability and ecosystem response. The questions of this type include the identification of the climate variability, whether the event or episode and/or the ecosystem response have an upper or lower limit, and whether the climate and/or ecosystem exhibit chaos. The discussion in the final chapter of the book (chapter 21) resequences the framework questions to better match the distinction between these two categories of questions.

Most of the questions that refer to the ecosystem are dependent on the scale of the particular ecosystem under consideration. On the other hand, the climate variability usually crosses multiple timescales and often has its root causes in other, larger, spatial scales. Both climate variability and ecosystem response, and the questions relating to them, cross multiple temporal scales. Beyond the scope of this book is the probability that the ecosystem effects may also be large spatial-scale effects and that the ecosystem effects may ultimately feed back on to the climate system at multiple scales.

Overview of Book

Two nonmutually exclusive sets of concepts emerge from our studies. The first set of concepts is that initial and intermediate cascade elements may act as gateways, filters, and/or catalysts to the climatic signal. Gateways can be open or closed; that is, they can either permit the passage of material, energy, or information or not. Filters may pass a variable amount of material, energy, or information along through the cascade. This amount varies from all to none and includes all the possibilities in between. Thus, the filters in the system help promote a buffering function to a climate disturbance. Catalysts occur where the presence of one component greatly enhances the effectiveness of two or more other components in the system. The second set of concepts deals with classes of ecosystem response to climate variability. There are at least three broad classes of interaction between systems and climate. First, the ecosystem buffers climate variability. Second, the ecosystem simply responds to individual climate events and episodes that exceed some threshold for response. Third, the ecosystem moves into resonance with the climatic variability with positive and negative feedbacks that produce a strong ecosystem response. These two sets of concepts will be discussed in the final chapter.

This first chapter of this book is an introduction to the general topic of climate variability and ecosystem response in the LTER program. We have also introduced our framework questions. Chapters 2–20, which form the body of the work, are organized into five parts, each one, except part V, dealing with the separate timescales at which we are looking. Each part, except part V, has its own introduction and a section synthesizing the material and results as they apply to the particular timescale being studied. Part I considers the short timescale ranging from an individual storm to a year or less. Part II focuses on the quasi-quintennial scale and concentrates on events that have a recurrence interval of about 5 years, such as the El Niño–Southern Oscillation. The group of chapters in part III addresses the timescale of several decades. Part IV treats climate variability at the century to millennial timescale. Individual chapters do not always fit with ease into one or the other divisions of timescales. Perhaps the best example of this is chapter 14, which deals with individual short period extratropical storms. The frequency of these storms is found to vary at a century timescale. Similarly, the Sevilleta chapter (chapter 15) could equally well fit into the quasi-quintennial or the decadal or even the century-scale section. Part V includes chapters from individual sites that cover the topic at several timescales. This material seeks to address the issue of climate variability and ecosystem response without being constrained to a particular scale. Chapter 21 is a review of the answers to our framework questions, concluding comments, and suggestions for further research.

18   Introduction

References

Baker, K. S., B. J. Benson, D. L. Henshaw, D. Blodgett, J. H. Porter, and S. G. Stafford. 2000. Evolution of a multisite network information system: The LTER information management paradigm. *BioScience* 50:963–978.

Barry, R.G., and R. J. Chorley. 1987. *Atmosphere, Weather and Climate.* 5th ed. New York: Routledge.

Bull, J. J., and R. C. Vogt. 1979. Temperature-dependent sex determination in turtles. *Science* 206:1186–1188.

Callahan, T. 1984. Long-Term Ecological Research. *BioScience* 34:363–367.

Chorley, R. J., and B. A. Kennedy. 1971. *Physical Geography: A Systems Approach.* London: Prentice-Hall International.

Franklin, J. F., C. S. Bledsoe, and J. T. Callahan. 1990. Contributions of the Long-Term Ecological Research Program. *BioScience* 40:509–523.

Greenland, D. 1999. ENSO-related phenomena at Long-Term Ecological Research sites. *Physical Geography* 20:491–507.

Greenland, D., and L. W. Swift, Jr., editors. 1990. *Climate Variability and Ecosystem Response.* USDA Forest Service. Southeastern Forest Experimental Station. General Technical Report SE-65. 90 pp.

Greenland, D. E., and L. W. Swift, Jr. 1991. Climate Variability and Ecosystem Response: Opportunities for the LTER Network. *Bulletin of the Ecological Society of America* 72:118–126.

Greenland, D., and T. G. F. Kittel. 2002. Temporal variability of climate at the U.S. Long-Term Ecological Research (LTER) sites. *Climate Research* 19(3):213–231.

Hayden, B. P. 1998a. Ecosystem feedbacks on climate at the landscape scale. *Philosophical Transactions of the Royal Society,* London B, 353:5–18.

Hayden, B. P. 1998b. Regional climate and the distribution of tallgrass prairie. Pages 19–34 in Knapp A. K., Briggs J. M., Hartnett D. C., Collins S. L., editors, *Grassland dynamics: Long-Term Ecological Research in tallgrass prairie.* New York: Oxford University Press.

IPCC. 2001. *Climate Change 2001: The scientific basis. Contribution of Working Group I to the Third Assessment Report of the Intergovernmental Panel on Climate Change.* Houghton, J. T., Y. Ding, D. J. Griggs, M. Noguer, P. J. van der Linden, X. Dai, K. Maskell, and C. A. Johnson, editors. Cambridge: Cambridge University Press.

Karl, T. R. 1985. Perspective on Climate Change in North America during the twentieth century. *Physical Geography* 6:207–229.

Likens, G. E., and F. H. Bormann. 1995. *Biogeochemistry of a Forest Ecosystem.* 2nd ed. New York: Springer-Verlag.

Likens, G. E., C. T. Driscoll, and D. C. Buso. 1996. Long-term effects of acid rain: Response and recovery of a forest ecosystem. *Science* 272:244–246.

Long-Term Intersite Decomposition Experiment Team (LIDET). 1995. *Meeting the challenge of long-term, broad-scale ecological experiments.* LTER Network Office, Seattle, Washington. Publication No 19. 23 pp.

LTER 1989. *1990s Global Change Action Plan Utilizing a Network of Ecological Research Sites. A Proposal from Sites Conducting Long-Term Ecological Research.* Workshop held in Denver, November 1989. Published by the LTER Network Office. University of Washington, College of Forest Resources, AR-10, Seattle, Washington.

LTER Network Office. 1998. *The International Long Term Ecological Research Network. 1998.* A summary of current activities in 15 countries. LTER Network Office, University of New Mexico, Albuquerque, New Mexico.

Magnuson, J. J. 1990. Long-term ecological research and the invisible present. *BioScience* 40:495–501.

Magnuson J. J., D. M. Robertson, B. J. Benson, H. Wynne, D. M. Livingstone, T. Arai, R. A. Assel, R. G. Barry, V. Card, E. Kuusisto, N. G. Granin, T. D. Prowse, K. M. Stewart, and V. S. Vuglinski. 2000. Historical trends in lake and river ice cover in the Northern Hemisphere. *Science* 289:1743–1746. Errata 2001. *Science* 291:254.

Merrens, E. J., and D. R. Peart. 1992. Effects of hurricane damage on individual growth and stand structure in a hardwood forest in New Hampshire, USA. *Journal of Ecology* 80(4):787–795.

Michener, W. K., J. H. Porter, and S. G. Stafford. 1998. *Data and information management in the ecological sciences: A resource guide.* LTER Network Office, University of New Mexico, Albuquerque, New Mexico. (http://www.lternet.edu/ecoinformatics/guide/frame.htm)

Parmesan, C., T. L. Root, and M. R. Willig. 2000. Impacts of extreme weather and climate on terrestrial biota. *Bulletin of the American Meteorological Society* 81:433–450.

Pielke, R. A., Sr., T. Stohlgren, W. Parton, N, Doesken, J. Money, L. Schell, and K. Redmond. 2000. Spatial representativeness of temperature measurements from a single site. *Bulletin of the American Meteorological Society* 81:826–830.

Risser, P., and J. Lubchenco. 1993. *Ten-year Review of the National Science Foundation Long Term Ecological Research (LTER) Program.* Commissioned by the Biological Sciences Directorate of the National Science Foundation. July 1993. NSF 94-96, National Science Foundation, Virginia.

Robertson, D. M., R. A. Ragotzkie, and J. J. Magnuson. 1992. Lake ice records used to detect historical and future climate changes. *Climatic Change* 21:407–427.

Schwartz, M. D. 1999. Advancing to full bloom: Planning phenological research for the 21st century. *International Journal of Biometeorology* 42:113–118.

Strahler, A. N. 1980. Systems theory in physical geography. *Physical Geography* 1:1–27.

Strayer, D., J. S. Glitzenstein, C. G. Jones, J. Kolasa, G. E. Likens, M. J. McDonell, G. G. Parker, and T. A. Pickett. 1986. Long-Term Ecological Studies: An Illustrated Account of Their Design, Operation, and Importance to Ecology. *Occasional Publication of the Institute of Ecosystem Studies.* Number 2. Millbrook, New York.

Swetnam, T. W., and J. L. Betancourt. 1998. Mesoscale disturbance and ecological response to decadal climate variability in the American Southwest. *Journal of Climate* 11:3128–3147.

Thomas, R. W., and R. J. Huggett. 1980. *Modelling in Geography: A Mathematical Approach.* Totawa, New Jersey: Barnes and Noble.

Van Cleve, K. and Martin, S. 1991. *Long-Term Ecological Research in the United States A Network of Research Sites 1991.* 6th ed., revised. LTER Publication No. 11. Long-Term Ecological Research Network Office, Seattle, Washington.

Webster, J. R., and Meyer, J. L., editors. 1997. Stream organic matter budgets. *Journal of the North American Benthological Society* 16:3–161.

Part I

Short-Term Climate Events

Introductory Overview

David Greenland

Television images of floods, hurricanes, tornadoes, snow and ice storms, and drought conditions are among the most vivid that leap into our minds when we think of short-term climatic events and their often obvious and direct ecosystem responses. The images are so striking that they tend to crowd out thoughts of longer term events. Yet, in many cases, even the longer term climatic events are often represented by the media as some manifestation of an individual severe weather event. The LTER sites have experienced a wide variety of severe weather events. Some of these are discussed in various chapters of this book. However, many other noteworthy instances are not treated in these pages. For example, we do not discuss the playa at the Jornada LTER site that has experienced a 100-year return period storm that filled the normally dry lake with water and brought to the fore many life forms that were surprising to Jornada investigators. Neither do we have room for the work at the Hubbard Brook LTER site by researchers who have documented in detail the effects on their trees of one of the most severe ice storms of the last century. Several other short-term climatic events, such as the 1996 flood at the Andrews rain forest, are discussed in the chapters of this book beyond this first section. In Part I the focus is on hurricanes, drought, and the short-term climatic events and ecosystem responses in the Arctic LTER site in Alaska.

Emery Boose of the Harvard Forest LTER in central Massachusetts introduces a Harvard Forest study on the effects of hurricanes on forest ecosystems in chapter 2. A strong hurricane passed over central New England in 1938 and left an indelible memory both in the minds of the inhabitants who experienced it and on the landscape. This stimulated Harvard Forest researchers to investigate the past history of hurricanes in their region and even to simulate a hurricane in their forest and study its effects on the ecosystem. The latter has become one of the legendary classic experiments of the LTER program. It has also been natural for Harvard Forest researchers to extend their interest to other LTER sites that have experienced hurricanes. They are interested particularly in the Luquillo LTER site in Puerto Rico and, to a lesser extent, the former North Inlet site in South Carolina that suffered immensely from the passage of hurricane Hugo in 1989. Boose provides an interesting comparison on the hurricanes in New England and Puerto Rico and their impact on the respective ecosystems.

Another important event in the 20-year annals of LTER history is the drought of 1988 that affected the entire United States in one way or another. Kloeppel and his coworkers examine the effects of this event in chapter 3 with respect to tree mortality at the Coweeta LTER site in the mountain forests of North Carolina. The same drought caused major detrimental responses to the agricultural ecosystems of the north central region of the United States, where the Kellogg Biological Station LTER is located. In chapter 4, Gage provides a penetrating analysis of the effects of this drought not only for the Kellogg LTER but also for a much larger region. Finally, in this chapter 5, we turn our attention to one of the younger LTER sites, the Arctic site in northern central Alaska where the climate record began in the early 1970s. Hobbie and his coworkers describe the effects of low temperatures in the air and soil, the 8-month snow cover, the reduced amount of light energy for photosynthesis, the completely frozen streams from mid-September until mid-May, and the long duration of the ice cover on lakes.

LTER sites are well set up to systematically observe the effects of short-term climatic events. The chapters in part I demonstrate the different ways in which LTER research on climate variability and ecosystem response operates at this timescale.

2

Hurricane Impacts in New England and Puerto Rico

Emery R. Boose

Introduction

Hurricanes have a profound effect on many coastal ecosystems. Direct impacts often include wind damage to trees, scouring and flooding of river channels, and salt-water inundation along shorelines (Simpson and Riehl 1981; Diaz and Pulwarty 1997). In some areas, secondary impacts may include landslides triggered by heavy rains (Scatena and Larson 1991) or catastrophic dry-season fires resulting from heavy fuel loading (Whigham in press). This chapter will focus on the long-term impacts of hurricane wind damage at two LTER sites, the Harvard Forest (HFR) in central New England and the Luquillo Experimental Forest (LUQ) in northeastern Puerto Rico. These two sites, both located in the North Atlantic hurricane basin and occasionally subject to the same storms, provide interesting examples of tropical and temperate hurricane disturbance regimes.

Wind damage from a single hurricane is often highly variable (Foster 1988). Damage to individual trees can range from loss of leaves and fine branches, which can significantly alter surface nutrient inputs (Lodge et al. 1991), to bole snapping or uprooting, which can significantly alter coarse woody debris and soil microtopography (Carlton and Bazzaz 1998a and b). At the stand level, damage can range from defoliation to individual tree gaps to extensive blowdowns, creating different pathways for regeneration (Lugo 2000). At landscape and regional levels, complex patterns of damage are created by the interaction of meteorological, topographic, and biological factors (Boose et al. 1994).

Adding to this spatial complexity is the fact that successive hurricanes are not necessarily independent in terms of their effects. A single storm lasting several hours may have effects that persist for decades (Foster et al. 1998). And forest susceptibility to wind damage is strongly influenced by composition and structure, which in turn are strongly influenced by previous disturbance history (Foster and Boose 1992). Thus, the impacts of a single hurricane may depend in part on the impacts of earlier storms as well as on other previous disturbances and land use. Hurricanes, like other disturbances, both create and respond to spatial heterogeneity (Turner et al. 2003).

To understand the long-term ecological role of hurricanes at a given site, we must consider these three sets of questions: (1) What is the hurricane disturbance regime? How do the disturbance events (hurricane wind damage) vary in space and time? (2) What is the ecosystem response to a single event? How does ecosystem response vary with disturbance intensity? (3) What is the ecosystem response to repeated events? How does hurricane disturbance fit into the overall disturbance regime? This chapter will highlight recent studies by the Harvard Forest and Luquillo LTER programs, focusing on the reconstruction of past hurricanes (first questions), summarizing results from field studies of ecosystem response to wind damage (second questions), and outlining possible directions for future research (third questions).

Study Regions

New England

The six New England states plus adjoining New York City and Long Island comprise a region of about 175,000 km². Topographic relief varies from coastal plains to mountains of 1000-m elevation (maximum 1900 m) to the west and north. The climate is temperate, with significant variation (especially in temperature) resulting from differences in elevation, latitude, and distance from the ocean. Major life zones include Northern Hardwoods-Spruce-Fir (northern New England), Transition Hardwoods (central New England), Central Hardwoods (southern New England), and Pitch Pine-Oak (Cape Cod and scattered sand plains) (Westveld 1956; Foster and Aber in press).

New England is affected by mature and late-stage Atlantic hurricanes that form at lower latitudes and approach from the south (figure 2.1a). Most hurricanes weaken by the time they reach New England, though an intensity of category 3 on the Saffir–Simpson scale (sustained wind speeds of 50–58 m/s) is not unusual. Because hurricanes derive most of their energy from warm ocean water, hurricanes that pass over inland areas to the south and west generally cause little wind damage in New England. Similarly, because the highest surface winds are normally located to the right of the storm track, storms that pass offshore to the east also tend to cause less wind damage. The greatest impacts result from hurricanes that travel northward over the warm waters of the Gulf Stream and pass directly over New England. Rapid forward motion helps to offset the effects of weakening as the storms come over land or over the cold waters of the Gulf of Maine (Smith 1946).

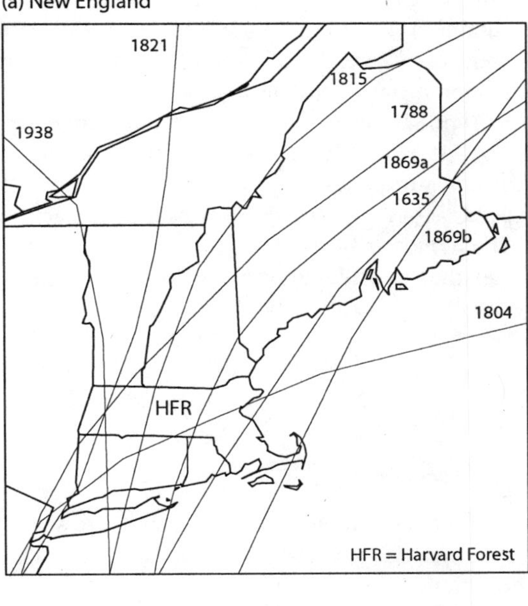

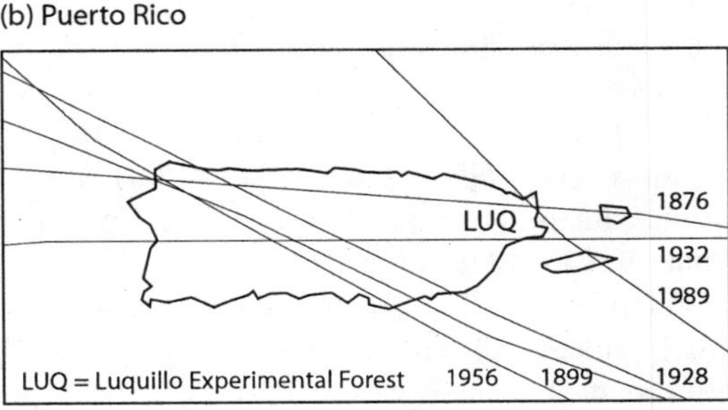

Figure 2.1 Tracks of hurricanes that caused F3 damage on the Fujita scale. (a) New England, 1620–1997. Reprinted with permission from *Ecological Monographs*. (b) Puerto Rico, 1851–1997 (adapted from Boose et al. 2001, in press). Adapted and reprinted with permission from *Ecology*.

Puerto Rico

Puerto Rico, the easternmost and smallest of the Greater Antilles, is a mountainous island roughly 55 by 160 km in size. The island is characterized by an east-west mountain ridge terminating in the northeast in the Luquillo Mountains (maximum elevation 1075 m). The climate is subtropical, with significant variation (especially in precipitation) caused by the interaction of topography and the prevailing north-easterly trade winds. Major life zones include Subtropical Dry Forest (southwest), Subtropical Moist Forest (elsewhere at lower elevations), Subtropical Wet Forest and Rain Forest (higher elevations), and Lower Montane Wet Forest and Rain Forest (near highest summits) (Holdridge 1946; Ewel and Whitmore 1973).

Hurricane frequency in Puerto Rico is among the highest in the North Atlantic basin. Most storms approach from the east and southeast (figure 2.1b). Though hurricanes sometimes originate close to the island, the most intense storms often form off the west coast of Africa and approach Puerto Rico at or near maximum intensity. On rare occasions hurricanes reach category 5 on the Saffir–Simpson scale (sustained wind speeds above 70 m/s), with devastating impacts on both human and natural systems. The mountainous topography of the island affects the overall intensity of hurricanes that make landfall as well as the extent of local protection from damaging winds (Boose et al. 1994, in press).

Methods

Reconstructing Historical Hurricanes

The frequency of hurricanes and the life span of trees are such that the long-term impacts of hurricanes on forests can be understood only at a scale of centuries. For much of the North Atlantic basin, the historical record provides evidence of past hurricanes over the last 300 to 500 years since European settlement. At the Harvard Forest we developed a method for interpreting this historical record using a combination of wind damage assessment and meteorological modeling. This historical-modeling method and its application to hurricanes in New England and Puerto Rico are outlined here (for more details, see Boose et al. 2001, in press). The computer models and historical data used in our analyses are available on the Harvard Forest web page (http://harvardforest.fas.harvard.edu).

The first task was to create a list of hurricanes in each study region for which there was historical evidence of wind damage. We relied on the works of other scholars to identify significant hurricanes during the early period (e.g., Salivia 1950; Ludlum 1963; Millas 1968). However, our assessment of the impacts of each storm was based, wherever possible, on contemporary accounts, mostly newspapers for the later period and letters, diaries, and government documents for the earlier period. As expected, the number of historical reports was greater for recent and/or severe hurricanes. Efforts focused on obtaining good regional coverage for each storm.

Actual wind damage in each hurricane was classified using Fujita's system (1971) for assessing wind damage in tornadoes and hurricanes. Fujita's damage classes extend from F0, minor damage caused by gale or storm force winds, to F5, extreme damage in the most severe tornadoes. Each F-scale (Fujita scale) class is defined by specified levels of damage to common cultural and biological features of the landscape. For New England we used a slightly modified version of Fujita's original system, whereas for Puerto Rico we made additional changes to account for different building practices and higher wind speeds (table 2.1). Though wind damage to exposed forests is strongly dependent on composition and structure (Foster and Boose 1992), as a general rule, F0 = loss of leaves and branches, F1 =

Table 2.1 The Fujita scale of wind damage, modified for application to New England and Puerto Rico.

	F0 Damage	F1 Damage	F2 Damage	F3 Damage
Sustained wind speed (m/s)[a]	18–25	26–35	36–47	48–62
Trees	Leaves and fruit off, branches broken, trees damaged	Trees blown down	Extensive blowdowns	Most trees down
Crops	Damaged or blown down			
Masonry buildings	Minor damage	Roof peeled, windows broken, chimneys down	Unroofed	Blown down or destroyed
Wood houses[b]	Minor damage	Roof peeled, windows broken, chimneys down	Unroofed or destroyed	3+ blown down or destroyed in same town
Unspecified buildings, wood-zinc houses[c]	Minor damage	Unroofed or damaged	Blown down or destroyed	50% or more blown down or destroyed in same town[d]
Barns, churches, town halls, cottages[e]	Minor damage	Unroofed, steeple blown down, damaged	Blown down or destroyed	
Shacks, sheds, outbuildings, warehouses	Minor damage	Unroofed, blown down, or destroyed		
Huts[f]	Damaged	Blown down or destroyed		
Furniture, bedding, clothes	Not moved	Blown out of building		
Masonry walls, radio towers, traffic lights	No damage	Blown down		
Utility poles	Wires down	Poles damaged or blown down, high-tension wires down		
Signs, fences	Damaged	Blown down		
Autos	No damage	Moving autos pushed off road	Stationary autos moved or pushed over	Heavy autos lifted and thrown

(continued)

Table 2.1 *Continued*

	F0 Damage	F1 Damage	F2 Damage	F3 Damage
Trains	No damage	Pushed along tracks	Boxcars pushed over	Trains overturned
Marinas, small airplanes	Minor damage	Destroyed		
Small boats	Blown off mooring	Sunk		
Missiles	None	None	Light objects, metal roofs	

Notes: PR = Puerto Rico. [a]Corresponding sustained wind speed values are derived from Fujita's equations (1971), assuming a wind gust factor of 1.5 over land. [b]Described as well constructed or owned by a wealthy person (PR); also municipal buildings (PR). [c]Constructed with light wood frame and metal roof (PR). [d]F2 assigned if buildings described as rural or poor (PR). [e]Also schools, sugar mills, commercial buildings, and military buildings (PR). [f]Constructed of palm leaves or similar materials (PR). Adapted from Boose et al. 2001, in press.

scattered blowdowns (small gaps), F2 = extensive blowdowns (large gaps), and F3 = most trees down.

Reports of wind damage were collected and indexed by town to create a database for each hurricane. Each report that contained sufficient information was assigned an F-scale value based on the highest level of damage reported. Care was taken to exclude coastal damage caused by the storm surge, valley damage caused by river flooding, and (in Puerto Rico) local damage caused by landslides. Regional maps of actual damage were then created for each hurricane using the maximum F-scale value assigned for each town. These maps provided a quantitative, spatial assessment of actual damage for each storm.

Meteorological modeling complemented the asssessment of actual wind damage by providing informed estimates for sites that lacked data as well as a complete regional picture of the impacts of each storm. The range and quantity of meteorological data for New England and Puerto Rico have, of course, increased dramatically since European settlement as a result of more widely distributed populations, better historical records, and steady improvements in technology (Ludlum 1963; Neumann et al. 1987). For hurricanes since 1851, our main source of meteorological data was the HURDAT (Hurricane Data) database maintained by the U.S. National Hurricane Center (NHC), which provides estimates of hurricane position and maximum sustained wind speed every 6 hours. HURDAT is available on the NHC web page (http://www.nhc.noaa.gov) (see Landsea et al. 2001 for information on current revisions). HURDAT values were modified in a few cases (see Boose et al. 2001, in press), including simulated weakening as hurricanes passed over the island of Puerto Rico (where actual damage maps showed a consistent pattern of storm weakening over the interior mountains). For New England, track and wind speed data for hurricanes before 1851 were reconstructed from contemporary accounts and from analyses by Ludlum (1963). Though actual measurements of wind speed are not available for the early period, observers often left careful records of wind speed (in qualitative terms) and direction (eight points of the compass) and noted the times of peak wind, wind shift, lulls, and changes in cloud cover and precipitation intensity. For Puerto Rico, reconstructions of hurricanes before 1851 were not attempted because of the lack of reliable estimates of hurricane tracks.

A simple meteorological model (HURRECON), based on published empirical studies of many hurricanes, was used to reconstruct the impacts of each storm (Boose et al. 2001, in press). HURRECON uses information on the track, size, and intensity of a hurricane, as well as the cover type (land or water), to estimate surface wind speed and direction. The model also estimates wind damage on the Fujita scale by using the correlation between maximum wind velocity and wind damage proposed by Fujita (1971; table 2.1). The model was parameterized and tested by comparing maps of actual and reconstructed F-scale wind damage for recent hurricanes, where the meteorological data used as input to the model were independent of the maps of actual damage created from historical accounts. The model was then used to reconstruct earlier storms. The resulting maps of reconstructed wind damage for each hurricane were compiled to create regional maps showing the number of storms at different damage levels (F0–F3) during the period of study. Model estimates were also compiled for individual sites to create time lines of hurricane damage and plots of wind damage as a function of peak wind direction. The latter were used in combination with a simple topographic exposure model (EXPOS) to create landscape-level maps of exposed and protected areas for individual hurricanes (Boose et al. 1994).

Field Studies of Ecosystem Response

Ecosystem response to individual hurricanes can be measured directly through field studies of actual or simulated wind damage. Research at the Harvard Forest has focused on two areas: (1) long-term studies of forest recovery in central New England after the 1938 hurricane, and (2) intensive studies of a simulated hurricane blowdown created at the Harvard Forest in October 1990. The 1938 hurricane, a category 3 storm at landfall, caused widespread F2 damage across much of central New England. Studies have focused on patterns of damage and long-term changes in forest composition and structure. In the 1990 blowdown experiment, mature trees in an upland, 0.8-ha 75-year-old red oak–red maple (*Quercus rubra–Acer rubrum*) stand were pulled over with a logging winch to closely approximate the effects of the 1938 hurricane. Studies have focused on vegetation mortality and regeneration, community dynamics, and ecosystem processes.

At the Luquillo Experimental Forest (LUQ), research has focused on the aftermath of Hurricane Hugo in September 1989. This storm, also a category 3 hurricane at landfall, caused widespread F2 damage across much of eastern Puerto Rico. Ecosystem response was investigated at various sites throughout the LUQ, with intensive studies at two sites. The first site, the Bisley watershed in the northeastern section of the LUQ, suffered extensive blowdowns. The second site is the Hurricane Recovery Plot at El Verde in the northwestern section of the LUQ. This area experienced widespread defoliation and branch break as well as scattered blowdowns. Both sites are located in the Tabonuco forest zone at elevations of 300–500 m. Studies have focused on patterns of damage, vegetation mortality and regeneration, community dynamics, ecosystem processes, and impacts on animal populations.

Results for New England

Historical Hurricanes

Sixty-seven hurricanes during the period 1620–1997 were selected for detailed analysis, including all hurricanes since 1851 that tracked within 200 km of New England according to HURDAT, and all earlier hurricanes for which Ludlum (1963) presented evidence of F1+ wind damage in the study region. Newspapers were often the best source of information for later hurricanes, especially the *Boston Globe* and the *New York Times* for storms since 1871, and various local newspapers, depending on the area of impact, for the period 1700–1870. Evidence for earlier storms (1620–1699) was drawn mainly from personal diaries and town histories (especially at the American Antiquarian Society, Worcester, Mass.). Primary sources cited in Ludlum (1963) were consulted wherever possible.

Our analysis of temporal variation in New England hurricanes was based on 52 hurricanes whose maximum actual wind damage in the study region equaled or exceeded F1. At a seasonal scale, 90% of such hurricanes (and all hurricanes that caused F3 damage) occurred during the months of August, September, and October. About 30% of these storms occurred in October and November, during or after leaf senescence in New England, when deciduous trees are much less likely to suffer wind damage. At an annual scale, there were 4 years with two hurricanes in the same year; in two of these years both storms caused F2 damage, whereas in one year both storms caused F3 damage, suggesting a higher than average hurricane intensity during these exceptionally active years. At a decadal scale, the number of hurricanes since 1851 varied from a minimum of 0 storms in the 1910s to a maximum of 4 storms in the 1950s, and evidence suggests that such multidecadal variation was present during the entire historical period (figure 2.2a). At a centennial scale, there was no clear trend in the timing of hurricanes causing F3 damage; the greatest number occurred in the nineteenth century. At lower levels of damage, fewer storms were recorded in the seventeenth and eighteenth centuries than in the nineteenth and twentieth centuries, probably because of improvements in meteorological observations and records since the early nineteenth century.

Our analysis of spatial variation in hurricane impacts was based on meteorological reconstructions of each storm using the HURRECON model. Results for F0+, F1+, and F2 damage were based on the periods 1871–1997, 1800–1997, and 1620–1997, respectively, to maximize the observation period while minimizing the likelihood that storms of a given magnitude escaped historical notice. The frequency of F0 events was probably underestimated, because F0 damage could result from storms not covered in this study. At a regional scale, the frequency and

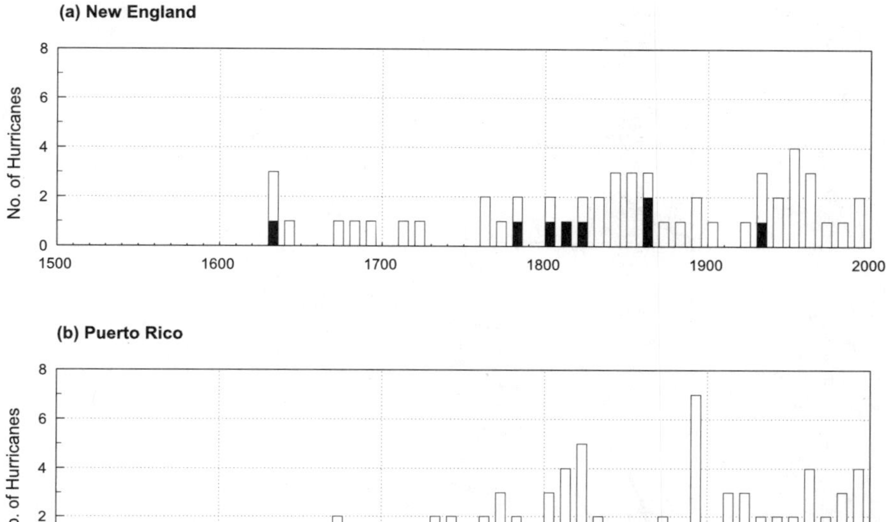

Figure 2.2 Number of hurricanes by decade with maximum reported damage equal to F1–F2 (white) or F3 (black). (a) New England, 1620–1997. (b) Puerto Rico, 1508–1997 (adapted from Boose et al. 2001, in press). Adapted and reprinted with permission from *Ecological Monographs* and *Ecology,* respectively.

intensity of hurricane wind damage decreased from southeast to northwest across New England (figure 2.3a). These gradients result from the consistent direction of the storm tracks, the shape of the coastline, and the tendency for hurricanes to weaken rapidly over land or over cold ocean water north of the Gulf Stream. At a site scale, estimated mean return intervals for F0+, F1+, and F2 damage at the Harvard Forest were 10 years, 20 years, and 125 years, respectively (figure 2.4a), with the highest winds from the southeast. At a landscape scale, there was a gradient of hurricane impacts across the town of Petersham, Massachusetts (the location of Harvard Forest), with reduced impacts in protected valleys and scattered lee hill slopes, though most of the gently rolling terrain was fully exposed to all storms.

Ecosystem Response

Results from field studies of the 1938 hurricane and the Harvard Forest blowdown experiment are summarized in this section.

Damage patterns. In the 1938 hurricane, wind damage on exposed sites increased with stand age and height and decreased with density. Conifer stands, mostly white and red pine (*Pinus strobus* and *P. resinosa*), were more susceptible to damage than hardwood stands. Fast-growing pioneer species were generally

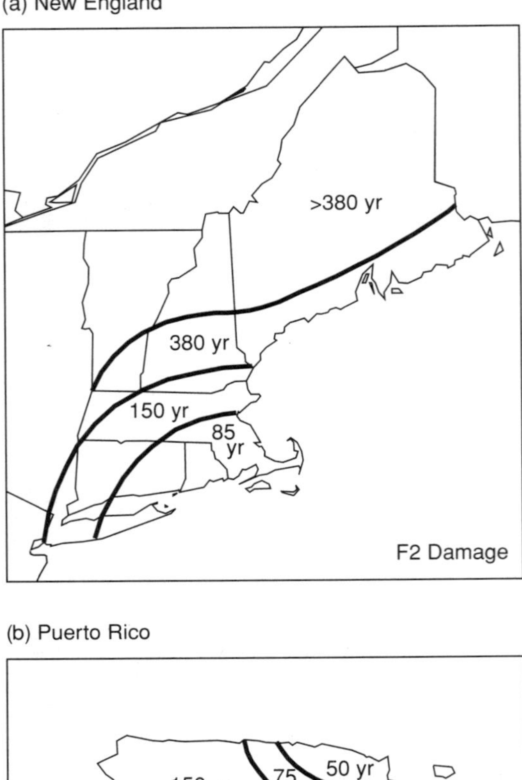

Figure 2.3 Regional gradients in reconstructed hurricane damage showing mean return intervals. (a) New England, F2 damage, 1620–1997. (b) Puerto Rico, F3 damage, 1851–1997 (adapted from Boose et al. 2001, in press). Adapted and reprinted with permission from *Ecological Monographs* and *Ecology,* respectively.

more susceptible than slower growing, shade-tolerant species. The number of damage patches decreased exponentially with patch size, with most patches < 2 ha (Foster 1988; Foster and Boose 1992). In the blowdown experiment, the initial rate of tree mortality was low. Survival of uprooted and broken trees exceeded 75% after the first year and remained above 40% after 4 years. Survival rates varied considerably with damage type and species. Standing trees had the same mortality (4%) as the control plot. Tree basal area and density declined initially by more than 70% (Foster et al. 1997; Cooper-Ellis et al. 1999).

Vegetation regeneration. In the 1938 hurricane, dramatic changes in forest

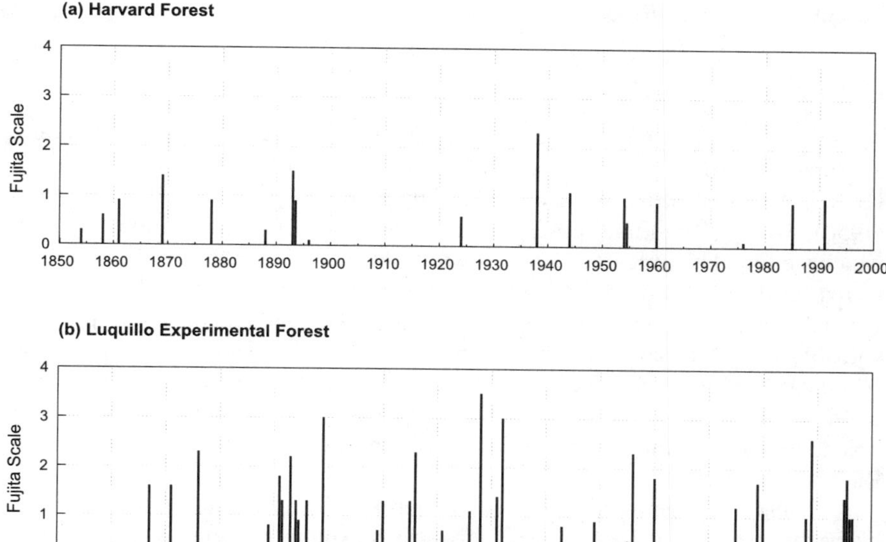

Figure 2.4 Reconstructed hurricane damage by year for 1851–1997. (a) Harvard Forest. (b) Luquillo Experimental Forest (adapted from Boose et al. 2001, in press). Adapted and reprinted with permission from *Ecological Monographs* and *Ecology,* respectively.

composition were probably the result of previous agricultural land use and extensive salvage operations after the hurricane (Foster and Boose 1995; Foster et al. 1997). In the blowdown experiment, canopy cover quickly reestablished through releafing and sprouting of damaged trees, increased growth in understory plants, and seedling establishment on disturbed microsites. Sapling and sprout numbers increased significantly, with only a slight change in composition. Increased light levels resulted in greater diameter and height growth of saplings and sprouts than in the control plot. Net ecosystem productivity declined. Litterfall decreased immediately after disturbance but returned to 71% of control levels after 4 years (Foster et al. 1997; Carlton and Bazzaz 1998a and b; Cooper-Ellis et al. 1999).

Soils and nutrient cycling. In the blowdown experiment, 8% of the soil surface was covered by new uproot mounds and pits. Soil moisture remained unchanged, and soil temperature increased only slightly. There was little or no change in nitrogen (N) cycling. Net N mineralization rates did not change during the first year. Net conversion of ammonium to nitrate increased, but absolute nitrification rates were low. Net fluxes of carbon dioxide and methane remained unchanged. Net nitrous oxide emissions were lower in the blowdown than in the control, though rates at both sites were extremely low (Bowden et al. 1993; Foster et al. 1997; Cooper-Ellis et al. 1999).

Results for Puerto Rico

Historical Hurricanes

A total of 143 hurricanes over the period 1508–1997 were investigated, including all hurricanes since 1851 that passed within 500 km of Puerto Rico according to HURDAT and all earlier hurricanes that impacted the island according to Salivia (1950), Millas (1968), and other scholars. Of these, a total of 85 hurricanes for which we found historical evidence of wind damage (F0+) in Puerto Rico were selected for detailed analysis. A wide range of Puerto Rican newspapers have provided direct accounts of hurricane impacts since 1876, and Salivia (1950) provided valuable secondary information. Evidence for earlier hurricanes was drawn from secondary studies by Salivia (1950), Millas (1968), and other scholars. Wherever possible these works were supplemented by primary sources (letters and other documents) from the University of Puerto Rico and the General Archives of Puerto Rico in San Juan.

Our analysis of temporal variation in Puerto Rican hurricanes was based on 73 hurricanes whose dates are known and whose maximum reported wind damage in Puerto Rico equaled or exceeded F1. At a seasonal scale, 84% of these hurricanes occurred during the months of August and September. At an annual scale, there were 7 years with two hurricanes in the same year, though none of these caused F3 damage. At a decadal scale, the number of hurricanes since 1851 varied from a minimum of 0 storms in the 1850s to a maximum of 7 storms in the 1890s, and evidence suggests that such multidecadal variation was present over the entire historical period (figure 2.2b). At a centennial scale, the number of F3 hurricanes was fairly constant during the historical period, with the greatest number in the nineteenth and twentieth centuries and the smallest number in the eighteenth century. At lower levels of damage, the number of F2 hurricanes increased steadily over time and the number of F1 hurricanes increased steadily until the nineteenth century. These trends are probably the result of improvements in meteorological observations and records over time.

Our analysis of spatial variation in hurricane impacts was based on meteorological reconstructions of storms since 1851 with the HURRECON model. The frequency of F0 events was probably underestimated, since F0 damage could result from storms not covered in this study. At a regional scale, the frequency of higher levels of wind damage decreased from east to west across the island, though the entire region was occasionally subject to F3 damage (figure 2.3b). These gradients result from the consistent direction of the storm tracks and the tendency for hurricanes to weaken as they pass over the island. At a site scale, estimated mean return intervals for F0+, F1+, F2+, and F3 damage in the LUQ were 4 years, 5 years, 20 years, and 50 years, respectively (figure 2.4b), with the highest winds from the northeast. At the landscape scale, the steep mountain topography produced striking differences in predicted impacts on the north and south slopes of the Luquillo mountains, with greatly reduced impacts on the south slopes.

Ecosystem Response

Results from field studies of Hurricane Hugo at various sites across the LUQ are summarized in this section. (For more details, see Walker et al. 1991, 1996.)

Damage patterns. Across the LUQ, wind damage was greatest on exposed northern slopes and in eastern sections closer to the hurricane track (Scatena and Larsen 1991). At El Verde, more than half of the trees were severely defoliated (56%), whereas some trees were snapped (11%) or uprooted (9%), though overall mortality was low (7%) (Walker 1991). At Bisley, aboveground biomass was reduced by 50%, and there were broad areas with 75–100% of trees blown down (Boose et al. 1994; Scatena et al. 1996). Nitrogen and phosphorus content in the leaf litter was about 30% and 100% greater at both sites, respectively, than prehurricane levels (Lodge et al. 1991).

Vegetation regeneration. Canopy cover was rapidly reestablished through releafing and sprouting of damaged trees and recruitment of pioneer species. Tree biomass and density decreased sharply and then returned to prehurricane levels, with no significant change in species richness. Aboveground net primary productivity rose and fell to prehurricane levels as a result of abundant regeneration of pioneer species. Temporary increases in light levels caused herb and woody seedling biomass to increase and then gradually decline. There was a transient rise and fall in forest floor biomass. Total fine litterfall decreased sharply and then increased, but did not reach prehurricane levels after 5 years. Fine root biomass decreased sharply, with little recovery after 5 years (Walker 1991; Zimmerman et al. 1994; Scatena et al. 1996; Zimmerman et al. 1996).

Soils and nutrient cycling. Soil temperature remained unchanged, and soil moisture increased only slightly. There were short-term increases in net N mineralization, net nitrification, and nitrous oxide fluxes, and short-term decreases in methane and carbon dioxide fluxes. There were also short-term increases in soil nutrient pools and nutrient concentrations in groundwater and streams that may have resulted from increased inputs, increased decomposition rates, increased leaching losses, and reduced nutrient uptake by biota. No significant change in soil organic matter was detected. High nutrient concentrations in pioneer species, reflected in increased aboveground pools of potassium and magnesium, helped to contain nutrient losses (Steudler et al. 1991; Scatena et al. 1996; Silver et al. 1996; Zimmerman et al. 1996).

Animal populations. Populations of many organisms declined immediately after the hurricane because of direct negative impacts on populations or because individuals migrated out of hurricane-damaged areas. Some populations increased and then returned to prehurricane levels because of improved resources, improved habitat, and/or a reduction in predator populations. Other populations declined sharply and then returned to prehurricane or above-hurricane levels, whereas still others declined with little recovery in 5 years (Waide 1991; Covich et al. 1996; Woolbright 1996; Zimmerman et al. 1996).

Discussion

The long-term ecological roles of hurricanes at the Harvard Forest and Luquillo Experimental Forest can be compared in terms of the spatial and temporal distribution of disturbance events (hurricane wind damage) and the corresponding ecosystem response. At a continental scale, the locations of New England and Puerto Rico relative to hurricane patterns in the North Atlantic basin account for historical differences in hurricane frequency and intensity. Hurricane frequency and maximum intensity are significantly higher in Puerto Rico, and average storm duration is longer (because storms move more slowly). As a result, hurricane impacts are both more frequent and more severe than in New England. At a regional scale, gradients of hurricane damage result from track patterns and the tendency for hurricanes to weaken as they pass over land (especially over mountains) or over cold ocean water. In New England these gradients extend from southeast to northwest, and in Puerto Rico from east to west. At a landscape scale, the interaction between local topography and constrained peak wind directions creates a landscape-scale gradient of impacts within the larger regional gradient. In the gently rolling terrain of central New England, only scattered areas were found to have long-term protection from the most damaging hurricane winds. In the mountainous terrain of the LUQ, more extended areas were found to have such protection. At smaller scales, the random nature of hurricane gusts contributes to spatial heterogeneity in patterns of wind damage.

Historical records for New England and Puerto Rico extend back roughly 400 and 500 years, respectively. Hurricane frequency and maximum intensity were higher in Puerto Rico than in New England, but otherwise the temporal distributions were similar. For example, we found no clear evidence of centennial-scale trends for either region. In both regions we found the same multidecadal variation that is well documented for North Atlantic hurricanes in general (Neumann et al. 1987), though the specific patterns were different in the two regions; there is a growing understanding of how such variation is linked to other global climatic factors (Gray et al. 1997). Individual hurricanes tended to be clustered in time, whereas the most extreme events were also the most rare.

A critical factor controlling the impact of hurricanes on ecosystems is the relative length of hurricane return intervals and vegetation life spans. For example, if hurricane damage at a given site occurred only at a millennial scale (comparable to tornadoes in southern New England; Fujita 1987), then we might expect those impacts to be negligible for most of the intervening period. On the other hand, if hurricane-force winds occurred every year (as is the case in some wind-swept alpine areas), then we might expect to find heath or shrub communities that were highly resistant to wind damage. In southern and coastal New England, most trees on exposed sites experience some wind damage from hurricanes during their lifetimes, whereas the maximum size and life span of susceptible species may be limited. In Puerto Rico, most trees on exposed sites experience significant wind damage from hurricanes during their lifetimes, and the maximum size and life span of many species, and possibly the distribution of some species, may be limited.

The mixed hardwood forests of central New England and the Tabonuco forests of the LUQ both exhibited remarkable resiliency to wind damage. In both cases, despite major structural reorganization, there was rapid regeneration of canopy cover through releafing, sprouting, or recruitment, which helped to reduce impacts on soil moisture, temperature, and nutrient cycling processes. A large number of damaged trees (even uprooted trees) survived, at least for a few years. Nutrient retention was high despite initial pulses. Long-term impacts on species composition depended on initial composition and the extent of damage. In general, the effects of individual storms remained visible longer in New England, where growth and decomposition rates are slower than in Puerto Rico. Because of their reduced stature, heavily damaged stands are naturally protected from subsequent wind damage for a period of years or decades.

Many interesting questions remain to be answered for these two regions; a few of these questions are outlined here.

Cumulative impacts of major damage. The cumulative impacts of major wind damage (bole snap or uprooting) are not well understood, and field studies are difficult because return intervals are often measured in decades. In this case a modeling approach that utilized meteorological data from historical storms, topographic data for the study site, damage data from past hurricanes, and information on ecosystem response might be useful. Such an approach could also be used to explore the effects of land use or climate change. The accuracy of such modeling efforts might well improve in the future with improvements in our understanding of topographic control of hurricane winds, the response of individual species to a range of wind speeds, and ecosystem response to wind damage in other forest communities.

Cumulative impacts of minor damage. The cumulative impacts of more frequent minor wind damage (defoliation and branch break) are also not well understood. The creation of new foliage, branches, and sprouts appears to be an important adaptive response in many species, but one that cannot be repeated indefinitely at short intervals. Presumably, the impacts of such damage are more significant when combined with other stresses such as drought, disease, or insect outbreak. Future field studies, especially in tree physiology, may shed light on this question.

Interactions with other disturbances. As mentioned previously, hurricane wind damage may be combined with other direct hurricane impacts (e.g., river floods or saltwater inundation) or secondary impacts (e.g., landslides or fires), whose effects sometimes rival or exceed wind damage, at least at a local scale. Hurricanes may also precipitate extensive logging operations whose long-term effects far surpass those of wind damage alone (Foster et al. 1997). Hurricane wind damage is also strongly dependent on previous disturbance history; for example, agricultural and logging activities in New England and Puerto Rico in recent centuries have strongly affected the impacts of hurricanes on forests in those regions (Foster et al. 1999; Boose et al. 2001). Many of these questions could be explored in future modeling efforts.

Prehistoric hurricane record. In general, the historical-modeling method described in this chapter provides a relatively high degree of accuracy and spatial resolution in its hurricane reconstructions. What it does not provide is millennial-scale data that could be invaluable for studying the frequency of the most intense (and rare) hurricanes, as well as the possible effects of climate change. Because our present understanding of hurricane meteorology is not sufficient to predict the effects of climate change on hurricane frequency and intensity on theoretical grounds alone (Emanuel 1997), there is a growing interest in investigating the relationship between hurricanes and climate in the past. Recently, several new techniques have emerged for studying past hurricanes on a millennial scale, for example, stratigraphic analyses of salt marsh deposits. The historical-modeling method can be used to help calibrate and test such methods for recent centuries (Donnelly et al. 2001).

Acknowledgments   The author thanks D. Foster, D. Greenland, and an anonymous reviewer for helpful comments on the manuscript. The research was supported by grants from the National Science Foundation (DEB-9318552, DEB-9411975, and DEB-9411973) and is a contribution from the Harvard Forest and Luquillo Long-Term Ecological Research Programs.

References

Boose, E. R., K. E. Chamberlin, and D. R. Foster. 2001. Landscape and regional impacts of hurricanes in New England. *Ecological Monographs* 71:27–48.

Boose, E. R., D. R. Foster, and M. Fluet. 1994. Hurricane impacts to tropical and temperate forest landscapes. *Ecological Monographs* 64:369–400.

Boose, E. R., M. I. Serrano, and D. R. Foster. In press. Landscape and regional impacts of hurricanes in Puerto Rico. *Ecology*.

Bowden, R. D., M. S. Castro, J. M. Melillo, P. A. Steudler, and J. D. Aber. 1993. Fluxes of greenhouse gases between soils and the atmosphere in a temperate forest following a simulated hurricane blowdown. *Biogeochemistry* 21:61–71.

Carlton, G. C., and F. A. Bazzaz. 1998a. Resource congruence and forest regeneration following an experimental hurricane blowdown. *Ecology* 79:1305–1319.

Carlton, G. C., and F. A. Bazzaz. 1998b. Regeneration of three sympatric birch species on experimental hurricane blowdown microsites. *Ecological Monographs* 68:99–120.

Cooper-Ellis, S., D. R. Foster, G. Carlton, and A. Lezberg. 1999. Forest response to catastrophic wind: Results from an experimental hurricane. *Ecology* 80:2683–2696.

Covich, A. P., T. A. Crowl, S. L. Johnson, and M. Pyron. 1996. Distribution and abundance of tropical freshwater shrimp along a stream corridor: Response to disturbance. *Biotropica* 28: 484–492.

Diaz, H. F., and R. S. Pulwarty, editors. 1997. *Hurricanes: Climate and socioeconomic impacts*. New York: Springer-Verlag.

Donnelly, J. P., S. S. Bryant, J. Butler, J. Dowling, L. Fan, N. Hausmann, P. N. Newby, B. Shuman, J. Stern, K. Westover, and T. Webb III. 2001. A 700-year sedimentary record of intense hurricane landfalls in southern New England. *Geological Society of America Bulletin* 113:714–727.

Emanuel, K. A. 1997. Climatic variations and hurricane activity: Some theoretical issues. Pages 55–65 in H. F. Diaz and R. S. Pulwarty, editors, *Hurricanes: Climate and socioeconomic impacts*. New York: Springer-Verlag.

Ewel, J. J., and J. L. Whitmore. 1973. Ecological life zones of Puerto Rico and the U.S. Virgin Islands. *USDA Forest Service Research Paper* ITF-18.

Foster, D. R. 1988. Species and stand response to catastrophic wind in central New England, U.S.A. *Journal of Ecology* 76:135–151.

Foster, D. R., and J. D. Aber, editors. In press. *Forests in time: The environmental consequences of 1000 years of change in New England.* New Haven, Connecticut: Yale University Press.

Foster, D. R., J. D. Aber, J. M. Melillo, R. D. Bowden, and F. A. Bazzaz. 1997. Forest response to disturbance and anthropogenic stress: Rethinking the 1938 Hurricane and the impact of physical disturbance vs. chemical and climate stress on forest ecosystems. *BioScience* 47:437–445.

Foster, D. R., and E. R. Boose. 1992. Patterns of forest damage resulting from catastrophic wind in central New England, U.S.A. *Journal of Ecology* 80:79–98.

Foster, D. R., and E. R. Boose. 1995. Hurricane disturbance regimes in temperate and tropical forest ecosystems. Pages 305–339 in M. P. Coutts and J. Grace, editors, *Wind and Trees.* Cambridge: Cambridge University Press.

Foster, D. R., M. Fluet, and E. R. Boose. 1999. Human or natural disturbance: Landscape-scale dynamics of the tropical forests of Puerto Rico. *Ecological Applications* 9:555–572.

Foster, D. R., D. H. Knight, and J. F. Franklin. 1998. Landscape patterns and legacies resulting from large, infrequent forest disturbances. *Ecosystems* 1:497–510.

Fujita, T. T. 1971. Proposed characterization of tornadoes and hurricanes by area and intensity. *SMRP Research Paper* 91. University of Chicago, Chicago, Illinois.

Fujita, T. T. 1987. U.S. Tornadoes: Part one, 70-year statistics. *SMRP Research Paper* 218. University of Chicago, Chicago, Illinois.

Gray, W. M., J. D. Sheaffer, and C. W. Landsea. 1997. Climate trends associated with multidecadal variability of Atlantic hurricane activity. Pages 15–53 in H. F. Diaz and R. S. Pulwarty, editors, *Hurricanes: Climate and socioeconomic impacts.* New York: Springer-Verlag.

Holdridge, L. R. 1946. A brief sketch of the Puerto Rican flora. Pages 81–83 in F. Verdoorn, editor, *Plants and plant science in Latin America.* Waltham, Massachusetts: Chronica Botanica.

Landsea, C, C. Anderson, N. Charles, G. Clark, J. Fernandez-Partagas, P. Hungerford, C. Neumann, and M. Zimmer. 2001. Atlantic hurricane re-analysis project. NOAA-Hurricane Research Division web page (http://www.aoml.noaa.gov/hrd/hurdat).

Lodge, D. J., F. N. Scatena, C. E. Asbury, and M. J. Sanchez. 1991. Fine litterfall and related nutrient inputs resulting from Hurricane Hugo in subtropical wet and lower montane rain forests of Puerto Rico. *Biotropica* 23:336–342.

Ludlum, D. M. 1963. *Early American hurricanes 1492–1870.* Boston, Massachusetts: American Meteorological Society.

Lugo, A. E. 2000. Effects and outcomes of Caribbean hurricanes in a climate change scenario. *The Science of the Total Environment* 262:243–251.

Millas, J. C. 1968. *Hurricanes of the Caribbean and Adjacent Regions.* Miami, Florida: Academy of the Arts and Sciences of the Americas.

Neumann, C. J., B. R. Jarvinen, and A. C. Pike. 1987. *Tropical cyclones of the North Atlantic ocean 1871–1986.* Third revised edition. NOAA-National Climatic Data Center, Asheville, North Carolina.

Salivia, L. A. 1950. *Historia de los temporales de Puerto Rico* (1508–1949). San Juan, Puerto Rico.

Scatena, F. N., and M. C. Larsen. 1991. Physical aspects of Hurricane Hugo in Puerto Rico. *Biotropica* 23:317–323.

Scatena, F. N., S. Moya, C. Estrada, and J. D. Chinea. 1996. The first five years in the reorganization of aboveground biomass and nutrient use following Hurricane Hugo in the Bisley Experimental Watersheds, Luquillo Experimental Forest, Puerto Rico. *Biotropica* 28:424–440.

Silver, W. L., F. N. Scatena, A. H. Johnson, T. G. Siccama, and F. Watt. 1996. At what temporal scales does disturbance affect belowground nutrient pools? *Biotropica* 28:441–457.

Simpson, R. H., and H. Riehl. 1981. *The Hurricane and its impact*. Baton Rouge, Louisiana: Louisiana State University Press.

Smith, D. M. 1946. Storm damage to New England forests. Master's Thesis. Yale University, New Haven, Connecticut.

Steudler, P. A., J. M. Melillo, R. D. Bowden, and M. S. Castro. 1991. The effects of natural and human disturbances on soil nitrogen dynamics and trace gas fluxes in a Puerto Rican wet forest. *Biotropica* 23:356–363.

Turner, M. G., S. L. Collins, A. L. Lugo, J. J. Magnuson, T. S. Rupp, and F. J. Swanson. 2003. Disturbance dynamics and ecological response: The contribution of Long-Term Ecological Research. *Bioscience* 53:46–56.

Waide, R. B. 1991. The effect of Hurricane Hugo on bird populations in the Luquillo Experimental Forest, Puerto Rico. *Biotropica* 23:475–480.

Walker, L. R. 1991. Tree damage and recovery from Hurricane Hugo in Luquillo Experimental Forest, Puerto Rico. *Biotropica* 23:379–385.

Walker, L. R., N. V. L. Brokaw, D. J. Lodge, and R. B. Waide, editors. 1991. Ecosystem, plant, and animal responses to hurricanes in the Caribbean. *Biotropica* 23:313–521.

Walker, L. R., W. L. Silver, M. R. Willig, and J. K. Zimmerman, editors. 1996. Long-term responses of Caribbean ecosystems to disturbance. *Biotropica* 28:414–613.

Westveld, M. 1956. Natural forest vegetation zones of New England. *Journal of Forestry* 54:332–338.

Whigham, D. In press. Impacts of hurricanes on the forests of Quintana Roo, Yucatán Peninsula, Mexico. In A. Gómez-Pompa, M. F. Allen, S. Fedick, and J. J. Jiménez-Osornio, editors, *Lowland Maya area: Three millennia at the human-wildland interface*. New York: Haworth Press.

Woolbright, L. L. 1996. Disturbance influences long-term population patterns in the Puerto Rican frog, *Eleutherodactylus coqui* (Anura: Leptodactylidae). *Biotropica* 28:493–501.

Zimmerman, J. K., E. M. Everham, R. B. Waide, D. J. Lodge, C. M. Taylor, and N. V. L. Brokaw. 1994. Responses of tree species to hurricane winds in subtropical wet forest in Puerto Rico: Implications for tropical tree life histories. *Journal of Ecology* 82:911–922.

Zimmerman, J. K., M. R. Willig, L. R. Walker, and W. L. Silver. 1996. Introduction: Disturbance and Caribbean ecosystems. *Biotropica* 28:414–423.

3

Drought Impacts on Tree Growth and Mortality of Southern Appalachian Forests

Brian D. Kloeppel

Barton D. Clinton

James M. Vose

Aaron R. Cooper

The Coweeta LTER Program represents the eastern deciduous forests of the southern Appalachian Mountains in the United States. Coweeta Hydrologic Laboratory was established in 1934 and hence has a long record of climate measurement and vegetation response to both natural and human disturbance (Swank and Crossley 1988). The general climate of the area is classified as marine humid temperate because of high moisture and mild temperatures (Critchfield 1966; Swift et al. 1988). These conditions have favored the evolution of high species diversity in organisms in the southern Appalachians at many levels.

In recent years, however, Coweeta has experienced several droughts that have caused significant tree growth reduction and increased mortality rates (Swift et al. 1990; Clinton et al. 1993; Vose and Swank 1994; McNulty and Swank 1995). In this chapter, we describe the general climate and features of Coweeta as well as the impact of droughts on tree growth and mortality. The timescale of this climate variability is annual, with the potential for preexisting soil moisture conditions either providing a buffer or further exacerbating the drought conditions.

Coweeta Hydrologic Laboratory

Climate

Mean annual precipitation at Coweeta Hydrologic Laboratory (latitude 35°14' N, longitude 83°26' W) varies from 1798 mm at the base climate station (686 m) to 2373 mm at the high-elevation Mooney Gap climate station (1364 m). Mean annual growing season precipitation, defined as May to October, is 782 mm at the

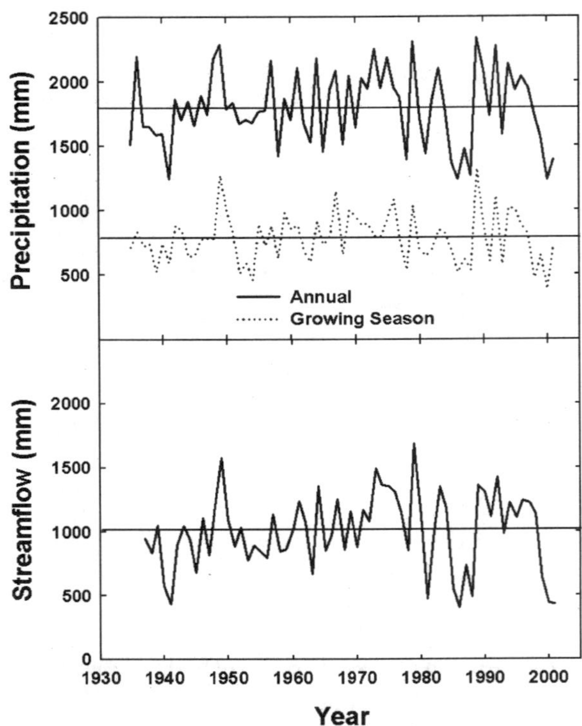

Figure 3.1 Precipitation from 1935 to 2001 (standard gauge 19) and streamflow from 1937 to 2001 (reference watershed 18) at Coweeta Hydrologic Laboratory near Otto, North Carolina, United States. The mean total annual precipitation (1798 mm), mean May–October growing season precipitation (782 mm), and mean streamflow (1011 mm) are shown by straight solid lines.

base climate station (figure 3.1). Mean annual streamflow from watershed 18, a low-elevation reference watershed, is 1011 mm or 56% of precipitation (figure 3.1).

Short-duration thundershowers at Coweeta are typical for midsummer and fall with occurrences of large rainfalls stimulated by tropical disturbances near the Atlantic or Gulf coasts. Forty-nine percent of the 133 storms each year have a total precipitation amount less than 5 mm, and 69% of the annual precipitation falls with an intensity less than 10 mm per hour. Snow is a minor part of the annual precipitation, averaging 2–5% depending on elevation. Snow cover rarely lasts for more than 3 or 4 days, even on the upper slopes. Compared with other mountain sites, wind speeds at Coweeta appear to be low and even imperceptible in the valley bottoms.

Periodic droughts occur in the southern Appalachians. The summer drought of 1925 in Asheville, North Carolina (May to August), generated only 32% of mean precipitation (Hursh and Haasis 1931). At Coweeta during the period 1985–1988, a severe drought occurred, totaling 1837 mm of precipitation deficit (mean minus annual precipitation), a 26% reduction, and 1849 mm of streamflow deficit (mean minus annual streamflow), a 46% reduction. A recent drought totals 1246 mm of precipitation deficit from 1998 to 2001, a 17% reduction, and 1349 mm of streamflow deficit from 1998 to 2001, a 34% reduction.

Physical Features

Coweeta Hydrologic Laboratory is located in the Nantahala Range of the southern Appalachian Mountains approximately 200 km north of Atlanta, Georgia, and 119 km southwest of Asheville, North Carolina. The laboratory comprises two adjacent, east-facing, bowl-shaped basins. The Coweeta Basin encompasses 1626 ha and has been the primary site for watershed experimentation, whereas the 559-ha Dryman Fork Basin has been largely held in reserve for future studies. More than 50 km of streams drain the area, including first- through fifth-order drainages. Ball Creek and Shope Fork are fourth-order streams draining the Coweeta Basin; they join within the laboratory boundary to form Coweeta Creek, a fifth-order tributary that flows 9 km east to the Little Tennessee River. Elevations range from 675 m in the administrative area to 1592 m at Albert Mountain. The diverse topography, including various aspects and slope positions distributed across the elevational gradient within the Coweeta Basin, creates a complex mosaic of environmental conditions that influence hydrologic, climatic, and biological characteristics of forest and stream ecosystems.

Since Coweeta Hydrologic Laboratory was established, numerous weirs have been installed on streams within the laboratory; currently 18 are operational. Stream gaging was initiated on most watersheds between 1934 and 1938. Relief in the watershed (weir to ridge top) averages 300 m on smaller catchments and 550 m on larger watersheds. Side slopes average about 50% and a variety of aspects are present within the basin. Eight Coweeta watersheds have remained relatively undisturbed since the establishment of the laboratory and serve as reference watersheds in paired watershed experiments. Over the 68-year history of Coweeta, a variety of watershed experiments have produced a diverse array of forest and stream ecosystems with respect to composition, structure, productivity, and successional state.

Eight long-term (60+ years) climatic stations are distributed across the basin, and bulk precipitation chemistry has been measured weekly at each station since 1971. In addition, stream water inorganic chemistry has been measured weekly for many of the watersheds since 1971. Over 400 permanent plots, established in 1934, remain undisturbed and provide a basis for assessing forest successional trends. This network of forested plots has been periodically resampled in 5- to 15-year increments to document changes in species composition, tree density, and tree basal area. Other long-term research on processes is facilitated by a five-site environmental gradient, canopy gap sites, riparian focused studies, and stream litter exclusion studies, to name a few.

Biological Features

Forests at Coweeta were traditionally classified as belonging to the oak–chestnut association. However, with the loss of chestnut (*Castanea dentata*) as the dominant canopy species, the area is more appropriately included in the oak–hickory or Appalachian oak association. The plant communities in the Coweeta Basin are distributed in a reasonably predictable mosaic over the highly varied topography in relation to complex moisture and elevational gradients (Bolstad et al. 1998). Generally, deciduous oak species are the dominant canopy species with an abundant evergreen understory component composed primarily of *Rhododendron maximum* and *Kalmia latifolia*. Four major forest types are recognized: (1) northern hardwoods, (2) cove hardwoods, (3) oak (–chestnut), and (4) oak–pine. These forest types exhibit successional change in response to historical disturbances (logging, fire, windstorm, drought, and chestnut blight). Generally, species that were codominants with chestnut at the time of the blight have increased in basal area. More opportunistic species, such as yellow birch, yellow poplar, and red maple, have also increased in relative basal area since the blight. The evergreen understory species, *Rhododendron* and *Kalmia*, have also increased in importance since the 1930s.

In addition to the variety of watershed experiments previously discussed are forest stands in a range of successional status and vegetation types within the Coweeta Basin. Two watersheds (WS 1 and WS 17) have 46-year-old eastern white pine (*Pinus strobus*) plantations (dating from 1957). Past treatments in other experimental watersheds in the basin have included light selection cutting, clearcutting without roads or products removed, commercial clearcutting with and without cable yarding, whole tree harvesting, a combination of thinning and clearcutting, and clearcutting followed by grass planting and then applying herbicide. These treatments have produced naturally regenerating forests ranging in age from 6 to 100 years, thus providing a unique opportunity to assess both the extent of ecosystem alteration following treatment and the patterns, rates, and mechanisms of post-treatment recovery. Superimposed on the environmental gradient in the Coweeta Basin are disturbed ecosystems that contain a wide range of biotic diversity.

Paralleling the diversity of forest ecosystems within the Coweeta Basin is a comparable diversity of stream ecosystems with a variety of bottom substrate types. These diverse headwater streams are characterized both by distinct biotic assemblages, particularly with reference to benthic invertebrates, and by distinct functional processes (e.g., organic matter processing).

Effects of Drought on Overstory Tree Species

Growth Rates

Observed values of species-specific basal area growth rates vary considerably at Coweeta Hydrologic Laboratory (figure 3.2). Because of the wide range in tree diameters, data are expressed on a relativized basis, annual basal area growth (cm^2)

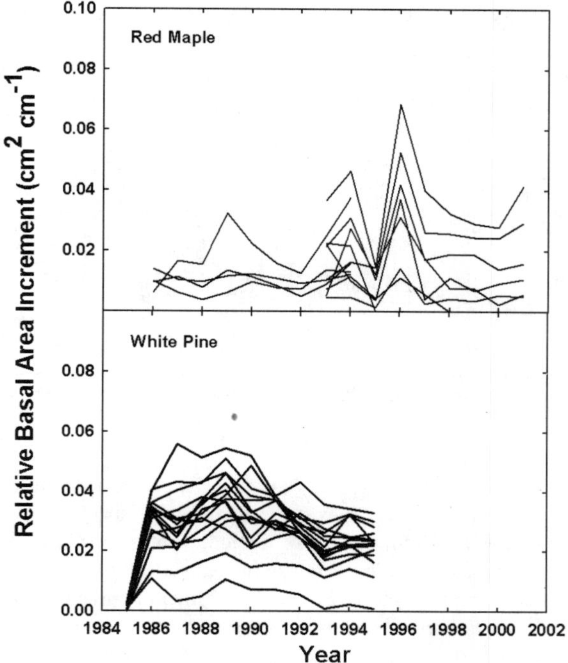

Figure 3.2 The relative basal area increment (cm² tree growth per cm tree diameter) of two selected species at Coweeta Hydrologic Laboratory near Otto, North Carolina. Each connected line represents a single tree over the measurement period. The deciduous red maple (*Acer rubrum*) and the evergreen white pine (*Pinus strobus*) exhibit wide variation of relative basal area increment between trees.

per diameter of the tree (cm). The species measured at Coweeta range from the greatest relative growth rate in *Acer rubrum* (see figure 3.2) to the lowest rate in *Quercus prinus* (data not shown). *Quercus rubra, Q. coccinea,* and the subcanopy species *Oxydendrum arboreum* are intermediate along with the combined "other" species. The understory *Rhododendron maximum* has widely varying growth rates, which likely depends on light availability, whereas the evergreen *Pinus strobus* (see figure 3.2) in watersheds 1 and 17 exhibits some of the greatest growth rates at Coweeta. *Acer rubrum* exists on a variety of sites with a wide range in moisture availability, whereas the *Quercus* species predominate on dry sites except for *Quercus rubra.*

A condensed summarization of the relativised growth rate data indicates that oaks maintained more consistent growth rates during dry and wet conditions, whereas white pine was more sensitive to moisture availability and hence exhibited a growth decline (figure 3.3; McNulty and Swank 1995). This observed difference between oaks and pines suggests that the oaks may be more deeply rooted than the white pines. Hence, the oaks were drawing from a deeper soil water resource that was more available during periods of precipitation decline.

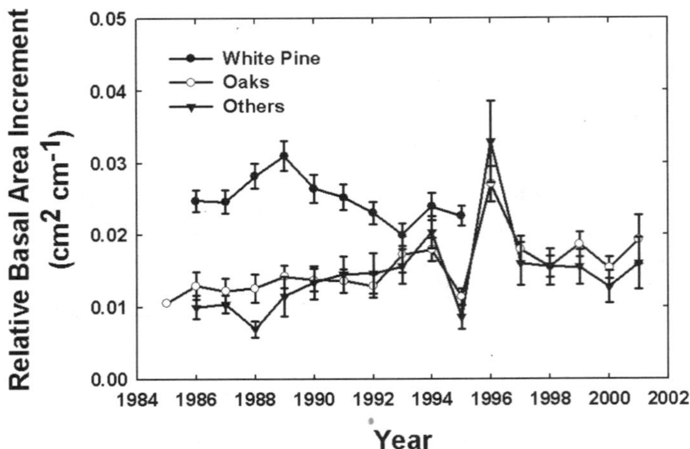

Figure 3.3 Mean (± standard error) relative basal area increment (cm² tree growth per cm tree diameter) for oaks, white pine (*Pinus strobus*), and other species at Coweeta Hydrologic Laboratory near Otto, North Carolina. "Oaks" include white oak (*Quercus alba*), scarlet oak (*Q. coccinea*), chestnut oak (*Q. prinus*), northern red oak (*Q. rubra*), and black oak (*Q. velutina*). "Others" include sweet birch (*Betula lenta*), hickory species (*Carya* spp.), yellow poplar (*Liriodendron tulipifera*), blackgum (*Nyssa sylvatica*), sourwood (*Oxydendrum arboreum*), and black locust (*Robinia pseudoacacia*).

Mortality Caused by Southern Pine Beetle

In the southern Appalachians, most pine species can be considered hosts for the southern pine beetle (SPB; *Dendroctonus frontalis*), although historically the SPB has been associated primarily with yellow pine species (subgenus *Diploxylon*). Yellow pine species native to the southern Appalachians include pitch pine (*Pinus rigida*), shortleaf pine (*P. echinata*), Virginia pine (*P. virginiana*), and table mountain pine (*P. pungens*), as well as small populations of planted and naturalized loblolly pine (*P. taeda*). More recently, eastern white pine (*P. strobus*), which occurs throughout the southern Appalachians, has also been under attack by the SPB, resulting in widespread mortality. There have also been reports of isolated SPB attacks on eastern hemlock (*Tsuga canadensis*).

SPB populations, as with most insect species, are cyclic. The coincidence of SPB outbreaks at epidemic levels with severe and sustained drought can result in significant impacts on pine populations, particularly the yellow pines. The primary mechanism used to combat bark beetle attack is the production of large quantities of oleoresins at the point of attack. Drought stress reduces oleoresin flow and pressure, disabling the trees' primary defense system (Lorio and Hodges 1977). During the period 1971–1988, the land area occupied by yellow pine species in the Coweeta Basin was reduced by 97% as a result of SPB attack (Smith 1991). More recently, watersheds planted in eastern white pine within the Coweeta Basin sustained heavy mortality caused by SPB attack. There are two monocultures of eastern white pine at Coweeta: one in the north-facing watershed 17 and another in the

Table 3.1 Eastern white pine density (stems ha-1) and basal area (m² ha⁻¹) (mean + SE) characteristics for the south-facing watershed 1 and north-facing watershed 17 at Coweeta Hydrologic Laboratory near Otto, North Carolina, USA.

Watershed	Variable	Total	Living	Dead	Beetle-Killed Dead[a]
1	Density	1064 ± 56	695 ± 36	369 ± 39	33 ± 22 (4.7)
	Basal area	66.2 ± 1.9	55.0 ± 2.4	11.2 ± 1.9	2.1 ± 1.4 (3.8)
17	Density	899 ± 34	600 ± 23	299 ± 19	7 ± 6 (1.2)
	Basal area	68.3 ± 2.0	59.3 ± 1.7	9.0 ± 0.8	0.6 ± 0.5 (1.0)

[a]Values in parentheses are the percentage of mortality caused by southern-pine beetle attack.

south-facing watershed 1. On the more exposed watershed 1, mortality from the SPB was estimated to be 5% of the density and 4% of basal area (table 3.1). By contrast, in north-facing watershed 17, only 1% of the density and 1% of the basal area was lost (table 3.1). The apparent difference in susceptibility to SPB attack between the two watersheds suggests the linkage between moisture availability (higher on the north-facing watershed) and susceptibility to attack.

Mortality Caused by Pathogens

The commonly occurring shoestring root rot fungus (*Armillaria mellea*) has been associated with oak mortality species (Wargo 1977) and has been implicated as the primary causal agent (D. J. Lodge, pers. comm.) in mortality observed in the southeast during the 1980s. Nonetheless, there is considerable speculation about whether primary or secondary causes of mortality can be assigned to a single vector (Wargo 1977). The effectiveness of the fungus in causing or contributing to mortality is related to an individual tree's condition, its degree of stress because of low moisture availability (Staley 1965), defoliation (Wargo 1977), or the presence of stem borers (*Agrilus bilineatus* Weber; Dunbar and Stephens 1975). To more efficiently support respiration and other metabolic processes during prolonged periods of severe moisture stress, carbohydrates stored in the root systems as starch are converted into simple sugars (Wargo 1977 and 1996). The fungus is better able to use simple sugars than complex starches and therefore depletes stored energy in roots much more quickly. The added stress on tree physiology often results in mortality or predisposes the individual to mortality during subsequent periods of stress.

By contrast, other species are susceptible to vectors whose optimum influence comes under much different conditions. Fungal pathogens tend to be more virulent under the cool, moist conditions characteristic of periods of abundant rainfall. For example, *Cornus florida* (flowering dogwood) has been under attack by the anthracnose fungus (*Discula destructiva*) since it was first observed in the northeastern United States in 1977 (Daughtrey and Hibben 1983), and at some sites the species is in serious decline. In 1992, the fungus was found in 144 southeastern U.S. counties, particularly in the southern Appalachians and the foothills of the Carolinas and Georgia. Some areas above 900 m in elevation have 100% mortality (Hofacker et al. 1992). The fungus attacks the dogwood's leaves, effectively severing communication between the leaf and the branch and disrupting the exchange of essential metabolites. Mortality results when the fungus moves from the leaves to the shoots and into the main stem, where stem cankers coalesce and girdle the tree. Chellemi et al. (1992) found that the disease was more prevalent and active on cooler, moister northeast-facing slopes than on drier southwest-facing slopes. Similarly, B. D. Clinton et al. (unpub. data, 1999) found that dogwood mortality rates were highest on north-facing slopes but were also higher during a wet period (1989–1998) than during a dry period (1983–1988), whereas oak mortality was higher during the dry period. Hence, in the dogwood example, drought has the effect of mitigating against mortality.

Interactions with Insects and Disease

The dogwood example raises an interesting question: Is an individual tree's susceptibility to drought-related mortality determined, at least in part, by the local conditions under which the individual developed (Waring 1987)? Gram and Sork (2001) have shown that under sufficient selection pressure, even within a localized area, some species can develop distinct genotypes that are associated with fine-scale microtopographic variation or with a specific set of resource availabilities. For example, Tainter et al. (1990) provided evidence suggesting that periods of prolonged moisture stress can result in differential within-species responses. In their study of the effects of drought on radial increments of trees, two populations emerged after a severe drought: a relatively healthy population and a declining population. In some species, gene switching during a fluctuating local climate—to compensate for periods of reduced resource availability (i.e., low moisture availability)—is a common drought-avoidance mechanism. Chang et al. (1996) demonstrated experimentally that genes with a variety of drought-avoidance functions are water-deficit inducible, particularly those that may fulfill a structural role either directly or through participating in the synthesis of cell wall components necessary for maintaining turgor. However, this mechanism may be ineffective where strong within-species genetic selection for specific resource conditions has occurred. That is, under widely fluctuating soil moisture conditions, the capacity for that form of gene expression to aid in the necessary adjustments in water-use efficiency may be exceeded.

Another mechanism responsible for variation in within-species responses to drought occurs when plants undergoing moisture stress incur increased levels of abscisic acid (ABA), which elicits a myriad of physiological responses, such as increased root/shoot ratios and regulation of stomatal function (Nilsen and Orcutt 1996). Long-term exposure to moisture stress, particularly during development, may result in greater sensitivity to ABA, as well as a more "hardened" physiological state, which would allow quicker responses to moisture stress and the maintenance of a higher level of drought resistance (Nilsen and Orcutt 1996). More research is needed to better explain spatial patterns of within- and among-species responses to stress.

A growing body of evidence in the literature supports the notion that the risk of tree death increases with a decreasing growth rate (Pedersen 1998). The rationale behind this assertion is that recovery from periods of stress becomes increasingly difficult and that the effects of repeated periods of stress compound problems of recovery (Pedersen 1999). Wyckoff (1999), through the use of various growth-mortality functions, showed that the probability of mortality increases with a decreasing growth rate. Specifically, dead trees of the two species he examined (*Cornus florida* and *Acer rubrum*) tend to have lower growth rates in the 5 years prior to mortality than their living cohorts. Conversely, fitted mortality functions show that the risk of death decreases with increasing growth for both species. Furthermore, he examined the effect of tree size on growth-mortality functions and found that when small trees and large trees are examined independent of one another, their respective mortality functions diverge, implying that their rates of mortality are driven by tree size.

Species-Specific Mortality

Mortality patterns during severe drought are often species specific (Tainter et al. 1984; Starkey et al. 1989; Clinton et al. 1993; Elliott and Swank 1994). For example, Clinton et al. (1993) found that the species most susceptible to drought-related mortality were members of the red oak group (particularly *Quercus coccinea*) and *Carya* spp. This pattern of mortality was observed across the southeastern region during the mid- to late 1980s (Starkey et al. 1989; Stringer et al. 1989; Oak et al. 1991). The same pattern of mortality was observed in other studies at Coweeta. B. D. Clinton et al. (unpubl. data, 1999) examined tree mortality on two opposing (north- and south-facing) mixed hardwood watersheds in the Coweeta Basin. The period of study covered 18 years and was generally split between an extremely dry period (1984–1988) and a period of above-average precipitation (1989–1997; table 3.2). During the dry period, annual precipitation averaged 20% less than the long-term (60+ years) mean, and, during the wetter period, precipitation averaged 12% above the long-term mean (table 3.2). In the study of Clinton et al., mortality varied considerably between watersheds and within species. On south-facing watershed 2 for the period 1983–1989, the highest mortality rates by species ranked *Carya* spp. > *Q. velutina* = *Oxydendrum arboreum* = *Acer rubrum*; in 1998, mortality rates for that watershed ranked *Cornus florida* > *A. rubrum* > *Liriodendron tulipifera* = *Carya* spp. = *O. arboreum*. On north-facing watershed 18, mortality rates ranked *C. florida* > *Q. prinus* > *A. rubrum* > *Q. velutina* > *Carya* spp. > *O. arboreum*. Even though watershed 2 was subject to the same meteorological variation over the sampling period, mortality rates were less significant for watershed 2 than for watershed 18. For example, aboveground woody net primary productivity (ANPP) for watersheds 2 and 18 for the period 1983–1998 were 3.4 and 2.1 Mg ha^{-1} yr^{-1}, respectively. The lower productivity for watershed 18 resulted from high rates of mortality following the earlier drought. For the oak species, variation in rates of mortality were considerable (table 3.3).

Drought as a Disturbance Regime

Severe drought has been implicated as a contributing factor to recent accelerated rates of tree mortality in the southeastern United States (Tainter et al. 1984; Starkey

Table 3.2 Comparison of rainfall between two sampling periods at Coweeta Hydrologic Laboratory near Otto, North Carolina, USA

Period	1984–1988	1989–1997
Years	5	9
Mean annual precipitation (mm)	1431	2010
Relative to long-term mean (%)	−20	+12
Mean growing season precipitation (mm)	634	913
Relative to long-term mean (%)	−14	+17
Number of consecutive growing season droughts and deficit range (%)	4 (−13 to −34)[a]	0 (−23 and −26)

[a]Values in parentheses for the number of consecutive growing season droughts represent the range of the deficit relative to the long-term mean.

Adapted from Clinton et al., unpubl. data, 1999.

Table 3.3 Mortality patterns of the red oaks (*Quercus rubra, Q. velutina, Q. coccinea*) for the two sampling periods by watershed for stems > 10 cm dbh at Coweeta Hydrologic Laboratory near Otto, North Carolina, USA

	South-Facing Watershed 2		North-Facing Watershed 18	
	1983–1989	1990–1998	1983–1989	1990–1998
Dead stems (# ha⁻¹)	8	2	38	16
Live stems (# ha⁻¹)	24	25	77	70
Mortality (%)	25	7	33	19
Total dead wood biomass (kg ha⁻¹)[a]	5615	8597	10777	14446
Total live wood biomass (kg ha⁻¹)[b]	45592	55273	87801	88640
Biomass lost (%)[c]	11	14	11	14
Aboveground woody net primary production (kg ha⁻¹ yr⁻¹)	10	1076	679	93

[a]This is the sum of standing and fallen dead red oak stems for a given sampling period.

[b]This represents total red oak wood standing crop.

[c]This is the percentage of the total for a given sampling period.

No adjustment for loss of wood density was made in the calculation of dead biomass.

Adapted from Clinton et al., unpubl. data, 1999.

et al. 1989; Stringer et al. 1989; Clinton et al. 1993). This pulse of mortality may have a long-term impact on stand structure and function (Clark et al. 2002). The structural pattern associated with drought-induced mortality (i.e., standing-dead snags) implies that important types of microhabitats are not produced. For example, species such as pitch pine that require large openings (Barden and Woods 1976) commonly associated with large-scale, wind-induced mortality or wildfire are at a distinct disadvantage. In addition, the lack of a pulse addition of coarse woody debris, typical of wind-induced gap formation, may reduce regeneration opportunities for species such as sweet birch (*Betula lenta*) and eastern hemlock (*Tsuga canadensis*), whose regeneration strategies include "nurse logs" as fresh substrate for seed germination (Burns and Honkala 1990). This is not to say that other important classes of microhabitat are not produced. The standing-dead tree and the shade it casts are an important microhabitat for many organisms and processes (Franklin et al. 1987). Thus, effects of drought-induced mortality may have important influences on micro- as well as macro-level processes (Mueller-Dombois 1987).

We must also begin to assess ecosystem-level impacts of such climatic alteration of the forest structure. Canopy openings and shifts in species composition alter microclimatic factors such as light, temperature, and moisture (B. D. Clinton, unpubl. data, 1999) that regulate nutrient cycling processes. For example, the response of the nitrogen-fixing black locust (*Robinia pseudoacacia*) in large gaps and shifts in litter quality or decomposition rates of leaves of different species are two potential manifestations. The long-term importance of increasing our understanding of drought impacts on forest structure and function is central to anticipating the full impacts of predicted long-term climate change.

Acknowledgments   Components of this work were funded by the National Science Foundation to the Coweeta Long-Term Ecological Research Program (Grant #9632854). We thank Jim Deal, Barry Argo, Sharon Taylor, and Susan Steiner for collection of the tree dendrometer band data. We thank the USDA Forest Service, Coweeta Hydrologic Laboratory, for the collection and management of the precipitation and streamflow data.

References

Barden, L. S., and F. W. Woods. 1976. Effects of fire on pine and pine-hardwood forests in the southern Appalachians. *Forest Science* 22: 399–403.

Bolstad, P. V., W. T. Swank, and J. M. Vose. 1998. Predicting southern Appalachian overstory vegetation with digital terrain data. *Landscape Ecology* 13: 271–283.

Burns, R. M., and B. H. Honkala. 1990. *Silvics of North America*. USDA Forest Service Agriculture Handbook 654, Volume 1. Conifers, 675 pp. Volume 2. Hardwoods, 877 pp.

Chang, S. J., J. D. Puryear, M. A. D. L. Dias, E. A. Funkhouser, R. J. Newton, and J. Cairney. 1996. Gene expression under water deficit in loblolly pine (*Pinus taeda*): Isolation and characterization of cDNA clones. *Physiologia Plantarum* 97: 139–148.

Chellemi, D. O., K. O. Britton, and W. T. Swank. 1992. Influence of site factors on dogwood anthracnose in the Nantahala Mountain Range of western North Carolina. *Plant Disease* 76: 915–918.

Clark, J. S., E. C. Grimm, J. J. Donovan, S. C. Fritz, D. R. Engstrom, and J. E. Almendinger. 2002. Drought cycles and landscape responses to past aridity on prairies of the northern great plains, USA. *Ecology* 83: 595–601.

Clinton, B. D., L. R. Boring, and W. T. Swank. 1993. Canopy gap characteristics and drought influences in oak forests of the Coweeta Basin. *Ecology* 74: 1551–1558.

Critchfield, H. J. 1966. *General Climatology*, second ed., Englewood Cliffs, New Jersey: Prentice Hall.

Daughtrey, M. L., and C. R. Hibben. 1983. Lower branch dieback, a new disease of Northeastern dogwoods. *Phytopathology* 73: 365–365.

Dunbar, D. M., and G. R. Stephens. 1975. Association of two-lined chestnut borer and shoe-string fungus with mortality of defoliated oak in Connecticut. *Forest Science* 21: 169–174.

Elliott, K. J., and W. T. Swank. 1994. Impacts of drought on tree mortality and growth in a mixed hardwood forest. *Journal of Vegetation Science* 5: 229–236.

Franklin, J. F., H. H. Shugart, and M. E. Harmon. 1987. Tree death as an ecological process. *BioScience* 37: 550–556.

Gram, W. K., and V. L. Sork. 2001. Associations between environmental and genetic hetero-geneity in forest tree populations. *Ecology* 82: 2012–2021.

Hofacker, T. H., R. F. Fowler, L. Turner, K. Webster, and M. Reiffe. 1992. Forest insects and disease conditions in the United States 1991. USDA Forest Service, Forest Pest Man-agement, AB-2S. 139 pp.

Hursh, C. R., and F. W. Haasis. 1931. Effects of 1925 summer drought on southern Ap-palachian hardwoods. *Ecology* 12: 380–386.

Lorio, P. L., and J. D. Hodges. 1977. *Tree water status affects induced southern pine beetle attack and brood production.* USDA Forest Service, Research Paper 20-135. Southern Forest Experiment Station, New Orleans, Louisiana. 7 pp.

McNulty, S. G., and W. T. Swank. 1995. Wood $\delta^{13}C$ as a measure of annual basal area growth and soil water stress in a *Pinus strobus* forest. *Ecology* 76: 1581–1586.

Mueller-Dombois, D. 1987. Natural dieback in forests. *BioScience* 37: 575–583.

Nilsen, E. T., and D. M. Orcutt. 1996. *The Physiology of Plants Under Stress: Abiotic factors.* New York: John Wiley and Sons. 689 pp.

Oak, S. W., C. M. Huber, and R. M. Sheffield. 1991. *Incidence and impact of oak decline in western VA*, 1986. USDA Forest Service, SEFES Resource Bulletin SE-123. 16 pp.

Pedersen, B. S. 1998. The role of stress in the mortality of midwestern oaks as indicated by growth prior to death. *Ecology* 79: 79–93.

Pedersen, B. S. 1999. The mortality of midwestern overstory oaks as a bioindicator of envi-ronmental stress. *Ecological Applications* 9: 1017–1027.

Smith, R. N. 1991. *Species composition, stand structure, and woody detrital dynamics asso-ciated with pine mortality in the southern Appalachians.* Masters thesis, University of Georgia, Athens, Georgia. 163 pp.

Staley, J. M. 1965. Decline and mortality of red and scarlet oak. *Forest Science* 11: 2–17.

Starkey, D. A., S. W. Oak, G. W. Ryan, F. H. Tainter, C. Redmond, and H. D. Brown. 1989. *Evaluation of oak decline areas in the South.* USDA Forest Service, Forest Protection Report R8-TR17.

Stringer, J. W., T. W. Kimmerer, J. C. Overstreet, and J. P. Dunn. 1989. Oak mortality in east-ern Kentucky. *Southern Journal of Applied Forestry* 13: 86–91.

Swank, W. T., and D. A. Crossley, Jr. 1988. *Forest Hydrology and Ecology at Coweeta.* Eco-logical Studies 66, New York: Springer-Verlag.

Swift, L. W., Jr., G. B. Cunningham, and J. E. Douglas. 1988. Climatology and Hydrology. Pages 35–55 in W. T. Swank and D. A. Crossley, Jr., editors, *Forest Hydrology and Ecology at Coweeta.* Ecological Studies 66, New York: Springer-Verlag.

Swift, L. W., Jr., J. B. Waide, and D. L. White. 1990. Application of the Z-T extreme event analysis using Coweeta streamflow and precipitation data. Pages 13–18 in D. Green-land, and L. W. Swift, Jr., editors, *Climate Variability and Ecosystem Response: Pro-ceedings of a Long-Term Ecological Research Workshop.* USDA Forest Service General Technical Report SE-65, Asheville, North Carolina.

Tainter, F. H., S. W. Fraedrich, and D. M. Benson. 1984. The effect of climate on growth, de-cline, and death of northern red oaks in the western North Carolina Nantahala Moun-tains. *Castanea* 49: 127–137.

Tainter, F. H., W. A. Retzlaff, D. A. Starkey, and S. W. Oak. 1990. Decline of radial growth in red oaks is associated with short-term changes in climate. *European Journal of Forest Pathology* 20: 95–105.

Vose, J. M., and W. T. Swank. 1994. Effects of long-term drought on the hydrology and growth of a white pine plantation in the southern Appalachians. *Forest Ecology and Management* 64: 25–39.

Wargo, P. M. 1977. *Armillaria mellea* and *Agrilus bilineatus* and mortality of defoliated oak trees. *Forest Science* 23: 485–492.

Wargo, P. M. 1996. Consequences of environmental stress on oak: Predisposition to pathogens. *Annales des Sciences Forestieres* 53: 359–368.

Waring, R. H. 1987. Characteristics of trees predisposed to die. *BioScience* 37: 569–574.

Wyckoff, P. H. 1999. *Growth and mortality of trees in the southern Appalachian Mountains.* Ph.D. dissertation, Duke University, Durham, North Carolina.

4

Climate Variability in the North Central Region: Characterizing Drought Severity Patterns

Stuart H. Gage

Introduction

This chapter examines the spatial and temporal variability and patterns of climate for the period 1972–1991 in the North Central Region of North America (NCR). Since the mid-1970s, climate has become more variable in the region, compared to the more benign period 1950–1970. The regional perspective presented in this chapter characterizes the general climatology of the NCR from 1972 to 1991 and compares the climate to a severe drought that occurred in 1988. This one-year drought was one of the most substantial in the region's recent history, and it had a significant impact on the region's agricultural economy and ecosystems. Petersen et al. (1995) characterize the 1988 drought with respect to solar radiation, and Zangvil et al. (2001) consider this drought from the perspective of a large-scale atmosphere moisture budget. A major reason for the seriousness of the drought in 1988 was the fact that May and June were unusually dry and hot (Kunkel and Angel 1989).

Drought is defined as a condition of moisture deficit sufficient to adversely affect vegetation, animals, and humans over a sizeable area (Warwick 1975). The condition of drought may be considered from a meteorological, agricultural, and hydrologic perspective. Meteorological drought is a period of abnormally dry weather sufficiently prolonged to a point where the lack of water causes a serious hydrologic imbalance in the affected area (Huschke 1959). Agricultural drought is a climatic digression involving a shortage of precipitation sufficient to adversely affect crop production or the range of production (Rosenberg 1980). Hydrologic drought is a period of below-average water content in streams, reservoirs, groundwater aquifers, lakes, and soils (Yevjevich et al. 1977). All of these drought conditions are mutually linked.

The objectives of this chapter are to (1) address the issues of climatic spatial scale to quantify variability of climate in the NCR, (2) examine the characteristics of the 1988 drought as it relates to characteristics of an ecoregion, (3) illustrate a means to quantify drought through a potential plant stress index, and (4) examine the link of regional drought to ecosystem processes. This analysis will provide background and methodology for ecologists, agriculturalists, and others interested in spatial and temporal characterization of climate patterns within large geographic regions.

Background

The North Central Region, one of several regions designated for administrative purposes by the U.S. Department of Agriculture (USDA), encompasses 12 midwestern states containing portions of the "Corn Belt." These are Ohio, Indiana, Illinois, Missouri, Michigan, Wisconsin, Minnesota, Iowa, North Dakota, South Dakota, Nebraska, and Kansas. The Corn Belt is one of the most productive agricultural regions in the world, and it includes six states bordering the Great Lakes (Ohio, Iowa, Michigan, Illinois, Wisconsin, Minnesota) and four states containing portions of the Great Plains (North Dakota, South Dakota, Nebraska, Kansas). Iowa is located in the central Corn Belt, and Missouri, with its agriculturally rich "boot heel," is the region's southern limit. Several great rivers, including the Ohio, the Mississippi, the Red River of the North, the Missouri, and the Platt, all flow within the region. The NCR spans several ecological and climatic zones. Row-crop agriculture, consisting of corn and soybean, dominates the region's agriculture.

The Corn Belt has been cropped intensively since the discovery of its rich soil base and a climate conducive for the production of row crops. These include government-supported crops such as corn, soybeans, and wheat. The intense cultivation of these crops (continuous planting of crops like corn) has required increased use of fertilizers, herbicides, insecticides, and irrigation to sustain high production levels. As a result, many of the region's fragile landscapes are facing critical issues, including reduction of soil fertility through monoculture cropping, loss of topsoil as the result of erosion, increased contamination of groundwater through leaching of chemical inputs, and depletion of subsurface water supplies through irrigation practices. Increasing pressures will be placed on the landscape not only to produce human food and livestock feed for U.S. consumption and foreign export, but also to generate biofuels to satisfy increasing energy demands in the United States (Gever et al. 1986).

The Kellogg Biological Station is one of the Long Term Ecological Research sites (KBS LTER) supported by the National Science Foundation (NSF) and is located in southwest Michigan within the eastern portion of the Corn Belt. One objective of the NSF LTER program is to enable research findings from an LTER site to represent larger ecosystems or management regimes as an aid to regional-scale decision making. "The environmental issues confronted in the second half of the 20th century approached the problem from the perspective of stressor, impact and mitigation. The environmental issues of the coming century will be resolved at the system level. Environmental problems within landscapes and ecosystems will, of necessity, be approached from a regional perspective. It is from a regional perspective that the LTER sites will become a distributed center for the strategic research required to meet these challenges . . ." (Bruce Hayden, pers. comm., 1998). Regional-scale issues include net primary productivity dynamics, land use and cover change, dynamics of crop productivity, and climate change and related plant and insect distribution change. The NSF LTER network has attempted to assist LTER scientists address the regional representation of LTER sites through three workshops that focused on regional analysis and synthesis (Helly et al. 1998; Fountain et al. 1999; Gage et al. 2000b).

The KBS LTER site, with its focus on row crop ecosystems, is representative of crops grown in the Midwest (corn, soybean, and wheat). Because the 12 NCR states produce about 80% of the U.S. corn and soybeans (NASS 2001), an analysis of climate impacts on crop production in the region may provide valuable information on the role of the changing climate on row-crop productivity and future trends of row-crop agriculture in the NCR.

The research objective at the KBS LTER site is to understand the ecological characteristics and drivers of row-crop agriculture with a goal of developing methods to replace chemical subsidies with appropriate ecological management systems. The focus on row-crop agriculture (corn, soybean, and wheat), and the emphasis of the effects of climate on crop production and on the ecological interrelationships associated with crop management, provide an opportunity to investigate the long-term climate dynamics in the NCR and how climate impacts crop productivity at regional scales. However, because research was initiated at KBS in 1989, the year after the 1988 drought, observations on crop production at the site were not available to incorporate into the analysis of this particular drought. Studies of the contribution of row-crop agriculture to greenhouse gas dynamics by Robertson et al. (2000) and of modeling regional crop productivity (Gage et al. 2000a, 2001) provide insight into the dynamics of regional agriculture.

There is an increasing need for an ecological perspective on current and future methods of farming and utilization of agricultural lands for agricultural production. These needs are exemplified by changing dynamics and trends in weather and climate, increasing urbanization of prime farmland, the need to predict future scenarios associated with management of sensitive ecosystems, appropriate management of bioengineered crops, and development of new crop varieties.

Patterns of Maize Yield in the Corn Belt

Corn (*Zea mays*) is a crop central to the region's economy. Crop production in the NCR is an important resource as a national supply of food and by-products as well as a component of the nation's export marketing strategy. The role of weather as a cause of the variability of crop production at local, regional, and national scales is a subject of considerable concern. Variability in the annual yield of crops such as corn is a useful indicator of regional climate patterns because plant growth and biomass accumulation are primarily dependent on weather (temperature, precipitation, and solar radiation). Other causes of variability include soils (texture, water-holding

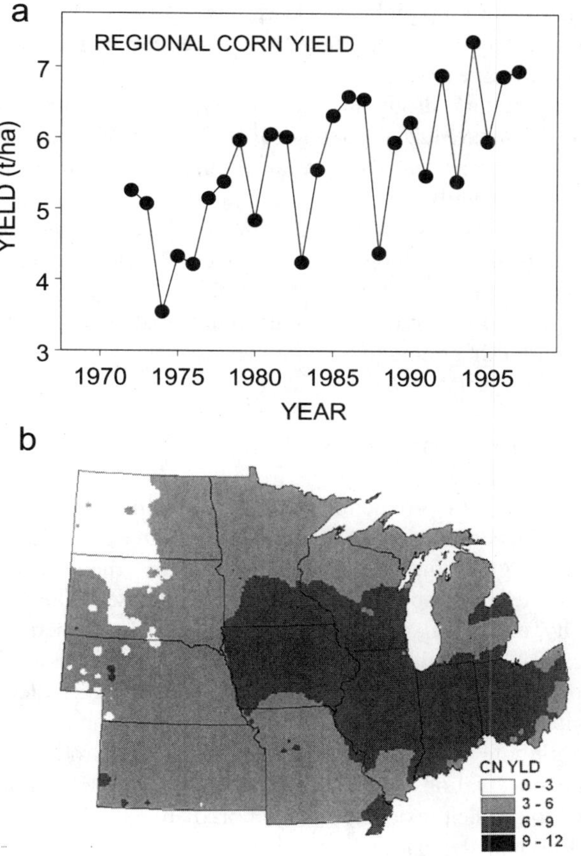

Figure 4.1 (a) Yield (t/ha) over time (1972–1997) for corn and (b) the spatial distribution of average corn yield (t/ha).

capacity) and technology (genetic manipulation, chemical subsidies, irrigation). In agricultural ecosystems, crop productivity or yield refers to the utilizable part of the plant (Tivy 1990). Four major ecological factors determine crop yield in agro-ecosystems: weather, water availability, "negative" biological factors (pests and diseases), and nutrients (Olson 1982; Tivy 1990).

Yield (t/ha) over time (1972–1997) for corn and the spatial distribution of average corn yield is shown in figure 4.1. The interannual variation is largely driven by meteorological factors, whereas the general increase in yield can be attributed to changes in technology (figure 4.1a). The average spatial distribution of corn yield (figure 4.1b) defines the boundaries of the Corn Belt and reflects a combination of the geographic distribution of prime soils and optimal climate for the growth and development of corn.

At regional scales (1,000,000 km^2), climate is the main driving variable of the ecological system (Burke et al. 1991; Bailey 1996). Water is the single most important limiting factors for crop yields worldwide (Tivy 1990). Water shortages cause varying levels of crop stress, contingent on the developmental stage of the plant (Doorenbos and Kassam 1979).

In 1988, an agricultural drought occurred in the North Central Region (Kunkel and Angel 1989; Kunkel 1992; Petersen et al. 1995; Zangvil et al. 2001). This one-year drought was a primary causal agent of a general crop failure in the Corn Belt. Corn yields were unusually low, resulting in a significant reduction in U.S. corn production. Regional mean corn yield in 1988 was 4.2 t/ha, approximately 2 t/ha below the previous and following years (see figure 4.1a). The drought was triggered by meteorological anomalies, including above-average temperatures associated with below-average rainfall that occurred unusually early in the growing season, compared with other years in the climate record.

Database Description Used for Analysis

A database containing crop production statistics and climate data was compiled and organized by members of the USDA Regional Committee (NC94) under the project title "Climate and Agricultural Landscape Productivity Analysis and Assessment." The NC94 committee developed a meteorological database, a database on historical crop productivity, and a soils database for the NCR. The objective of developing these databases was to provide research communities with the ability to map and link annual crop production and monthly weather and soil variables to gain knowledge of the temporal and spatial characteristics of crop production in the region and the effects that climate has on regulating patterns of crop production. The climate database of daily observations associated with each of the 1055 counties in the NCR was constructed to enable the development of predictive models of major row crops in the region. A climate database was developed as part of a long-term study to characterize patterns of climate so that the patterns could be used to investigate linkages to agricultural productivity in the states that comprise the Corn Belt.

The meteorological database consists of daily weather records for a 20-year time period at 1055 locations in the NCR. Daily meteorological data are maximum and minimum temperatures and total precipitation. This contiguous database is composed of interpolated measurements of maximum and minimum temperature and precipitation based on National Weather Service (NWS) cooperative observation network in the region for the period 1972–1991. This climate database contains over 7 million records.

Analysis of Climate Patterns

Spatial Organization

Trends and patterns of climate in the NCR were examined and linked to the ecological regions classified by Bailey (1996). Bailey (1996) used a hierarchy of scales to define ecoregions, including (from larger to smaller scales) domain, division,

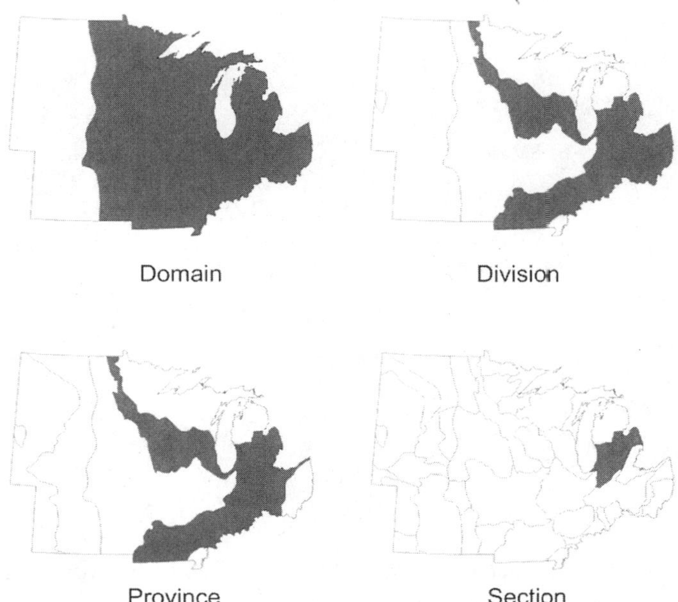

Figure 4.2 Bailey ecoregion classification. The lighter area in each map shows the ecoregion represented by the KBS LTER.

province, and section. The Bailey system of classification was used in this analysis because this hierarchy is primarily based on climatic criteria. Figure 4.2 illustrates distribution of the Bailey ecoregion classification. A weather data set was geographically located at the center of each of the 1055 counties in the region (figure 4.3). Each location, therefore, comprises a 20-year set of observations of daily estimates of temperature and precipitation. These data, in addition to derived variables (see below), were organized and entered into a relational database to provide ease of manipulation and computation.

Monthly summaries of variables were computed from the daily data for each of the 1055 locations in the NCR, as described previously. Variables in the data tables consist of location name (state-county code), mean maximum monthly temperature (°C), mean minimum monthly temperature (°C), and accumulated monthly precipitation (mm) (MPP). Monthly degree-day accumulation (MDD) and a ratio of (MDD/MPP), called HPR, were derived variables. The method for computing degree-days (base 10°C) was that of Baskerville and Emin (1969). The HPR (Heat/Precipitation Ratio) is an index of potential plant stress, assuming that a large accumulation of heat during one month associated with a small accumulation of precipitation during the same month will result in greater potential plant stress.

The interpretation of drought induced plant stress is a focus of the analysis in this chapter. Several indexes have been developed to estimate the severity of drought (Loomis and Connor 1992). Although the Palmer Drought Severity Index (Alley 1984) is widely used, a simpler Heat/Precipitation Ratio (HPR), proposed

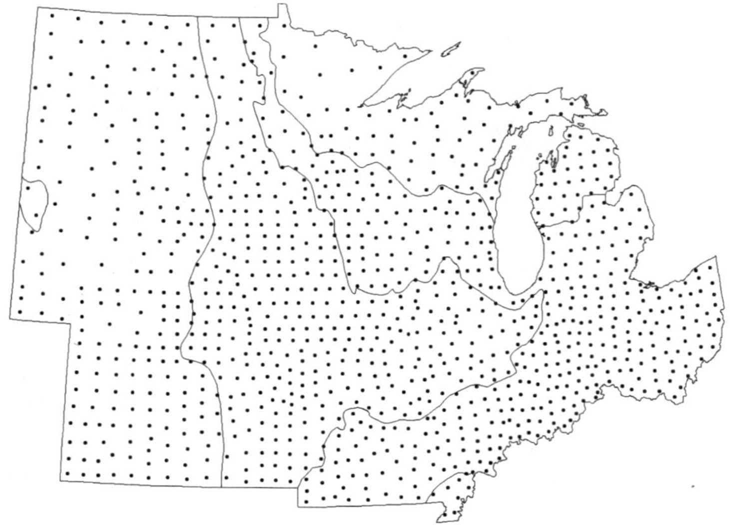

Figure 4.3 Each weather data set is located at the center of each of the 1055 counties in the region.

by Gage and Mukerji (1977), was used in this analysis as an index to characterize potential plant stress. The HPR index is estimated by the equation HPR = MDD/ (MPP + 1), where MDD is the number of heat units accumulated during a period and MPP is the amount of precipitation accumulated during the same interval. This ratio was developed to study (1) the response of grasshoppers to combinations of heat and moisture in arid environments and (2) the subsequent crop loss caused by these insects (Gage and Mukerji 1977, 1978). Although several methods to characterize drought have been developed (Alley 1984; Harouna and Carlson 1994), using HPR to indicate potential plant stress (Gage and Mukerji 1977, 1978) avoids the need for continuous evapotranspiration observations for a large numbers of stations over a long time period—observations that are not available. High HPR values indicate that high heat accumulations are associated with low amounts of precipitation. When this occurs at a monthly timescale, high potential plant stress may occur. For example, if 80 mm of precipitation accumulate in the presence of 400 units of heat >10°C during a month, the HPR would be 400/80 = 5. However, if 40 mm of precipitation were to occur in the presence of 400 units of heat >10°C, the HPR would be 400/40 = 10. In this chapter, the analysis will focus on the patterns of the HPR ratio over time.

Table 4.1 shows the mean maximum HPR for key growing season months for the 20-year period (1972–1991) and the average monthly HPR for 1988 alone. In all cases, the average HPR in1988 is greater than the average HPR for the period of record and was 6.85-fold greater in June. The high plant stress (high HPR) in June 1988 stands out as an unusual event compared to other years in the period of record. The 1988 HPR in May, July, and August was approximately 2.5-fold above average, demonstrating that 1988 was indeed a summer of high plant stress.

Table 4.1 Mean maximum HPR (heat/precipitation ratio) during the period 1972–1991 and HPR values in 1988 in the North Central Region

Month	Mean maximum 1972–1991	Mean maximum 1988	1988/mean
May	11.89	25.42	2.14
June	17.17	117.61	6.85
July	38.68	104.78	2.70
August	26.09	73.78	2.83

Ecological Scaling of Climate Variables

To scale the climate data to ecological regions, each of the ecological categories defined by Bailey (1996) (section, province, division, and domain) was associated with each of the 1055 counties in the NCR. Table 4.2 shows the data points, located at the county centers, in the database of daily weather and annual corn yields in the Bailey ecoregion classification scheme over the 20-year period. This analysis focuses on those ecoregions associated with the LTER in southwest Michigan (Kalamazoo County). The lighter areas in the maps shown in figure 4.2 represent these ecoregions. This hierarchy provides a comparative method to evaluate weather and corn yield patterns associated with the measurements made at the KBS LTER. Although the KBS-LTER was not initiated until 1989, this scaling provides a template for evaluation of the performance of corn yields under a variety of ecological management regimes.

Figure 4.4 shows the monthly patterns of mean temperature (C), precipitation (mm), Heat Units (>10 C), and HPR for the period 1972–1991. Each variable is presented for each of the five levels of regional hierarchy corresponding to the ecological scales plus the county scale (see table 4.2 and figure 4.2). Mean temperature and degree-days (>10°C), as expected, follow a regular pattern with peak occurrence in July (month 7), whereas the variability in precipitation, and hence HPR, is more stochastic.

Table 4.2 Hierarchical organization of Bailey ecoregions and associated weather station locations in the North Central Region

Bailey Ecoregion	Ecoregion Code	Ecoregion Name	Counties (Weather)
County	26077	Kalamazoo	1
Section	200010	South Central Great Lakes	42
Province	2000	Eastern Broadleaf Forest	361
Division	200	Hot Continental	399
Domain	2	Humid Temperate	842

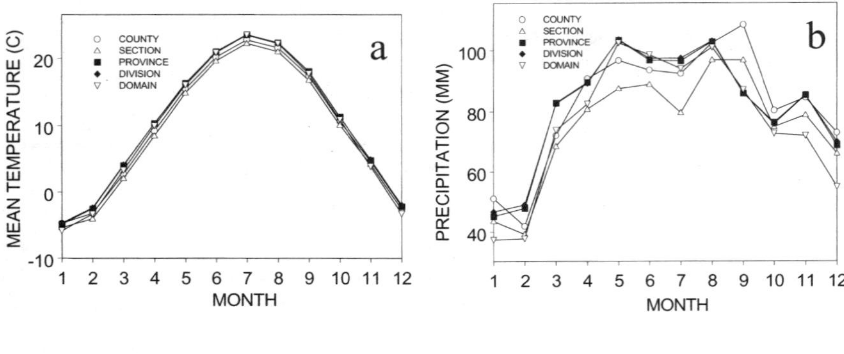

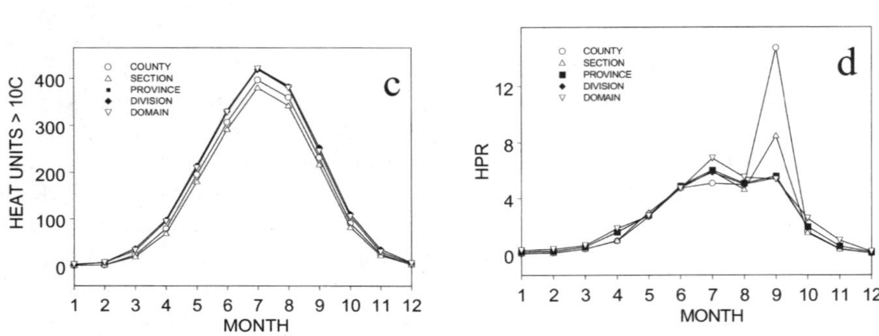

Figure 4.4 Monthly patterns of mean temperature (C), precipitation (mm), Heat Units (>10C), and HPR for the period 1972–1991.

The HPR exhibited the most irregular patterns where the single county observations (Kalamazoo County) had the highest value of HPR, followed by the Section (see figure 4.4d). Because of the focus on the evaluation of the 1988 drought, we anticipated that the HPR values would be most indicative of potential stress to plant communities. The heat/precipitation ratio can be a useful indicator of potential stress to biotic communities.

Patterns of Drought During the Growing Season (May–August)

To examine the annual patterns of HPR during the 20-year period of record, the average HPR was computed for May, June, July, and August, based on all 1055 locations in the climate database. These months were selected because they encompass months when plant stress will have a significant impact on plant productivity, particularly in agricultural crops. The monthly patterns of HPR, computed for May–August in the North Central Region, are provided in figure 4.5. In May, relatively high HPR values occurred in 1977, 1980, and 1988, indicating high potential

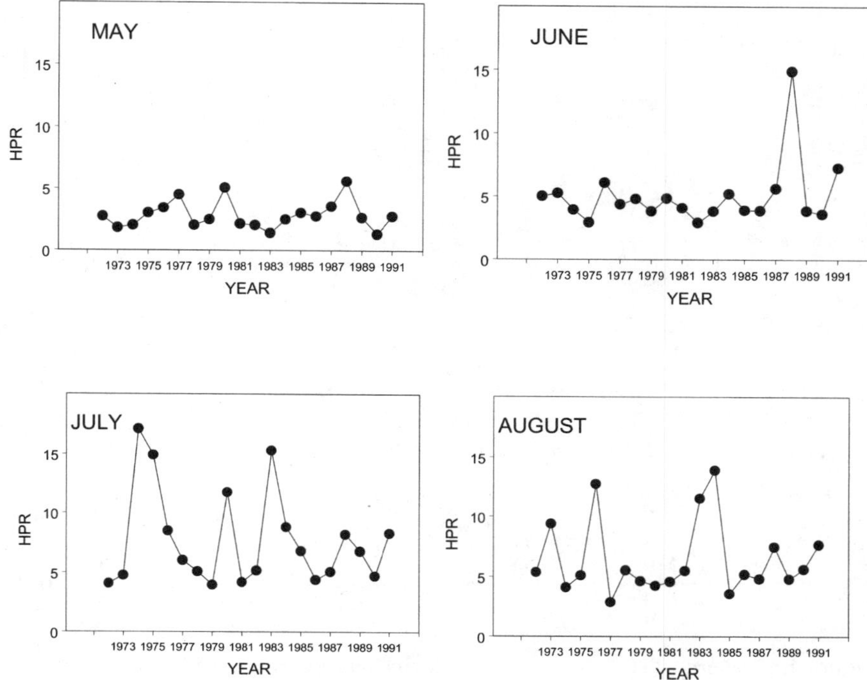

Figure 4.5 Calendar month time series of HPR for May–August 1972–1991.

plant stress early in the growing season. In June, the value of HPR was greater than the mean only in 1988. In July, values of HPR were above average in 1974, 1975, 1980, and 1983, whereas patterns of HPR in August were above average in 1973, 1976, 1983, and 1984. In 1988, both May and June exhibited high values of HPR. These early season periods are important for the seedlings, which require moisture for root growth and thus are more vulnerable to mortality because they have a lower tolerance for stress.

Further examination of the HPR, calculated at different spatial scales (county, section, division, and domain) reveals important patterns (figure 4.6). In figure 4.6a–d, the average HPR for the time interval (1972–1991) within each of the ecoregion classifications is contrasted with the HPR values in 1988. The co-occurrence of high values of HPR in May and June within each of the ecological classifications demonstrated a general phenomenon (May–June peak) across all levels of the spatial hierarchy. The HPR values were less extreme at the County level (Kalamazoo County) with a value of ~12 and greatest in magnitude (~ 22) at the ecological section level (South Central Great Lakes). Had this analysis not been restricted to ecoregions that are associated with the LTER, plant stress potential would have been even more extreme, particularly in western portions of the NCR where the ecoregion division is classified as short-grass prairie (Temperate Steppe).

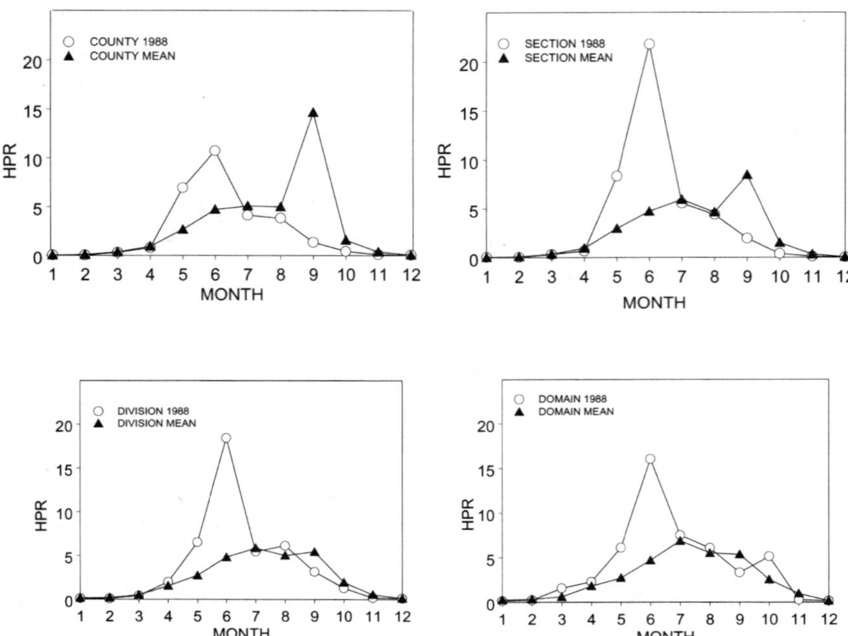

Figure 4.6 Patterns of HPR at ecoregion scales (county, section, division, and domain) contrasting the average HPR for the time interval (1972–1991) with the HPR values in 1988.

Spatial Characteristics of Drought

To characterize the spatial distribution of drought throughout the North Central Region, the HPR was associated with the geographic center of each county. ArcView GIS software (ESRI 1999) was used to interpolate a surface of HPR based on the spatial association between points using the inverse distance weighting (IDW) algorithm and a spatial moving average by associating 12 nearest neighbors to interpolate values of HPR between the 1055 points. Surface grid maps of HPR were developed based on monthly values of HPR at each location, and maps of cumulative HPR from May to July were produced. Two maps based on HPR were developed using this method: one shows the 20-year mean (figure 4.7b) and one shows only 1988 values (figure 4.7a). The higher HPR values (i.e., HPR > 15), based on the 20-year mean, occurred primarily in the western third of the region. However, in 1988, high values of HPR (HPR >30) occurred in central Michigan, Illinois, and Iowa, and in parts of Minnesota and North Dakota.

The magnitude of the difference between the 20-year mean HPR for each month (May–August) and the corresponding1988 HPR values is expressed spatially in figure 4.8 by computing the percent difference between the two spatial data set grids. The following equation was used to compute the difference:

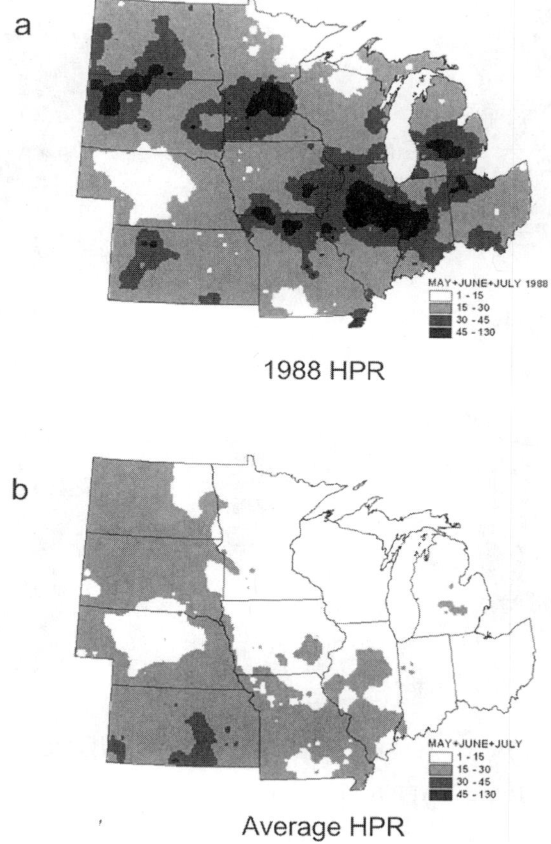

Figure 4.7 Spatial distribution of (a) the 1988 May–July HPR for the time interval (1972–1991) and (b) the average May–July HPR.

$$\frac{\text{HPR}_{Mean} - \text{HPR}_{1988}}{\text{HPR}_{Mean}} * 100$$

This computation was applied to grid maps for May, June, July, and August. Differences, expressed as a percentage, illustrate the patterns of the 1988 HPR deviation from the 20-year mean, with the darker portions of the maps representing the maximum percent differences for May (600%), June (1600%), July (700%), and August (600%). Note the corresponding areas for each of the classes. The month of June 1988 shows the greatest potential plant stress, particularly in the eastern portions of the NCR.

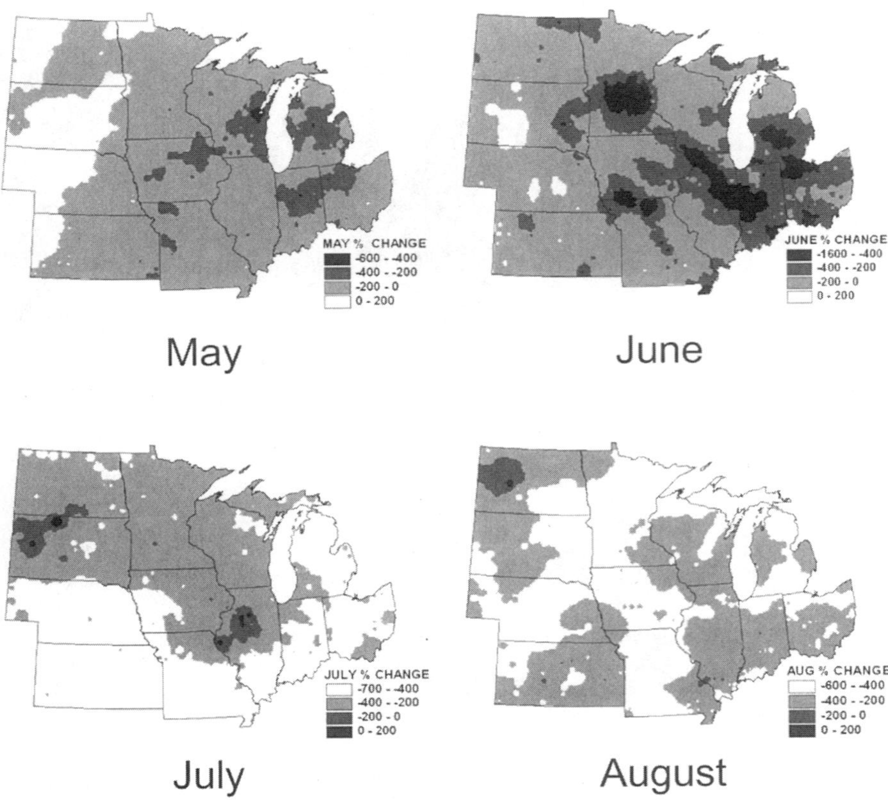

Figure 4.8 Percent change in HPR relative to 1988.

Plant Sensitivity to Drought

Regional Crop Production Patterns

When drought begins, the agricultural sector is usually the first to be affected because of its dependence on stored soil water, which is rapidly depleted during extended dry periods. Corn is particularly sensitive to summer droughts because of its high physiological requirements for moisture during the growing phase. In the North Central Region, planting generally occurs when the land is dry enough to support planting machinery and when the soil warms to about 10°C, which is usually during late April in the southern Corn Belt through mid-May in the north. During May and June, seedling and early root growth are vulnerable to periods without moisture, especially if temperatures are high. This can cause desiccation, particularly when vegetation is young. The corn crop experienced a significant period of stress throughout the Corn Belt in 1988, when the particular climatic anomaly took place. Figure 4.9 contrasts the average corn yield during the 20-year period (1972–1991) with the yield distribution in 1988. The month of June was a period of exceptionally high stress throughout the region (see figure 4.5). In June 1988 the

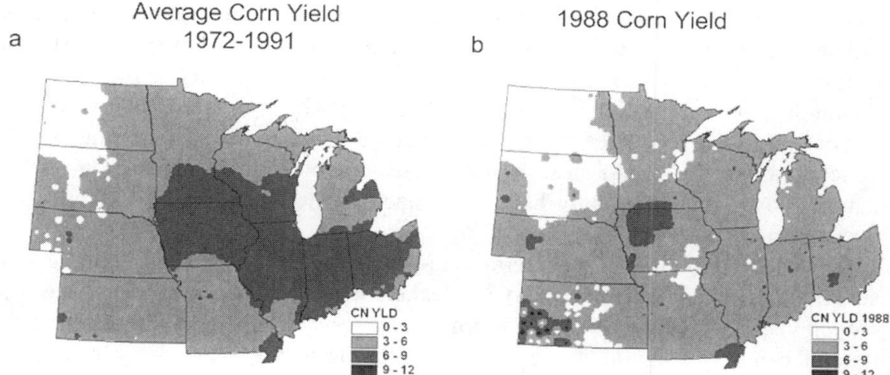

Figure 4.9 Average distribution of corn yield during the period 1972–1991 (t/ha) and the distribution of corn yield in 1988 (t/ha).

HPR (> 15 HPR) was three times greater than the mean HPR for other June months in the 20-year period. In addition, the HPR value for May 1988 (see figure 4.5) was also above average (> 5 HPR). High values of HPR in May and June 1988 indicate a long period of unusual stress during the early period of corn growth. Farmers attempt to override stress through use of irrigation. Evidence of this is shown in figure 4.9, where corn yields in the southwestern area of the NCR were above average in 1988, although only USDA data on nonirrigated yields were selected for analysis. One can speculate that additional irrigation was used in an attempt to gain economic advantage in 1988 and may not have been reported.

Although July stress was above average (HPR = 7) in 1988, the physiological stress that occurred during May and June was enough to significantly reduce the yield (t/ha) of the corn crop across the region. The general spatial patterns of corn yield in figure 4.9a–b provide a comparison between the 20-year corn yields (t/ha) and corn yields in 1988. Because corn is severely affected by drought, primarily through a deficit of moisture, its sensitivity to heat in the absence of moisture (provided by the HPR) makes it a valid indicator of drought stress. Other crop and plant communities were significantly affected by the 1988 drought. Although soybeans are generally planted later than corn (late May or June), because of their shorter growing season requirement, soybean yields were also well below average in 1988.

Long-Term Effects of Drought on Ecosystems

Several studies characterize the ecological effect of the drought of 1988. Tilman and Downing (1994) provided documentation of drought effects on plant communities other than those in agriculture. They characterized the influence of the 1988 drought on plants at the Cedar Creek LTER in Minnesota, and they measured the effects of drought and the dynamics of recovery from drought against a known baseline. Indeed, it was not until 1993, the fifth year after the 1988 drought and the twelfth year of the Cedar Creek LTER, that the effects of the drought on the species richness in successional grasslands were no longer discernible. However, the effects of the 1988 drought were still clearly evident in the oak savanna complex in 1993. About 30% of mature pin oaks died during the drought, compared to only 10% of bur oaks. Most of these dead trees are still standing. Tilman and Downing (1994) concluded that this major shift in oak species composition and reduction in oak canopy cover that will likely have an impact on these savanna ecosystems for decades to come.

In wooded vegetation a short-term (one-year) drought, depending on timing and severity, may cause plant mortality or weaken the plant system, predisposing the crop to insect herbivory or disease. In row-crop ecosystems, a short-term (one-year) drought can have a significant effect on yield, thus reducing productivity in the year of the anomaly. The long-term effects of a short-term drought on annual rotational agronomic systems are generally minimal. A drought may be local or regional. A drought that occurs over a large geographic region for a relatively short time period (several months compared with several years), such as the 1988 drought (May and June), can be economically devastating but not ecologically catastrophic. However, as Tilman and Downing (1994) illustrate, even a short-term drought can have cascading ecological consequences. Other unanticipated ecological consequences may occur. For instance, a drought may stimulate the need for irrigation as a means to override the effects of the drought. The addition of irrigation can affect ecosystem function by adding water to soils in dry ecosystems, thus stimulating changes in ecosystem flows and functions; irrigation can also affect the water table by mining groundwater and river systems to provide water to crops, particularly in ecosystems unaccustomed to large amounts of water. Crops such as corn that require high amounts of water. Corn grown in the western portion of the region is under stress in most years (see figure 4.7b), and this was especially the case in 1988 (figure 4.7a). Thus it is necessary to override the climate (via irrigation) each year to sustain "profitable" corn yields.

When a short-term drought transitions into a prolonged agricultural drought, this can have a cascade of effects by modifying ecosystem function. Poor agricultural practices and dry soil, coupled with wind, can erode soils, resulting in their redistribution. This agricultural management practice in the 1920s and 1930s caused soil erosion at regional scales, resulting in widespread ecological and economic consequences. Herbivore population fluctuations are associated with moisture and heat cycles. For instance, grasshopper egg survivorship and development of the eggs is intimately associated with soil moisture regimes and temperature (Mukerji and Gage 1978), and grasshopper populations respond to combinations of warmer than average spring temperatures (early hatch) and warmer than average temperatures in September, especially in northern regions. These longer periods of warm weather provide maximum potential for increased numbers of eggs to be laid in the soil. Because eggs can develop after oviposition in fall, they hatch earlier in spring, resulting in increased crop loss potential due to herbivory of crops during their early growth stages. Thus climate is a major contributing factor to pest outbreaks (Gage and Mukerji 1977).

Climate Change Implications

The drought that occurred in 1988 was clearly different from other droughts during the 20-year period analyzed (1972–1991). The 1988 drought occurred in May and June, whereas in other years, significant periods of stress occurred later in the crop-growing season (July–August). The severity of the stress and subsequent loss in productivity was due to the inability of young seedlings to tap soil moisture reserves prior to a stress period. Long periods of intense heat without precipitation during early phases of plant growth resulted in crop mortality. In the western Corn Belt (see figure 4.9b), rain-fed corn did not survive. However, the existence of the Ogallala Aquifer and the irrigation infrastructure developed to support row-crop agriculture in this short-grass prairie ecosystem enabled irrigation to override the stress induced by the 1988 drought in some parts of the region. In these areas, yields in 1988 were comparable to years when moisture was adequate for good corn production (see figure 4.9b, bottom left). However, the ecological costs and subsequent economic costs of depleting aquifer resources have not been fully evaluated. The 1988 drought was a one-year drought compared to the multiyear "dust bowl" drought that occurred in the 1930s. During the past 60 years, a significant multiyear drought has not occurred in the NCR, thus the probability of such a multiyear drought is high.

There has been considerable debate regarding the effect that a changing climate will have on agricultural productivity in the United States. In an assessment of the adaptation of agriculture to climate change, Rosenberg (1992) argues that agriculture may be both negatively and positively impacted by a changing climate and that additional information is needed at regional scales to provide a more complete assessment. More recent assessments of the impacts of climate change (NAST 2000) address the potential effects on agriculture in the Midwest and the Great Plains. The NAST (2000) report suggests that crop productivity may increase as a result of enhanced CO_2 in the northern reaches of the Midwest (eight of the NCR states) but may decline in southern portions of the Midwest. Four of the NCR's 12 states are also in the Great Plains. The NAST (2000) report predicts that higher evapotranspiration will result in decreased water availability, a problem for both the Midwest and the Great Plains.

The analysis presented in this chapter does not address whether climate change will have an impact on agriculture in the NCR. Instead, it shows that if significant changes in temperature and precipitation regimes occur, then organisms such as plants and insects, which depend on these variables for growth and survivorship, will respond based on ecological principles. An increase in temperature associated with a decrease in precipitation will result in a larger HPR and thus will cause an increase in stress to most biological communities, inducing them to adapt. More complete and higher quality data are needed to improve our ability to make more accurate ecological assessments. The LTER climate network, data archives, and associated ecological observations will satisfy part of that need.

Acknowledgments I express my appreciation to Manuel Colunga and Brian Napoletano for earlier reviews of the manuscript. Gene Safir made many helpful suggestions and contributions. Work associated with this chapter was accomplished with support from the NSF-LTER program (DEB 98-10220), the Michigan Agricultural Experiment Station, and the members of the USDA Regional Research Committee NC94. I also greatly appreciate the review, provision of additional literature, and suggested revisions to the chapter made by Peter Lamb.

References

Alley, W. M. 1984. The Palmer Drought Severity Index: Limitations and assumptions. *J. Climate and Applied Meteorology* 23: 1100–1109

Bailey, R. G. 1996. *Ecosystem geography*. New York: Springer-Verlag.

Baskerville, G., and P. Emin. 1969. Rapid estimation of heat accumulation from maximum and minimum temperature. *Ecology* 50: 514–517.

Burke, I. C., T. G. F. Kittel, W. K. Lauenroth, P. Snook, C. M. Yonker, and W. J. Parton. 1991. Regional analysis of the central Great Plains. *Bioscience* 41: 685–692.

Doorenbos, J., and A. H. Kassam. 1979. *Yield response to water*. Irrigation and Drainage Paper 33. Food and Agriculture Organization (FAO), Rome.

Environmental Systems Research Institute (ESRI). 1999. *ArcView GIS desktop mapping program*. Redlands, Calif.: ESRI.

Fountain, T, J. Helly, R. Waide, and S. H. Gage. 1999. Biological scale process modeling. 2nd Workshop on Modeling Ecosystem Processes at Regional Scales. http://www.sdsc.edu/sdsc-lter/modeling.html.

Gage, S. H., M. Colunga-Garcia, J. J. Helly, G. Safir, and A. Momin. 2001. Structural design for management and visualization of information for simulation models applied to a regional scale. *Computers and Electronics in Agriculture* 33: 77–94.

Gage S. H., J. J. Helly, and M. Colunga-Garcia. 2000a. A framework to integrate analytical and visual applications to regional models. In B. O. Parks, K. M. Clarke, M. P. Crane, editors. *Proceedings of the 4th international conference on integrating geographic information systems and environmental modeling: Problems, prospects, and needs for research*. 2000. Boulder, Colorado. Boulder: University of Colorado, Cooperative Institute for Research in Environmental Science. www.colorado.edu/research/cires/banff/pubpapers/78/.

Gage, S. H., J. Helly, D. Ojima, and W. Parton. 2000b. Fundamental questions that define regional analysis. *In 3rd Workshop on Modeling Ecosystem Processes at Regional Scales*. LTER All Scientists Meeting. Snowbird, Utah. http://www.lternet.edu/allsci2000/abstract.html#ls5.

Gage, S. H., and M. K. Mukerji. 1977. A perspective of grasshopper population distribution in Saskatchewan and interrelationship with weather. *Environ. Entomol.* 6: 469–479.

Gage, S. H., and M. K. Mukerji. 1978. Crop losses associated with grasshoppers in relation to economics of crop production. *J. Econ. Entomol.* 71: 487–498.

Gever, J., R. Kaufmann, D. Skole, and C. Vorosmarty. 1986. *Beyond oil: The threat to food and fuel in the coming decades*. Cambridge, Mass.: Ballinger. 304 pp.

Harouna, S., and R. E. Carlson. 1994. Analysis of an Iowa aridity index in relation to climate and crop yield. *Jour. Iowa Acad. Sci.* 101: 14–18.

Helly, J., T. Fountain, S. H. Gage, and R. Waide. 1998. Biological scale process modeling. *1st Workshop on Modeling Ecosystem Processes at Regional Scales*. http://www.sdsc.edu/sdsc-lter/modeling.html.

Huschke, R. E., editor. 1959. *Glossary of meteorology*. Boston, Mass.: American Meteorological Society.

Kunkel, K. E. 1992. *Measurements and estimates of evaporation in the Mid-western United States during the 1988 drought*. Proceedings of Workshop on the 1988 U.S. drought, College Park, Maryland. Center for Ocean-Land-Atmosphere Interactions and Department of Meteorology, University of Maryland, College Park, p. 83–95.

Kunkel, K. E., and J. R. Angel. 1989. Perspective on the 1988 midwestern drought. *EOS* 36: 817–819.

Loomis, R. S., and D. J. Connor. 1992. *Crop ecology: Productivity and management in agricultural systems*. Cambridge: Cambridge University Press.

Mukerji, M. K., and S. H. Gage. 1978. A model for estimating hatch and mortality of grasshopper egg populations based on soil moisture and heat. *Ann. Ent. Soc. Amer.* 71: 487–498.

National Agricultural Statistics Service (NASS). 2001. *Agricultural statistics*. Washington, D.C.: U.S. Government Printing Office.

National Assessment Synthesis Team (NAST). 2000. *Climate change impacts on the United States: The potential consequences of climate variability and change*. Cambridge: Cambridge University Press.

Olson, R. A. 1982. Soil fertility and plant productivity. Pages 85–101. In M. Rechcigl Jr., editor. *Handbook of agricultural productivity*. Volume I. Boca Raton, Florida: CRC Press Inc.

Petersen, M. S., P. J. Lamb, and K. K. Kunkel. 1995. Implementation of a semiphysical model for examining solar-radiation in the Midwest. *Journal of Applied Meteorology* 34: 1905–1915.

Robertson, G. P., E. A. Paul, and R. R. Harwood. 2000. Greenhouse gases in intensive agriculture: Contributions of individual gases to the radiative forcing of the atmosphere. *Science* 289: 1922–1925.

Rosenberg, N. J. 1992. *Adaptation of agriculture to climate change*. Climatic Change 21: 385–405.

Rosenberg, N. J., editor. 1980. *Drought in the Great Plains—Research on impacts and strategies*. Proceedings of the Workshop on Research in Great Plains Drought Management Strategies, University of Nebraska, Lincoln. Littleton, Colorado: Water Resources Publications.

Tilman, D., and J. A. Downing. 1994. Biodiversity and stability in grasslands. *Nature* 367: 363–365.

Tivy, J. 1990. *Agricultural ecology*. New York: Longman Scientific Technical.

Warwick, R. A. 1975. *Drought hazard in the United States: A research assessment*. University of Colorado, Institute of Behavioral Science, Monograph no. NSF/RA/E-75/004. Boulder, Colorado.

Yevjevich V., W. A. Hall, and D. J. Salas, editors. 1977. *Proceedings of the Conference on Drought Research Needs*. Colorado State University, Fort Collins, Colorado.

Zangvil, A., D. H. Portis, and P. J. Lamb. 2001. Investigation of the large-scale atmospheric moisture field over the Midwestern United States in relation to summer precipitation. Part I: Relationships between moisture budget components on different timescales. *Journal of Climate* 14: 582–597.

5

Climate Forcing at the Arctic LTER Site

John E. Hobbie

Neil Bettez

Linda A. Deegan

James A. Laundre

Sally MacIntyre

Steven Oberbauer

W. John O'Brien

Gaius Shaver

Karie Slavik

Introduction

The Arctic LTER site is located at 68°38'N and 149°43'W, at an elevation of 760 m in the northern foothills of the Brooks Range, Alaska. The location, 208 km south of Prudhoe Bay, was chosen for accessibility to the Dalton Highway, which extends along the Trans-Alaska Oil Pipeline from north of Fairbanks to Prudhoe Bay on the Arctic Ocean (figure 5.1). The rolling foothills at the site are covered with low tundra vegetation (Shaver et al. 1986a), which varies from heaths and lichens in dry sites to sedge tussocks on moist hillslopes to sedge wetlands in valley bottoms and along lakes. Riparian zones often have willow thickets up to 2 m in height. Small lakes are frequent; the best studied such lake is the 25-m-deep Toolik Lake (O'Brien 1992), the center of the LTER research site. Some 14 km from Toolik Lake, the Dalton Highway crosses the fourth-order Kuparuk River, the location of much of the LTER stream research (Peterson et al. 1993).

Climate records at Toolik Lake have been kept since the early 1970s when a pipeline construction camp was established. On completion of the road in 1975, climate stations were set up by the U.S. Army Cold Regions Research Laboratory (CRREL, climate reported in Haugan 1982 and Haugen and Brown 1980). Since 1987, the LTER project has maintained climate stations at Toolik Lake (http://ecosystems.mbl.edu/arc/) whereas the Water Resources Center of the University of Alaska has continuous records beginning in 1985 from nearby Imnavait Creek. An

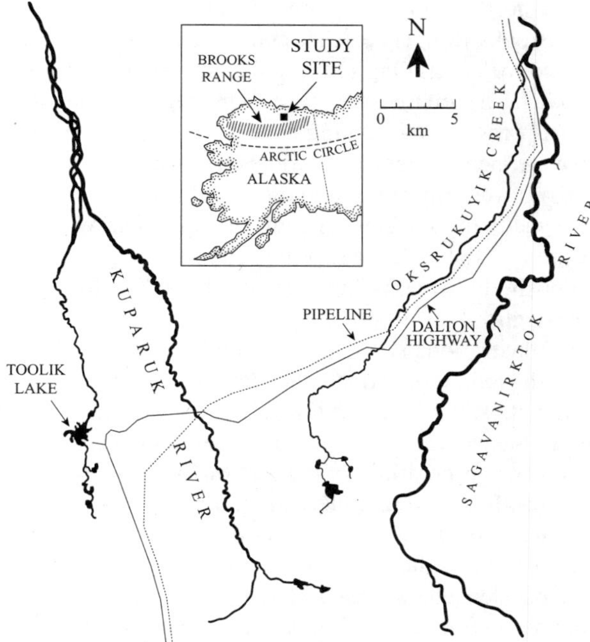

Figure 5.1 Location of Arctic LTER site at Toolik Lake, Alaska.

automatic station at Imnavait now reports every few hours to the Natural Resources Conservation Service–Alaska of the U.S. Dept. of Agriculture.

The characteristics of the climate in northern Alaska are summarized by Zhang et al. (1996), who pointed out the strong influence of the ocean during both summer and winter months. They reported that the mean annual air temperature is coldest at the coast (–12.4°C), where there are strong temperature inversions in the winter, and warmest in the foothills (–8.0°C). At Toolik Lake, snow covers the ground for about eight months, and some 40% of the total precipitation of 250–350 mm falls as snow. However, snowfall can occur on any day of the year. Summer temperatures at Toolik Lake are significantly warmer than at Prudhoe Bay on the coast.

Important climate factors that affect the ecology of the ecosystems at Toolik Lake include (1) the low temperatures in the air and soil that affect the metabolism of all the biota but especially cause a reduction in the microbial decomposition rates, (2) the 8-month snow cover that allows only a very short growing season for plants, (3) the reduced amount of light energy for photosynthesis because plant growth does not begin until after half of the annual radiant energy input has occurred, (4) the completely frozen streams from mid-September until mid-May that reduce the fish diversity to one species, and (5) the long duration of the ice cover of lakes (from the end of September until mid-to-late June) that reduces the light available for photosynthesis.

One consequence of the climate at Toolik Lake is the lack of significant trees in the vegetation. The northern limit for spruce is a hundred kilometers away on the south side of the Brooks Range. Another consequence is the presence of permafrost, frozen soil and rock extending some 200 m into the ground. On top of the permafrost is the active layer, soil that thaws each summer to a depth of 30–50 cm. Not only does permafrost restrict the rooting zone of plants to the active layer, it also seals the soils to water penetration. The result is that water from snowmelt and rain is held in the active layer, especially in the organic matter–rich upper 10–20 cm, thus the soils are usually moist despite the low precipitation. When there is enough precipitation to saturate the soil, the resulting runoff is "flashy"—that is, there is a quick peak of flow in the streams, but there is little water storage so the peak decreases quickly (Stieglitz et al. 1999).

The preceding illustrations show how the fundamental ecology at the Arctic LTER site is set by the long-term climate that determines such things as the makeup of the plant communities, the length of the growing season in tundra, streams, and lakes, and the hydrologic cycle. But important clues about ecosystem function and controls also arise from observations of the ecosystem response to short- and long-term climate changes. In the Arctic, there are many aspects of short-term climate variability, including year-to-year snow cover duration, the variation in lake temperatures from year to year, the effects of air and soil temperature changes from year to year, the ecosystem changes caused by stream flow and stream temperature differences from one summer to another, and the changes within a lake related to irregular stream flows caused by rain events. Of the possible long-term changes in climate, an increase in air and permafrost temperatures is the only one detected thus far.

Variability of Climate and Related Physical Factors

The 11-year climate record for the Toolik Lake site (table 5.1) indicates a mean daily air temperature of −8.8°C and a total annual precipitation of 315 mm. Monthly means are above freezing for 3 months, and most of the precipitation occurs from June through September. In figure 5.2, the year-to-year variability for two biologically important indices, the annual degree-days above 0°C and the summer rainfall, illustrate the nearly twofold difference from year to year.

Solar radiation is a very important physical factor that affects characteristics such as the stratification and heating of the surface layers of lakes as well as the depth of thaw of soils. Biological processes are also affected, particularly photosynthesis in terrestrial plants and plankton. Solar radiation varies at different timescales from the minute-to-minute variation caused by passing clouds to the interannual changes between summers differing in cloudiness.

Although the air temperatures and their sum, the degree-days shown in figure 5.2, are useful indicators of possible effects on the aboveground parts of plants, a better indicator of soil temperatures and its effect on soil roots and microbes is the depth of thaw of the active layer of the soil (figure 5.3). The thaw depth is mainly affected by the amount of insulating plant material on the surface of the tundra, by the air temperature, and by the soil moisture. For example, 10 years after a single

Table 5.1 Eleven-year summary of climate at Toolik Lake, 1989–2000

Month	Mean Daily Air Temperature, °C	Maximum Average Temperature, °C	Minimum Average Temperature, °C	Total Monthly Precipitation, mm
1	−24.2	−20.3	−31.7	8.2
2	−21.4	−8.1	−31.3	13.3
3	−19.9	−12.5	−26.6	9.4
4	−11.4	−4.5	−17.5	9.2
5	−0.9	4.3	−9.6	16.7
6	8.5	9.6	6.2	44.6
7	11.6	14.1	9.2	67.8
8	7.3	11.3	3.7	67.1
9	−0.8	3.3	−8.8	37.3
10	−12.7	−6.5	−16.7	18.3
11	−19.7	−12.3	−29.2	9.8
12	−22.6	−17.1	−29.7	13.0
Long-term average or sum	−8.8			314.7

dose of fertilizer, the amount of plant litter was greater and the soil temperature was significantly lower in a treated tundra plot than in a nearby untreated plot (LTER data of G. Shaver and J. Laundre). Despite these complications, the degree-days (figure 5.2) are roughly correlated with the thickness of the active layer (figure 5.3). The two periods with the thickest depth of thaw, 46 cm in 1993 and 1997, occurred during warm summers but not the warmest.

Another indicator of the effect of climate variability is the 26-year record of July temperatures of the surface waters of Toolik Lake (figure 5.4). These temperatures are affected by the air temperature but most of all by the amount and timing of the

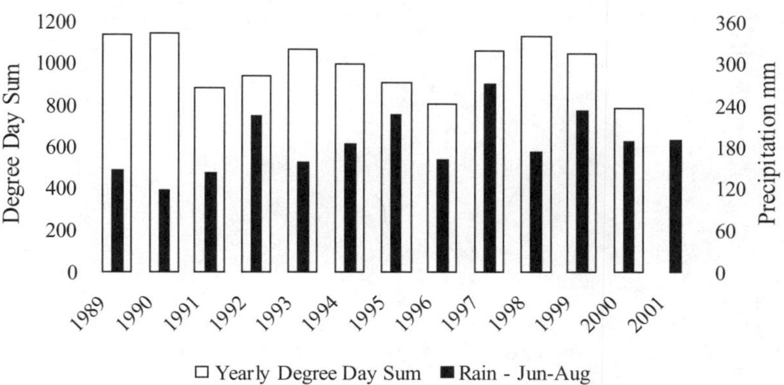

Figure 5.2 The annual degree-days (sum of daily average temperatures above 0°C) and the June through August rainfall at Toolik Lake, 1989–2000.

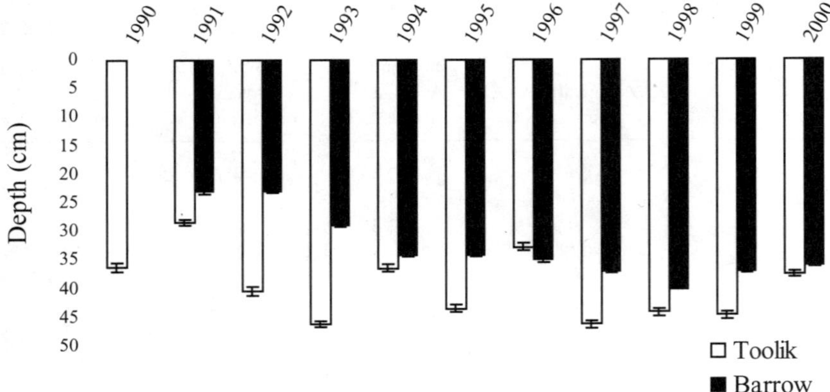

Figure 5.3 The mean thickness of the active layer at the end of the summer season at the Toolik Lake LTER site (68° 37' N, 149° 36' W) and the Barrow CRREL site (71° 19' N, 156°35' W), Alaska (data from the Circumpolar Active Layer Monitoring network) (Brown et al. 2000).

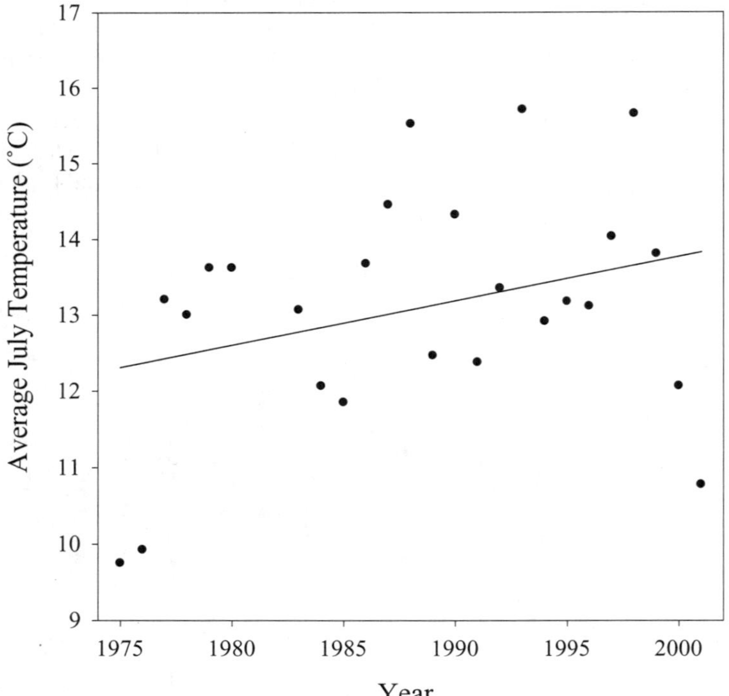

Figure 5.4 Toolik Lake average temperatures during July at 1 m depth.

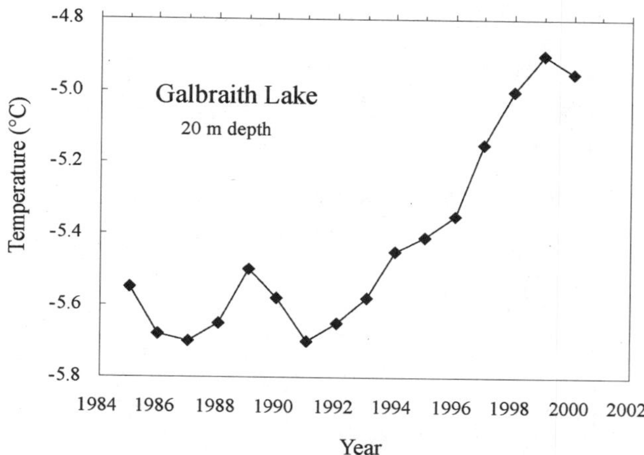

Figure 5.5 Average temperature at 20 meters in an 80-m-deep borehole near Galbraith Lake, Alaska (68° 28' N, 149° 29' W) (pers. Comm. T. Osterkamp and V. Romanovsky, University of Alaska, Fairbanks. 4/2/02).

solar radiation. When the ice stays on the lake until 1 July, then the monthly mean water temperature will be cooler than a summer with ice out in mid June. Given the variation in the dates of the ice melt on the lake, it is remarkable that there is any trend at all. The one obvious trend is the long-term increase of about 2°C. Other studies (e.g., Chapman and Walsh 1993) have pointed out that there is a continuing 30-year warming of air temperature in northern Alaska. A possible integrator of air temperature is the temperature in the upper levels of the permafrost. T. E. Osterkamp and V. Romanovsky of the University of Alaska Fairbanks (pers. comm., 4/2/02) found that at Galbraith Lake, 20 km south of Toolik, the temperatures at a depth of 20 m in a borehole driven into the soil showed an impressive warming of 0.8°C since 1991 (figure 5.5). Unfortunately for the perfect integrator theory, recent analysis (Marc Stieglitz, Lamont Doherty Earth Observatory, Columbia University, pers. comm., 1/8/03) points out that the change is very likely caused by two factors: a warming of air temperatures and an increase in the amount of snow during the winter. Only about half of the permafrost warming is due to an increase in air temperature. While the permafrost temperatures may not tell us about the air temperatures, the analysis illustrates changes in snow cover can be just as important as changes in air temperature in regulating soil temperatures and can even amplify below-ground temperatures. For long-term predictions, the alteration in winter precipitation must be better understood.

The year-to-year variability is especially important for stream ecology. Figure 5.6 illustrates two extremes in the Kuparuk River. In 1990 the flow was very low, with only one high flow event (or spate) after the spring runoff; temperatures fell mostly between 10 and 15°C. In 1995 there were some nine spates during the summer, and temperatures fell between 5 and 10°C. The inlet stream to Toolik Lake showed similar variability in the number and temperature of spates.

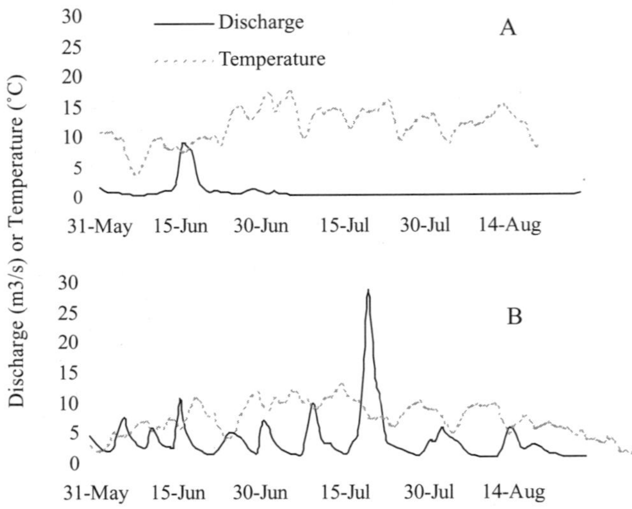

Figure 5.6 The Kuparuk River discharge (m³/s) and water temperature (°C) during (A) a low flow year (1990) and (B) a high flow year (1995).

Day-to-Day Variability in Solar Radiation and Photosynthesis

The most important short-term control of photosynthesis of tundra plants at Toolik Lake is the amount of solar radiation. This is illustrated in figure 5.7 by the process-based model of net ecosystem production (NEP) developed by Williams et al. (2000). The model, the solid line in the top panel of this figure, is driven by the hourly amount of photosynthetically active radiation (PAR), by hourly air temper-

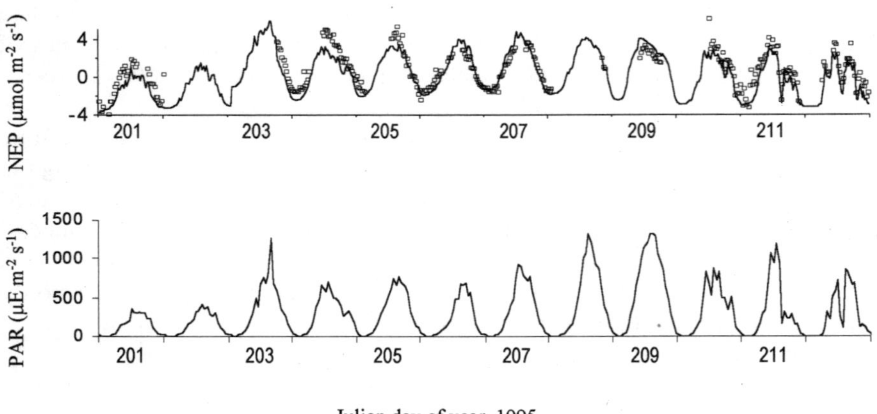

Figure 5.7 The top panel is the net ecosystem production (NEP) for acidic tundra at Toolik Lake measured by eddy covariance and modeled (continuous line) by Williams et al. (2000, 2001). The bottom panel is the photosynthetically active radiation (PAR) at the same site (Chapin data base at National Snow and Ice Data Center).

atures, and by the leaf nitrogen content. Only a small change in photosynthesis occurred over the temperature ranges for the 12 days in summer, and the leaf nitrogen did not change appreciably. Accordingly, the changes in photosynthesis are mostly due to the changes in PAR, shown in the bottom panel. How close is the model to the real world? We were able to compare the model runs for acidic tussock tundra with actual measurements of NEP by the eddy covariance method of measuring CO_2 exchange with the atmosphere. These measurements, shown as small open squares in the top panel, were carried out on the acidic tundra at Toolik Lake by Eugster et al. (1997) and corroborate the model. Note especially the low values of NEP and PAR on days 201 and 202, followed by high values of both for day 204. Also, the variable PAR values of days 211 and 212 are closely reflected in the multiple peaks of photosynthesis for the eddy covariation data.

Within-Season Variability in Stream Flow and Lake Ecosystems

Recent studies have shown that the mixing and stratification of Toolik Lake may be greatly changed when one or more high-discharge spates in the inlet stream occur during the summer; this has important consequences for algal primary productivity. The incoming stream water may be differentiated from the lake water through small differences in temperature and conductivity.

One example of the impact of a spate occurred in mid-July 1999. Prior to the spate (figure 5.8A), the subbasins of Toolik Lake were strongly stratified, with the surface water at 17°C and the hypolimnion at 4°C. Much of the primary productivity occurred in the deep water as evidenced by a chlorophyll maximum at 7 m and below. Algal counts showed that the chlorophyll reflected an increase in abundance of the same algal species, small flagellates such as cryptophytes, found throughout the surface layers. When the spate occurred and flow rates exceeded 12 m³/s (figure 5.8B), the two subbasins nearest to the inflow point quickly mixed with the inflow water at 11°C. The chlorophyll maximum was dispersed, and the chlorophyll content of the upper waters increased. Primary productivity of much of the lake increased. The most likely explanation for the increase in productivity, based on detailed measurements of NH_4^+, of the ratio of C:N in particulate matter, and of the nutrient limitation of algal photosynthesis, is that nutrient-replete algae from the chlorophyll-maximum region of the lake became mixed throughout the upper layers and were able to obtain sufficient light to make use of excess nutrients they had stored.

Year-to-Year Variability in Lake Heating and Fish Habitat

The course of changes in temperature and stratification has been described many times for temperate lakes in the spring. The water temperature immediately beneath the ice is 0°C and warms to 3–4°C in the depths of the lake. When the ice leaves the lake, wind action circulates the entire water column before the surface waters warm and stratification begins again. During the circulation period, the lake

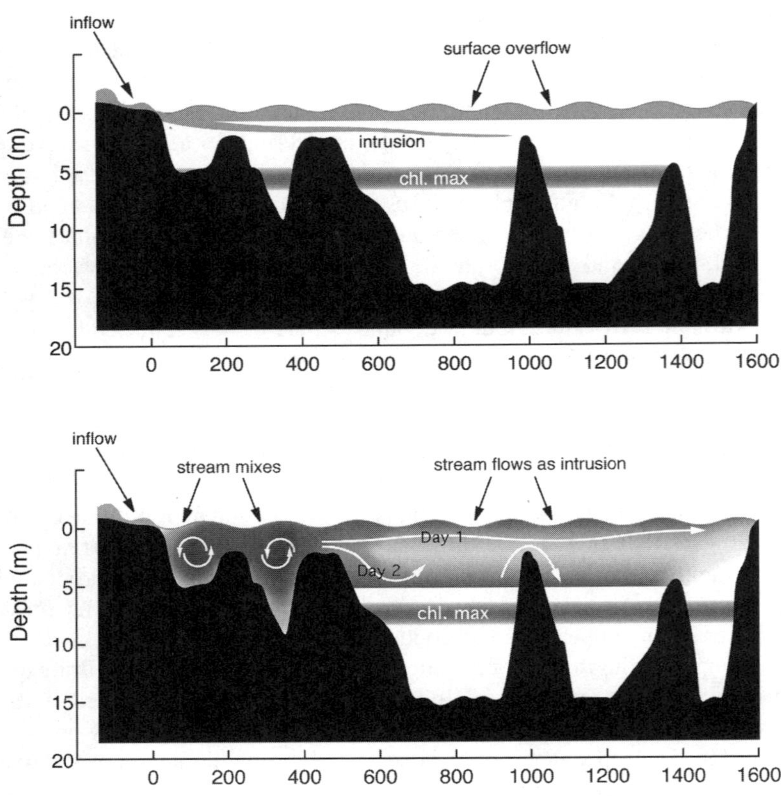

Figure 5.8 Schematic diagrams of the flow path of stream inflows into Toolik Lake based on unpublished data of S. MacIntyre and G. Kling. (Top) In early July, there are low inflow rates of water less dense than upper waters of the lake; the water flows in as a subsurface intrusion. The stability of the stratification allows a deepwater layer of algae to develop (chlorophyll maximum). (Bottom) In mid-July, the first days of a spate show strong mixing of the stream waters with the lake water; the mixing is vigorous enough to destroy the chlorophyll-maximum layers in the basins close to the inflow point.

waters absorb oxygen to replace the amount lost during winter respiration. The situation is often different in Toolik Lake. During some years there may be no circulation period after the ice leaves the lake; instead, the summer stratification begins immediately as a result of calm conditions and high amounts of solar radiation in the crucial few days after the ice leaves the lake. The ecological result is that the deep waters of the stratified lake begin the summer with reduced concentrations of oxygen, and the deepwater oxygen can be further reduced by the breakdown of plankton settling from the surface layers. Although the oxygen is eventually restored during the long-lasting fall circulation of the water column, the impact of lowered levels of oxygen can eventually reduce the habitat for fish. For example, lake trout, the dominant predator in lakes of the arctic foothills, require 3 mg O_2

L^{-1}, a level approached during a several-year experimental addition of low-level of nutrients to a lake next to Toolik Lake. In Hobbie et al. (1999), a model of lake circulation and temperature predicts that too low oxygen concentrations and too warm temperatures in the surface waters will drastically reduce lake trout habitat if a 4°C change in the annual air temperature occurs.

Year-to-Year Climate Variability and Grayling Growth

As seen in figure 5.6, there is a large amount of annual variability in streamflow and temperature in the Kuparuk River as a direct result of summer precipitation and air temperatures. The growth of the arctic grayling, the only species of fish in this river and in a nearby smaller stream called Oksrukuyik Creek, is strongly affected by the flow rate and by the temperature as a result of its life cycle.

The arctic grayling (*Thymallus arcticus*) is found in North America from Minnesota north to northern Canada and northern Alaska. Adults reach approximately 35 cm in length and live as long as 20 years. Grayling adults live in pools within streams and exclusively feed on drifting insects, such as immature stages of mayflies, stoneflies, black flies, and other stream insects. Spawning takes place in the spring as soon as the adults return from lakes where they have overwintered to their summer territories in the pools. The young live in shallow water along the edge of the streams where they feed on tiny insect larvae and other invertebrates.

Deegan et al. (1999) measured the growth of young-of-the-year and adult fish for a number of years (figure 5.9), and they have used the variability of the flow and water temperatures to make correlations with physical factors controlling growth. The young were measured in field samples collected throughout the summer, and their average weight in grams at the end of the summer was taken as the amount of growth. At the beginning of each summer, adults were caught on a barbless hook, weighed, tagged (in fact, most of the population is already tagged), and released. At the end of the summer the same fish are recaught, weighed, and released. In this way, the growth for each summer can be calculated as the grams added or lost per day. Measurements were made in control sections of the river as well as in sections where the primary productivity was increased by daily fertilization with phosphorus (details in Peterson et al. 1993).

The young-of-the-year fish grew best in warm and relatively dry summers when the stream discharge was low. This makes ecological sense because the young fish would find suitable habitat in small pools and the warm temperatures would encourage the growth of their tiny prey. In high-flow years the small pools are absent, and the young fish have to live in the stream itself.

In contrast, the adult fish grew best in cool summers with high flow rates. Under these conditions, a greater number of drifting insects move downstream compared with the number in low-flow years. In addition, the low temperatures result in a lower rate of metabolism for the fish that could lead to improved growth for the same amount of food eaten. In two relatively warm summers, the adults in the Oksrukuyik Creek actually lost weight over the summer.

The adults of this population of grayling appear to grow reasonably well most sum-

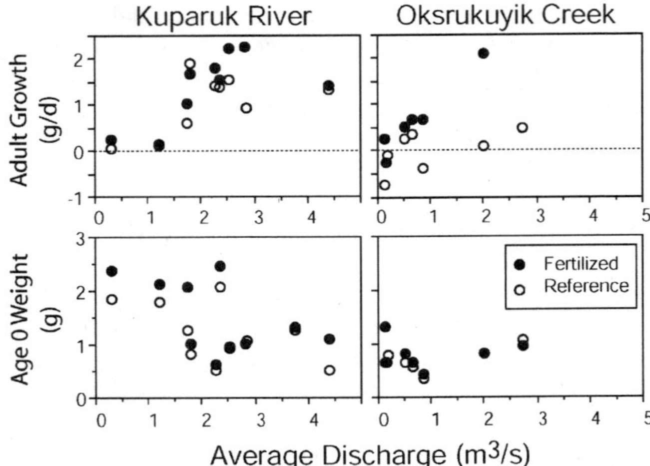

Figure 5.9 Interannual variability of adult and age 0 arctic grayling is related to river discharge and nutrient availability.

mers. The young survive and grow well only during exceptionally warm and dry summers with low flow. Thus, a part of the population does well no matter what the environmental conditions are during that year. However, the survival of the populations depends on the exceptional year classes that occur infrequently, every 6 or 7 years. The population as a whole survives only because the fish are so long lived—they can live for up to 20 years. It is likely that the great amount of climate variability at this site would not allow fish that only lived 6 or 7 years to survive as a population.

Year-to-Year Climate Variability and Plant Flowering

Every summer since 1980, G. Shaver has counted the average number of flowers of the cotton grass (*Eriophorum vaginatum*) at 38 sites along the highway from Fairbanks to Prudhoe Bay (Shaver 1986b) (figure 5.10). The variability from year to year was amazing; the average ranged from 1% to 46% (figure 5.11). Even more remarkable was the synchrony of flowering along the entire transect that covered ~650 km. This is evident in figure 5.12, where the mean inflorescence for a year is compared, for each site, with the long-term mean. Years of above-average flowering, that is, those above the 0.0 line, have above-average flowering almost everywhere along the transect, whereas years of below-average flowering are below average everywhere.

The environmental cause of the synchronous flowering must be linked to events of the past one or two years, because the flower buds are set at the end of the previous summer. When the detailed climate record at Toolik Lake was examined, there was a good correlation between the number of flowers per plot and the cumulative degree-days above 0°C at a depth of 20 cm in the soil (figure 5.13). The

Figure 5.10 *Eriophorum vaginatum* (arctic cotton grass) in bloom at Toolik Lake.

best correlation was found when the period for the degree-days was the 12 months beginning in the fall and extending to the end of the summer in the year before the counting. For example, the number of flowers counted in July 2000 correlated well ($R^2 = 0.69$) with the degree-days in the fall of 1998 and the spring and summer of 1999.

One hypothesis that arises from these findings is that the plants flower after

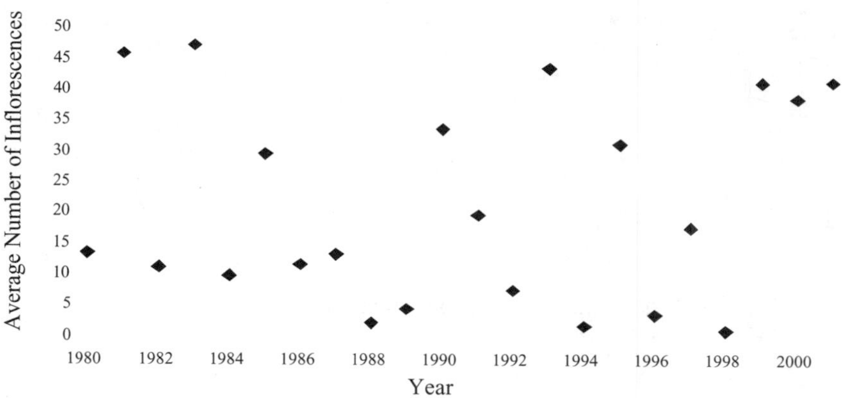

Figure 5.11 The yearly average number of inflorescences for 38 plots along the highway from Fairbanks to Prudhoe Bay from 1980 to 2001.

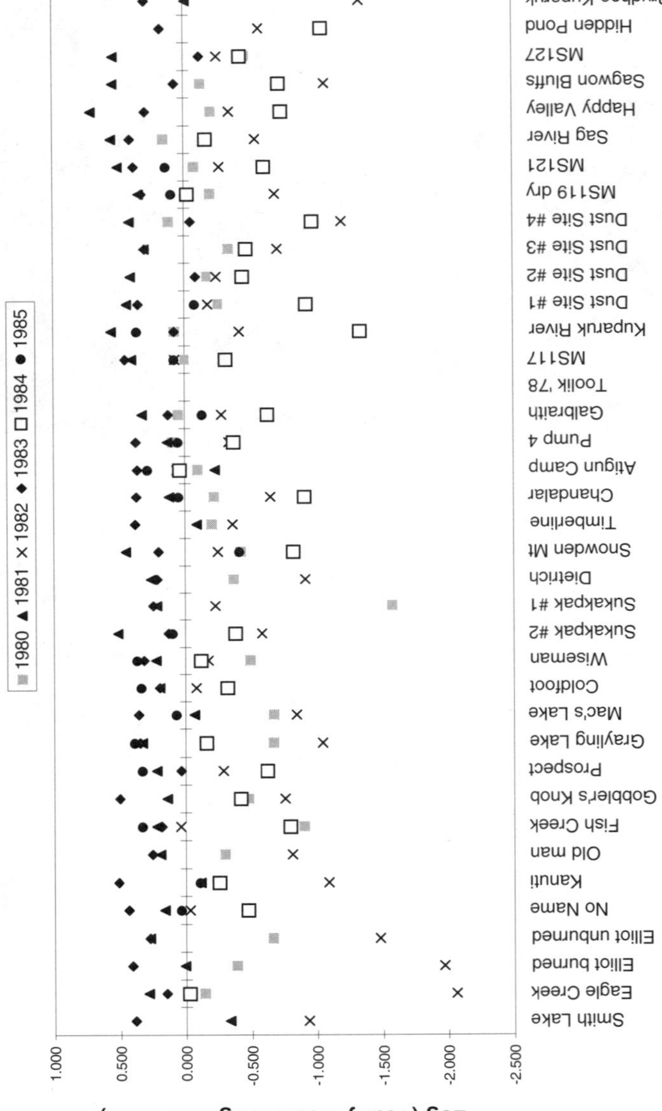

Figure 5.12 The mean inflorescence counts relative to the long term mean for sites from Fairbanks to Prudhoe Bay. Only 5 years of data are shown.

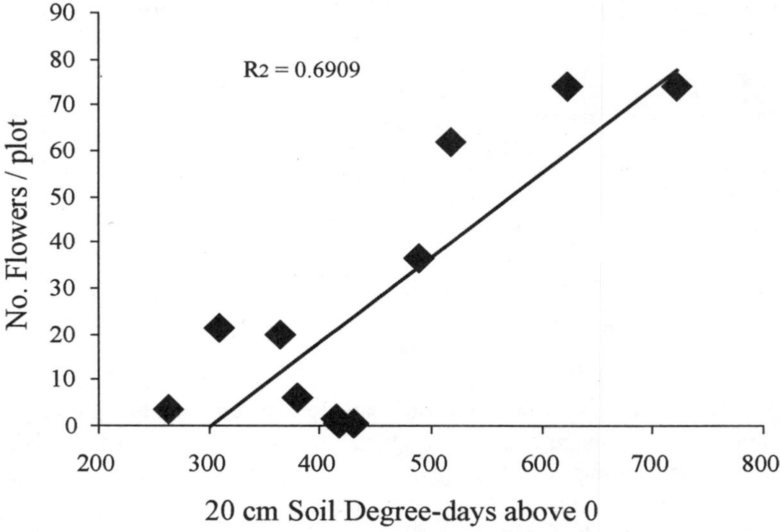

Figure 5.13 Number of flowers of *Eriophorum vaginatum* per 2-m by 2-m plot at the Toolik site plotted against the cumulative degree-days above 0°C at a 20-cm depth in the soil. The degree-days cover one complete year beginning 22 months before the counting time.

they have accumulated a threshold quantity of nitrogen from the soil. The nitrogen is made available to the plants through the process of decomposition in the soil. This microbial process will be slightly faster during years when the soils are warmer.

Year-to-Year Climate Variability and Plant Phenology, Physiology, and Ecosystem Exchange

The date of the snowmelt near the LTER greenhouse plots varied by 26 days over the last 7 years; the earliest date was 13 May in 1995 and the latest was 9 June in 2000 (table 5.2). A longer record from nearby Imnavait Creek had a range of 30 ± 10 days (sd) over the past 17 years (Kane et al. 2000).

The timing of snowmelt has a significant effect on soil temperatures and thickness of the active or thawed layer (figure 5.3). On experimental plots where snow was removed two to three weeks before the snow melted on nearby control plots, soil temperatures averaged more than 1°C higher than in control plot soils (Oberbauer et al. 1998). Such differences undoubtedly affect belowground processes such as nutrient mineralization rates and root growth.

Plant photosynthesis begins at snowmelt. Obviously, late snowmelt leads to a shorter growing season, but more subtly, late snowmelt shifts the window of peak leaf performance further toward the declining sun angles and lower irradiances that follow the summer solstice. The weather after snowmelt can also be important; for

Table 5.2 Date of complete snowmelt and snow depth for a site near Toolik LTER

Year	Date of snowmelt	Snow depth on 2 May
1995	13 May	52.8
1996	29 May	71.3
1997	3 June	65.4
1998	24 May	48.6
1999	22 May	49.5
2000	9 June	62.2
2001	8 June	65.0

example, very low spring temperatures may slow development of photosynthetic capacity or even damage overwintering leaves. Evergreens are the exception— they very quickly ramp up to full photosynthetic capacity after snowmelt (Oberbauer et al. 1996).

Snowmelt also marks the beginning of the accumulation of forcing temperatures required for bud break of deciduous and evergreen shrubs and for initiation of leaf expansion of graminoids and forbs. The date of bud break is strongly dependent on spring temperatures after snowmelt (Pop et al. 2000). Consequently, the effects of late snowmelt can be partially offset by warm spring temperatures. Variation in the timing of snowmelt in conjunction with spring temperatures has a strong effect on when and how much leaf area develops. In experimentally manipulated plots, Oberbauer et al. (1998) found that plants in the experimental plot with early snow removal developed leaf area earlier and to a greater extent than plants in control plots. These increases in leaf area translate to higher gross ecosystem uptake capacity, though they may be partially offset by higher respiration resulting from higher soil temperatures. Also, for some species, early bud break or leaf initiation leads to early senescence (Oberbauer et al. 1998; Starr et al. 2000).

Variation in the timing of fall freezes and development of snow cover at the end of the season has less impact on ecosystem uptake capacity because shoots of most plant species near Toolik become dormant in response to declining photoperiod. Early freezes unquestionably do accelerate the end-of-season leaf senescence (McGraw et al. 1983). However, perhaps of more importance is the timing of snow cover with respect to hard freezes; late development of snow cover exposes evergreens to potentially damaging hard freezes from which they otherwise would be protected. Die-offs of evergreens noted at the beginning of the growing season in some years are probably a result of such conditions.

Long-Term Climate Variability and Lake Water Chemistry

The 25-year record of water chemistry at the Arctic LTER site has documented a doubling of the average alkalinity or acid neutralizing capacity of Toolik Lake (figure 5.14). This change in alkalinity is balanced primarily by changes in calcium and

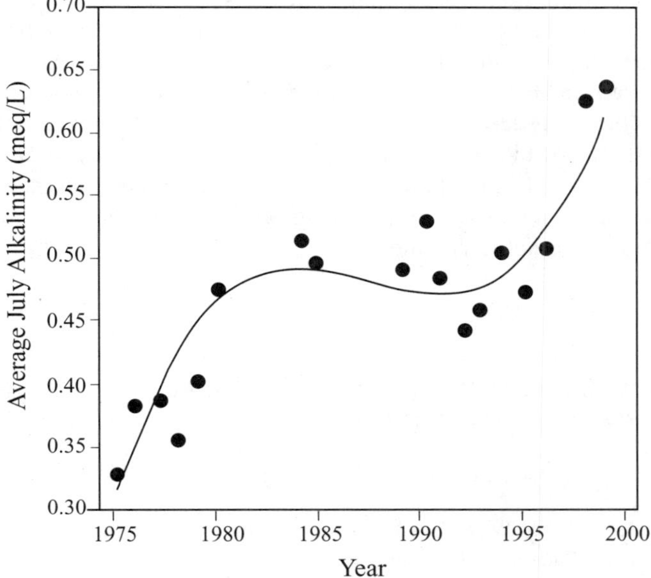

Figure 5.14 The annual July average alkalinity (1 m depth) in Toolik Lake, Alaska.

magnesium. There are no corresponding changes in the chemistry or amount of the precipitation that would account for these changes. One suggestion is that dust from the road is causing the changes. However, similar changes in alkalinity have been found in streams and lakes quite distant from the road.

The most reasonable explanation is that the lake alkalinity is an indicator of changes in soil chemistry and groundwater chemistry. These same changes are also found in the stream waters. It is possible that small increases in thaw depth expose new soil material to weathering, material that has been frozen for millennia. Another possibility is that new material is weathering because the active layer has thickened in riparian zones where more water is flowing. Given the large year-to-year variations in thaw depth, these processes are difficult to sort out, but, there can be no doubt that the air and permafrost temperatures have warmed over the past 15 years in northern Alaska (see previous discussion in the section "Variability of climate and related physical factors").

Conclusions

Ecosystems at the Arctic LTER site at Toolik Lake, Alaska, respond to variations in climate at a number of temporal scales. At the scale of a single season, changes in PAR directly affect NEP, whereas rainfall events change stream flow and associated lake stratification and ecosystems. At the scale of year-to-year variation, warming lake temperatures may reduce the habitat suitable for lake trout, and the growth of grayling is positively correlated with wet and cold summers. For land plants, soil warming caused by early snow melt or by warm summers causes the synchronous flowering across northern Alaska of arctic cotton grass and an increase in leaf area. Finally, at the scale of multiyear variation in air temperatures, the alkalinity in lakes and steams near the Toolik Lake Field Station has doubled in the last decade. The exact reasons for this change are unknown, but they are likely related to long-term trends in climate.

References

Brown, J., G. W. Kling, K. M. Hinkel, L. D. Hinzman, F. E. Nelson, V. E. Romanovsky, and N. I. Shiklomonov. 2000. Arctic Alaska and Seward Peninsula. Pages 182–187 in J. Brown, K. M. Hinkel, and F. E. Nelson, editors. The circumpolar active layer monitoring (CALM) program: Research designs and initial results. *Polar Geography* 24 (3): 165–258.

Chapman, W. L., and J. E. Walsh. 1993. Recent variations of sea ice and air temperatures in high latitudes. *Bulletin of the American Meteorological Society* 74(1): 33–47.

Deegan, L. A., H. E. Golden, C. J. Harery, and B. J. Peterson. 1999. Influence of environmental variability on the growth of age-0 and adult Arctic grayling. *Transactions of the American Fisheries Society* 128: 1163–1175.

Eugster, W., J. P. McFadden, and F. S. Chapin III. 1997. A comparative approach to regional variation in surface fluxes using mobile eddy correlation towers. *Boundary-Layer Meteorology* 85: 293–307.

Haugan, R. K. 1982. Climate of remote areas in north-central Alaska 1975–1979. Summary. *Cold Regions Research and Engineering Laboratory Report* 82–35. Hanover, New Hampshire.

Haugan, R. K., and J. Brown. 1980. Coastal-inland distributions of summer air temperature and precipitation in Northern Alaska. *Arctic and Alpine Research* 12(4): 403–412.

Hobbie, J. E., B. J. Peterson, N. Bettez, L. A. Deegan, W. J. O'Brien, G. W. Kling, and G. W. Kipphut. 1999. Impact of global change on biogeochemistry and ecosystems of an arctic freshwater system. *Polar Research* 18: 207–214.

Kane, D. L., L. D. Hinzman, J. P. McNamara, Z. Zhang, and C. S. Benson. 2000. An Overview of a nested watershed study in Arctic Alaska. *Nordic Hydrology* 31: 245–266.

McGraw J. B., A. L. Chester, and L. Stuart. 1983. A note on July senescence in tundra plants at Eagle Creek, Alaska, U.S.A. *Arctic and Alpine Research* 15: 267–269.

O'Brien, J., editor. 1992. Toolik Lake: Ecology of an aquatic ecosystem in arctic Alaska. *Developments in Hydrobiolgia* 78: 1–269.

Oberbauer, S. F., W. Cheng, B. Ostendorf, A. Sala, R. Gebauer, C. T. Gillespie, R. A. Virginia, and J. D. Tenhunen. 1996. Landscape patterns of carbon gas exchange in tundra ecosystems. Pages 223–257 in J. F. Reynolds and J. D. Tenhunen, editors. *Landscape Function and Disturbance in the Arctic*. New York: Springer Verlag.

Oberbauer, S. F., G. Starr, and E.W. Pop. 1998. Effects of extended growing season and soil warming on carbon dioxide and methane exchange of tussock tundra in Alaska. *Journal of Geophysical Research* 103: 29075–29082.

Peterson, B. J., L. A. Deegan, J. Helfrich, J. E. Hobbie, M. Hullar, B. Moller, T. E. Ford,

A. E. Hershey, A. Hiltner, G. Kipphut, M. A. Lock, D. M. Fiebig, V. McKinley, M. C. Miller, R. Westal, R. Ventullo, and G. Volk. 1993. Biological responses of a tundra river to fertilization. *Ecology* 74: 653–672.

Pop, E. W., S. F. Oberbauer, and G. Starr. 2000. Predicting vegetative bud break in two arctic deciduous shrub species, *Salix pulchra* and *Betula nana*. *Oecologia* 124: 176–184.

Shaver, G. R., F. S. Chapin III, and B. L. Gartner. 1986a. Factors limiting seasonal growth and peak biomass accumulation in *Eriophorum vaginatum*. *Journal of Ecology* 74: 257–278.

Shaver, G. R., N. Fetcher, and F. S. Chapin III. 1986b. Growth and flowering in *Eriophorum vaginatum*: Annual and latitudinal variation. *Ecology* 67: 1524–1525.

Starr, G., S. F. Oberbauer, and E. W. Pop. 2000. Effects of extended growing season and soil warming on phenology and physiology of *Polygonum bistorta*. *Global Change Biology* 6: 357–369.

Stieglitz, M., J. Hobbie, A. Giblin, and G. Kling, 1999. Hydrologic modeling of an arctic watershed: Towards Pan-Arctic predictions, *Journal of Geophysical Research-Atmospheres* 104(D22): 27507–27518.

Williams, M., W. Eugster, E. B. Rastetter, J. P. McFadden, and F. S. Chapin III. 2000. The controls on net ecosystem productivity along an Arctic transect: A model comparison with flux measurements. *Global Change Biology* 6(1): 116–126.

Williams, M., E. B. Rastetter, G. R. Shaver, J. E. Hobbie, E. Carpino, and B. L. Kwiatkowski. 2001. Primary production of an Arctic watershed: An uncertainty analysis. *Ecological Applications* 11: 1800–1816.

Zhang, T., T. E. Osterkamp, and K. Stamnes. 1996. Some characteristics of the climate in northern Alaska. *Arctic and Alpine Research* 28: 509–518.

Short-Term Climatic Events—Synthesis

David Greenland

S hort-term climatic events produce some of the most dramatic ecosystem responses. Sometimes the responses may last for a long time into the future. Three themes will be emphasized in this synthesis. The first is that short-term climatic events have both short- and long-term responses in the ecosystem. Second, the timing of short-term events is important in partially determining the kind and degree of ecosystem response that might occur. A third theme is the importance of putting short-term variability into a long-term context.

Short- and long-term responses in the ecosystem

The chapters about hurricanes and droughts in this section demonstrate that short-term climatic events may have short- and long-term responses in the ecosystem. Both the short- and long-term responses are important. The short-term responses have noteworthy economic influences in the agricultural ecosystem. One could argue that the ecosystems containing species with short life spans such as grasslands are able to respond and recover from a short-term climatic disturbance more quickly than those ecosystems with longer lived species such as trees. Corn responds quickly to variability in precipitation during important parts of the growing season. Gage believes the long-term effects of a short-term drought on annual rotational agronomic systems are generally minimal. Other LTER studies have documented strong relationships between annual precipitation and grassland aboveground net primary productivity (Knapp et al. 1998). Conversely, the Coweeta study brings to our attention the insidious, long-term effects of drought that quietly kills trees and leaves their dead necromass on the landscape for decades into the future. However, each ecosystem is responding at its own characteristic timescale. Boose notes that the mixed hardwood forests of central New England and the Tabonuco forests of Puerto Rico both exhibit remarkable resiliency to wind damage. In both cases, despite major structural reorganization after a hurricane, there was rapid regeneration of canopy cover through releafing, sprouting, or recruitment, which helped to reduce impacts on soil moisture, temperature, and nutrient cycling processes. Nevertheless, some signs of the hurricane impact are present for decades, although less so in Puerto Rico where decomposition and regeneration rates are much faster than in New England. Gage quotes work from the Cedar Creek LTER in Minnesota that also exhibits a difference in the time response to the 1988 drought between grasslands and the semiarboreal ecosystem. He reminds us that it was not until 1993, the fifth year following the 1988 drought, that there were no longer discernible effects of drought on species richness in successional grasslands. However, the effects of the 1988 drought were still clearly evident in the oak savanna complex in 1993. About 30% of mature pin oaks died during the drought, compared to only 10% of bur oaks. Most of these dead trees are still standing. Tilman and Downing (1994) concluded that this major shift in oak species composition and reduction in oak canopy cover will likely impact these savanna ecosystems for decades to come. Comparable to the Cedar Creek case is the Kloeppel et al. report that drought effects in the Appalachian forests are species specific and the necromass remains on the landscape.

One of the implications of these dynamics relates to the stability of the ecosystem. On the one hand, following a climatic event giving rise to a short-term response the ecosystem might return to approximately the state that it was in before. On the other hand, some of the long-term responses, especially those related to human activities such as increasing irrigation systems or logging, may change the ecosystem in important, semipermanent and permanent ways. The Cedar Creek LTER work pointed to another pivotal finding on ecosystem operation with respect to the 1988 drought. At this site plant communities with the greatest biodiversity were the least susceptible to year-to-year fluctuations in total plant growth and also were the least susceptible to nutrient loss from the soil by leaching (Tilman and Downing 1994). Cedar Creek researchers point out that this conclusion was the first experiment showing a clear connection between biodiversity and the functioning of ecosystems since Darwin first suggested it over a century ago.

Investigators of short-term climate events and related ecosystem responses are also concerned with the return period of the event. Boose comments that a critical factor controlling ecosystem impact is the relative length of hurricane return intervals and vegetation life spans. However, it is important that there was no clear evidence of centennial-scale trends for either region. In both regions Harvard Forest investigators found the same multidecadal variation that is well documented for North Atlantic hurricanes in general.

Timing of Short-Term Climatic Events

The chapters in this section support the hypothesis that the timing of short-term events is important in partially determining the kind and degree of ecosystem response that might occur. At the Arctic LTER site, early snow melt is an issue. Hobbie et al. report that every few years soil warming, caused by early snow melt or by warm summers, results in both a synchronous flowering of arctic cotton and an increase in leaf area that occurs across the entire area of northern Alaska. The corn study particularly illuminates the importance of timing of soil moisture deficit conditions in determining the degree of impact of drought on agriculture. During May and June, seedling and early root growth are vulnerable to periods without moisture, especially if temperatures are high. The severity of the stress in the early growing season and subsequent loss in productivity was due to the inability of young seedlings to tap soil moisture reserves prior to a stress period. In another instance, Gage reports that depending on timing and severity, drought may cause plant mortality or weaken the plant system, predisposing the crop to insect herbivory or disease. In row-crop ecosystems, a short-term (one-year) drought can have a significant effect on yield, thus reducing productivity in the year of the anomaly. Timing is important in the forest ecosystem as well. Kloeppel and his colleagues claim Southern Pine Beetle (SPB) populations, as with most insect species, are cyclic, but the coincidence of SPB outbreak or epidemic populations and stress brought on by severe and sustained drought can have monumental impacts on pine populations.

In subsequent chapters, we will see several other examples of the importance of timing of a climatic event at other time scales. This issue relates, in part, to the framework question, What preexisting conditions will affect the impact of the climatic event or episode? In many cases the exact state or stage of seasonal development of an ecosystem will create a preexisting condition that will control the intensity of the effect of the climatic event or episode.

Short-Term Variability in a Long-Term Context

Another aspect of short-term climatic variability and ecosystem response, partially and implicitly treated above, is the need to place these items into a longer term perspective. Boose points out the need for paleohurricane record studies as the next step in our understanding of the forest response to such storms. The hurricane study also shows how important preexisting conditions are, especially with regard to the passage of previous hurricanes and to human land-use patterns that may have been set decades to centuries previously. Kloeppel et al. note that the long-term importance of increasing our understanding of drought impacts on forest structure and function is central to anticipating the full impacts of suggested long-term climate change. The Arctic analysis demonstrates that the short-term climate variability and more or less direct response must be seen against a backdrop of steadily increasing temperatures at the decadal scale, as shown by borehole records. This situation foreshadows strong, nonlinear ecosystem responses if ground temperatures rise high enough to melt the permafrost. Gage's analysis uses long-term data to establish that the drought in 1988 was clearly different from other droughts that occurred during the 20-year period analyzed (1972–1991). The 1988 drought occurred in May and June, whereas in other years, significant periods of stress occurred later in the crop-growing season (July–August).

These comments end on an ominous note. Gage reminds us that the 1988 drought was a one-year drought compared to the multiyear "dust bowl" drought that occurred in the 1930s. During the past 60 years, a significant multiyear drought has not occurred in the North Central Region of the United States, and thus the probability of such a multiyear drought is high. The LTER program is well prepared to monitor and study the ecosystem responses to even more severe climatic events and episodes than those that occurred during the first two decades of the program.

References

Knapp, A. K., J. M. Briggs, J. M. Blair, and C. L. Turner. 1998. Patterns and controls of aboveground net primary production in tallgrass prairie. Pages 193–221 in *Grassland dynamics: Long-Term Ecological Research in tallgrass prairie.* New York: Oxford University Press.
Tilman, D., and J. A. Downing. 1994. Biodiversity and stability in grasslands. *Nature* 367: 363–365.

Part II

The Quasi-Quintennial Timescale

Introductory Overview

Raymond C. Smith

The El Niño–Southern Oscillation (ENSO) is a coupled ocean–atmosphere phenomena that has a worldwide impact on climate. An aperiodic phenomena that reoccurs every 2 to 7 years, the ENSO is second only to seasonal variability in driving worldwide weather patterns. As Greenland notes in chapter 6, the term "quasi-quintennial" is chosen to recognize that climatic events other than ENSO-related events might occur at this timescale, although it is widely recognized that ENSO contributes the lion's share of the higher frequency variability in paleorecords of the past several thousand years. In this section, we consider variability with cycles of 2 to 7 years and the resulting ecological response. Although we emphasize the ENSO timescale in this section, there is growing evidence that this phenomena is neither spatially nor temporally stable over longer time periods. Indeed, Allan (2000) suggests the ENSO climatic variability must be viewed within the context of climate fluctuations at decadal to interdecadal timescales, which often modulate the higher frequency ENSO variability. As a consequence, results in this and the next section often display overlapping patterns of variability, and their separation is not sharply defined.

An important theme in this section is the worldwide influence of ENSO-related climate variability. Greenland (chapter 6) provides an LTER network overview with an analysis of ENSO-related variability of temperature and precipitation records for many LTER sites from the Arctic to the Antarctic. He discusses the general nature of ENSO and its climatic effects, summarizes previous climate-related work in the LTER network, and provides a cross-site analysis of the correlations between the Southern Oscillation Index (SOI) and temperature and precipitation at LTER sites. His results are consistent with the expected patterns of the geography of ENSO effects on the climate. Greenland's cross-site analysis provides the basis for studying climate variability and ecosystem response within the context of the series of framework questions that form an underlying theme for this volume.

Brazel and Ellis (chapter 7) provide an excellent analysis of climate-related parameters within the context of ENSO indices. Reporting on the Central Arizona and Phoenix (CAP) LTER urban-rural ecosystem, these authors provide a comprehensive analysis linking water-related parameters to climate forcing, as indicated by these indexes. Their studies show a strong connection between ENSO and winter moisture in Arizona, perhaps making it possible to forecast impending conditions. This arid to semiarid ecosystem is strongly dependent on water resources, and Brazel and Ellis provide several excellent examples of ecological response to ENSO-related climate variability. Their examples show that both the natural and human components of the CAP ecosystem are substantially affected at the ENSO timescale. They also discuss how climate responses potentially result in complex cascades within this ecosystem. For example, drought periods lead to dust storms, wildfires, vegetation change, water quantity and quality changes, and the subsequent consequences of these changes.

A study of a Puerto Rican tropical rainforest (Luquillo Experimental Forest LTER, LUQ) provides dramatic contrast to the arid southwestern United States. Schaefer (chapter 8) studies the effects of ENSO and the North Atlantic Oscillation (NAO) on extreme rainfall events and finds the effects of those oscillations to be minor. Schaefer focuses on extreme rainfall events because of the highly nonlinear response of the system to precipitation whereby 75% of the sediment export occurs during only 1% of days with the greatest rainfall. These major sediment exports have important, and generally nonreversible, ecological effects on both the watersheds and downstream ecosystems. This study provides an excellent example of the high sensitivity of an ecosystem to the extreme nonlinearity of the process such that more regular variability may be overwhelmed.

The western Antarctic Peninsula (site of the Palmer LTER, PAL) is now a recognized "hot spot" with respect to a global warming trend (IPCC 2001). Smith and coworkers (chapter 9 and Synthesis) show that there is a significant correlation between air temperature of the western Antarctic Peninsula (WAP) and the SOI. Further, there is a strong anticorrelation between sea ice extent in the area and the SOI. These observations are further evidence for ENSO-related teleconnections to high latitudes. This Antarctic marine ecosystem is dominated by sea ice, and these researchers show that sea ice extent in the WAP has trended down and the sea ice season has shortened. Although ecological responses to this climate variability are evident at all trophic levels, Smith and coworkers show that changes are most clearly seen in a shift in the population size and distribution of penguin species with different affinities to sea ice. Analogous to, but in distinct contrast to, the extreme nonlinearity of sediment processes for a tropical rainforest, this study also emphasizes the importance of nonlinear processes. At the PAL site, the fine balance of temperature with respect to the phase transition between ice and water is such that warming trends may remove large areas of this ice-related habitat with significant consequences for this marine environment.

Welch and coworkers (chapter 10) report on studies of the driest and coldest deserts on the planet (McMurdo Dry Valleys, MCM). These workers show that the key climatic parameters influencing ecosystem structure and function are the ones that affect the state of water. The ecosystem is very sensitive to relatively small climatic variations because the change between ice and liquid water is delicately balanced. Thus, small changes in temperature and/or radiant energy are amplified by large, nonlinear changes in the hydrologic budgets that can cascade through the system. Indeed, this cold dry ecosystem provides outstanding examples of how small climatic shifts cascade with impacts on stream, lake, and soil ecosystems. Further, these small variations can have a significant multiyear impact. The record from Taylor Valley is too short to discern statistically significant long-term trends or ENSO-related variability, although a few paleorecords show dominant periodicities coincident with the SOI.

The five chapters in this section examine dramatically different ecosystems that often represent extremes with respect to temperature and/or precipitation. Interestingly, a common theme for such ecosystems is a high sensitivity whereby relatively small changes are amplified and cascade through the system.

References

Allan, R. J. 2000. ENSO and climatic variability in the past 150 years. Pages 3–55 in H. F. Diaz and V. Markgraf, editors, *El Niño and the Southern Oscillation.* Cambridge University Press, Cambridge, UK.

IPCC. 2001. *Climate Change 2001, Synthesis Report.* A contribution of Working Groups I, II and III to the Third Assessment Report of the Intergovernmental Panel on Climate Change [Watson, R. T. and the Core Writing Team, editors]. Cambridge University Press, Cambridge, UK.

6

An LTER Network Overview and Introduction to El Niño–Southern Oscillation (ENSO) Climatic Signal and Response

David Greenland

P art II of this book deals with the quasi-quintennial timescale that is dominated by the El Niño–Southern Oscillation (ENSO) phenomenon. During the last 50 years, ENSO has operated with a recurrence interval between peak values of 2–7 years. The term *quasi-quintennial* is chosen to recognize that climatic events other than ENSO-related ones might occur at this timescale. The general significance of the ENSO phenomenon lies in its influence on natural and human ecosystems. It has been estimated that severe El Niño–related flooding and droughts in Africa, Latin America, North America, and Southeast Asia resulted in more than 22,000 lives lost and more than $36 billion in damages during 1997–1998 (Buizer et al. 2000). The specific significance of ENSO within the context of this book is that it provides fairly well-bounded climatic events for which specific ecological responses may be identified.

In the other chapters in part II, we first look at the U.S. Southwest. The Southwest is home to an urban LTER site, the Central Arizona-Phoenix (CAP) site. Tony Brazel and Andrew Ellis describe the clear ENSO climatic signal at this site and identify surprising responses that cascade into the human/economic system. Ray Smith, Bill Fraser, and Sharon Stammerjohn provide more details of the fascinating ecological responses of the Palmer Antarctic ecosystem to ENSO. World maps of ENSO climatic signals do not usually show the Antarctic, and the LTER program provides some groundbreaking results at this location, with Smith and coworkers (see the Synthesis at the end of this part) providing such maps (figures S.1 and S.2). Kathy Welch and her colleagues present equally new discoveries related to freshwater aquatic ecosystems from the other Antarctic LTER site at the McMurdo Dry Valleys.

This chapter gives a general introduction to ENSO and its climatic effects. However, these general patterns may mask the detailed responses that occur at individual locations. This is one reason for presenting the principal results of previous findings related to El Niños and LTER sites and one particular analysis focused on LTER sites. This analysis for the period 1957–1990 investigates the response of monthly mean temperature and monthly total precipitation standardized anomaly values to El Niño and La Niña events as indicated by the Southern Oscillation Index (SOI) (Greenland 1999). The chapter then reviews some of the ENSO-related responses occurring at LTER sites. Some of these responses are treated in more detail in other chapters in this section and this book. The goal here is to provide an introduction to climate variability and ecosystem response at the quasi-quintennial scale. Finally, this chapter addresses some of the framework questions of this book.

The General Nature of ENSO and Its Climatic Effects

ENSO is the acronym for El Niño–Southern Oscillation despite the fact that the Southern Oscillation is composed of swings between El Niño and the almost opposite La Niña events. El Niño is a warming of the Pacific Ocean between South America and the international date line, centered on the equator, and typically extending several degrees of latitude to either side of the equator. La Niña exists when cooler than usual ocean temperatures occur in the same area (Trenberth 1997; Kelly Redmond, pers. comm., 2000). Both El Niño and La Niña affect the atmosphere as well as the ocean. There are many ways of measuring ENSO variability. One of the most common is the use of the Southern Oscillation Index (SOI), which represents the standardized Tahiti-Darwin sea level pressure (SLP) anomaly. Data on this and other climatic variables may be found at http://www.cpc.ncep.noaa.gov/products/.

The general climatological effects of El Niños and La Niñas worldwide have been well established. El Niños in North America are generally associated with higher than average precipitation in the Southwest and Southeast of the country and lower than average precipitation in the Pacific Northwest (Ropelewski and Halpert 1986; Kiladis and Diaz 1989,1992). Trenberth and Caron (2000) have updated the climatology and provide new global maps of ENSO effects on climate. Among other things, such maps and high resolution data emphasize the fact that the LTER sites by no means provide a high-resolution spatial climatology. Rather, the role of the LTER sites is to provide detailed ecological responses. There are many internet web sites that describe the ENSO phenomenon and give almost worldwide coverage of its climatic results (e.g., http://www.pmel.noaa.gov/tao/elnino/nino-home.html). Occasionally, very intense super El Niños occur as in 1982–1983 and 1997–1998. El Niños and La Niñas can be forecasted a season or two in advance; because they tend to have similar climate effects in known parts of the world, this permits climate forecasts to be made. Both the National Oceanographic and Atmospheric Administration's (NOAA) Climate Prediction Center (CPC) and the International Research Institute for Climate Prediction (IRI) provide these forecasts. Model nested approaches can now provide accurate seasonal climate and streamflow forecasts for some areas that include LTER sites. Model nested results have been given for the Pacific Northwest (PNW) (Leung et al. 1999). General forecasts are also available during intense El Niño events for the coterminous United States (Barnston et al. 1999) and globally (Mason et al. 1999). These forecasts will be very useful for the design of future ecological experiments at LTER sites and are already being used in many areas of human societal systems (Buizer et al. 2000). However, K. T. Redmond (pers. comm., 2000) makes the following points with respect to the predictability of ENSO-related climate. Not every El Niño produces the same climatic effect. La Niñas have a more consistent signal, in general, than El Niños. The relations between an ENSO "cause" and a climatic "effect" are not perfect, in part because other things are happening in the climate system. Super El Niños sometimes do not show the expected resulting climatic patterns in some locations. Resulting climatic patterns for large El Niños may differ in some ways from those of typical El Niño patterns. The relationship between extreme ENSO events and climatic results is lagged. In general, the best associations are between the summer/autumn SOI and the following winter climate.

Previous Work at LTER Sites

Within the LTER network, there has been ongoing interest in the ENSO-related phenomena. A workshop, held in 1993, concentrated on the effect of El Niños and La Niñas at LTER sites (Greenland 1994a). Since the LTER network is spread across the North American and Antarctic continents (figure 1.1), it is natural that ENSO climatic signals should be stronger and more marked for some LTER sites than others. The line of LTER sites, from New Mexico through Colorado to the Pacific Northwest and into Alaska (JRN, SEV, NWT, SGS, AND, BNZ), follows the inverse influence of El Niño-related above-average precipitation in the Southwest to below-average precipitation in the Northwest. However, the intensity of the ecological response differs. The effect of low streamflows on the ecosystems in the Northwest is less marked than the large hydrologic and ecological impact documented by workers at the Seviletta (SEV) LTER site in New Mexico (Molles and Dahm 1990; Dahm and Molles 1992). Wetter than usual winters during El Niños have large effects on the aquatic and terrestrial ecosystems at the Sevilleta site. Plants, invertebrates, rodents, and rabbits all react to the increase in autumn and spring moisture associated with El Niño. Additionally, Dahm and Moore (1994) showed a series of dry La Niña episodes in the late 1940s and mid-1950s led to significant dieback of pinyon pine (*Pinus edulis*) and juniper (*Juniperus monosperma*) at the site. Ecosystem responses at SEV are treated in detail in chapter 15. At the NWT site, Woodhouse (1994) showed that wet springs occur the year after an El Niño and dry springs follow a La Niña. An ecological response is visible in tree ring chronologies, which indicated a more marked response for La Niñas than for El Niño. Tree growth response to SOI values varies over time at this site.

The first El Niño workshop (Greenland 1994a) demonstrated that the ENSO signal could be geographically subtle in its effects on ecosystems. For example, sophisticated analyses were made for the North Temperate Lakes (NTL) LTER sites in Wisconsin. Robertson et al. (1994) showed that although during El Niño events ice breakup dates were earlier and spring air temperatures were warmer than long-term averages across the whole state of Wisconsin, the effect was more extreme for southern lakes than northern lakes. Robertson and coworkers attribute the difference to the average breakup dates for the southern lakes being in late March, directly following a period when air temperatures are strongly related to El Niño events. In contrast, average breakup dates for northern lakes are in mid- to late April after a period when air temperatures are not significantly related to El Niño events. Later work by individual site investigators has identified an ENSO signal at other sites. An investigation by Schaefer (see chapter 8) suggests the Luquillo (LUQ) site in Puerto Rico has increased rainfall in May and decreased rainfall in October in El Niño years, whereas in La Niña years this pattern is reversed, although, in both cases, the signal is weak. The Andrews Forest (AND) site in the Pacific Northwest often suffers from below-average precipitation in El Niño years, whereas many severe floods and above-average precipitation years are associated with La Niñas (Greenland 1994b; see also chapter 19).

Until 1999 the effect of ENSO phenomena on LTER sites had not been investigated with a standardized methodology across all sites—a situation that made intersite comparison difficult. This was remedied by a systematic analysis on a common set of climatic data applicable to 17 of the LTER sites and designed to identify the relative strength of the El Niño and La Niña signal across the LTER network (Greenland 1999).

A Cross-Site Analysis

The Pearson product moment correlation coefficients for the period 1957–1990 between the SOI values for a given month and the standardized temperature or precipitation anomalies do not show very high values (tables 6.1 and 6.2) because the data are inherently noisy. However, most of the values shown are statistically significant, partly because of the large number of pairs of observations (397–408) in the analyses. Another reason that the correlation coefficients are low is that all SOI values during the period are used, and thus both El Niño (extreme negative SOI values) and La Niña (extreme positive SOI values) events occur in the series along with intermediate values. This approach differs from one where correlations are found between one or more climate variables of El Niño years only and some other ecosystem variable. Many studies of this kind tend to deal only with extreme ENSO values instead of all the data (e.g., Cayan et al. [1999] use the 90th percentile ENSO events). In addition, Sardeshmukh et al. (2000) have noted that away from the tropical Pacific Ocean, an ENSO event is associated with relatively minor changes of the probability distributions of atmospheric variables. Nonetheless, it is important to estimate the changes accurately for each ENSO event, because even small changes of means and variances can imply large changes in the likelihood of extreme values. Wolter et al. (1999) have also quantified similar relationships. The higher correlation coefficients in Greenland (1999) are used to indicate sites where the climatic variable displays an association to both El Niño and La Niña events.

Table 6.1 Correlation coefficients between monthly SOI values and LTER site temperatures

Lag	0	1	2	3	4	5	6	7	8	9	10	11
AND	−0.196	−0.226	−0.193	−0.187	−0.15	−0.089	−0.088	−0.085	−0.065	−0.041	−0.014	0.005
ARC	−0.082	−0.046	−0.07	−0.073	−0.077	−0.043	0.057	0.014	0.028	0.011	−0.028	−0.002
BNZ	−0.081	−0.098	−0.108	−0.105	−0.104	−0.096	−0.014	−0.042	−0.024	0.011	−0.025	−0.03
CDR	−0.09	−0.011	0.025	−0.028	−0.041	−0.023	−0.095	−0.064	−0.075	−0.066	−0.078	−0.064
CWT	0.019	0.093	0.142	0.037	0.003	−0.028	−0.005	−0.037	−0.062	−0.047	0.005	−0.062
HBR	0.004	−0.055	−0.006	−0.013	−0.022	−0.018	−0.073	−0.103	−0.093	−0.035	−0.07	−0.049
HRF	0.007	−0.03	−0.025	−0.027	−0.052	−0.056	−0.106	−0.137	−0.11	−0.05	−0.093	−0.081
JRN	−0.059	−0.089	−0.079	−0.093	−0.131	−0.136	−0.095	−0.088	−0.045	−0.061	−0.022	0.011
KBS	−0.127	−0.103	−0.063	−0.046	−0.017	−0.001	−0.041	−0.029	−0.049	−0.075	0.045	0.083
KNZ	−0.007	0.065	0.116	0.091	0.037	0.044	0.009	−0.08	−0.043	−0.017	−0.002	0.014
LUQ	−0.157	−0.199	−0.279	−0.238	−0.223	−0.267	−0.252	−0.234	−0.152	−0.157	−0.15	−0.141
NTL	−0.092	−0.087	−0.023	−0.049	−0.082	−0.128	−0.142	−0.178	−0.156	−0.069	−0.068	−0.066
NWT	0.005	0.054	0.093	0.023	0.032	0.083	0.019	−0.01	0.03	0.03	−0.015	0.036
PAL	0.166	0.208	0.202	0.217	0.185	0.169	0.156	0.164	0.121	0.154	0.135	0.133
SEV	−0.005	0.026	0.05	0.032	0.036	0.031	−0.002	−0.062	−0.085	−0.132	−0.165	−0.117
SGS	−0.096	−0.029	0.012	−0.01	−0.014	−0.004	−0.022	−0.072	−0.051	−0.03	−0.011	0
VCR	0.059	0.065	0.096	0.062	0.011	0.013	−0.022	−0.067	−0.1	−0.023	−0.008	−0.048

[a]Trailing numbers represent lags by month of correlation (i.e., 0 = no lag, 1 = SOI value correlated against the temperatures of the following month).

Table 6.2 Correlation coefficients between monthly SOI values and LTER site precipitation

Lag	0	1	2	3	4	5	6	7	8	9	10	11
AND	0.024	0.086	0.069	0.139	0.072	0.02	0.068	0.01	−0.024	−0.024	0.013	−0.027
ARC	0.02	0.019	0.019	0.004	−0.056	−0.027	0.006	−0.011	−0.006	0.003	−0.011	0.022
BNZ	0.097	0.1	0.1	0.051	0.013	0.029	0.013	0.063	−0.021	−0.027	−0.002	0.008
CDR	0.037	0.079	−0.006	−0.087	−0.034	−0.055	−0.051	−0.069	−0.027	−0.052	−0.058	−0.051
CWT	0.075	0.005	0.057	0.002	0.074	0.081	0.067	0.061	0.036	0.059	0.04	0.045
HBR	0.05	0.023	−0.01	0.017	−0.009	0.036	−0.015	0.033	0.032	0.034	0.018	−0.015
HRF	0.008	0.001	−0.007	0.021	0.036	0.081	0.046	0.016	0.014	0.051	0.01	−0.038
JRN	0.003	−0.031	0.058	0.081	0	−0.028	0.027	−0.011	−0.043	−0.091	−0.049	−0.111
KBS	0.094	0.069	0.07	0.032	−0.005	0.057	0.002	0.014	−0.024	−0.02	−0.032	0.001
KNZ	−0.069	−0.05	−0.037	−0.042	0.017	−0.007	−0.028	−0.015	−0.041	−0.038	0.001	−0.027
LUQ	0.012	0.022	−0.042	−0.037	−0.089	−0.025	−0.041	−0.079	−0.073	−0.019	−0.035	−0.048
NTL	0.034	0.036	−0.032	−0.012	−0.008	−0.059	−0.071	−0.108	−0.1	−0.111	−0.1	−0.086
NWT	−0.028	−0.019	−0.061	−0.008	−0.026	−0.033	−0.025	−0.039	−0.055	−0.06	0.012	−0.025
SEV	−0.106	−0.093	−0.159	−0.172	−0.124	−0.094	−0.088	−0.043	−0.068	−0.015	0.015	0.051
SGS	−0.082	−0.062	−0.084	−0.085	0.006	0.019	0.007	0.005	−0.04	−0.046	−0.042	−0.044
VCR	−0.024	−0.083	−0.105	−0.104	−0.05	−0.028	0.028	−0.005	0	−0.054	0.017	−0.017

aTrailing numbers represent lags by month of correlation (i.e., 0 = no lag, 1 = SOI value correlated against the precipitation of the following month).

Table 6.3 Classification of sites by signal strength and duration of signal

Very strong signal (r>0.20)	Site and climate variable	Signal duration in lag months
	ANDt[a]	1
	LUQt	2,3,4,5,6,7
	PALt	2,3,4

Signal (r=0.1 to 0.2)	Site and climate variable	Signal duration in lag months
	ANDp[b]	3
	BNZt	2,3,4
	BNZp	1,2
	CWTt	2
	HBRt	7
	HRFt	6.7
	JRNt	4,5
	JRNp	11
	KBSt	0,1
	NTLt	5,6,7,8
	NTLp	7,8,910
	SEVt	9,10,11
	SEVp	0,2,3,4
	VCRt	8
	VCRp	2,3

No signal (r<0.1)		
	ARCt	ARCp
	CDRt	CDRp
	CWTp	
	HBRp	
	HRFp	
	KBSp	
	KNZt	KNZp
	LUQp	
	SGSt	SGSp
	NWTt	NWTp

[a] t represents temperature.

[b] p represents precipitation.

Unless otherwise specified, I use the term ENSO in the subsequent discussion to refer to both of these extremes. Although there are a few exceptions, the results of the current study are consistent with the expected patterns of the geography of ENSO effects on the climate as illustrated in the Synthesis to this section.

Given these facts, it is appropriate to classify the LTER sites in terms of their ENSO responses into three categories (table 6.3). The first category consists of three data series (ANDt, LUQt, and PALt) that show a strong response (r > 0.2). Here, and in the following, I use the abbreviation letters of the LTER sites (chapter 1, table 1.1) and a lowercase "t" or "p" to indicate temperature and precipitation, respectively. The second category consists of 15 data series that display a detectable signal (r = 0.1 to 0.2). The third category is where there is no signal according to the definition (r < 0.1). Some of the data series in this category do come close to the cutoff r value and might well have an ENSO signal by other definitions. The duration in months of the ENSO signal is also shown in table 6.3. In some data series the signal lasts for up to 6 months, whereas in other series the signal occurs only in a single month. The general types of patterns in the data are represented by the data series for PALt, SEVp, and ARCp. The ENSO signal at the Palmer site in the Antarctic is strong and actually lasts in some form for the whole 12 months. Although the analysis does not extend to longer periods, the signal may be found before the beginning and after the end of the 12 months considered here. The SEVp ENSO signal is the strongest in the second and third month after the extreme SOI value and then gradually decreases in strength, becoming insignificant after the seventh month. The ARCp series is an example of a data series with no ENSO signal.

Some interesting details and implications arise from these results. At some sites, such as LUQ, higher or lower than average precipitation values are more ecologically important than higher or lower than average temperature values. The LUQ, Puerto Rico, site shows strong and long-lasting higher than normal temperatures associated with an El Niño occurrence and lower than normal temperatures with a La Niña. The site also displays drier than normal conditions for zero- and one-month lags between the SOI value and the rainfall. But, with a higher number of monthly lags, the LUQ precipitation is higher when the SOI has indicated El Niño conditions, or at least negative SOI values two or more months previously. This is generally consistent with Schaefer's findings (chapter 8). However, he deals only with true El Niño years when SOI values are negative and large, rather than with all months of SOI values. Specifically, Schaefer finds that although El Niño years have higher precipitation than average this effect occurs only in May. The wetter or drier than normal conditions actually have more effect on the ecosystem than do the warmer than average conditions. Temperatures are usually high at LUQ, and a little increase will not make much ecological difference. However, the precipitation at this site is the highest of all LTER sites (2470 mm annually), and the variability around this value can be very large. Between 1961 and 1990, the wettest year (1979) had 3955 mm of precipitation, whereas the driest year (1967 following the 1966 El Niño) had 1540 mm. The increase or decrease of precipitation, according to Schaefer, has a large ecological impact on streamflows and their sediment load and water chemistry. This can have an even greater ecological impact when the below-normal precipitation occurs in the dry season between January and March or during large storm events at this site. Gianinni et al. (2000) have demonstrated a complex geography of Caribbean climate in response to El Niño, with some parts of the region wetter and others drier than average.

At the Colorado alpine site (NWT), the ENSO precipitation response is also important. This site had one of its highest precipitation years (1581 mm) in the super El Niño year of 1983. Net primary productivity was above average during this year. However, the El Niño–related precipitation signal at NWT is not strong in the correlation analysis described here. Most likely, the high precipitation of 1983 cannot be explained by the 1982–1983 El Niño alone.

SEV is another site where the importance of the ENSO-related precipitation appears more important than the temperature signal. The ENSO-related precipitation increase is manifested shortly after the occurrence of an extreme SOI value. The effect of the increase in winter precipitation in the case of El Niños has been well documented (Molles and Dahm 1990; Dahm and Molles 1992). After the forecast of the 1997–1998 El Niño event, workers at the SEV site issued a warning to New Mexico residents to be particularly careful not to allow the buildup of household and other waste. Such waste would add to the natural increased rainfall-derived accumulation of vegetative material on which rats and other small mammals feed. By issuing the forecast, scientists hoped to decrease the possibility of outbreaks of the ratborne hantavirus. Such outbreaks did not occur during the months after the 1997–1998 El Niño, but it is not possible to assess the direct effectiveness of the warning (Robert Parmenter, pers. comm., 2000).

At sites such as NTL, El Niño–related temperatures are more important than precipitation. El Niño occurrence is associated with higher than normal temperatures at the NTL site in Wisconsin, and La Niña corresponds to lower than average temperatures. However, the greater effect is found in the following summer for a wintertime maximum or minimum SOI value. Investigations have not yet been made to see whether this has effects on the ecosystem. The work of Robertson et al. (1994) focused on air temperatures during the spring melt of lake ice, and this had obvious ecosystem effects. The ENSO effect for El Niños also is manifested by increasing NTL precipitation values in summer and into fall. This may affect the atmosphere/groundwater water input ratio to the lakes that, in turn, affects the water chemistry and has a cascading effect through the ecosystem.

At the Antarctic site, PAL, the ENSO-related signal in temperature is extremely strong in the context of the present study. Smith et al. (1996) suggested the lag may extend to 19 months. During El Niño events, temperatures at PAL tend to be colder than average. Smith et al. (1996) have noted that El Niño occurrence is associated with above-average ice extents in the Western Antarctic Peninsular area. Here the effect at the quasi-quintennial timescale somewhat offsets the strong warming trend that has been noted at this site during the last 40 to 50 years. An important ecological linkage is associated with penguins in this location. Optimum sea ice conditions no longer exist for Adélie penguins in the Western Antarctic Peninsular because of the lack of sea ice as the result of long-term warming. In contrast, Chinstrap penguin populations are increasing because they do better in open-water conditions (Fraser et al. 1992). Thus, at the longer timescale of five decades, the smaller timescale El Niños give a "momentary" respite to the Adélie penguins at the expense of the Chinstrap penguins, whereas La Niñas may have the opposite effect.

The result of the analysis of the climatic response to the 1982–1983 super El Niño compared to more normal-size warm events was not clear-cut. The LTER sites that had shown the highest response in the previous analysis to El Niños were examined. The 1982–1983 El Niño was certainly larger in terms of its SOI value than those of 1958, 1965, 1972, and 1987. However, the responses to these five ENSO events are not altogether consistent. At AND, in the Pacific Northwest, the 1982–1983 temperature anomaly was larger than for any of the other El Niño years. This was also true for the NTL, Wisconsin, temperature anomaly of the following summer. But with respect to temperature at LUQ, Puerto Rico, and precipitation at SEV, New Mexico, the 1982 to 1983 El Niño led to a smaller response than in some of the other El Niño years. The PAL temperatures in the Antarctic showed a larger positive anomaly for the 1982–1983 event than other years, but the pattern is confounded by the large negative anomaly for the 1958 event. Thus super El Niños might give rise to larger climatic responses than "normal" El Niños at some of the LTER El Niño–sensitive sites but not necessarily all of them.

Application of Framework for Investigating Climate Variability and Ecosystem Response

In this chapter, we have concentrated only on a periodic type of climatic variability—the ENSO. We have seen that any particular site may exhibit a variety of responses, ranging from long-lasting responses of several months, to short-lived responses of one month, to no response at all. At some sites, precipitation anomalies have the greatest effect on the ecosystem, whereas at others temperature anomalies are more important.

Although it is not possible to go into the details of the framework questions for all LTER sites, some examples are appropriate. Having identified the nature of ENSO as a climatic signal, we present the next part of the framework questions.

1. *Are there any preexisting conditions that will affect the impact of the climatic event or episode?*

One of the benefits of asking this question is that it will usually stimulate new research questions. An example relates to the effect of ENSO events that are themselves superimposed on a trend of longer warming and less ice at the Antarctic PAL site. The new question arises, At what stage does the environmental condition pass a threshold, for example, a change from pack ice to open ocean, that might lead to a fast decline or increase in the penguin populations? The issue of preexisting conditions further raises questions about the relationship between climate events at one timescale and those at another. It is possible, for example, that La Niña years set the stage in the Pacific Northwest for increased, short-term, rain-on-snow flood events at the Andrews rainforest. That the ENSO scale can be related to the individual storm scale has been shown in at least two cases. The first case is the increase of Atlantic hurricane frequency and damage during La Niña periods (Pielke and Landsea 1999). The second case is the January 1998 ice storm in the northeastern United States that had a documented impact on the HBR LTER site and was attributed to the presence of an El Niño event (Barsugli et al. 1999).

2. *Is the effect of climate direct or does it cascade? If it cascades, how many levels does it have, and is the interaction between each level linear or nonlinear?*

A direct climatic effect on an ecosystem is exemplified by a windthrow event in which trees are severely broken; yet even this sets off a sequence of ecosystem responses at a relatively small spatial scale. Climate effects on ecosystems, however, most often go into cascades. So, for example, the increased El Niño–related precipitation at SEV increases the water in the aquatic systems and also sets the stage for increased primary productivity on the terrestrial systems. The latter, in turn, provides increased forage for small mammals, which provide transportation for the hantavirus. This example of a terrestrial ecosystem simplistically identifies a three-level system. Quantification of this system would help determine the degree to which the various stages were linear or not.

The subtle timing of the ecological response of the southern Wisconsin lakes to the El Niño signal illustrates that sometimes whether a cascade results depends on exactly how the signal and response are coupled. Much more work is needed to identify the coupling mechanisms and their temporal and spatial aspects in various ecosystems.

At this point, it is appropriate to address the question of whether correlation analysis should be used at all in attempts to relate ENSO to its climate signal and later to a potential ecological response. At the LTER All Scientist's Workshop on Climate Variability and Ecosystem Response (2000), this question was raised on the ground that meteorological, climatological, and, to a lesser extent, ecological data are inherently autocorrelated both in time and space and thus violate the assumptions of the statistical methods being used. This may be viewed as part of a larger question that has recently been discussed by Nicholls (2001), who points out many criticisms of null hypothesis significance testing in atmospheric science in general. As an alternative, Nicholls suggests we focus on the strength of the effect rather than on its significance. The correlation coefficients used in this chapter do just that. The correlation coefficients used here should be regarded as an index of the strength of the relationship between ENSO and temperature and precipitation rather than being viewed within the context of central tendency statistics.

3. *Is the primary ecological effect completed by the time of the next climatic event or episode (or part thereof)? If the effect is complete, we may consider the next part of the cascade (if any). If the primary ecological effect is not complete (i.e., reaches a new constant level), is it still of sufficient magnitude to have an effect on the rest of the ecosystem? If so, we should pass the effect along the cascade.*

The answer to the first question depends on the "characteristic time scale" of the ecosystem. On the one hand, at NTL the ecological effect of early ice melt during an El Niño is completed by the time of the next El Niño event. When an early melt occurs, primary productivity can have an early start, and, at least in hypothesis, there can be more productivity during the growing season at all higher trophic levels. The higher trophic levels represent the next part of the cascade. On the other hand, within the ENSO context, the ecological effect may not be complete for ecosystems, such as forests, acting at long timescales. In most of these kinds of cases, an individual El Niño will not have a measurable effect except possibly on the aquatic parts of the system, as in the case of LUQ. Apart from forest ecosystems, however, there will be few examples at the ENSO time scale where the ecological effect is not complete by the time of the next event.

4. *Does the climatic event or episode have an identifiable upper or lower limit? If a limit exists, we can stop the consideration if necessary at the limit but keep the cascade going until it reaches limits that may exist in later parts of the cascade.*

The ice melt at the NTL represents a more or less linear change. The change has a limit because there is always a finite amount of ice to melt. After the melt, the cascade of the energetics of the lake ecosystem will continue through the various trophic levels. Most changes will have limiting values. It will often be important to identify what these limiting values are for both the climatic system and the ecosystem. We might assume, in the case of the climate system, that the climatic effects of the super El Niño represent a set of limiting values. However, we showed that this does not necessarily lead to the identification of unequivocal limiting values. We also have to consider how limiting values might change as one moves across timescales and specifically how past and future climate change might affect the limiting values. In this context we must remember that the ENSO–intensity time series itself is not stationary (Torrence and Webster 1999).

5. *Does the climatic event or episode reverse to some original state (i.e., is it periodic, homeostatic, etc)? If so, what timescales are involved? Does the climate state go back to the original position or beyond? Do cascades reverse? Can we identify the timing of these events?*

The first of these five questions has a relatively easy answer as far as ENSO climatic phenomena are concerned. The climatic variation is quasi periodic and returns more or less to its original position. Atmospheric cascades do not usually reverse. The timing of ENSO events is at a quintennial timescale in terms of the usual occurrence of the event, yet when an ENSO event occurs it does so at a seasonal and monthly timescale. These answers also apply to many El Niño–influenced ecosystems. An increase in Adélie penguin populations during a greater than average ice year associated with an El Niño will be reversed by the occurrence of a La Niña if we assume La Niña has the opposite climatic effect at the Palmer site. Energy flow through trophic levels is not reversed. The energy flow is always from the primary producers to the top carnivores. Therefore, in this sense, the flow of food along the food chain cannot be reversed. Regarding the final question in this series, the timing of many events in the Palmer, Antarctic, ecosystem is well established.

6. *After the climatic event or episode, do the values of the climatic variables return along their outward path or is there hysteresis or some other trajectory in operation? If the latter, how does this affect the cascade?*

Changes in the atmospheric part of the ENSO system tend to return along their outward path at least as far as the values of the climatic variables are concerned. This generally applies to atmospheric pressures in the Pacific Ocean source areas of the events and the values of temperature and precipitation in the affected climates of the world. The energy transfers related to the ENSO phenomenon in the Pacific Ocean do not return along the same path because of the operation of the second law of thermodynamics. El Niño–related ecosystem changes such as the increase of populations in the NTL lake ecosystem will often reverse themselves along the same or similar pathway after the El Niño event. El Niño–related changes such as the loss of aquatic species in the LUQ aquatic ecosystem conceivably may take some time to reverse, and a hysteresis effect might come into play. An extreme example of this is the episode in the late 1940s and mid-1950s of a series of dry La Niña events that led to significant dieback of pinyon pine and juniper at the SEV site.

All the above relates to a deterministic, nonchaotic system. A consideration of chaotic systems is beyond the scope of this chapter. Many, if not most, of our atmospheric systems and ecosystems display some degree of chaos, and it will be essential to address this topic in the future.

The application of this framework to the ENSO case of climate variability has been very effective in raising further research questions and providing a manner in which they can be posed. Although some of the answers to the framework questions yield nothing new and are sometimes even trivial, the realization that the climatic ENSO signal has to be specifically connected to some part of the ecosystem to be effective provides a great stimulus for further investigations. We have learned that the timing of, or trigger of a sensitive nonlinear mechanism by, the climate signal is critical for the effectiveness of the signal. Also, the particular climatic variable in which the ENSO signal is found has to be one with a direct link to the ecosystem. The existence of a coupling mechanism between the climate variability signal and an ecosystem-driving function is therefore critically important. The same would also be true for climate variability relations with human systems.

Conclusions

Clearly, the idea of a simple forcing event and its direct response must be extended when considering ecosystems. The example of an ENSO event has been a useful, and relatively simple, one for illustrating the utility of the framework questions of this book. The LTER sites that manifest strong, detectable, and weak or no climatic signals to ENSO events have been identified. We have learned that the timing of the ENSO and the identification of an ecosystem-coupling mechanism are critical for this particular form of climate variability to have an effect. A statistically significant climate signal at an LTER site does not necessarily mean there will be an ecologically significant response. ENSO signals in the temperature series at the AND, LUQ, and PAL sites are the strongest statistically. Of these, only the signal at the PAL site has an important direct ecological effect. Somewhat less statistically strong ENSO signals at NTL and SEV do have important ecological effects. The results of the analysis of the climatic response to the 1982–1983 super El Niño compared to more normal-size warm events were not clear-cut, although in some cases the effects of the super El Niño were more pronounced.

The framework questions about climate variability and ecosystem response have allowed us to at least begin a thorough consideration of ecosystem response to a climatic phenomenon. The framework must also be applied in a quantitative fashion. In other sections of this book, we apply the framework to climatic forcing functions at other timescales ranging from an individual storm to a major glacial period. Only after many such applications will we begin to see some of the important basic principles relating climate variability and ecosystem response.

Acknowledgments   This study was supported by NSF Grant DEB 9416820 and the Crystal Harmony.

References

Barnston, A G., A. Leetmaa, V. E. Kousky, R. E. Livezy, E. A. O'Lenic, H. Van den Dool, A. J. Wagner, and D. A. Unger. 1999. *Bulletin of the American Meteorological Society* 80:1829–1852.

Barsugli, J. J., J. S. Whitaker, A. F. Lughe, P. D. Sardeshmukh, and Z. Toth. 1999. The effect of the 1997/98 El Niño on individual large-scale weather events. *Bulletin of the American Meteorological Society* 80:1399–1411.

Buizer, J. L., J. Foster, and D. Lund. 2000. Global Impacts and Regional Actions: Preparing for the 1997–98 El Niño. *Bulletin of the American Meteorological Society* 81:2121–2141.

Cayan, D. R., K. T. Redmond, and L. G. Riddle. 1999. ENSO and Hydrologic Extremes in the Western United States. *Journal of Climatology* 12:2881–2893.

Dahm, C. N., and M. C. Molles., Jr., 1992. Streams in semi-arid regions as sensitive indicators of global change. pp. 250–260. In P. Firth and S. Fisher, editors. *Troubled Waters of the Greenhouse Earth*. New York: Springer-Verlag.

Dahm, C. N., and D. I. Moore. 1994. The El Niño/Southern Oscillation Phenomenon and the Sevilleta Long-Term Ecological Research Site. Pages 12–21. In D. Greenland, editor. *El Niño and Long-Term Ecological Research (LTER) Sites*. LTER Publication No. 18. LTER Network Office. University of Washington. College of Forest Resources. AR-10. Seattle, Wash. 98195. 57 pp.

Fraser, W. R., W. Z. Trivelpiece, D. G. Ainley, and S. G. Trivelpiece. 1992. Increases in Antarctic penguin populations: Reduced competition with whales or a loss of sea ice due to environmental warming? *Polar Biology* 11:525–531.

Giannini, A., Y. Kushnir, and M. A. Cane. 2000. Interannual Variability of Caribbean Rainfall, ENSO, and the Atlantic Ocean. *Journal of Climate* 13:297–311.

Greenland, D., editor. 1994a. *El Niño and Long-Term Ecological Research (LTER) Sites*. LTER Network Office Publication No. 18. LTER Network Office, Univesity of Washington. College of Forest Resources. AR-10. Seattle, Wash. 98195. 57 pp.

Greenland, D. 1994b.The Pacific Northwest regional context of the climate of the H.J. Andrews Experimental Forest Long-Term Ecological Research Site. *Northwest Science* 69(2):81–96.

Greenland, D. 1999. ENSO-related phenomena at Long-Term Ecological Research sites. *Physical Geography* 20:491–507.

Kiladis, G. N., and H. E. Diaz. 1989. Global climatic anomalies associated with the extremes of the Southern Oscillation. *Journal of Climate* 2:1069–1090.

Kiladis, G. N., and H. E. Diaz. 1992. Atmospheric teleconnections associated with the extreme phase of the Southern Oscillation. Pp. 7–28 in *El Niño: Historical and paleoclimatic aspects of the Southern Oscillation*. H. F. Diaz and V. Markgraf, editors. Cambridge: Cambridge University Press. 474 pp.

Leung, L. R., A. F. Hamlet, D. P. Lettenmaier, and A. Kumar. 1999. Simulations of the ENSO hydroclimate signals in the Pacific Northwest Columbia river basin. *Bulletin of the American Meteorological Society* 80:2313–2339.

Mason, S. J., L. Godard, N. E. Graham, E. Yulaeva, L. Sun, and P. A. Arkin. 1999. The IRI seasonal climate prediction system and the 1997/98 El Niño event. *Bulletin of the American Meteorological Society* 80:1853–1873.

Molles, M. C., Jr., and C. N. Dahm. 1990. A perspective on El Niño and La Niña: Global implications for stream ecology. *Journal of the North American Benthological Society* 9:68–76.

Nicholls, N. 2001. The insignificance of significance testing. *Bulletin of the American Meteorological Society* 82:981–986.

Pielke, R. A., Jr., and C. N. Landsea. 1999. La Niña, El Niño, and Atlantic Hurricane Damages in the United States. *Bulletin of the American Meteorological Society* 80:2027–2033.

Robertson, D. M., W. Anderson, and J. J. Magnuson. 1994. Relations between El Niño/Southern Oscillation events and the climate and ice cover of lakes in Wisconsin. Pages 48–57. in D. Greenland, editor. *El Niño and Long-Term Ecological Research (LTER) Sites*. LTER Publication No. 18. LTER Network Office. University of Washington. College of Forest Resources. AR-10. Seattle, Wash. 98195. 57 pp.

Ropelewski, C. F., and M. S. Halpert. 1986. North American precipitation and temperature patterns associated with the El Niño/Southern Oscillation (ENSO). *Monthly Weather Review* 114:2352–2362.

Sardeshmukh P. D., G. P. Compo, and C. Penland, 2000. Changes of Probability Associated with El Niño. *Journal of Climate* 13:4268–4286.

Smith, R. C. S. E. Stamerjohn, and K. S. Baker. 1996. Surface air termperature variations in the Western Antarctic Peninsula Region. Pages 105–121 in *Foundations for Ecological Research West of the Antarctic Peninsular*. Antarctic Research Series. Vol. 70. American Geophysical Union. Washington, D.C.

Torrence, C., and P. J. Webster. 1999. Interdecadal Changes in the ENSO–Monsoon System. *Journal of Climate* 12:2679–2690.

Trenberth, K. E. 1997. The Definition of El Niño. *Bulletin of the American Meteorological Society* 78:2771–2777.

Trenberth, K. E., and J. M. Caron. 2000. The Southern Oscillation Revisited: Sea Level Pressures, Surface Temperatures, and Precipitation. *Journal of Climate* 13:4358–4365.

Wolter, K., R. M. Dole, and C. A. Smith. 1999. Short-Term Climate Extremes over the Continental United States and ENSO. Part I: Seasonal Temperatures. *Journal of Climate* 12:3255–3272.

Woodhouse, C. A. 1994. Tree-growth response to ENSO events near Niwot Ridge in the central Colorado Front Range: An extended abstract. Pages 22–28 in D. Greenland, editor. 1994. *El Niño and Long-Term Ecological Research (LTER) Sites*. LTER Network Office Publication No. 18. LTER Network Office, Univesity of Washington. College of Forest Resources. AR-10. Seattle, Wash. 98195. 57 pp.

7

The Climate of the Central Arizona and Phoenix Long-Term Ecological Research Site (CAP LTER) and Links to ENSO

Anthony J. Brazel
Andrew W. Ellis

Introduction

The Central Arizona and Phoenix LTER (CAP LTER) is one of two urban LTERs in the world network (Grimm et al. 2000; see http://caplter.asu.edu). Many LTER sites display a detectable climatic signal related to the El Niño–Southern Oscillation (ENSO) phenomenon (Greenland 1999). The purpose of this chapter is twofold: (1) to provide some insight into the role of the tropical Pacific Ocean as a driver of several climatic (and thus, ecologically related) variables in the CAP LTER location of central Arizona, and (2) to suggest the linkages of ENSO events to selected ecosystem processes near and within the geographical region of CAP LTER (figure 7.1a).

From past studies, it is clear that the seasonal and annual climate regimes of the southwestern United States, particularly water-related parameters, are linked to the periodicities and anomalies of what is known as the Multivariate ENSO Index (MEI) and Southern Oscillation Index (SOI) (e.g., Wolter 1987; Molles and Dahm 1990; Redmond and Koch 1991; Woolhiser and Keefer 1993; Wolter and Timlin 1993; Cayan and Redmond 1994; Redmond and Cayan 1994; Cayan et al. 1999; Redmond and Cayan 1999; Simpson and Colodner 1999; Redmond 2000; and Mason and Goddard 2001). In Arizona, and especially in the CAP LTER region, precipitation is bimodal during the year with peaks in winter (mostly midlatitude-derived frontal storms) and in mid-to-late summer, mostly in the form of convective thunderstorms during the North American monsoon season. Recent studies show a strong connection between ENSO and winter moisture in Arizona, such that it is even possible to forecast impending conditions in advance (Pagano et al. 1999). These studies have established relationships between the climate of the southwest-

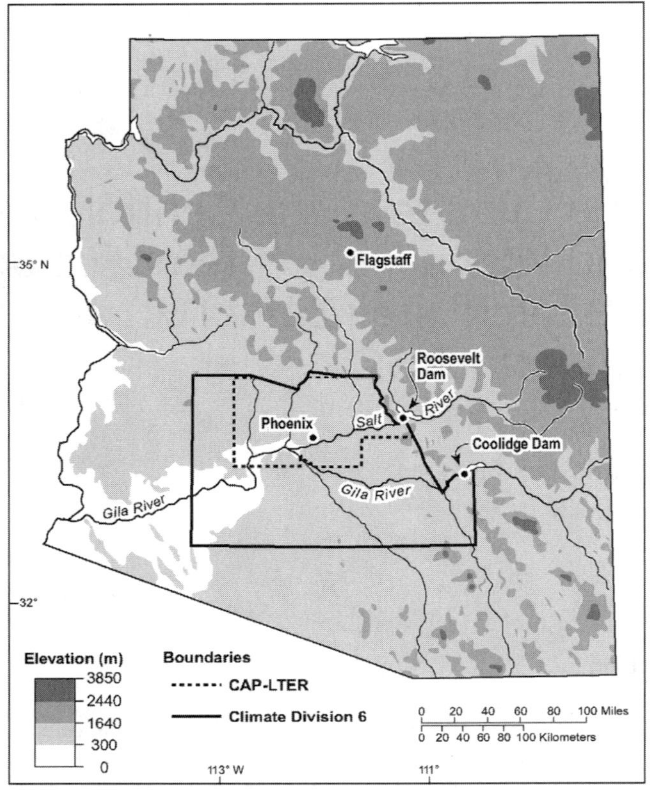

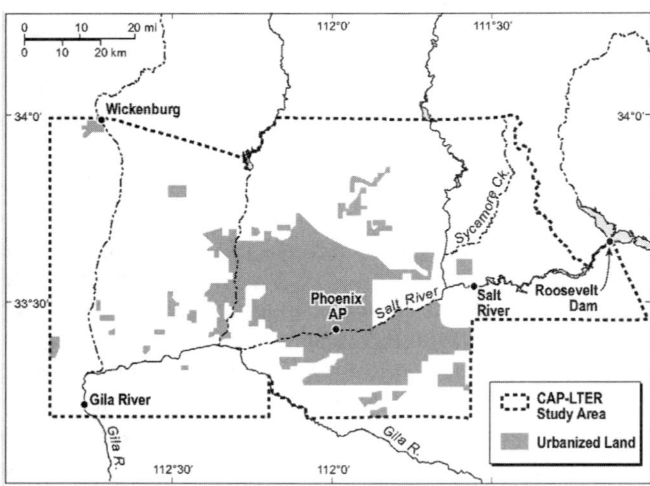

Figure 7.1 Geographical region of the Central Arizona and Phoenix Long-Term Ecological Research site (CAP LTER). (a) CAP LTER boundaries within Arizona. (b) CAP LTER study area.

ern United States and ENSO by demonstrating monthly and daily timescale effects on inputs of moisture and resultant streamflow in Arizona (e.g., Molles and Dahm 1990; Cayan et al. 1999; and Simpson and Colodner 1999). The synoptic- and large-scale circulation patterns associated with anomalies of MEI/SOI in the southwestern United States provide additional insight into regional forces that drive the CAP-LTER climate (e.g., Redmond and Koch 1991). Generally, when the warm phase of the tropical Pacific Ocean occurs (El Niño, thus negative SOI, positive MEI), across the Southwest precipitation is generally anomalously high. Conversely, when the cool phase occurs (La Niña, thus positive SOI, negative MEI), an input of moisture that is less than normal appears to be the case in the Southwest (e.g., Cayan et al. 1999). Generally, when neither El Niño nor La Niña occurs, it is unclear what the precipitation regime across the Southwest will be in relation to ENSO.

Increased daily, monthly, seasonal, and annual-to-decadal moisture or extended dry periods have important implications for the Southwest and the CAP LTER ecosystem. We suggest several potential linkages, and review three specific examples in this chapter: (1) studies related to the hanta virus and ENSO that have been conducted for the nearby Four Corners Area and New Mexico by researchers of the Sevilleta LTER site (e.g., Parmenter et al. 1999) and others; (2) our interpretation of ENSO phases in relation to past and ongoing stream ecological research on Sycamore Creek at the CAP LTER (analyzing data from Grimm 1993); and (3) possible impacts of ENSO on concentrations of river constituents routinely observed by the U.S. Geological Survey for the Phoenix region. (CAP LTER researchers have recently constructed a detailed nitrogen budget for this region; see Baker et al. 2001). The first example suggests linkages in a trophic cascade from inputs of moisture, to increased vegetative cover and insects, to abundance of fleas and mammals, to human plague incidences perhaps spanning over more than a year for the cascade. The second suggests linkages from inputs of moisture to a typical individual desert stream and its ecological conditions spanning short periods to months. The third example suggests linkages from seasonal inputs of moisture to larger river systems, and observed winter variations of many stream constituents over a quarter of a century upstream and downstream of the built-up urbanized and agricultural sector of the CAP LTER area.

ENSO effects are most obvious on the winter climate, but also anomalous conditions in the tropical Pacific Ocean in winter may influence the summer monsoon season in Arizona. Breaks and bursts in the monsoon and overall monsoon seasonal intensity are strongly related to flash-flood risks, local storm damage, dust storm frequencies, the urban heat island, human comfort and energy demand, and vegetation green-up and biomass. Thus, it is likely that climate impacts on natural and human components of CAP LTER are substantial at the timescale of MEI/SOI variations. Scientists in CAP LTER are just beginning to study ecosystem response and feedback to a host of natural and human-induced processes. The human dimension acts as a driver to ecosystem change and, in turn, is affected by these changes (Brazel et al. 2000; Collins et al. 2000; and Grimm et al. 2000).

This chapter certainly does not purport to explain all of the cascading effects on CAP LTER at a quasi-quintennial timescale. However, we explicitly relate indices of the warm and cool phase of the tropical Pacific Ocean to several climatic variables in CAP LTER. The work is intended to outline those relationships that deserve further study to expand our knowledge of cascading climate effects on the urban ecosystem. In this analysis, data are expressed at the monthly timescale for the period 1951–1999 using thermal and moisture climatic variables. The variables include (1) maximum, minimum, mean, and range of temperature in rural and urban locales, (2) regional temperature and precipitation, (3) evaporation at reservoirs in or near CAP LTER, (4) snowfall at the upper end of a major watershed important to CAP LTER, (5) simulated soil moisture surpluses and deficits in CAP LTER, and (6) streamflow within a representative natural stream in CAP LTER (Sycamore Creek). This analysis hopefully will assist researchers in (1) the development of hypotheses for retrospective analyses of urban ecosystem dynamics, (2) the recognition of the climate context of field experiments conducted at CAP LTER, (3) the climate context of a repetitive 3- to 5-year snapshot ecosystem survey of 200 points at CAP LTER in rural-urban locales, and (4) the illustration of external drivers on the local ecosystem, thus making links from CAP LTER to large scale (global and regional) change more explicit. Our analysis also provides composite views of regional atmospheric circulation features that are expected with anomalies in the SOI and MEI. Thus, regional explanations of local area effects are more easily facilitated and understood.

Method of Analysis

Teleconnection Indexes

In examining the linkages between Pacific teleconnections and CAP LTER, two teleconnection indexes were correlated with climate characteristics (thermal and moisture) across central Arizona. Monthly values of the Southern Oscillation Index (SOI) and the Multivariate El Niño–Southern Oscillation Index (MEI) were collected for the 49-year period 1951–1999 (e.g., March in figure 7.2). The beginning date of the study period is confined by the SOI and the MEI records, whereas records of climatic data confined the ending date. SOI values, representing differences in monthly sea level pressure values across the southern Pacific Ocean, were obtained from the U.S. Climate Prediction Center (CPC). MEI values were obtained directly from K. Wolter of the Climate Diagnostics Center (CDC) of the National Oceanographic and Atmospheric Administration (NOAA).

The MEI is explicitly used in our analysis. The MEI is built from six observed variables across the tropical Pacific. The variables are (1) sea-level pressure (P), (2) zonal (U) and (3) meridional (V) components of the surface wind, (4) sea surface temperature (S), (5) surface air temperature (A), and (6) total fraction of the sky covered by cloud (C). The MEI is calculated separately for each of twelve moving bimonthly seasons. After spatially filtering the individual fields into clusters, the MEI is calculated as the first unrotated Principal Component (PC) of all six observed fields combined. In doing this, the total variance of each field is first normalized prior to the extraction of the first PC on the covariance matrix of the com-

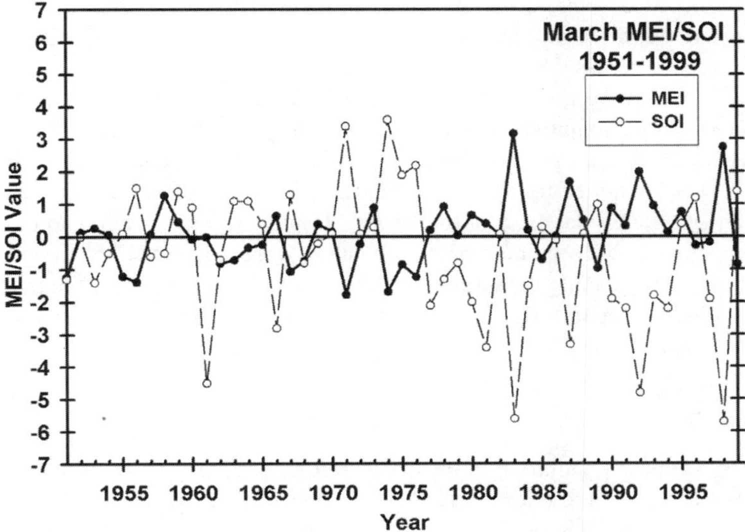

Figure 7.2 Southern Oscillation Index (SOI) and Multivariate El Niño–Southern Oscillation Index (MEI) values during March for the period 1951–1999.

bined fields. Finally, the computed MEI values are standardized with respect to the 1950–1993 reference period. Negative values of the MEI represent the cold ENSO phase, or La Niña, whereas positive values represent the warm ENSO phase, or El Niño. The sea level pressure (P) loadings characterize the Southern Oscillation. For example, negative MEI values (La Niña) are derived from negative pressure anomalies in the west and positive pressure anomalies in the east. The latitudinal (U) component of the surface wind corresponds to east-west wind direction anomalies along the equator near the international dateline. The meridional (V) component of the surface wind corresponds to north-south wind direction anomalies north of the equator across the Pacific Ocean, largely reflecting oscillation of the Inter-Tropical Convergence Zone (ITCZ). Sea (S) and air (A) surface temperatures indicate the typical ENSO pattern of temperature anomalies from the western South American coastline to the date line. Finally, total cloudiness (C) across the central equatorial Pacific versus over the Philippines and north of Australia indicate the migration of convective activity.

Thermal Data

To examine the covariance of MEI with near-surface air temperatures across the area, U.S. climate division data for central Arizona (division 6; figure 7.1a) were obtained from the National Climatic Data Center (NCDC; NOAA 1983a, b). Monthly values represent mean monthly temperatures as calculated from all regional stations at which daily maximum and minimum near-surface air temperatures are recorded (over 30 stations in the division 6 region). As such, the data represent the temporal variation within the general monthly lower atmospheric temperature record across the region as a whole.

To more closely examine MEI associations with temperature within the Phoenix urban area, daily maximum and minimum temperature values for a central Phoenix station (Sky Harbor Airport, Phoenix AP in figure 7.1b) and rural Wickenburg (figure 7.1b) were obtained for the period of study. Data were taken as a subset of the *Summary of the Day* database of the NCDC. Using daily maximum and minimum temperature data, daily temperature range values were calculated (maximum minus minimum), as were daily rural-urban differences (urban minus rural) in maximum and minimum temperature. All daily temperature values were translated into monthly means.

Moisture Data

With a burgeoning population in the desert setting of CAP LTER, water resources are of constant concern (Carter et al. 2000). Given the general convective nature of regional precipitation, and therefore large spatial inhomogeneity, precipitation data from the climate divisional records were obtained from NCDC to represent the variation in monthly mean precipitation across the CAP LTER region as a whole, again using over 30 sites. Since the water resources of the area are also dependent on spring snowmelt across the higher terrain to the north, daily snowfall values for Flagstaff in northern Arizona (figure 7.1a) were extracted from the *Summary of the Day* database of NCDC. Daily values were summed to monthly totals through the period of study to correlate with MEI values.

To translate monthly thermal and moisture variables into aspects of the climatic water conditions for the region, monthly divisional temperature and precipitation means were used to calculate mean monthly soil moisture values. The Thornthwaite-Mather climatic water budget technique (Thornthwaite and Mather 1955; Mather 1978) was used as a first approximation to produce monthly soil moisture surplus and deficit values, of which only deficit values were considered because of the infrequency of soil moisture surpluses in the CAP LTER area. To further represent the temporal variability in the water resources, daily streamflow values for Sycamore Creek (important to CAP LTER objectives; Grimm 1993; figure 7.1b) were totaled to monthly values for the period of study. Streamflow data were obtained from the U.S. Geological Survey stream gauge database (www.usgs.gov). Later we illustrate the links of SOI/MEI and streamflow oscillations to processes of stream ecology (Grimm 1993; Grimm et al. 1997).

To represent the temporal variability of evaporation from an open water surface (e.g., reservoir)—a parameter of extreme interest to water managers—data for daily pan evaporation at two reservoirs close to the CAP LTER area were obtained from the *Summary of the Day* database of the NCDC. Daily pan evaporation totals at Roosevelt Dam northeast of Phoenix and San Carlos Dam east of Phoenix (figure 7.1a) were totaled to monthly values for the period of the study.

Quality Assurance of Data

The SOI and MEI data are quality controlled and complete for the period 1951 through 1999 to form a comprehensive data set for the study period (figure 7.2). Likewise, U.S. climate division data are complete for the full period of study. Daily temperature data for Phoenix and Wickenburg, snowfall data for Flagstaff, and evaporation data at the Roosevelt and San Carlos Dams extend through the period of study, but are not entirely complete. Daily streamflow data are available for the period 1961–1997. In processing the incomplete records of daily data, for each month of the period of study a threshold value of 90% coverage of daily data was required. Otherwise, the monthly data value was labeled as missing. Subsequently, 90% coverage of monthly data was required for inclusion in covariance calculations with MEI values, a threshold value we accept as representative based on the work of Stooksbury et al. (1999).

As MEI values are representations of monthly deviations in standard values, all climatic variables were standardized to monthly Z-scores. In calculating Z-scores, the mean of each distribution was subtracted from each observation and then subsequently divided by the standard deviation of the distribution. The products were distributions of monthly climatic variables through the period of study (1951–1999) where each distribution has a zero mean and number units of standard deviations.

Analysis of Covariance

To assess the covariance between the SOI and MEI, simple correlation coefficients (r) were calculated to determine the extent of the covariance of monthly values of each index for the period of study. Likewise, to assess the extent of the covariance between each of the teleconnection indices and variables representing the climate of the CAP LTER area, correlation coefficients were calculated. Concurrent relationships were tested as well as lagged relationships; a monthly MEI value was correlated with each climate variable for each month as well as for each month of the subsequent 11-month period (in a manner similar to Greenland 1999). Finally, for each calculated correlation coefficient, a t-test for the significance of correlation was determined to highlight relationships of significance.

Physical Forcings

Lastly, to gain a physical understanding of any significant statistical relationships between MEI values and CAP LTER climate variables, the characteristics of the larger background synoptic atmosphere were diagnosed. For those intra-annual periods exhibiting strong statistical relationships, synoptic atmospheric composites and anomalies were constructed using only data from years with extreme MEI values. An extreme year was identified as possessing an MEI value in either the 10th or 90th percentile, which is to say those years possessing one of the five highest and five lowest MEI values for the 49-year period of study.

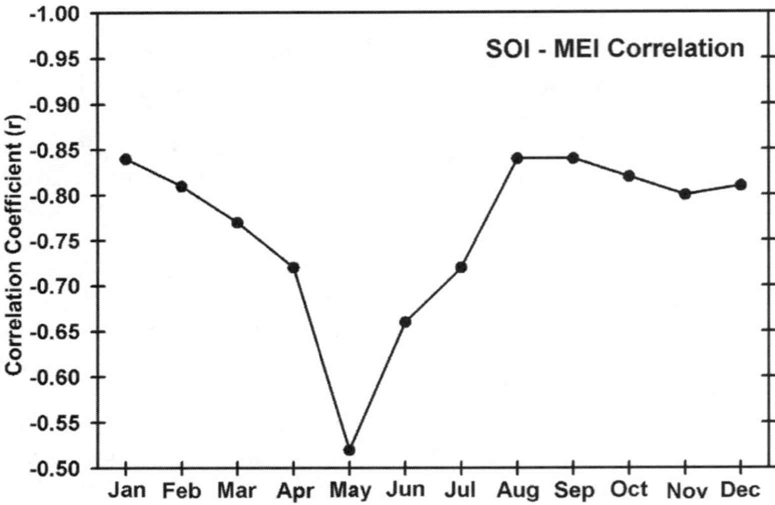

Figure 7.3 Correlation between monthly values of the SOI and MEI indices. Each correlation is significant at the 99% level.

Using data from the National Center for Environmental Prediction (NCEP) re-analysis data set (Kalnay et al. 1996), simple composites of the synoptic atmosphere on a 2.5°-latitude by 2.5°-longitude spatial resolution were constructed. Composites of 500-mb geopotential height (large-scale atmospheric flow), 850-mb air temperature (regional thermal conditions), and 850-mb specific humidity (regional moisture conditions) were created. The purpose of these specific composites is to illustrate the anomalies in the synoptic atmosphere affecting the CAP LTER climate, and, in turn, the ecosystem, and driven by the remote atmospheric anomalies represented by the MEI.

Results

SOI-MEI Covariance

Correlation coefficients measuring the significance of the covariance between SOI and MEI values (figure 7.3) indicate a rather significant inverse relationship throughout the annual period. The covariance is highly significant during late summer through early spring. However, the correlation between the two indexes decreases dramatically in May and June just before the typical onset of the monsoon in the southwestern United States. There is a good agreement between the two indexes. However, the MEI is correlated slightly better with the climate parameters of central Arizona than is the SOI, particularly during May and June. This seems reasonable, since more descriptive parameters of activity in the tropical Pacific Ocean are included in the MEI. For this reason, discussion from this point forward will be confined to the relationship between the MEI and the variability of climate in the CAP LTER area.

MEI-Temperature Associations

It is clear that there is very little association between MEI values and mean monthly temperature across CAP LTER (table 7.1). Inverse relationships are common in fall through early spring (October–March), whereas positive relationships exist from spring through late summer (April–September). Inverse relationships are indicative of decreased (increased) temperatures during El Niño (EN) [La Niña (LN)] events of eastern tropical Pacific Ocean warming (cooling). From spring through late summer, positive relationships indicate increased (decreased) temperatures during EN (LN) conditions. Still, there are no significant associations between the MEI and mean monthly CAP LTER temperatures.

We examined MEI associations with daily maximum and minimum temperatures for an urban (Phoenix) and a rural (Wickenburg) location. This provides greater insight into associations between the MEI and CAP LTER temperatures. It is apparent that MEI mean temperature relationships are weakened by the fact that the relationship between the MEI and maximum daily temperature tends to be opposite to that between MEI and minimum daily temperature. This is evidenced by the significance of the associations between MEI values and monthly means of daily temperature range (table 7.1). High (EN) [low (LN)] MEI values are associated with decreased (increased) maximum temperatures during the period October through March in Phoenix, and for every month of the year, but during August at rural Wickenburg. February and March relationships are significant at each location, as is the November relationship at Wickenburg. At Phoenix, a positive relationship exists from April through September and is significant during July, the typical month of the commencement of the monsoon circulation. However, from spring through summer, EN (LN) conditions are associated with increased (decreased) maximum temperatures.

Throughout the year, and most obvious in spring (March–June) and fall (October–November), a significant positive relationship exists between the MEI and minimum temperatures at Phoenix, whereby EN (LN) conditions are associated with higher (lower) minimum temperatures. The same positive relationship between MEI values and minimum temperatures exists at Wickenburg from the middle of the monsoon season (August) through early spring (March) and is significant during the monsoon season (August–September). However, the relationship weakens in spring and early summer, and it reverses significantly just prior to the monsoon (June–July). During this period, minimum temperatures at Wickenburg are inversely associated with the MEI, where EN (LN) conditions are associated with lower (higher) minimum temperatures.

Within the MEI-temperature range correlation, the products of the associations between the MEI and maximum and minimum daily temperatures can be seen. An inverse relationship between MEI values and temperature range in Phoenix exists throughout the year and is most significant (October–June) outside the monsoon season. An inverse relationship suggests that EN (LN) conditions are associated with decreased (increased) daily temperature range. The inverse associations exhibit a similar intra-annual pattern at Wickenburg (August–May), but are not quite as strong, with significant inverse relationships occurring only during the periods

Table 7.1 Correlation between monthly MEI values and monthly mean regional temperature and monthly means of daily maximum and minimum temperature, daily temperature range, and temperature difference for and between Phoenix and Wickenburg

	Regional	Phoenix			Wickenburg			Phoenix-Wickenburg	
Month	Mean	Max	Min	Range	Max	Min	Range	Max	Min
January	−0.04	−0.12	0.19	−0.28*	−0.10	0.07	−0.14	0.01	0.22
February	−0.11	−0.27*	0.20	−0.49**	−0.31*	0.21	−0.51**	0.14	0.06
March	−0.13	−0.26*	0.27*	−0.61**	−0.29*	0.16	−0.45**	0.18	0.20
April	0.10	0.07	0.36**	−0.49**	−0.05	−0.07	−0.01	0.16	0.42**
May	0.12	0.03	0.37**	−0.42**	−0.08	0.04	−0.15	0.35**	0.39**
June	0.01	0.01	0.31*	−0.44**	−0.16	−0.27*	0.15	0.22	0.48**
July	0.09	0.25*	0.22	−0.02	−0.04	−0.29*	0.25*	0.27*	0.39**
August	0.21	0.10	0.20	−0.15	0.05	0.23*	−0.17	0.10	−0.01
September	0.13	0.09	0.21	−0.19	−0.08	0.24*	−0.29*	0.20	−0.02
October	−0.03	−0.05	0.25*	−0.29*	−0.17	0.11	−0.24*	0.19	0.22
November	−0.10	−0.22	0.26*	−0.52**	−0.30*	0.13	−0.40**	0.21	0.18
December	−0.02	−0.14	0.21	−0.33**	−0.16	0.19	−0.26*	0.09	0.09

*Significance level of 95%.

**Significance level of 99%.

September through December and February through March. However, a positive relationship between the MEI and temperature range exists at Wickenburg during July, when EN (LN) conditions are associated with larger (smaller) temperature ranges.

In examining the correlation between MEI values and urban-rural differences in daily temperature (table 7.1), a typical method used in urban heat island studies and important in the energy service sector (Brazel et al. 1993), the greatest association of the MEI is with urban-rural minimum temperature differences (time of day when heat islands are more pronounced). High (low) MEI values associated with an EN (LN) situation are correlated with large (small) differences between Phoenix and Wickenburg minimum temperatures year round, but most significantly during the spring and summer period of April through July. Because urban minimum temperatures in Phoenix are nearly always milder than those at surrounding rural locations, the positive relationship suggests that minimum temperatures are greater than usual during EN and lesser during LN at Phoenix than at Wickenburg. This is supported by the stronger relationship between the MEI and minimum temperature at Phoenix than at Wickenburg, where the relationship actually reverses in June and July (table 7.1).

For much of the year a positive relationship between MEI values and urban-rural maximum temperature differences exists, most significantly in May and July (table 7.1). Positive correlation between MEI and urban-rural maximum temperature differences suggests that high (low) MEI values associated with EN (LN) conditions are associated with large (small) differences in temperature. Although not always the case, Phoenix maximum temperatures are typically warmer than those at surrounding moist rural locations. As such, the relationship suggests that during EN (LN) situations, especially in May and July, Phoenix maximum temperatures are generally larger than those at Wickenburg by an amount that is greater than (less than) usual. The significance to the urban ecosystem of the MEI/SOI forcers has not previously been demonstrated for CAP-LTER. Currently, these urban-rural climate differences and their impacts on a host of processes (e.g., human stress, heat stress on plants, energy consumption, arthropod abundance, cotton and dairy production) are the focus of a "feedbacks" subgroup of CAP-LTER researchers (L. A. Baker et al., unpubl. data, 2002).

MEI-Precipitation Associations

Correlations between MEI values and mean CAP LTER (climate division 6) monthly precipitation indicate a positive relationship during the fall through spring, most significantly during the months November–December, February–March, and May (table 7.2). During these periods, high (low) MEI values corresponding to EN (LN) conditions are associated with greater (small) amounts of precipitation across the CAP LTER area. The same is true of the relationship between the MEI and snowfall in Flagstaff in late winter (February–March; table 7.2). A significant inverse relationship between MEI values and mean CAP LTER precipitation exists in July. This indicates that during the month in which the monsoon season typically begins, EN (LN) conditions are associated with a(n) decrease (increase) in precipitation.

Table 7.2 Correlation between monthly MEI values and monthly CAP LTER area precipitation, Flagstaff snowfall, and CAP LTER area soil moisture deficit, pan evaporation, and streamflow

Month	Regional Precipitation	Flagstaff Snowfall	Soil Moisture Deficit	Roosevelt Evaporation	San Carlos Evaporation	Sycamore Creek Stream-flow
January	0.09	0.05	−0.17	0.07	−0.23	0.18
February	0.60**	0.26*	−0.39**	−0.12	−0.52**	0.36**
March	0.53**	0.27*	−0.41**	−0.38**	−0.54**	0.40**
April	0.06	−0.05	−0.29*	−0.41**	−0.29*	0.45**
May	0.36**		−0.32*	−0.50**	−0.56**	0.32*
June	−0.10		−0.14	−0.44**	−0.26*	0.23*
July	−0.37**		0.33**	0.26*	0.12	0.15
August	0.09		−0.03	0.13	−0.09	0.07
September	−0.07		0.11	0.02	−0.16	−0.17
October	0.22			−0.13	−0.12	0.24*
November	0.30*	−0.16	−0.31*	−0.03	−0.37**	0.43**
December	0.32*	−0.05	−0.26*	0.03	−0.12	0.25

*Significance level of 95%.

**Significance level of 99%.

MEI-Climatic Water Associations

In translating MEI associations with temperature and precipitation within the CAP LTER area into associations with climatic water variability (table 7.2), it is evident that the MEI is significantly associated with climatic water parameters during the period late fall through early summer. Soil moisture across the CAP LTER area exhibits a significant inverse relationship with MEI values during the period November through May, excluding January (table 7.2). The inverse relationship indicates that when MEI values are high (low), indicating EN (LN) conditions, soil moisture deficit values are low (high). In other words, under EN (LN) conditions, when precipitation tends to be increased (decreased) and temperatures tend to be decreased (increased), the soil moisture deficit typical of the region is decreased (increased). The relationship between MEI and the soil moisture deficit becomes significantly positive in July (table 7.2), indicating that EN (LN) conditions at the inception of the monsoon season are associated with increased (decreased) soil moisture conditions.

As in the case of the soil moisture deficit, an inverse relationship exists between the MEI and pan evaporation at the Roosevelt (significant March–June) and San Carlos (significant November, February–June) reservoirs (table 7.2). High (low) MEI values associated with EN (LN) conditions are linked to decreased (increased) evaporative loss from an open water surface. As with soil moisture deficit, MEI-evaporation relationships become positive in July, significantly so at the Roosevelt reservoir (table 7.2). As such, EN (LN) conditions during July are associated with increased (decreased) evaporative rates.

Taken together, MEI relationships with temperature, precipitation, soil moisture deficit, and evaporative loss lead to an association with streamflow. For nearly the entire year, a positive relationship exists between MEI values and streamflow at Sycamore Creek, significantly so during the period October through June, excluding January (table 7.2). High (low) monthly MEI values, indicating EN (LN) conditions, are associated with higher (lower) monthly streamflow. During the monsoon season, the relationship weakens (July–August) and actually reverses direction (September). The implications for these relationships are discussed in the section entitled Sycamore Creek Stream Ecology.

Summary

The correlation between MEI and various CAP LTER climate variables examined within this study indicate that the strongest relationships occur in late winter and spring, most significantly in March (table 7.3). In general during this period, high MEI values indicative of EN conditions are associated with (1) decreased maximum temperatures, (2) increased minimum temperatures, (3) decreased temperature ranges, (4) increased urban-rural temperature differences, (5) increased precipitation (including snowfall over higher terrain to the north), (6) increased soil moisture, (7) decreased evaporative losses, and (8) increased streamflow. Low MEI values (LN conditions) are associated with opposite responses. MEI-climate associations are high in midautumn as well, with generally the same strength of the relationships as in spring. Of additional interest is the reversal in the nature of the MEI relationships with precipitation and many of the climatic water variables in July, the month during which the annual monsoon typically begins. In this case, high MEI values indicative of EN conditions are associated with decreased precipitation and increases in soil moisture deficits and pan evaporation rates (table 7.3).

Atmospheric Dynamics

To better understand the physical forcing that drives the variation in CAP LTER climate with EN and LN conditions, synoptic atmospheric composites of March and July were constructed for the five strongest EN years (highest MEI values) and the five strongest LN years (lowest MEI values). For March EN, these are the years 1958, 1983, 1987, 1992, 1998 (July: 1972, 1982, 1983, 1987, 1997); for March LN, 1951, 1956, 1971, 1974, 1976 (July: 1954, 1955, 1956, 1964, 1971). In March, it is apparent that the mean 500-mb height pattern is shifted more to the south during EN years than during LN years (figures 7.4a–c). The ridge/trough pattern (figures 7.4a,b) is very similar, however the magnitudes are considerably different, such that during LN (EN) years the Pacific ridge is strengthened (weakened). LN (EN) years seem to be associated with higher (lower) 500-mb heights across the southwestern United States. The strengthened (weakened) Pacific ridge during LN (EN) years is

Table 7.3 Direction of monthly correlation between MEI values and CAP LTER climatic parameters. Positive (P) relationships indicate an increase (decrease) in the variable under EN (LN) conditions, whereas inverse relationships (N) indicate a decrease (increase) under EN (LN) conditions.

	J	F	M	A	M	J	J	A	S	O	N	D
Temperature												
Mean regional	N	N	N	P	P	P	P	P	P	N	N	N
Phoenix max	N	**_N_**	**_N_**	P	P	P	**_P_**	P	P	N	N	N
Wickenburg max	N	**_N_**	**_N_**	N	N	N	N	P	N	N	**_N_**	N
Phoenix min	P	P	**_P_**	**_P_**	**_P_**	**_P_**	P	P	P	**_P_**	**_P_**	P
Wickenburg min	P	P	P	N	P	**_N_**	**_N_**	**_P_**	**_P_**	P	P	P
Phoenix range	**_N_**	**_N_**	**_N_**	**_N_**	**_N_**	**_N_**	N	N	N	**_N_**	**_N_**	**_N_**
Wickenburg range	N	**_N_**	**_N_**	N	N	P	**_P_**	N	**_N_**	**_N_**	**_N_**	**_N_**
Phoenix-Wickenburg max	P	P	P	P	**_P_**	P	**_P_**	P	P	P	P	P
Phoenix-Wickenburg min	P	P	P	**_P_**	**_P_**	**_P_**	P	N	N	P	P	P
Precipitation												
Mean regional	P	**_P_**	**_P_**	P	**_P_**	N	**_N_**	P	N	P	**_P_**	**_P_**
Flagstaff snowfall	P	**_P_**	**_P_**	N							N	N
Climatic Water Variables												
Soil moisture deficit	N	**_N_**	**_N_**	**_N_**	**_N_**	N	**_P_**	N	P	N	**_N_**	**_N_**
Pan evaporation-Roosevelt	P	N	**_N_**	**_N_**	**_N_**	**_N_**	**_P_**	P	P	N	N	P
Pan evaporation-San Carlos	N	**_N_**	**_N_**	**_N_**	**_N_**	**_N_**	P	N	N	N	**_N_**	N
Streamflow-Sycamore Creek	P	**_P_**	**_P_**	**_P_**	**_P_**	**_P_**	P	P	N	**_P_**	**_P_**	**_P_**

Significant (95%) correlation is bold and underlined.

likely to be associated with a more northerly (southerly) storm track, and is associated with a relatively warmer (cooler) and drier (moister) lower atmosphere in March (figure 7.5), accounting for the associations between the MEI and CAP LTER-area climate variables.

LN (EN) conditions during July are associated with a strengthened (weakened) 500-mb ridge across the western United States and 500-mb trough across the eastern Pacific Ocean (figure 7.6). July marks the beginning month of the monsoon season in the southwestern United States, and it is initiated by northward displacement of the subtropical ridge (Adams and Comrie 1997). The stronger (weaker) western U.S. 500-mb ridge during LN (EN) years is associated with a warmer (cooler), but moister (drier) lower atmosphere (figure 7.7). This is opposite to the drier (moister) atmosphere associated with LN (EN) conditions during March (figure 7.5).

Discussion of Results

Our results and others (e.g., Simpson and Colodner 1999) point to significant climate connections between the southwestern United States and periodicities of ENSO as represented by the MEI and SOI. In fact, climate responses may be

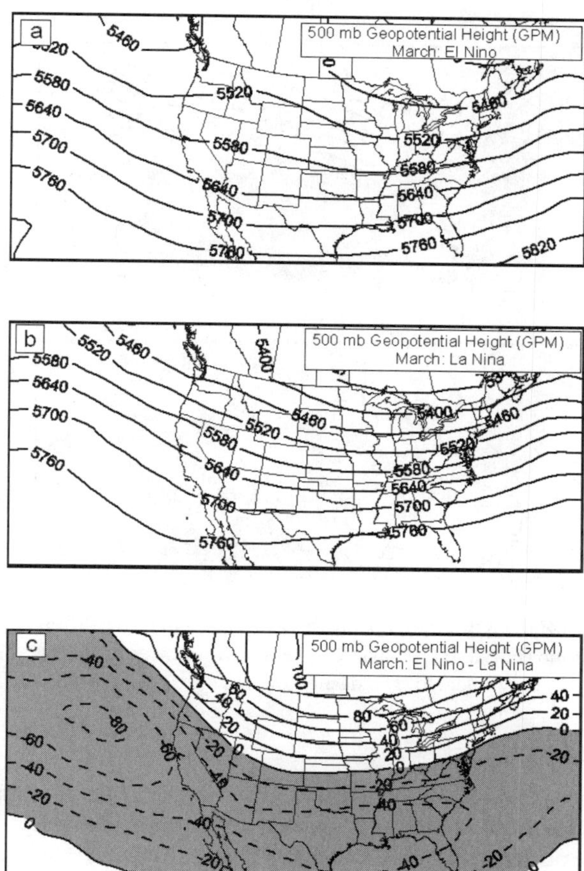

Figure 7.4 Mean March 500-mb height during the five strongest years of (a) El Niño and (b) La Niña. (c) their differences taken as El Niño (EN) minus La Niña (LN); negative differences are indicated by a dashed line.

greatly predictable at seasonal timescales (Pagano et al. 1999). The cascading-like effects through the climate system—from the Pacific tropical ocean temperatures, to southwestern U.S. circulation dynamics, to central Arizona seasonal thermal and hydrological regimes—are quite pronounced for the fall/winter/spring time frame. Weak inverse connections to EN are even evident for the North American monsoon regime of summer (EN yields dry summer; LN yields active, wet summer). At the upper level of what might be viewed as a local CAP LTER climate cascade, therefore, there exist variable processes of moisture (precipitation, evaporation, and soil moisture), local storms, and clear/cloudy day frequencies, for example. These components are likely to have strong connections to the main driver variables analyzed in this chapter (e.g., MEI yields distinct variations in thermal/moisture inputs).

Month-to-month lag effects (not shown) are also pronounced for initial MEI

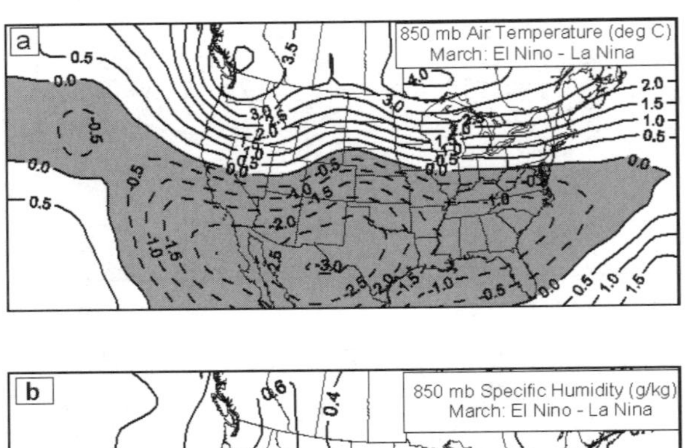

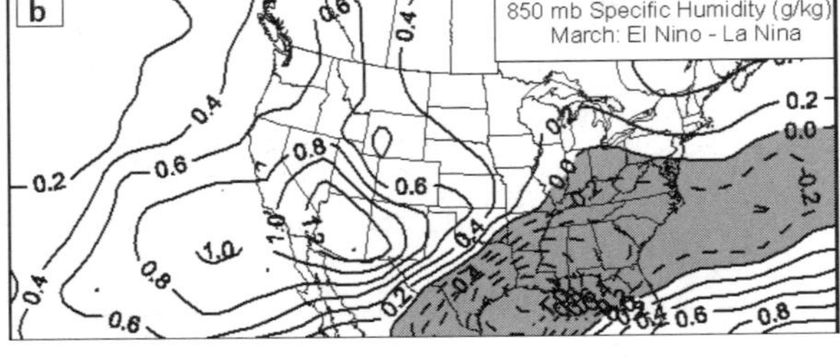

Figure 7.5 Mean March 850-mb temperature (a) and moisture (b) differences between the five strongest El Niño (EN) and La Niña (LN) years. Differences are taken as EN minus LN. Negative differences are indicated by a dashed line.

anomalies and persistence of local climate responses. Typically, longer lags are most evident for moisture variables from early winter through late spring. Thus, for example, a large positive MEI anomaly is felt from early winter through late spring as increased snow packs in the high country of Arizona, rising stream flows, increasing soil moisture, and reductions in evaporative losses. However, no association with mean temperature variations exists. Daily mean temperature shows little relation to large MEI anomaly years. As stated previously, this appears to be explained by offsetting responses of regional maximum and minimum temperatures. Thus, there is a marked variance in the temperature range: large for negative MEI, and small for positive MEI anomalies. This is likely important to ecosystem components that are sensitive to threshold values of temperature, not just to mean temperature (e.g., growing degree-day accumulation for plants; cooling degree-day accumulation for energy consumption).

A measure of the urban heat island effect (using Phoenix airport minus Wickenburg, Arizona—an urban minus rural site used previously; Balling and Cerveny 1987) shows a surprisingly significant relation to MEI anomalies. When positive anomalies occur (i.e., EN), larger urban-rural differences (bigger heat islands) are evident for the late spring and early summer months. Because there is an inverse relationship between positive MEI (EN events) and summer moisture (drier), it most likely means more clear nights and lower humidity values result. These are

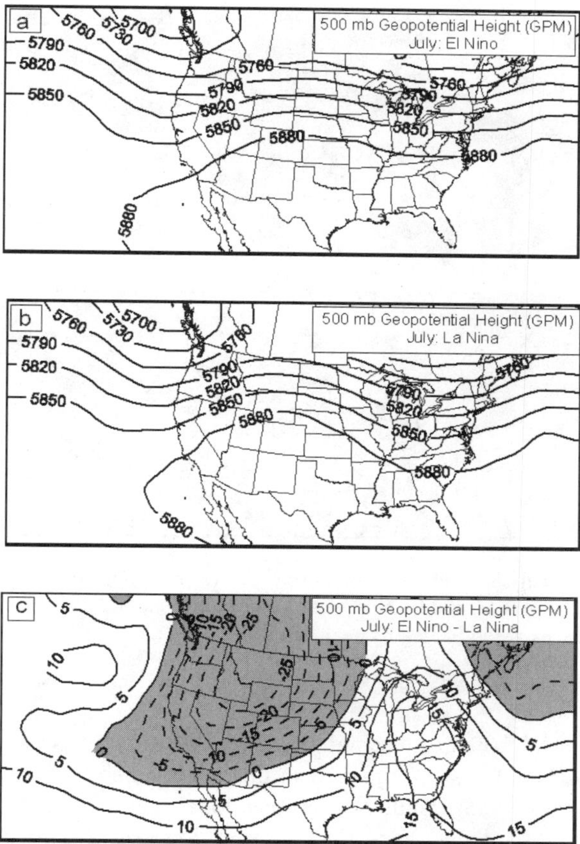

Figure 7.6 Mean July 500-mb heights for the five strongest years of (a) El Niño and (b) La Niña (c) their differences, taken as EN minus LN; negative differences are indicated by a dashed line.

the sorts of local conditions that promote chances for intense heat island development in CAP LTER urban locales, especially because the heat island is predominantly a nighttime phenomenon (Brazel et al. 2000).

Ecosystem Examples

These climate responses potentially result in more complex cascades in the CAP LTER ecosystem. For example, dust storms (Brazel 1987), wildfires (Swetnam and Betancourt 1990), vegetation change (Li and Kafatos 2000), and water quantity and quality (Carter et al. 2000) are all driven by surface processes that are a combination of natural and human-impacted environmental conditions. Detailed linkages

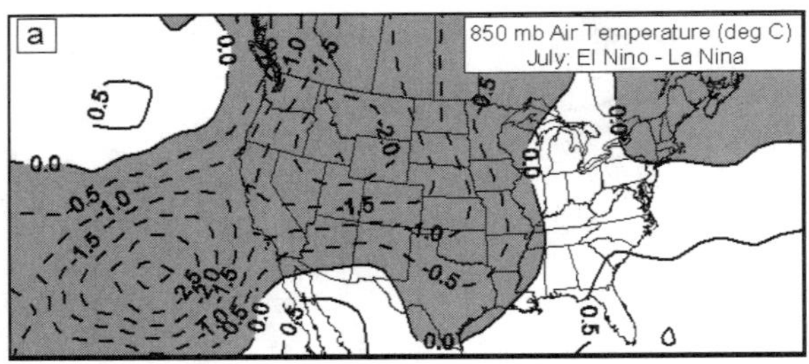

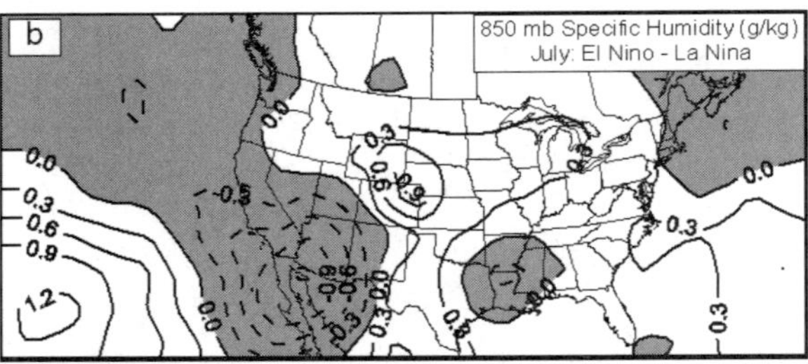

Figure 7.7 Mean July 850-mb (a) temperature and (b) moisture differences between the five strongest El Niño (EN) and La Niña (LN) years. Differences are taken as EN minus LN. Negative differences are indicated by a dashed line.

have yet to be established to disentangle natural from human controls for many themes such as health risks, human comfort levels, energy demand variations, transportation impacts, air quality variations, local urban flooding, and the variability of water uses at local scales. On transportation issues, for example, personal correspondence with Arizona Department of Transportation Office officials and independent analysis of precipitation intensity and traffic data in the local CAP LTER area by A. Ellis (unpubl. data, 2001) have revealed that precipitation events are strongly related to urban-area traffic accident frequencies in a nonlinear fashion; that is, light rainfall initially stimulates higher accidents rates, moderate rainfall, lessening rates, and very high rainfall rates, high accident rates. Part of this pattern relates to driver behavior and levels of risk perception. Subtle differences in rainfall intensity rates that relate to accident variability may or may not be significantly related to phases of ENSO at daily-to-seasonal timescales. More research is needed on this issue. Transportation is also disrupted by blowing dust in central Arizona (Brazel 1991). An analysis of the period 1948–1984 revealed a strong link of incidences of dust storms to lack of antecedent fall/winter precipitation, little surface vegetative armoring, human disturbance of dust source areas, enhanced entrainment due to exceedance of threshold wind speeds, and subsequent incidences of blowing-dust-related accidents on the major interstates and other roads in central Arizona (Brazel et al. 1986; Brazel and Nickling 1987; Brazel 1991). Those authors did not relate this pattern to ENSO per se, but in retrospect it is clear that the arid antecedent years were associated with LN event years. Three more specific examples of significant ENSO impacts are provided here.

Hantavirus and ENSO

One of the startling findings recently in the health risk area in the southwestern United States (primarily Four Corners Area and New Mexico) is that of the link of the hantavirus to environmental moisture parameters, and thus possibly to ENSO phases and climate change (e.g., Hjelle and Glass 2000; and Parmenter et al. 1999; and Sprigg and Hinkley 2000). Much of this work has been conducted under the auspices of the Sevilleta LTER site in New Mexico. Increased precipitation apparently creates a trophic cascade wherein small mammal abundance (related to increased plants and insects) leads to an increase in plague hosts, which in turn results in higher hantavirus incidences. The recent 1990s EN events of 1991–1992, 1993–1994, and 1997–1998 have been linked to subsequent accelerated virus incidences (Hjelle and Glass 2000). Parmenter et al. (1999) of the Sevilleta LTER explain the cascade in a three-stage scale analysis of moisture (ENSO, regional, local), emphasizing the strong relationship to, and need to understand, local precipitation processes. They illustrated an insignificant, yet suggestive correlation of plague case rates to the previous winter moisture conditions using the SOI index. We reanalyzed Hjelle and Glass's (2000; figure 1) 1990s data, and found a statistically significant relation to a previous winter's MEI ($r^2 = 0.42$). The cascade illustrates large moisture lag effects in this case, larger than an annual period from MEI variability to hantavirus events.

Sycamore Creek Stream Ecology

In an analysis of aquatic ecosystems related to climate change, Grimm et al. (1997) illustrate the sensitivity of a range of western U.S. streams to a number of environmental variables, among them precipitation/runoff and net basin supply, in addition to anthropogenic variables (e.g., diversions, withdrawal, and consumptive use). As indicated previously in this chapter, streamflow is correlated to MEI variability for Sycamore Creek. Specific to Sycamore Creek is Grimm's analysis (Grimm 1993) of hydrological characteristics of extreme wet and dry years, annual runoff, number of floods and other stream-specific conditions (wetness and dryness durations relative to an 11-m^3/s peak discharge threshold). We reanalyzed Grimm's (1993) data set of 5 wet and 5 dry years in terms of the MEI index to shed light on the more regional and hemispheric climate connections to this local stream system. We found that, on average, the set of wet years was associated with a December–January winter mean value of MEI = 1.31 (on the EN side of the teleconnection), whereas the dry years on average yielded a value of −0.22 for MEI (toward LN conditions). Two more specific and important hydrological characteristics for Sycamore Creek are the "days in succession" (number of days in the water year ≤ 30 days since a spate of ≥ 11-m^3/s peak discharge occurred) and "days in drying" (number of days in the water year ≥ 200 days since a spate of ≥ 11-m^3/s peak discharge occurred). These two hydrologic parameters specific to Sycamore Creek turn out to be strongly correlated to the MEI Index ($r^2=0.61$ for MEI vs. days in succession—a direct relation showing that higher MEI values indicative of EN relate to higher incidences of lessening time spans between spates; $r^2 = 0.48$ for MEI vs. days in dryness—an indirect relation in which lower MEI, non-EN periods, are associated with increases in periods between spates). Grimm (1993) also analyzed a biotic control factor for the stream, a time during which biotic interactions predominate. She found that the percentage of time during which neither postflood succession nor drying were occurring was relatively constant (mean = 177 days or 48% of the time during the year) even across differing hydrologically wet and dry years. Thus, links of ENSO to local biotic controls as opposed to disturbance controls in the stream would be relatively weak, and the cascade from ENSO to other related biotic factors would be limited. The ENSO cascade here is seasonal and appears to be restricted ecologically to the more disturbance-related aspects of the stream rather than to biotic aspects.

CAP LTER Area River Constituents

A third cascade example specifically related to the strong moisture signal of ENSO at CAP LTER is the cascade of inputs of moisture, stream runoff, and resultant dissolved and mineral river constituents upvalley and downvalley of the metropolitan Phoenix area. Several CAP LTER scientists have focused considerable efforts on creating a composite nitrogen budget for the CAP LTER region based on data for 1988–1996 (Baker et al. 2001; Lauver and Baker 2000). They suggest that there is a hydrologic control, especially for unusually high flow years such as 1993 when an ENSO event occurred at the midpoint of their analysis period. Overall, however, nitrogen fluxes in streams are relatively small in the budget (e.g., riverine export is low, about 3% overall of total input to the ecosystem of CAP LTER; Baker et al. 2001). They also indicate, however, that the N concentration was about 20 times higher in the outflow than in the inflow, reflecting N gained from agricultural drainage, urban runoff, and wastewater, and that N export from the ecosystem via the Gila River was twice as high as the surface-water input. Based on this, we accessed the U.S. Geological Survey's database on stream constituents (web site: www.usgs.gov) and used data for the past 25-year period to develop direct correlations of MEI index values with selected stream constituents for the winter months when MEI correlates highest with inputs of moisture (February through May). Table 7.4 presents the results for an upstream site of the metropolitan region below Bartlett Dam on the Salt River (upvalley of urban; USGS 09502000) and for a site at Gillispie Dam on the Gila River downstream of the metropolitan region (downvalley of urban; USGS 09518000). Many significant correlations are evident in table 7.4 (assuming a standard significance level of 0.05, for example, shown in parentheses). For the Salt River, oxygen, pH, solids, dissolved solids, calcium, sodium, magnesium, chloride, and sulfate all show a significant relation to MEI. For the Gila River, pH, solids, dissolved solids, calcium, sodium, magnesium, chloride, sulfate, total nitrogen, nitrogen nitrite and nitrite total, and nitrogen nitrite and nitrite dissolved are all significantly related to variations in the MEI index. R values are shown to illustrate the directionality of the relationships, positive or negative, versus the MEI index. Generally, the greater the MEI (EN phase), the less the concentration of constituents becomes per volume of water. Two differences between the Salt River site (upvalley of urban) and the Gila River site (downvalley of urban) emerge. (1) Dissolved solids are higher per volume with more runoff at Gila, whereas they

Table 7.4 Correlations (with significance levels in parentheses) of MEI versus February–May monthly concentrations over the 25–year period 1971–1995

Parameter	Gila River[a]	Salt River[b]
Water temperature	0.08 (0.38)	0.02 (0.889)
Turbidity	0.16 (0.19)	−0.05 (0.756)
Oxygen	0.10 (0.36)	0.38 (0.016)
pH	0.23 (0.06)	0.23 (0.002)
Arsenic	−0.03 (0.81)	0.20 (0.254)
Solids	−0.35 (0.004)	−0.47 (0.001)
Dissolved solids	0.23 (0.030)	−0.54 (0.000)
Dissolved calcium	−0.50 (0.000)	−0.44 (0.001)
Dissolved sodium	−0.54 (0.000)	−0.38 (0.003)
Dissolved magnesium	−0.51 (0.000)	−0.36 (0.007)
Dissolved chloride	−0.54 (0.000)	−0.44 (0.002)
Dissolved sulfate	−0.52 (0.000)	−0.37 (0.010)
Total nitrogen	−0.18 (0.250)	Little Data
Nitrogen nitrite + nitrate total	−0.40 (0.000)	−0.07 (0.970)
Nitrogen nitrite + nitrate dissolved	−0.49 (0.000)	0.10 (0.950)

[a] Gila River site is USGS station 09518000 (diversions at Gillespie Dam at 33°13'45" lat., 112°46'00" long).

[b] Salt River site is USGS station 09502000 (below Stewart Mountain Dam at 33°33'10" lat., 111°34'33" long).

are lower at the Salt River site. (2) Nitrogen-related parameters show no correlation at the Salt River site where there is little input, whereas there is a significant MEI climate signal at the Gila River site, where flushes of nitrogen elements occur more readily. With flood releases and significant variations in river discharges that reach downvalley of the urbanized region, the input of nitrogen-related constituents emanating from the urban/agricultural lands and released sediments varies significantly in relation to MEI at a seasonal timescale. Immediately downstream of Bartlett dam, only elemental constituents and not N-related constituents appear to be significantly related to the MEI variability. Hence, it appears that the human-modified urban/agricultural ecosystem in CAP LTER tends to create a positive feedback, or amplification, to the climate signal-stream constituent relationship.

The Future

Spigg and Hinkley (2000) have suggested that global warming may increase the frequency of EN events in the future. It has been hypothesized that a major impact of continued global warming might be an increased frequency of EN events in the Southwest desert area. Should this occur, increases in moisture inputs may result (presumably doubling of moisture in some areas). This could have many positive and negative benefits for the southwestern United States and CAP LTER. For example, more water may be available for the rapidly growing central Arizona area from increased snowpacks in the mountains and runoff into critical reservoirs of central Arizona. But possible negative impacts may occur in the form of reservoir releases, flood risks, ecosystem disturbance, and damage to urban areas. These scenarios are critical for the populace of this region (e.g., Carter et al. 2000). As illustrated in the three ecosystem case examples above, enhanced frequencies of EN may also cause intensification of disease risks, disturbance in streams, and severe variability in river constituents and sediment transfers. However, there remains considerable uncertainty in the global warming–enhanced EN scenarios, and researchers in the LTER network certainly share common goals in unraveling the science and ecology of possible shifts in the climate system that will cascade into important local site effects.

Acknowledgments   We acknowledge David Greenland for encouraging us to make a contribution for CAP LTER; anonymous reviewers; and Nancy Grimm and Charles Redman, principal investigators of CAP LTER for encouraging us to pursue the climate aspects of CAP LTER with support from NSF and grant number DEB 9714833. We would like to acknowledge the influence of K. Wolter in making us aware of the MEI index and for sharing its database. We also thank Barbara Trapido-Lurie for some of the cartographic work.

References

Adams, D.K., and A.C. Comrie, 1997. The North American monsoon. *Bulletin of the American Meteorological Society*, Vol. 78, 2197–2213.

Baker, L. A., D. Hope, Y. Xu, J. Edmonds, and L. Lauver, 2001. Nitrogen balance for the Central Arizona-Phoenix (CAP) ecosystem. *Ecosystems*, Vol. 4, 582–602.

Balling, R. C., Jr., and R.S. Cerveny, 1987. Long-term associations between wind speeds and the urban heat island of Phoenix, Arizona. *Journal of Climate and Applied Meteorology*, Vol. 26, 712–716.

Brazel, A. J. 1987. Dust and climate in the American Southwest. In *Paleoclimatology and Paleometeorology: Modern and Past Patterns of Global Atmospheric Transport*. Edited by M. Leinen and M. Sarinthein. NATO ASI Series, Kluwer Academic Pubs., 65–96.

Brazel, A. J., 1991. Blowing dust and highways: The Case of Arizona, U.S.A. In *Highway Meteorology*. Edited by A. Parry and L. Symons. Taylor and Francis Books Ltd., London, 131–161.

Brazel, A. J., and W. G. Nickling, 1987. Dust storms and their relation to moisture in the Sonoran-Mojave desert region of the Southwestern United States. *Journal of Environmental Management*, Vol. 24, 279–291.

Brazel, A. J., W. G. Nickling, and J. Lee, 1986. Effect of antecedent moisture conditions on dust storm generation in Arizona. In *Aeolian Geomorphology*. Boston: Allen and Unwin. 261–271.

Brazel, A. J., N. Selover, R. Vose, and G. Heisler, 2000. The tale of two climates — Baltimore and Phoenix urban LTER sites. *Climate Research*, Vol. 15, 123–135.

Brazel, A. J., H. J. Verville, and R. Lougeay, 1993. Spatial-temporal controls on cooling degree hours: An energy demand parameter. *Theoretical and Applied Climatology*, Vol. 47, 81–92.

Carter, R. H., P. Tscharert, and B. J. Morehouse, 2000. *Assessing the sensitivity of the Southwest's urban water sector to climatic variability: Case studies in Arizona.* Climate Assessment for the Southwest (CLIMAS), Report Series CL1-00, University of Arizona.

Cayan, D. R., and K. T. Redmond, 1994. ENSO influences on atmospheric circulation and precipitation in the Western United States. In K. T. Redmond and V. L. Tharp, *Proceedings of the Tenth Annual Pacific Climate (PACLIM) Workshop*, April 4–7, 1993.

California Dept. of Water Resources, Interagency Ecological Studies Program, Tech Report 36, pp. 5–26.

Cayan, D. R., K. T. Redmond, and L. G. Riddle, 1999. ENSO and hydrological extremes in the Western United States. *Journal of Climate*, Vol. 12, 2881–2893.

Collins, J. P., A. Kinzig, N. B. Grimm, W. F. Fagan, D. Hope, J. Wu, and E. T. Borer, 2000. A new urban ecology. *American Scientist,* Vol. 88, 416–425.

Greenland, D., 1999. ENSO-related phenomena at long-term ecological research sites. *Physical Geography,* Vol. 20, No. 6, 491–507.

Grimm, N. B., 1993. Implications of climate change for stream communities. In *Biotic Interactions and Global Change.* Edited by P. M. Kareiva, J. G. Kingsolver, and R. B. Huey, Sinauer Associates Inc. Pubs., Massachusetts, USA, pp. 293–313.

Grimm, N. B., A. Chacon, C. N. Dahm, S. W. Hostetler, O. T. Lind, P. L. Starkweather, and W. W. Wurtsbaugh, 1997. Sensitivity of aquatic ecosystems to climatic and anthropogenic changes: The basin and range, American Southwest and Mexico. *Hydrological Processes*, Vol. 11, 1023–1041.

Grimm, N. B., J. M. Grove, S. T. A. Pickett, and C. L. Redman, 2000. Integrated approaches to long-term studies of urban ecological systems. *BioScience*, Vol. 50, No. 7, 571–584.

Hjelle, B., and G. E. Glass, 2000. Outbreak of Hantavirus infection in the Four Corners region of the United States in the wake of the 1997–1998 El Niño-Southern scillation. *The Journal of Infectious Diseases*, Vol. 181, 1569–1573.

Kalnay, E., M. Kanamitsu, R. Kistler, W. Collins, D. Deaven, L. Gandin, M. Iredell, S. Saha, G. White, J. Woollen, Y. Zhu, A. Leetmaa, and R. Reynolds, 1996. The NCEP/NCAR reanalysis 40-year project. *Bulletin of the American Meteorological Society*, Vol. 77, 437–471.

Lauver, L., and L. A. Baker, 2000. Mass balance for wastewater nitrogen in the central Arizona-Phoenix ecosystem. *Water Research*, Vol. 34, No. 10, 2754–2760.

Li, Z., and M. Kafatos, 2000. Interannual variability of vegetation in the United States and its relation to El Niño/Southern Oscillation. *Remote Sensing of the Environment.* Vol. 71, 239–247.

Mason, S. J., and L. Goddard, 2001. Probabilistic precipitation anomalies associated with ENSO. *Bulletin of the American Meteorological Society,* Vol. 82, No. 4, 619–638.

Mather, J. R., 1978. *The climatic water budget in environmental analysis.* Lexington, Mass.: Lexington Books. 239 pp.

Molles, M. C., Jr., and C. N. Dahm, 1990. A perspective on El Niño and La Niña: Global implications for stream ecology. *Journal North American Benthological Society*, Vol. 9, No. 1, 68–76.

NOAA, 1983a. State, regional and national monthly and average temperatures, weighted by area, January 1931–December 1983. Historical Climatology Series 4-1, National Climatic Data Center, Asheville, NC, 68 pp.

NOAA, 1983b. State, regional and national monthly and average precipitation, weighted by area, January 1931–December 1983. Historical Climatology Series 4-2, National Climatic Data Center, Asheville, NC, 68 pp.

Pagano, T., H. Hartman, S. Sorooshian, and R. Bales. 1999. *Advances in seasonal forecasting for water management in Arizona: A case study of the 1997–98 El Niño.* Department of Hydrology and Water Resources, University of Arizona.

Parmenter, R. R., E. P. Yadav, C. A. Parmenter, P. Ettestad, and K. L. Gage, 1999. Incidence of plague associated with increased winter-spring precipitation in New Mexico. *American Journal of Tropical Medicine and Hygiene*, Vol. 61, No. 5, 814–821.

Redmond, K. T., 2000. Climate monitoring: Taking the long view. *Water Resources Impact,* Vol. 2, No. 4, 7–10.

Redmond, K. T., and D. R. Cayan, 1994. El Niño/Southern Oscillation and western climate variability, Preprint, *Sixth Conference on Climate Variations*. American Meteorological Society, Nashville, Tenn., January 23–28, Paper 3.5, 141–145.

Redmond, K. T., and D. R. Cayan, 1999. ENSO phase and precipitation persistence in the western U.S., Preprint, 11th Conference on Applied Climatology, American Meteorological Society, Dallas, Texas, January 10–15, Paper J2.3.

Redmond, K. T., and R. W. Koch, 1991. Surface climate and stream flow variability in the Western United States and their relationship to large-scale circulation indices. *Water Resources Research,* Vol. 27, No. 9, 2381–2399.

Simpson, H. J., and D. C. Colodner, 1999. Arizona precipitation response to the Southern Oscillation: A potential water management tool. *Water Resources Research*, Vol. 35, No. 12, 3761–3769.

Sprigg, W.A., and T. Hinkley. 2000. *Preparing for a Changing Climate the Potential Consequences of Climate Variability and Change.* Report of the Southwest Regional Assessment Group for the U.S. Global Change Research Program, September 2000, Institute for the Study of Planet Earth, 60 pp.

Stooksbury, D. E., C. D. Idso, and K. G. Hubbard, 1999. The effects of data gaps on calculated monthly mean maximum and minimum temperatures in the continental United States: A spatial and temporal study. *Journal of Climate*, Vol. 12, 1524–1533.

Swetnam, T. W., and J. L. Betancourt, 1990. Fire-Southern oscillation relations in the Southwestern United States. *Science.* Vol. 249, 1017–1020.

Thornthwaite, C.W., and J. R. Mather. 1955. The water balance. *Publications in Climatology*, Vol. 8, 1–104.

Wolter, K., 1987. The Southern Oscillation in surface circulation and climate over the tropical Atlantic, Eastern Pacific, and Indian Oceans as captured by cluster analysis. *J Climate Appl. Meteor.*, 26, 540–558.

Wolter, K., and M. S. Timlin, 1993. Monitoring ENSO in COADS with a seasonally adjusted principal component index. *Proc. of the 17th Climate Diagnostics Workshop*, Norman, OK, NOAA/N MC/CAC, NSSL, Oklahoma Clim. Survey, CIMMS and the School of Meteor., Univ. of Oklahoma, 52–57.

Woolhiser, D. A., and T. O. Keefer, 1993. Southern Oscillation effects on daily precipitation in the Southwestern United States. *Water Resources Research,* Vol. 29, No. 4, 1287–1295.

8

Watershed Hydrological and Chemical Responses to Precipitation Variability in the Luquillo Mountains of Puerto Rico

Douglas Schaefer

Introduction

Variations in temperature and precipitation are both components of climate variability. Based on coral growth rates measured near Puerto Rico, the Caribbean was 2–3°C cooler during the "Little Ice Age" during the seventeenth century (Winter et al. 2000). At the millennial scale, temperature variations in tropical regions have been inferred to have substantial biological effects (such as speciation and extinction), but not at the multidecadal timescales considered here. My focus is on precipitation variability in particular, because climate models examining effects of increased greenhouse gases suggest greater changes in precipitation than in temperature patterns in tropical regions.

Some correspondence between both the El Niño–Southern Oscillation (ENSO) and the Northern Atlantic Oscillation (NAO) and average temperatures and total annual precipitation have been reported for the LTER site at Luquillo (Greenland 1999; Greenland and Kittel 2002), but those studies did not refer to extreme events. Based on climate records for Puerto Rico since 1914, Malmgren et al. (1997) found small increases in air temperature during El Niño years and somewhat greater total rainfall during the positive phase of the NAO. Similar to ENSO, the NAO index is characterized by differences in sea-level atmospheric pressure, in this case based on measurements in Iceland and Portugal (Walker and Bliss 1932). Its effects on climate have largely been described in terms of temperature and precipitation anomalies in countries bordering the North Atlantic (e.g., Hurrell 1995).

Puerto Rico is in the North Atlantic hurricane zone, and hurricanes clearly play a major role in precipitation variability. The association between extreme rainfall events and hurricanes is discussed in detail in this chapter. I examine the degree to

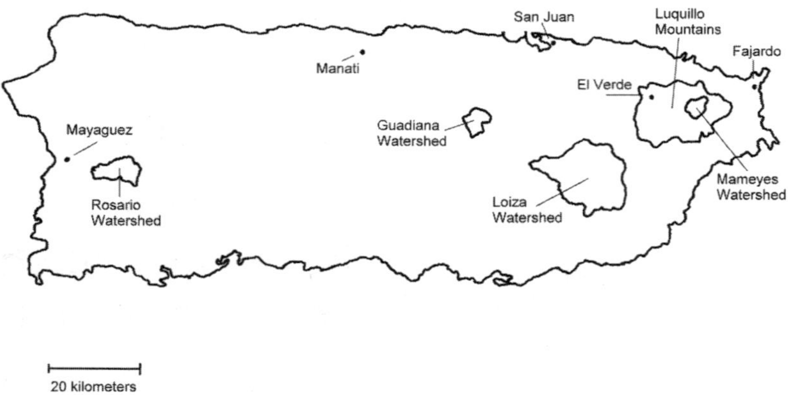

Figure 8.1 The island of Puerto Rico (centered approximately 18°15' N and 66°30' W), showing the locations of the sites mentioned in the text.

which extreme rainfall events are associated with hurricanes and other tropical storms. I discuss whether the occurrence of these extreme events has changed through time in Puerto Rico or can be linked to the recurrent patterns of the ENSO or the NAO. I examine the 25-year daily precipitation record for the Luquillo LTER site, the 90-year monthly record from the nearest site to Luquillo with such a long record, Fajardo, and those of the two other Puerto Rico stations with the longest daily precipitation records, Manati and Mayaguez (figure 8.1).

To explore the relationship between extreme rainfall events and sediment export, I used daily sediment export data from the Mameyes watershed in the Luquillo Mountains, and two other watersheds with the longest available records, Loiza and Rosario (figure 8.1). The Loiza watershed (23,260 ha) is in the north-central area and provides much of the inflow to the Carraizo Reservoir, the major water supply for San Juan, Puerto Rico's largest city. The rapid sedimentation of this reservoir (as well as others in Puerto Rico) is being intensively studied (Gellis 1993; Larsen et al. 1993; Morris and Fan 1998; Larsen et al. 1999). The Loiza watershed is currently experiencing substantial land clearing at the lower elevations as a result of urban expansion (Larsen et al. 1999). The Rosario watershed (4,740 ha) is in far western Puerto Rico. Sediment production has been modeled there using a combination of ground and remotely sensed data (Cruise and Miller 1994). As with much of the island outside the Luquillo Mountains, both the Loiza and Rosario watersheds were largely deforested for agriculture and pasture prior to 1940 and have since experienced secondary forest regrowth (Birdsey and Weaver 1987). The Mameyes watershed (1,782 ha) is in an area of largely undisturbed forest in the Luquillo Mountains in northeastern Puerto Rico. The analyses of daily sediment export data from Puerto Rico watersheds presented here indicates that 75% of the export occurs during the 1% of the days that have the greatest rainfall. Of course, this is an intuitive conclusion, but the extreme nonlinearity of the response deserves consideration. This type of nonlinearity in sediment delivery has been reported for the continental United States as well. Meade and Parker (1984) found that 50% of sediment export takes place in 1% of days, and 90% in 10% of days.

Variability of Extreme Rainfall Events

Hurricane Variability through Time

Since the first anecdotal record in the year 1515, 51 hurricanes have passed over the island for an average of about 1 per 9.5 years. The reliability of this record may be questioned prior to 1851, the first year of Atlantic hurricane tracks published by the National Oceanic and Atmospheric Administration, so that recognition of long-term patterns may be uncertain. At the same time, there is evidence for substantial variability over the last five centuries. From 1616 to 1737 (122 years), no hurricanes were reported to pass over the island. At the other extreme, there were 9 hurricanes from 1804 to 1819 (16 years). More recently, 3 hurricanes passed over Puerto Rico from 1989 through 1998 (10 years). The approximately 16 hurricanes that have passed over Puerto Rico since 1851 constitute too small a sample set to subject to an analysis of temporal variability.

The total number of North Atlantic hurricanes does show multidecadal trends during the last 150 years, but connections with Earth's climate system remain controversial. Just as controversial is the possibility that future global climate change will affect the number or intensity of hurricanes, or the length of the annual hurricane season. In any case, the island of Puerto Rico (figure 8.1) is only 175 by 55 km and thus presents a small "target" for the hurricanes (99% of them go elsewhere), so hurricane visits to the island must always have a large random component regardless of climate variability.

Other Extreme Rainfall Events

The Luquillo LTER has kept daily rainfall records since 1975. Data on all the days with 100 mm or more rainfall at Luquillo are presented in table 8.1. Even though hurricanes may be seen as the most extreme events because of the damaging winds, they are not the only source of extreme rain events in Puerto Rico. Since 1975, 65 rain events of 100 mm or more have occurred in Luquillo. Of those, 12 (or about 18%) were from hurricanes (including instances where the "hurricane eye" did not pass over the island directly), tropical storms, or tropical depressions (the latter being essentially precursor stages to the formation of hurricanes). The other 53 (about 82%) of the events occurred in the absence of cyclonic storms, either from localized convective cells or stalled low-pressure systems. Therefore, in Puerto Rico, most of the extreme rainfall events and sediment export occur in the absence of hurricane development. The annual average precipitation at this site is 3400 mm, so these extreme events represent about 13% of the total rainfall. So 75% of the sediment export was caused by 13% of the total rainfall, on less than 1% of the days. These extreme events have occurred with no unidirectional trend through time, so that during this period, no connection with unidirectional climate change

Table 8.1 Dates of all rainfall events in the Luquillo Mountains of Puerto Rico (1975–1999) with more than 100 mm rainfall

Date	H/TS/TD >100 mm	Name	Others > 100 mm
16–Sep–75[a]	**243**	**TS. Eloise**	
20–Nov–75			112
10–Dec–75			132
23–Apr–77			147
18–May–77			121
24–Nov–77			247
07–Mar–78			106
10–Apr–78			228
27–May–78			130
26–Oct–78	**160**	**TS. Kendra**	
19–Jan–79			111
14–Feb–79			292
30–May–79			103
30–Aug–79	**309**	**H. David**	
04–Sep–79	**197**	**TS. Fredric**	
14–May–80			155
06–Dec–80			139
21–Apr–81			118
13–Dec–81			222
09–May–82			117
11–May–82			106
27–May–82			149
12–Sep–82	**128**	**TD. Debby**	
21–Apr–83			127
06–Jul–83			150
02–Dec–83			133
15–May–85			161
18–May–85			104
20–Jul–85			104
13–Sep–85			118
06–Oct–85	**243**	**TD. Isabel**	
13–Nov–85			102
16–Nov–85			120
02–May–86			117
14–May–86			130
14–Aug–86			117
08–Oct–86			104
12–Apr–87			389
19–Jun–87			101
27–Nov–87			318
08–Dec–87			501
11–Aug–88			109
25–Aug–88	**153**	**TD. Chris**	
17–Feb–89			124
03–Jun–89			120
18–Sep–89	**170**	**H. Hugo**	
14–Aug–90			122
17–Oct–90			107
08–Nov–91			114

Table 8.1 *Continued*

Date	H/TS/TD >100 mm	Name	Others > 100 mm
30–Dec–92			110
16–May–95			104
16–Sep–95	**173**	**H. Marilyn**	
04–Apr–96			120
07–Sep–96	**540**	**H. Bertha**	
23–Nov–96			258
22–Aug–97			104
26–Nov–97			122
07–Mar–98			420
16–Apr–98			160
22–Aug–98			211
19–Sep–98	**377**	**H. Georges**	
02–Dec–98			119
17–Jun–99			130
21–Oct–99	**101**	**TS. Jose**	
02–Dec–99			254
Count	**12**		53
Average mm	**233**		159
Maximum mm	**540**		501

[a] Hurricanes (H), tropical storms (TS), and tropical depressions (TD) and their dates and rainfall amounts are shown in bold.

can be detected (table 8.2). These events can occur during any month but are not common during the January–March dry season (table 8.3).

In terms of monthly total precipitation, a weak ENSO signal can be discerned at stations throughout Puerto Rico. The longest monthly precipitation time series near the Luquillo Mountains comes from Fajardo on the northeastern coast, starting in 1909, and it is used here as an example. Specifically, El Niño years have more rainfall and La Niña years have less rainfall than other (index) years, but during the month of May only (figure 8.2). Although it has long been recognized that precipitation totals for the month of May are more variable than any other month (F. Scatena, pers. comm., May 3, 1997), the connection between this variability and the ENSO cycle has not previously been recognized.

In light of this pattern, the 12 extreme rainfall events in May that are shown in table 8.3 for Luquillo merit further attention. Only two of them (in 1978 and 1995) occurred during El Niño years. Thus, El Niño years had 0.22 extreme rainfall events during May per year, whereas the other (index and La Niña) years had 0.30 May events per year. From the limited 25 years of daily data examined here, I cannot conclude that the ENSO cycle alters the occurrence of extreme rainfall events in Luquillo. Longer-term (100 years) daily rainfall data from other Puerto Rico stations are considered subsequently.

The cumulative frequency distributions of daily rainfall for the wettest year of record (1987), the driest (1994), and all other years together are compared in figure

Table 8.2 Annual occurrence of extreme rainfall events at the Luquillo LTER site in Puerto Rico

Year	Number of rain events >100 mm
1975	3
1976	0
1977	3
1978	4
1979	5
1980	2
1981	2
1982	4
1983	3
1984	0
1985	7
1986	4
1987	4
1988	2
1989	3
1990	2
1991	1
1992	1
1993	0
1994	0
1995	2
1996	3
1997	2
1998	5
1999	3

Table 8.3 Seasonality of extreme rainfall events at the Luquillo LTER site in Puerto Rico

Month	Number of rain events greater than 100 mm
Jan	1
Feb	2
Mar	1
Apr	7
May	12
Jun	3
Jul	1
Aug	7
Sep	8
Oct	5
Nov	7
Dec	7

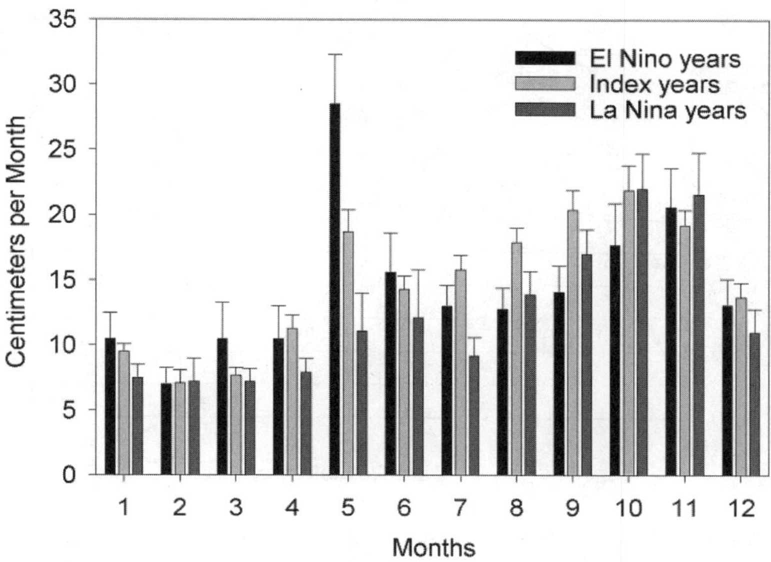

Figure 8.2 Average monthly precipitation at Fajardo, Puerto Rico (northeastern coast), 1909–1996. Data are segregated into El Niño (solid black bars), La Niña (dark shaded bars), and Index years (light shaded bars) as described in the text. Errors are one standard deviation above and below the means. During May, El Niño years have significantly greater precipitation than Index years, and La Niña years have less. During other months, the differences were not significant.

8.3. In the "other years" category, 1% of days have 100 mm or more of rain, and this is the daily rainfall threshold associated with 75% of the sediment export in the Luquillo Mountains. The year 1987 had 2% of such days, and 1994 had none. During that drought year (1994), the Mameyes watershed had only 28% of its 6-year average sediment export.

The longest daily records available for the island are Manati (north-central coast) and Mayaguez (far western coast) both beginning in 1900. The latter data were obtained in digital form from the National Climate Data Center. Whereas 1% of the days at El Verde have 100 mm rain or more, that value is reached at 54 mm at both Manati and Mayaguez, based on their entire records. To determine whether such extreme days are associated with the ENSO cycle, data were segregated into El Niño, La Niña, and Index (all other) years. For this analysis El Niño years were 1919, 1926, 1940–1942, 1952, 1958, 1964, 1966, 1973, 1978, 1983, 1987, 1992–1994, and 1998. La Niña years were 1918, 1939, 1950, 1951, 1956, 1971, 1974, 1976, 1989, and 1999. The ENSO data and categorical segregation followed Trenberth (1984) with more recent data taken from the University Center for Atmospheric Research web site (www.ucar.edu). For these three stations, there is no clear association between the state of ENSO and the proportion of days with greater than the threshold amounts of rain (table 8.4).

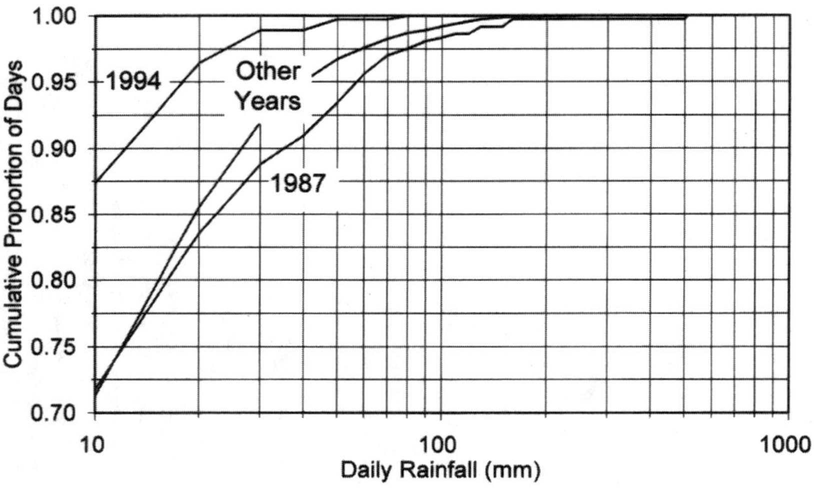

Figure 8.3 Cumulative frequency distribution of daily rainfall at the Luquillo LTER site, 1975–1999. The highest curve represents 1994, a drought year, when no days had more than 100 mm of rainfall. The lowest curve represents 1987, a flood year, when about 2% of the days had more than 100 mm of rainfall. The intermediate curve includes data from all other years, for which about 1% of days exceeded 100 mm of rainfall. Note that most days (having less than 10 mm of rainfall) are outside the range of this figure.

To determine whether such extreme days are associated with the NAO cycle, NAO negative years were 1918, 1919, 1966–1972, 1979, 1980, and 1981. Positive NAO years were 1907–1910, 1914, 1922–1927, 1934–1939, 1946, 1954, 1955, and 1990–1995, and index years were all others. This segregation followed Hurrell (1995) with more recent NAO data obtained from the UCAR web site listed previously. In this case both Mayaguez and El Verde had less extreme rainfall days when NAO was negative than in index or negative years (table 8.5). But we must recall that in the short El Verde record, both drought years (1994 and 1995) were NAO positive years. Whether this relationship is casual (as opposed to causal) cannot be determined from a record of this length. Droughts in Puerto Rico during future NAO-positive periods would constitute very important observations.

Table 8.4 The El Niño–Southern Oscillation (ENSO) has not altered the frequencies (percentage of days) of extreme rainfall events in Puerto Rico.

ENSO State	Manati (1900–1998) **54**[a]	Mayaguez (1900–1998) **54**	El Verde (1975–1999) **100**
La Niña (cold phase)	0.86	1.11	0.82
Index years (all other years)	0.89	0.85	1.10
El Niño (warm phase)	0.87	0.87	1.10

Periods of record are in parentheses and rainfall thresholds (mm/day) used for each site are in **bold**.

Table 8.5 The Northern Atlantic Oscillation (NAO) may have altered the frequencies (percentage of days) of extreme rainfall events in Puerto Rico

NAO State	Manati (1900–1998) **54**	Mayaguez (1975–1999) **54**	El Verde (1975–1999) **100**
Negative	0.94	1.52	1.37
Index years (all other years)	0.98	1.13	1.28
Positive	0.85	0.83	0.37

[a] Periods of record are in parentheses and rainfall thresholds (mm/day) are in **bold**.

Finally, the two longest records were compared during consecutive 20-year periods to search for unidirectional trends. Manati was indifferent, whereas extreme rainfall events in Mayaguez may have been increasing since 1920 (table 8.6). Whether a developing "heat island" in Mayaguez could be related to more extreme rain events (as suggested for continental U.S. cities by J. Luvall, pers. comm., January 17, 1996) is at present simply an interesting hypothesis.

Effects of Extreme Rainfall Events

Extreme rainfall events, regardless of whether they are from hurricanes, have pervasive ecological effects on this tropical forest. The historical data used for this analysis do not lead to the conclusion that global climate variability will necessarily lead to increased numbers of such extreme rainfall events. Yet it must be stressed that because of the extreme nonlinearity of sediment export versus precipitation amounts, even a small increase in the number of extreme rainfall events would have a large impact on sediment flux, as well as on its associated ecological responses, as mentioned elsewhere. Longer term daily precipitation data from other tropical sites could provide additional insight into this issue.

Table 8.6 Multidecadal trends in the frequencies (percentage of days) of extreme rainfall events in Puerto Rico

Interval	Manati (1900–1998) **54**	Mayaguez (1900–1998) **54**
1900–1919	1.14	1.15
1920–1939	0.72	0.90
1940–1959	0.97	0.91
1960–1979	1.01	1.20
1980–1998	0.87	1.35

[a] Periods of record are in parentheses and rainfall thresholds (mm/day) are in **bold**.

Effects on Stream Water Chemistry

Since Luquillo became an LTER site in 1988, two of the three Puerto Rico hurricanes have passed close enough to the forest to cause substantial wind damage (Hugo in 1989 and Georges in 1998). Stream water chemical export is one of the ecosystem responses to these events that we have examined. In brief, stream water export of potassium and nitrate ions increased markedly after Hurricane Hugo and remained elevated for 18 to 24 months until the canopy leaf cover returned. Other ions were much less affected (Schaefer et al. 2000). Hurricane Georges caused much less forest damage, as its eye passed south of the Luquillo Mountains, and our north-facing research areas were sheltered. Consequently, the effects on stream water chemistry were much smaller than after Hugo. One conclusion from these observations is that biogeochemical cycling in this forest is resilient to the moderate hurricane damage caused by Hugo (near-total canopy defoliation and 7–11% stem mortality; Walker 1991). If there is a higher damage threshold for biogeochemical cycling in this forest, it has apparently not been crossed during the last century.

Because of the positive concentration versus discharge relationships for potassium and nitrate ions (e.g., Schaefer et al. 2000), most of the stream export for these ions occurs during high flows associated with extreme rainfall events as well. However, those relationships appear to saturate at the highest stream flows, unlike those for sediment export, so watershed export of those dissolved ions will be less sensitive to the number of extreme rainfall events than is sediment export.

Effects of Sediment Export

Watershed sediment export is also studied at the Luquillo LTER and by the U.S. Geological Survey (USGS) at watersheds elsewhere in Puerto Rico. Stream sediment export is a particularly tangible example of several coupled ecosystem responses to extreme precipitation events. Where reservoirs are located downstream, sediment fluxes reduce reservoir capacity over time. Morris and Fan (1998) found that average sediment yields from the watersheds supplying the 14 major reservoirs in Puerto Rico ranged from 7 to 27 t ha^{-1} yr^{-1}. These values represent an annual loss of reservoir capacity from 0.3 to 1.3%. According to Soler-López (2001), the 14 major water-supply reservoirs in Puerto Rico have lost an average of 35% of their storage capacity since construction, which is twice the rate projected at their time of construction. With regard to the near-shore marine environment, Rogers (1990) summarized negative effects of sediment discharge on coral reefs and sea grass beds. Sediment export can carry adsorbed toxins such as heavy metals, pesticides, and other organic compounds (Meade and Parker 1984). With gully formation in particular, soil erosion is linked to the loss of arable cropland (Lal 1994). Riverine biota may be negatively impacted by sediment exports via reduction in food supplies, clogging of feeding structures, mechanical scouring, suffocation, and downstream relocation during the high flows (Newcombe and MacDonald 1991). The sediment fluxes during extreme rainfall events in the Luquillo Mountains of Puerto Rico are also associated with landslides (Larsen and Simon 1993). Their data indicate that 24-hour rainfall totals of 200 mm or greater trigger landslides.

Table 8.7 Annual sediment fluxes for three Puerto Rico watersheds in the years of record and the number of extreme events (defined as those with 75% of the total recorded export) occurring in each year

Water Year (Oct–Sep)	Loiza extreme t ha[-1] yr[-1]	days	Rosario extreme[a] t ha[-1] yr[-1]	days	Mameyes extreme[b] t ha[-1] yr[-1]	days
1983–1984	10.5	3	—	—	—	—
1984–1985	20.1	7	—	—	—	—
1985–1986	33.3	7	—	—	—	—
1986–1987	4.5	3	7.0	3	—	—
1987–1988	32.1	9	21.1	9	—	—
1988–1989	8.2	3	10.9	6	—	—
1989–1990[c]	—	—	—	—	—	—
1990–1991	4.9	1	4.0	0	—	—
1991–1992	19.1	6	10.3	2	—	—
1992–1993	7.2	1	13.8	7	2.1	3
1993–1994	2.7	1	2.4	1	0.6	1
1994–1995	3.8	2	5.1	1	0.9	1
1995–1996	23.5	2	3.4	1	4.7	5
1996–1997	3.3	0	2.1	0	2.0	4
1997–1998	26.0	4	83.1	5	8.7	15
Average	14.2	3.6	14.8	3.4	3.2	4.9
CV (%)	128	75	157	78	100	108

[a] Data begin in 1986.

[b] Data begin in 1992.

[c] Data unavailable.

Quantifying Sediment Export in Puerto Rico

Daily sediment export is currently being measured in more than 20 rivers in Puerto Rico (Diaz et al. 1984–1998). For this study, daily streamflow and sediment fluxes in the Loiza, Rosario, and Mameyes watersheds were taken from the USGS annual water resources data reports for Puerto Rico (Diaz et al. 1984–1998). Daily data were summed for annual totals, and those days totaling 75% of the sediment export (extreme days) were identified (table 8.7). In all three watersheds, 75% of the sediment export occurs in approximately 1% of the days. Sediment exports from these three watersheds average 3.2 t ha[-1] yr[-1] (Mameyes 1993–1998), 14.2 t ha[-1] yr[-1] (Loiza 1984–1998), and 14.8 t ha[-1] yr[-1] (Rosario 1987–1998).

By comparison, sediment yields in the continental United States range from 0.15 to 150 t ha[-1] yr[-1] (Vanoni 1975), and tropical rates range from 1 to more than 100 t ha[-1] yr[-1], with the higher rates being associated with severe land degradation (Lal 1990). In a compilation of data from 280 global rivers, Milliman and Syvitski (1992) provide annual sediment yields from the 16 largest tropical rivers (draining land areas of 430,000 km[2] and greater) ranging from 0.1 to 14 t ha[-1] yr[-1]. Their data for 73 smaller tropical rivers range from 0.1 to 360 t ha[-1] yr[-1]. Sediment yields in excess of 300 t ha⁻¹ yr⁻¹ are observed in Taiwan (Li 1976) that exhibit an unfortunate combination of extensive land clearing, heavy rainfall (in part from tropical typhoons, the Pacific Ocean equivalent of hurricanes), and steep lands resulting from rapid rates of geological uplift.

Based on 1984–1990 data for the Loiza watershed, Morris and Fan (1998) reported that 65% of the sediment was delivered in 10 days (0.3% of the total days) and 17% of the total occurred in the single largest event (290,000 tons on 13 May 1986). That sediment discharge has since been exceeded by the 2-day total that occurred during Hurricane Georges (352,200 tons during 21–22 September 1998) and 396,000 tons during Hurricane Bertha (10 September 1996).

In a related study, the revised universal soil loss equation was applied to the Guadiana watershed in central Puerto Rico by Del Mar Lopez et al. (1998). They reported that, within that watershed, areas of open forest eroded at 26 t ha⁻¹ yr⁻¹ and closed forest at 7 t ha⁻¹ yr⁻¹. Based on their sediment delivery ratio (the fraction of eroded soil material that reaches the river channel; Trimble 1975) of 0.17, these cover classes yield 4.42 and 0.12 t ha⁻¹ yr⁻¹ sediment to the river, respectively.

My selection of the 75% level of total stream sediment export was arbitrary, but convenient in that, based on 25 years of daily precipitation records at the Luquillo site, this occurred in precipitation events of 100 mm or more.

Effects of Numbers of Extreme Events on Sediment Exports

For all three watersheds, the extreme events, defined as producing 75% of the total sediment exports, were identified. In the Loiza watershed, there were 50 such events in 14 years (average 3.6 yr⁻¹), of which 12 were associated with hurricanes, tropical storms, or tropical depressions (24% of total). In the Rosario watershed, there were 37 such events in 11 years (average 3.4 yr⁻¹), of which 4 were associated with hurricanes, tropical storms, or tropical depressions (11% of total). In the Mameyes watershed, there were 29 such events in 6 years (average 4.8 yr⁻¹), of which 3 were associated with hurricanes, tropical storms, or tropical depressions (10% of total). Analyses of how increasing the number of these extreme events could increase sediment exports were conducted as follows: Annual exports were calculated without these events, and the annual sediment export attributed solely to those extreme events was regressed against the number of events in each year (figures 8.4–8.6).

Sediment export from the Rosario watershed during Hurricane Georges appeared to be an outlier in this analysis, as the r^2 of that regression was increased from 0.14 to 0.74 by its exclusion. For that reason the regressions for all three watersheds were performed both with and without Hurricane Georges, and both versions are presented in figures 8.4–8.6 (showing Loiza, Rosario, and Mameyes, respectively). This modification slightly improved the Loiza r^2 (from 0.65 to 0.74), and reduced the Mameyes r^2 (from 0.95 to 0.70). Based on these models, the effect of increasing numbers of extreme events was explored and those results are presented in table 8.8. For these analyses, all extreme events were treated as equals, even though they vary greatly in terms of sediment production. This simplification may be justified in two ways. First, the relationships between rainfall amounts or

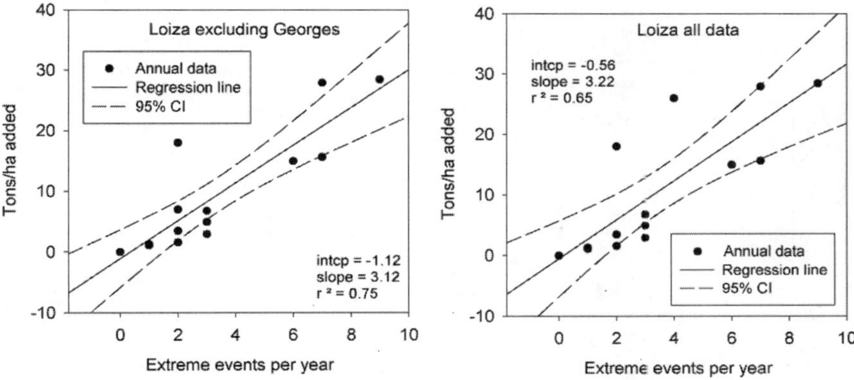

Figure 8.4 Regressions of the annual numbers of extreme events on the sediment exports during those events in the Loiza watershed of Puerto Rico. Filled circles represent annual data, solid lines are regressions, and dashed curves are 95% confidence intervals. Regressions are performed with all data and excluding the 1998 Hurricane Georges (see text).

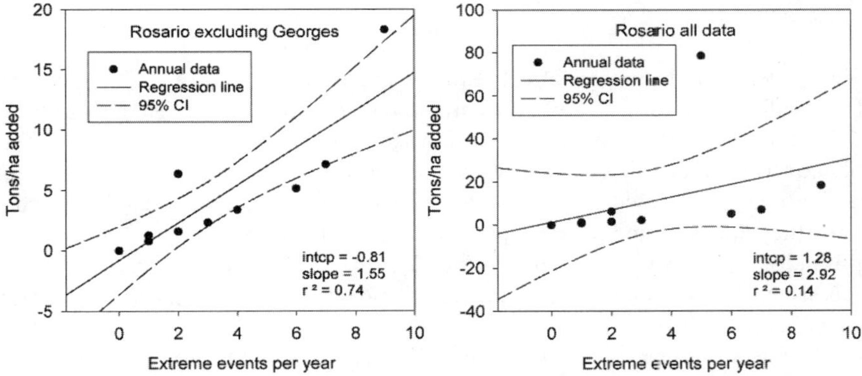

Figure 8.5 Regressions of the annual numbers of extreme events on the sediment exports during those events in the Rosario watershed of Puerto Rico. Filled circles represent annual data, solid lines are regressions, and dashed curves are 95% confidence intervals. Regressions are performed with all data and excluding the 1998 Hurricane Georges (see text).

intensities and sediment production are complex. Second, it was not deemed fruitful to specify the sizes of extreme events that would be added for these analyses.

Although climate variations (whether natural or anthropogenic) could alter the number of such extreme events, the analyses performed on long-term rainfall data from Puerto Rico provide scant evidence that this has occurred to date. Rather, this exercise demonstrated the sensitivity of total sediment export to the number extreme events that may occur. In Puerto Rico, one additional event per year could add 13 to 24% to the total sediment export, and doubling the number of extreme events occurring could increase sediment export by 61 to 95%.

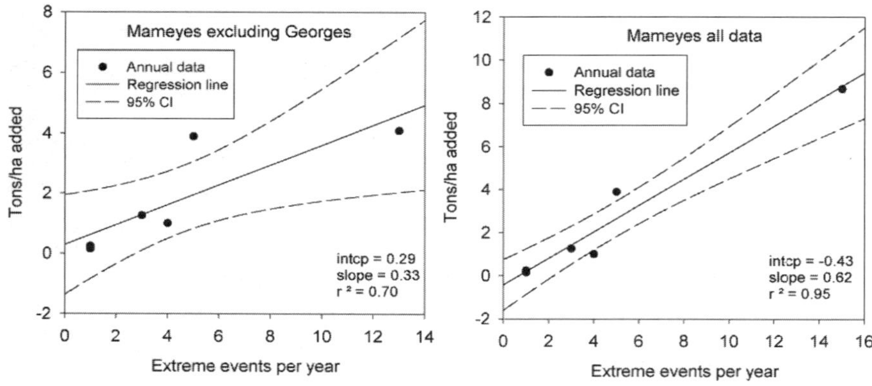

Figure 8.6 Regressions of the annual numbers of extreme events on the sediment exports during those events in the Mameyes watershed of Puerto Rico. Filled circles represent annual data, solid lines are regressions, and dashed curves are 95% confidence intervals. Regressions are performed with all data and excluding the 1998 Hurricane Georges (see text).

Conclusions

What preexisting conditions affect sediment export during extreme rainfall events? The condition of saturated soil is important, but not required. Surface flow of water is the minimum necessary condition for sediment mobilization. This can occur whenever the hydraulic conductivity of the surface soil is exceeded by the rainfall rate. When unsaturated, surface soils have higher hydraulic conductivities than when they are saturated (by previous rainfall). In effect, sediment mobilization begins at lower rainfall rates when soils are saturated (Lal 1990). The rainfall thresholds selected in this study exceed this requirement because of their observed effects on sediment export.

The lower limit was arbitrarily selected as being one of the larger rainfall events associated with the export of 75% of the sediment totals. In the Luquillo Mountains, this rainfall threshold is approximately 100 mm, and it is about 50 mm in the lower elevation rain records (Mayaguez and Manati) considered here. Although there is no fixed upper limit to the amount of rain that can fall within a 24-hour period, there are no records that it has exceeded 600 mm in Puerto Rico. The largest recorded flood (5–10 October 1970; Haire 1972) in Puerto Rico exhibited a maximum 1-day total of about 460 mm (5-day total of 970 mm), and the 2-day total for Hurricane Bertha (1996) in El Verde was 540 mm.

With respect to soil erosion and sediment deposition, conditions generally do not return to a previous state. The biota in the sediment source areas, the watershed itself, may persist (or not) in the remaining soil resource. River channels and coastal systems may be cleared of their sediments by subsequent water flow, but a persistent biological signal (perhaps best expressed as long-term variation in coral growth rates) may remain. In the very long term, geologic uplift and pedogenesis may keep pace with soil erosion—or it may not.

Table 8.8 Projection of the effects of adding one extreme sediment export event (defined as one of the class that produces 75% of the total export recorded) and of doubling the annual number of such events on the average annual sediment export (t ha^{-1} yr^{-1}) in three Puerto Rico watersheds

	Watersheds		
	Loiza (14)	Rosario (11)	Mameyes (6)
Case 1. All data[a]			
Average export (t ha^{-1} yr^{-1})	14.2	14.8	3.2
Average export w/one	17.4	17.7	3.8
extreme event added (t ha^{-1} yr^{-1})	(+23%)	(+20%)	(+19%)
Average export with	25.8	24.7	6.2
extremes doubled (t ha^{-1} yr^{-1})	(+82%)	(+67%)	(+95%)
Case 2. Hurricane Georges excluded[b]			
Average export (t ha^{-1} yr^{-1})	13.2	7.9	2.6
Average export w/one	16.3	9.4	3.0
extreme event added (t ha^{-1} yr^{-1})	(+24%)	(+20%)	(+13%)
Average export with	24.4	13.2	4.2
extremes doubled (t ha^{-1} yr^{-1})	(+85%)	(+67%)	(+61%)

[a] Analyses performed on all the total data sets.

[b] Analyses excluded Hurricane Georges (21–22 September 1998) sediment fluxes.

There is weak evidence, at best, that climatic cycles (ENSO and NAO) influence the occurrence of extreme rainfall events in Puerto Rico. One of the century-long daily rain records examined here (Mayaguez) may indicate an increase over the last several decades, and thus may merit further study. Sediment export records for other LTER sites (and elsewhere) also merit examination with respect to climate cycles and decadal (or longer) trends. The hypothetical analysis performed here on possible increases in extreme rainfall events is intended to highlight the high sensitivity of the sediment export process to these events. Doubling the number of events would not cause annual rainfall totals to fall outside the historic range of variation, but it could increase sediment export to unusually high levels.

Finally, even though hurricanes and their precursor stages (tropical depressions and tropical storms) constitute a portion of the extreme rainfall events, they are by no means the majority. If either natural climate variability or anthropogenic effects increase the number of extreme rainfall events, this could occur with or without changes in hurricane frequency or intensity.

Acknowledgments This research was supported by the U.S. National Science Foundation Long Term Ecological Research Program (BSR-8718396, BSR-8811902 and DEB-9411973) and a NASA Institutional Research Award (NAGW-4059) to the University of Puerto Rico.

References

Birdsey, R. A., and P. L. Weaver. 1987. Forest area trends in Puerto Rico. *U.S. Department of Agriculture Forest Service Research Note* SO-331.

Cruise, J. F., and R. L. Miller. 1994. Hydrologic modeling of land processes in Puerto Rico using remotely sensed data. *Water Resources Bulletin* 30:419–428.

Del Mar Lopez, T., T. M. Aide, and F. N. Scatena. 1998. The effect of land use on soil erosion in the Guadiana watershed in Puerto Rico. *Caribbean Journal of Science* 34:298–307.

Diaz, P. L., Z. Aquino, C. Figueroa-Alamo, R. J. Vachier, and A. V. Sanchez. 1984–1998. Water Resources Data for Puerto Rico and the U.S. Virgin Islands, U.S. Geological Survey, San Juan, Puerto Rico.

Gellis, A. 1993. The effect of Hurricane Hugo on suspended-sediment loads in the Lago Loíza basin, Puerto Rico. *Earth Surface Process and Landforms* 18:505–517.

Greenland, D. 1999. ENSO-related phenomena at long-term ecological research sites. *Physical Geography* 20:491–507.

Greenland, D., and T. G. F. Kittel. 2002. Temporal variability of climate at the U.S. Long-term ecological research (LTER) sites. *Climate Research* 19:213–231.

Haire, W. J. 1972. Flood of October 5–10, 1970 in Puerto Rico. *U.S. Geological Survey Water Resources Bulletin* 12. San Juan, Puerto Rico.

Hurrell, J. W. 1995. Decadal trends in NAO, regional temperatures and precipitation. *Science* 269:676–679.

Lal, R. 1990. *Soil erosion in the topics: Principles and management.* McGraw-Hill, New York.

Lal, R. 1994. Soil erosion by wind and water: problems and prospects. In R. Lal, editor. *Soil erosion research methods.* Soil and Water Conservation Society, Ankeny, Iowa.

Larsen, M. C., P. D. Collar, and R. F. Stallard. 1993. Research plan for the investigation of water, energy, and biogeochemical budgets in the Luquillo Mountains, Puerto Rico. U.S. Geological Survey Open-File Report 92-150.

Larsen, M. C., and A. Simon. 1993. Rainfall intensity-duration threshold for landslides in a humid tropical environment, Puerto Rico. *Geografiska Annaler* 75A:13–21.

Larsen, M. C., A. J. Torres-Sánchez, and I. M. Concepción. 1999. Slopewash, surface runoff, and fine-litter transport in forest and landslide scars in humid-tropical steeplands, Luquillo Experimental Forest, Puerto Rico. *Earth Surface Processes and Landforms* 24:481–506.

Li , Y. H. 1976. Denudation of Taiwan island since the Pliocene epoch. *Geology* 4:105–107.

Malmgren, A., A. Winter, and D. Chen. 1997. El Niño–southern oscillation and North Atlantic oscillation control of climate in Puerto Rico. *Journal of Climate* 11:2713–2718.

Meade, R. H., and R. S. Parker. 1984. Sediment in rivers of the United States. *National water summary.* United States Geological Survey, Reston, Va.

Milliman, J. D., and J. P. M. Syvitski. 1992. Geomorphic/tectonic control of sediment discharge to the ocean: The importance of small mountainous rivers. *Journal of Geology* 100:525–544.

Morris, G. L., and J. Fan. 1998. *Reservoir sedimentation handbook: Design and management of dams, reservoirs and watersheds for sustainable use.* McGraw-Hill, New York.

Newcombe, C. P., and D. D. MacDonald. 1991. Effects of suspended sediments on aquatic ecosystems. *North American Journal of Fisheries Management* 11:72–82.

Rogers, C. S. 1990. Response of coral reefs and reef organisms to sedimentation. *Marine Ecology Progress Series* 62:185–202.

Schaefer, D. A., W. H. McDowell, F. N. Scatena, and C. E. Asbury. 2000. The effects of hurricane disturbance on long-term stream water concentrations in eight tropical forest watersheds of the Luquillo Experimental Forest, Puerto Rico. *Journal of Tropical Ecology* 16:189–207.

Soler-López, L. R. 2001. Sedimentation survey results of the principal water-supply reservoirs of Puerto Rico. In W. F. Sylva, editor. *Proceedings of the sixth Caribbean Islands Water Resources Congress.* Mayagüez, Puerto Rico, February 2001.

Trenberth, K. E. 1984. Signal versus noise in the southern oscillation. *Monthly Weather Review* 112:326–332.

Trimble, S. W. 1975. Denudation studies: Can we assume steady state? *Science* 188:1207–1208.

Vanoni, V. A. 1975. *Sedimentation engineering.* American Society of Civil Engineers, New York.

Walker, G. T., and E. W. Bliss. 1932. The North Atlantic oscillation. *Memoirs of the Royal Meteorological Society* 44:53–83.

Walker, L. R. 1991. Tree damage and recovery from Hurricane Hugo in the Luquillo Experimental Forest, Puerto Rico. *Biotropica* 23:379–385.

Winter, A., H. Ishioroshi, T. Watanabe, T. Oba, and J. Christy. 2000. Caribbean Sea surface temperatures: Two-to-three degrees cooler than present during the Little Ice Age. *Geophysical Research Letters* 27:3365–3368.

9

Climate Variability and Ecological Response of the Marine Ecosystem in the Western Antarctic Peninsula (WAP) Region

Raymond C. Smith

William R. Fraser

Sharon E. Stammerjohn

Introduction

The Antarctic Peninsula, a relatively long, narrow extension of the Antarctic continent, defines a strong climatic gradient between the cold, dry continental regime to its south and the warm, moist maritime regime to its north. The potential for these contrasting climate regimes to shift in dominance from season to season and year to year creates a highly variable environment that is sensitive to climate perturbation. Consequently, long-term studies in the western Antarctic Peninsula (WAP) region, which is the location of the Palmer LTER (figure 9.1), provide the opportunity to observe how climate-driven variability in the physical environment is related to changes in the marine ecosystem (Ross et al. 1996; Smith et al. 1996; Smith et al. 1999).

This is a sea ice–dominated ecosystem where the annual advance and retreat of the sea ice is a major physical determinant of spatial and temporal change in its structure and function, from total annual primary production to the breeding success and survival of seabirds. Mounting evidence suggests that the earth is experiencing a period of rapid climate change, and air temperature records from the last half century confirm a statistically significant warming trend within the WAP during the past half century (King 1994; King and Harangozo 1998; Marshall and King 1998; Ross et al. 1996; Sansom 1989; Smith et al. 1996; Stark 1994; van den Broeke 1998; Weatherly et al. 1991). Air temperature–sea ice linkages appear to be very strong in the WAP region (Jacka 1990; Jacka and Budd 1991; King 1994; Smith et al. 1996; Weatherly et al. 1991), and a statistically significant anticorrelation between air temperatures and sea ice extent has been observed for this region. Consistent with this strong coupling, sea ice extent in the WAP area has trended down

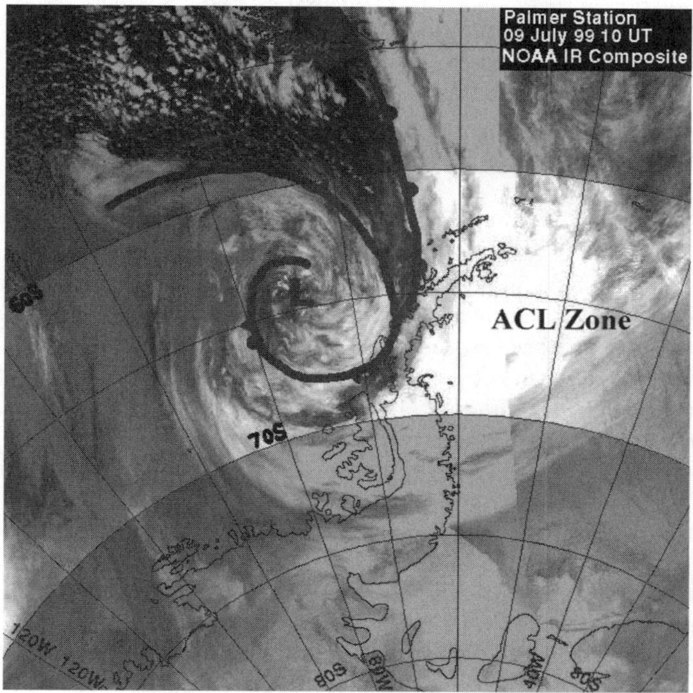

Figure 9.1 Satellite (NOAA infrared composite) image of the Antarctic Peninsula. A low-pressure system (with corresponding warm and cold fronts illustrated) is to the west of the peninsula. Palmer Station (64°41′ S, 64°03′ W), on Anvers Island is positioned roughly under the apex of the schematic outline of the frontal system. Also illustrated as an overlay on the image is the Antarctic Convergence Line (ACL), the mean position of the circumpolar low-pressure trough surrounding Antarctica. The ACL undergoes a semiannual cycle, whereby, on average, it is nearest the continent when the ice edge is near its extreme equatorward (spring) or poleward (autumn) position. The ACL, on average, is farther equatorward when the ice edge is at an intermediate position (winter, summer) (van Loon, 1967). The Palmer LTER regional sampling grid is along the western Antarctic Peninsula (WAP).

during this period of satellite observations, and the sea ice season has shortened. In addition, both air temperature and sea ice have been shown to be significantly correlated with the Southern Oscillation Index (SOI), which suggests possible linkages among sea ice, cyclonic activity, and global teleconnections.

Ecological responses to this climate variability are evident at all trophic levels, but are most clearly seen in a shift in the population size and distribution of penguin species with different affinities to sea ice. In the text that follows, we update both air temperature and sea ice records for the WAP to demonstrate their continued statistical significance and to place the related ecological and environmental observations into a long-term context that shows how the WAP region is responding to an increasing maritime, as opposed to continental, influence. We further show the correlation of these environmental variables to the Southern Oscillation Index (SOI), address issues of seasonal timing, and discuss the broad implications of these changes to the ecosystem.

Climate and Ecological Data

Surface Air Temperature

The British Antarctic Survey (BAS) has a long and distinguished history of scientific research in Antarctica, and their meteorological observations at Faraday/Vernadsky Station have been especially useful to WAP research because of their length (5+ decades), consistency, and quality control. In this chapter, we update and augment earlier studies (Smith et al., 1996) with data from the 1990s. Figure 9.2 shows the Faraday/Vernadsky annual average air temperatures from 1945 to 2000 (N = 56). The solid line is the least-squares regression line, which shows a statistically significant warming trend over the last 56 years. The dotted lines indicate the ± 1 standard deviation (s.d.) from the regression line and has been used as a designator for defining "high" (above 1 s.d.) or "low" (below 1 s.d.) temperature years.

After accounting for serial correlation present in this 56-year record (for method, see Smith et al., 1996), we found the trend to be statistically significant at a >99% confidence level. These annual results are further supported by a monthly and seasonal analysis (see table 1 in Smith et al., 1996) showing that the warming trend in Faraday/Vernadsky air temperatures is strongest during the midwinter months and peaks in June at 0.11°C/year. This represents about a 6°C increase in June temperatures over the 56-year record. Spring and summer trends, however, are not as pronounced. The record from Rothera (further south on the WAP) shows a strong temporal coherence (King 1994; Smith et al., 1996) to Faraday/Vernadsky, displaying similar trends but with mean annual temperatures that average a few degrees cooler. This evidence suggests there is a north-south temperature gradient along the WAP and that observed trends are coherent throughout the region.

The annual progression of temperatures and the amount of variability associated with those temperatures have also changed over the last half century. Figure 9.2 shows that the last two decades (1980s and 1990s) were warmer than the previous several decades. The seasonal variability associated with this change is illustrated in figure 9.3a, where we have plotted the annual curves of monthly mean air temperature for Faraday/Vernadsky for the following periods: the full instrument record, 3/44 to 12/99 (solid); the period 1/78–12/89 (hereafter called the 1980s, dotted); and the period 1/90–12/99 (the 1990s, dashed). The curves in figure 9.3a also illustrate that the largest temperature changes have occurred in winter (Jun–Aug), in contrast to less change in spring and early summer (Sept–Dec).

Figure 9.3c shows the standard deviations of the monthly mean surface air temperatures shown in figure 9.3a. Several observations in air temperature variability are apparent. First, there is significantly higher variation from May through September during all periods. Second, during the summer, when ice-free conditions are increasingly typical and maritime conditions prevail, there is relatively lower vari-

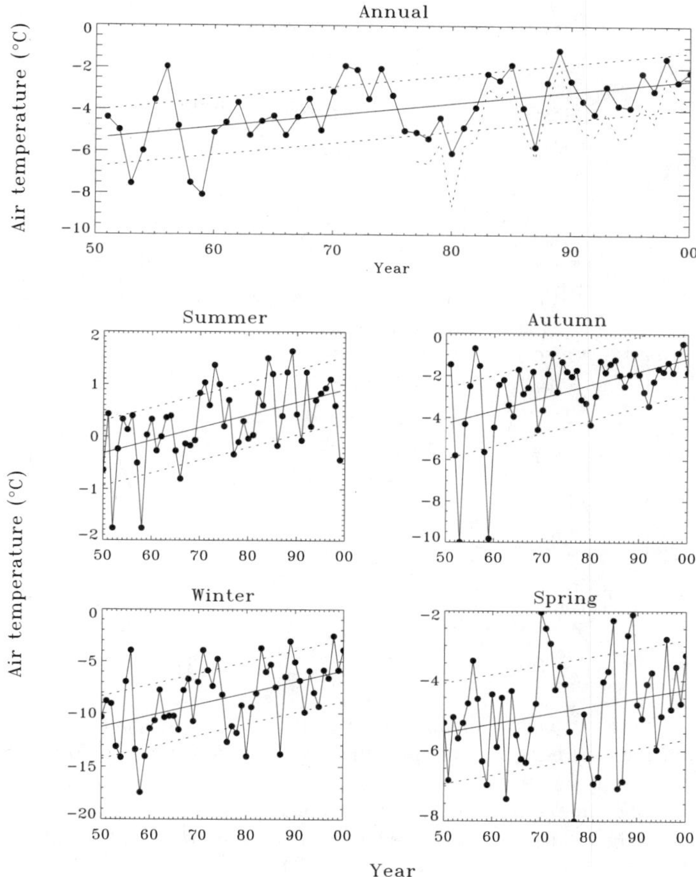

Figure 9.2 Faraday/Vernadsky (65° 15' S, 64° 15' W) annual average air temperatures from 1945 to 2000 (N=56). The solid line is the least-squares regression line with a gradient of 0.052°C/year, and the dotted lines indicate ± 1 standard deviation from this line. A linear regression model shows the warming trend over this period to be significant at greater than the 99% confidence level. The shorter-period Rothera (67° 34' S, 68° 08' W) annual temperature is plotted as a dotted line. Temperature data for Faraday/Vernadsky and Rothera kindly supplied by the British Antarctic Survey.

ability in air temperatures. Third, the high midwinter (July) variation during the 1980s is caused by greater extremes between warm and cold winters. These changes in the annual progression of temperature and the amount of variability associated with those temperatures suggests a climate shift, in which continental influences are giving way to increasing maritime influences along the WAP. Smith and Stammerjohn (2001) have detailed why these observations are consistent with the characteristics of a maritime environment in which temperatures are moderated by the open ocean.

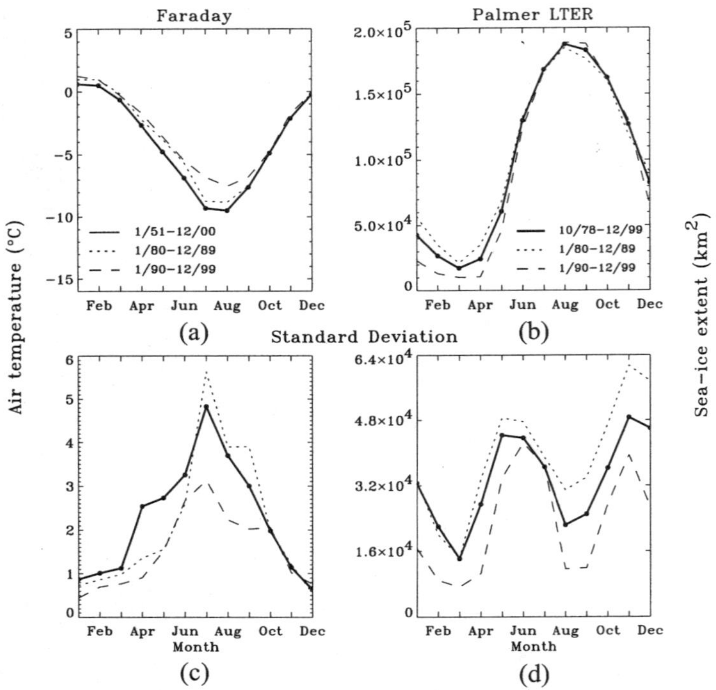

Figure 9.3 (a) Annual curves of monthly mean surface air temperatures for Faraday/ Vernadsky for the total period of the instrument record (3/44–12/99, bold line with solid dots), the decade of the 1980s (1/80–12/89, dotted line), and the decade of the 1990s (1/90–12/99, dashed line). (b) Annual curves of monthly mean sea ice extent for the Palmer LTER region for the full period of satellite passive microwave data (10/78 –12/99, bold line with solid dots), the decade of the 1980s (10/80–12/89, dotted line), and the decade of the 1990s (1/90–12/99, dashed line). Sea ice data supplied by the National Snow and Ice Data Center. (c) Standard deviations of the monthly mean surface air temperatures for the same periods shown in part (a). (d) Standard deviations of the monthly mean sea ice extent for the periods shown in part (b).

Although the mechanistic processes linked to these WAP temperature trends are still being debated, the role of the mean position of the circumpolar atmospheric low-pressure trough (i.e., the Atmospheric Convergence Line (ACL), figure. 9.1) bears close inspection as a possible causal mechanism. The Antarctic Peninsula is the only area in Antarctica where the ACL crosses land. The seasonal cycle displayed in temperature, pressure, wind, and precipitation (Schwerdtfeger 1984; van Loon 1967) is linked to both increased cyclonic activity and a southward shift of approximately 10° of latitude of the ACL during spring and autumn. The relative position of the ACL influences not only the semiannual cycle of climate variables but also the timing and distribution of sea ice. Van Loon suggested that this seasonal temperature cycle is associated with enhanced meridional flow from middle to high latitudes during winter.

Indeed, more recent work by Meehl (1991) confirms that transient eddy heat flux likely contributes to this seasonal cycle in the Antarctic coastal zone. King and coworkers (King 1994; King and Harangozo 1998; Marshall and King 1998) also show a strong correlation between surface air temperature and meridional sea-level pressure indexes calculated for the WAP area. Their results demonstrate that increased boundary-layer winds, flowing from the northwest sector toward the WAP, are associated with increased cyclonic activity and warm air advection from lower latitudes. The increase in surface temperatures associated with the increase in northerly winds consequently produces an environment with more maritime (warm and moist) characteristics, as opposed to the continental environment (cold and dry) that would result from the effects of southerly winds and colder temperatures. Stammerjohn et al. (2003) have discussed in detail the responses of sea ice and drift dynamics to synoptic forcing in the WAP region and suggested that, with longer term shifts in the mean position of the ACL, these synoptic-scale systems may provide a mechanism for longer term climate variability.

Sea Ice

Also shown in figure 9.3 are the mean annual cycles of sea ice extent (figure 9.3b) and the standard deviations of the monthly means (figure 9.3d). Means for the full period of the passive microwave satellite record (1978–1999, solid) and for the 1980s (dotted) and the 1990s (dashed) are included. Methods we used when working with passive microwave satellite data are described in Stammerjohn and Smith (1996) and Smith et al. (1998). Several observations can be made with respect to figure 9.3. First, the winter seasonal cycle of air temperature (figure 9.3a) is inversely related to the winter seasonal cycle of sea ice extent (figure 9.3b), but the summer sea ice extent minimum lags the summer air temperature maximum by 2 to 3 months. Second, summer (Jan–Mar) and fall (Apr–May) sea ice extent in the 1990s is below that for the 1980s. Third, spring (Sept–Dec) also follows this pattern, with the 1990s showing less sea ice on average than the 1980s. Fourth, the earlier retreat and later advance of sea ice in the 1990s (as compared with the 1980s) translates into a shorter sea ice season by roughly two weeks. The variance also changed (figure 9.3d); the 1980s, when contrasted to the 1990s, have a higher variance because of the seasonal persistence of anomalies during April to September.

Within the period of satellite multichannel microwave records (1978 to present), anomalies in WAP air temperature and sea ice extent (King 1994; Smith et al. 1996; Weatherly et al. 1991) have been shown to be significantly anticorrelated. Figure 9.4a shows monthly standard deviates of Faraday/Vernadsky air temperature versus Palmer LTER sea ice extent smoothed with a 5-month running average. Standard deviates are the normalized anomalies determined by dividing the anomaly (for the month and year in question) by the standard deviation of the anomaly (for the month in question). However complex the mechanisms linking air temperature and sea ice trends are, these data show that since 1978 these two parameters behave almost as mirror images within the WAP. During the 1980s, when anomalies in sea ice extent showed strong persistence, so did air temperature, but during the 1990s this persistence gave way to greater month-to-month variability in both parameters.

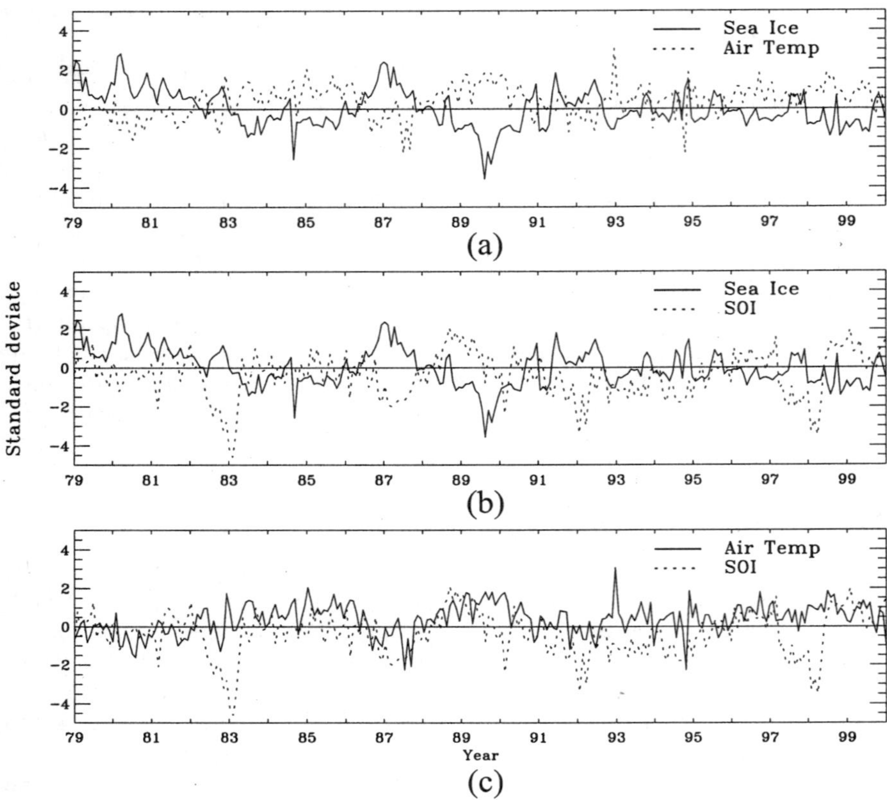

Figure 9.4 Monthly standard deviates (smoothed by a 5-month running mean) from January 1979 to December 1999. (a) Faraday/Vernadsky air temperature (dotted line) and Palmer LTER sea ice extent (solid line). (b) Palmer LTER sea ice extent (solid line) and Southern Oscillation Index (dotted). (c) Faraday/Vernadsky air temperature (solid line) and Southern Oscillation Index (dotted). The SOI data were obtained digitally (http://www.cpc.ncep. noaa.gov/data/indices/soi) from the Climate Prediction Center (Department of Commerce, NOAA).

As expected from the relationships discussed previously, but in contrast to the Southern Ocean as a whole, the annual mean sea ice extent has trended down in the WAP region (figures 9.5). Here the mean annual sea ice extent for the WAP region (a) and the Southern Ocean (inset) are presented along with mean seasonal data for summer (b), autumn (c), winter (d), and spring (e). The annual trend is due mostly to the decreasing trend in summer sea ice, which was also inferred from figure 9.3. Given the relatively short satellite record and high interannual variability, these trends are not statistically significant. However, the trends are suggestive, and less summer sea ice is consistent with increased maritime influence in the WAP region as noted previously. During the 1980s over half the annual means are greater than ± 1 s.d. from the regression line, in contrast to the 1990s when all the annual means

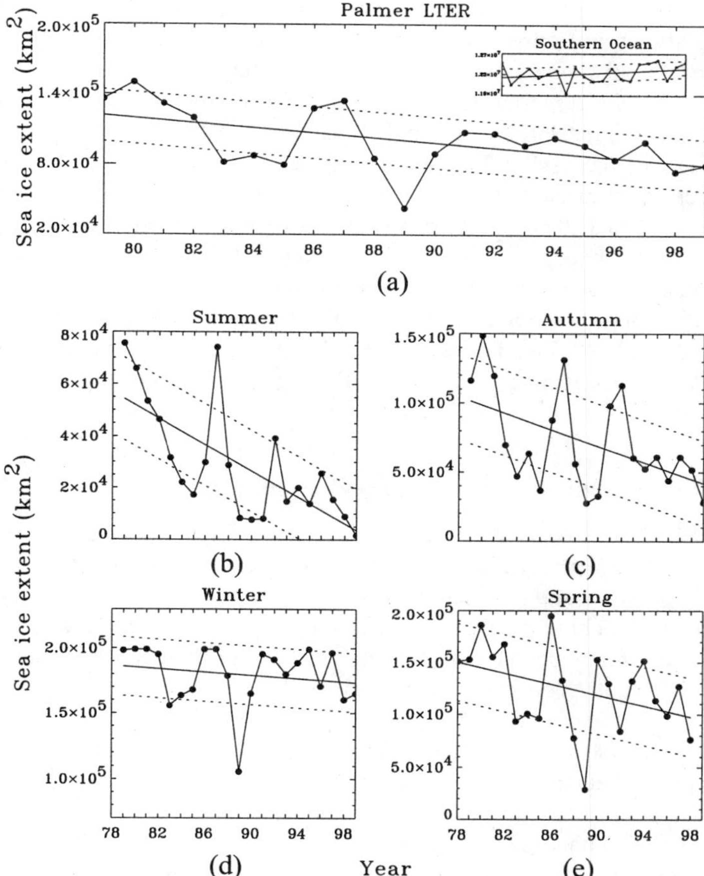

Figure 9.5 Mean annual sea ice extent for the Southern Ocean (insert) and the Palmer LTER region (a). See Stammerjohn and Smith (1996) for details on the satellite data used. Mean annual sea ice extent for the Palmer LTER region for summer (b), autumn (c), winter (d), and spring (e) are shown to illustrate that the annual trend in the Palmer LTER region is due mostly to the decreasing sea ice trend during summer.

are within ± 1 s.d. Also, during the earlier decade the periods of anomalously high (1979–1981 and 1986–1987) and low (1983–1985 and 1988–1990) sea ice extents stand out clearly. We expect the ecosystem to respond to these anomalies.

Links to the Southern Oscillation Index

Monthly standard deviates of Palmer LTER sea ice extent and Faraday/Vernadsky air temperature versus the Southern Oscillation Index (SOI) (which is determined by the standardized sea level pressure difference between Tahiti and Darwin, Australia) are shown in figures 9.4b and c, respectively. Figure 9.4b shows an anticorrelation between Palmer LTER sea ice extent and SOI. As expected based on the relationship shown in figure 9.4a, figure 9.4c shows a correlation between Faraday/Vernadsky air temperature and SOI. Smith and coworkers (1996) have discussed this relationship previously and we include an updated figure here to show that the relationships continue to hold throughout the 1990s. These relationships support the idea of possible linkages among sea ice, cyclonic activity and global teleconnections (Carleton 1988; Mo and White 1985; van Loon and Shea 1985; van Loon and Shea 1987; White and Peterson 1996; White et al. 1998; Yuan and Martinson 2000). In particular, the semiannual oscillation (SAO, the twice-yearly contraction and expansion of the atmospheric low-pressure trough around Antarctica) is an important component of the Southern Hemisphere climate regime and has been shown to be linked to variability in air temperature and cyclonic activity in the WAP and elsewhere in the Antarctic (Meehl 1991; van den Broeke 2000; van Loon 1967).

Pygoscelid Penguins, Upper Trophic Level Predators

High variability and long-term change constitute the setting in which this polar marine ecosystem has evolved. Solar radiation, atmospheric and oceanic circulation, and air temperature and sea ice cover are the physical forcing mechanisms that drive variability in biological processes at all trophic levels. The extreme seasonality of these forcing mechanisms in conjunction with the seasonal timing of ecologically important events in the life histories of key species from each trophic level provides a conceptual model for understanding WAP trophic interactions (Smith et al. 1995, figure 4). Figure 9.6 presents annual time lines of selected physical and biological components in the WAP region with emphasis on the variability of sea ice and the life histories of three sympatric, congeneric penguins, the Adélie (*Pygoscelis adeliae*), chinstrap (*P. antarctica*), and gentoo (*P. papua*). Adélie penguins are obligate inhabitants of the winter pack ice, whereas chinstraps and gentoos are almost exclusively associated with ice-free Antarctic and sub-Antarctic waters (Fraser et al. 1992). These three species are closely related and have a similar breeding cycle of courtship, egg laying, incubation, brooding, and fledging. However, as illustrated in figure 9.6, the Adélie breeding cycle begins roughly 3 weeks earlier than that of the other two species. The timing associated with these relatively fixed breeding chronologies, in association with interannual variability in sea ice cover and in the life histories of primary and secondary producers, provides the ecological context that determines penguin breeding success and recruitment.

The basis for understanding the possible causal factors associated with WAP penguin population trends originated with the hypothesis that a decrease in the number of cold years with heavy winter sea ice because of climate warming produced habitat conditions more suitable for the ice-intolerant, as opposed to the ice-dependent, species (Fraser et al. 1992). Figure 9.7 shows the changes in Adélie and chinstrap penguin populations near Palmer Station during the past two decades, and for gentoo penguins since founder colonies became established in the area during the early 1990s. These trends clearly support this ice reduction hypothesis. Chinstrap and gentoo penguins, the more ice-intolerant species, have increased,

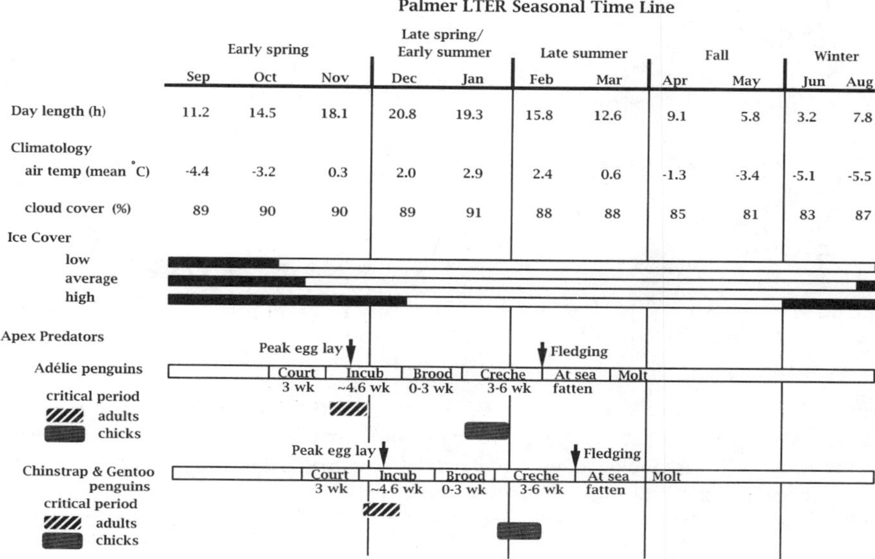

	Early spring			Late spring/ Early summer		Late summer		Fall		Winter	
	Sep	Oct	Nov	Dec	Jan	Feb	Mar	Apr	May	Jun	Aug
Day length (h)	11.2	14.5	18.1	20.8	19.3	15.8	12.6	9.1	5.8	3.2	7.8
Climatology air temp (mean °C)	-4.4	-3.2	0.3	2.0	2.9	2.4	0.6	-1.3	-3.4	-5.1	-5.5
cloud cover (%)	89	90	90	89	91	88	88	85	81	83	87

Figure 9.6 Annual time lines of selected physical and biological components in the WAP region: day length (h), mean monthly air temperature (°C), cloud cover (%), ice cover variability, Adélie penguins, Chinstrap and Gentoo penguins (include peak egg lay, critical periods for adults and chicks, incubation, brood, creech, fledging, molt periods). This schematically illustrates how the variability in sea ice cover relates to the relatively fixed breeding chronology of selected upper level predators.

whereas the ice-dependent Adélie penguins have decreased. Moreover, the causal mechanisms suggested by this hypothesis have now been implicated as key factors affecting penguin demography at a range of spatial and temporal scales in both paleoecological and demographic studies (Baroni and Orombelli 1991; Baroni and Orombelli 1994; Denton et al. 1991; Emslie 1995; Emslie et al. 1998; Fraser and Patterson 1997; Smith et al. 1999; Taylor et al. 1990). The emerging evidence is that penguin distributions are undergoing a fundamental reorganization in the WAP and other regions of Antarctica (see Fraser and Trivelpiece 1996) as the result of climatic factors that appear to influence long-term recruitment.

Discussion and Summary

Several comments can be made with respect to the air temperature and sea ice data. First, to place the more recent observations within the context of the past half century, it is important to recall that the decade of the 1990s is the warmest for the entire period of the instrument record (figure 9.2). Second, the strong inverse relationship between air temperature and sea ice extent continues to be clearly evident. Further, in contrast to earlier periods, departures from the mean during the decade

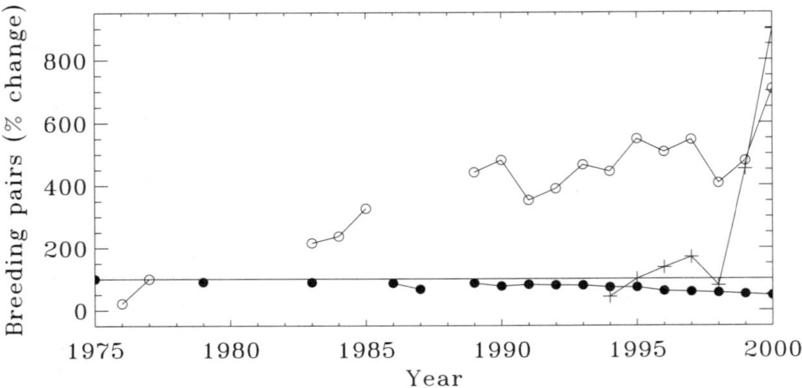

Figure 9.7 Twenty-five-year trends in Adélie and chinstrap penguin populations at Arthur Harbor (Palmer Station) and for gentoo penguins since founder colonies became established in the early 1990s. Adélie penguins (solid dots) are normalized to 100% in 1975 when the record began. Chinstrap (open circles) and gentoo (plus signs) penguins are normalized to 100% in 1977 and 1995, respectively, one year after founding colonies were established.

of the 1990s are relatively low. Third, the trends in the WAP area are such that there are fewer high sea ice years, and the seasonal progression of sea ice, although highly variable from year-to-year, is such that the average ice-free period is roughly 2 weeks longer than it was 5 decades ago. Fourth, climate warming in the Antarctic Peninsula has, in some areas, raised the mean annual temperature above the suggested climate limit (–5°C) for ice shelf stability, leading to the complete disintegration of some shelves (Skvarca et al. 1999; Vaughan and Doake 1996). The removal of large areas of this ice-related habitat illustrates the role that temperature plays in the phase transition between ice and water, which has important consequences for this marine ecosystem.

King and Harangozo (1998) have discussed the trends in climate change in the WAP and identified two possible factors as causes for the interannual variability in the temperature record: changes in atmosphere-ice-ocean interactions, and variability in maritime versus continental control on climate. The increased maritime influence during recent decades is relatively clear from the data, whereas the mechanisms underlying atmosphere-ice-ocean interactions and the causative factors involved remain to be elucidated. The variability of the Antarctic Convergence Line (ACL, figure 9.1), both semiannual and long-term, with its corresponding influence on climatic conditions in the WAP, appears to play a significant role at temporal scales that range from synoptic to long-term. We can thus hypothesize that maritime conditions are likely to become the prevailing climatic regime in the WAP region, and this, in turn, will force a restructuring of the marine ecosystem from a more polar to a more maritime state.

Climate variability along the peninsula holds the potential to cascade through the ecosystem through a variety of mechanisms. Recent work (Dierssen et al. 2002) has shown the potential influence of glacial meltwater, as distinct from the usual meltwater from sea ice, on the hydrography of the WAP ecosystem. Glacial meltwater freshens and warms coastal surface waters, leading to enhanced water column stability and increased primary productivity. The influence of glacial meltwater on the space-time variability of the system is currently under investigation, but the potential of this mechanism to act as a catalyst influencing both the magnitude and timing of primary production and to cascade this influence to higher trophic levels is clear. Further, the amount of meltwater may have important secondary effects on the ecosystem by influencing the timing of sea ice formation the following fall.

The life histories of various polar marine species are synchronized with the seasonality of the sea ice (Ross et al. 1996; Smith et al. 1995). For example, Ackley and Sullivan (1994) have proposed a conceptual model of the seasonal cycle of sea ice with the following characteristics: (1) autumn formation entrains phytoplankton as a seed population within the sea ice matrix; (2) entrained sea ice communities grow and develop during winter as the sea ice evolves; and (3) sea ice decay in the spring releases a potential bloom inoculum of particulate organic matter into the water column. Palmer LTER multiyear observations on phytoplankton biomass and production variability support this hypothesis because several factors controlling abundance and distribution of phytoplankton biomass, often dominated by diatom blooms, have been shown to be modulated by sea ice (Smith et al. 1998; Smith and Stammerjohn 2001). Further up the food web, the Antarctic krill (*Euphausia superba* Dana), a major herbivore responsible for the transfer of energy within the ecosystem, has a life history that is closely coupled to sea ice (Quetin et al. 1996). It has been hypothesized that the wintertime survival of larval krill depends on sea ice to provide a habitat and an algal food source. Further, recent evidence supports the hypothesis that maximum krill growth rates are only possible during diatom blooms and that year-class success in Antarctic krill is limited by both food quantity and quality (Ross et al. 2000). This suggests strong linkages among sea ice, phytoplankton, and krill. Continued significant warming will reduce the dominance of sea ice in the WAP ecosystem with subsequent changes and/or shifts in primary and secondary production.

For higher trophic predators such as penguins, variability in sea ice concentrations can affect foraging ecology directly through its effects on krill recruitment and abundance (Fraser and Hofmann 2003) or indirectly through habitat changes that mediate the availability of krill (Fraser et al. 1992; Fraser and Trivelpiece 1996; Fraser and Patterson 1997). A conceptual model that is roughly analogous to the intermediate disturbance model (Connell 1978) was proposed by Fraser and Trivelpiece (1996) and Smith et al. (1999) to account for the direction of change in Adélie penguin populations in the Ross Sea and WAP regions in relation to climate warming and a decline in the frequency of heavy sea ice years. Penguin breeding colonies are located on coastal sites that offer an optimal combination of foraging and nesting habitats. Such sites, we are now beginning to understand, are associated with environmental conditions that ensure some level of predictability in the availability of prey at ecological time scales, here associated with the presence or absence of sea ice via its controlling effects on primary and secondary production.

Although the mechanisms that control these conditions are not fully understood, significant progress has recently been made in understanding cause and effect between some of the basic linkages circumscribed by weather, ice, primary and secondary production, and predator population responses. This has provided a better perspective on the magnitude of the changes induced by the current warming trend and on the linearity or nonlinearity of the associated processes. A most interesting observation based on the paleoecological record is that the presence of chinstrap and gentoo penguins in the Palmer Station area is unprecedented in the 600-year fossil record, which is entirely dominated by Adélie penguin remains (Emslie et al. 1998). This pattern stands in sharp contrast to trends evident 250 km north of the Palmer area, where the relative dominance of Adélie and chinstrap penguins has changed cyclically in response to multicentury cooling and warming periods (Emslie 1995). That chinstrap and gentoo penguins have invaded the Palmer region thus seems to affirm the unusual nature of this twentieth-century WAP warming event. However, that founder colonies of these species have increased so dramatically—and, conversely, that Adélie penguins have decreased so substantially—in roughly 25 years (figure 9.7) strongly suggests that causal processes are more linear than nonlinear, involve fewer potentially diffusive links, and may impinge directly on key aspects of the life history of penguins and/or their prey.

Evidence supporting this perspective stems from recent studies by Fraser and Hofmann (2003), who analyzed changes over a period of 30 years in the diets of Adélie penguins. Their results show that there is a direct, causal relationship between variability in ice cover and krill recruitment, krill abundance, and predator foraging ecology. Of particular relevance is the observation that time lags between sea ice formation and changes in the responses of Adélies foraging on krill are short, less than 12 months during some years, and can simultaneously affect parameters such as chick fledgling weight that have longer term consequences to recruitment (Salihoglu et al. 2001). Moreover, there is some evidence that the coupling strength between these interactions shows a strong 4–5 year periodicity. This periodicity is consistent with the periodicity of the Antarctic Circumpolar Wave (ACW) (White and Peterson 1996) and is coherent with the development of cold temperatures and heavy ice years in the WAP. Possible teleconnections between the ACW and the SOI were previously discussed.

Several studies (Fraser and Hofmann 2003; Smith et al. 1996; White et al. 1998; Yuan and Martinson 2000) strongly suggest that ENSO-type events govern key biophysical interactions in the WAP that affect all trophic levels, but the unprecedented characteristics of the current warming trend make it difficult to envisage an "end scenario" to these climate-induced ecosystem changes. In light of present sea ice trends, however, it is not inconceivable that Adélie penguins will continue to decline in the Palmer Station area and that the locus of their distribution will be forced farther south along the WAP, while chinstrap and gentoo penguins emerge as the dominant top predators. The fossil record already supports such a scenario at more northern sites along the WAP, where there is also evidence that squid and fish replaced krill as the dominant component in penguin diets as the climate warmed (Emslie 1995; Emslie et al. 1998). This would imply that, at least within the confines of some spatial and temporal scales, climate-induced ecological effects were complete (defined as one food web replacing another) before new climate events restored the marine ecosystem to its previous state. This scenario thus argues in favor of cyclical as opposed to absolute changes in WAP ecosystems in response to climate change. The fault with this argument, of course, is that previous changes in climate were in all probability unrelated to anthropogenic forcing.

Acknowledgment   This work was supported by NSF Office of Polar Programs grants OPP-9632763 (RCS and WRF) and OPP-9505596 (WRF). This is Palmer LTER contribution # 214.

References

Ackley, S. F., and C. W. Sullivan. 1994. Physical controls on the development and characteristics of Antarctic sea ice biological communities—A review and synthesis. *Deep-Sea Research* I 41: 1583–1604.

Baroni, C., and G. Orombelli. 1991. Holocene raised beaches at Terra Nova Bay, Victoria Land, Antarctica. *Quaternary Research* 36: 157–177.

Baroni, C., and G. Orombelli. 1994. Abandoned penguin rookeries as Holocene paleoclimatic indicators in Antarctica. *Geology* 22: 23–26.

Carleton, A. M. 1988. Sea ice-atmosphere signal of the Southern Oscillation in the Weddell Sea, Antarctica. *Journal of Climate* 1: 379–388.

Connell, J. H. 1978. Diversity in tropical rainforests and coral reefs. *Science* 199: 1302–1310.

Denton, G. H., J. G. Bockheim, S. C. Wilson, and M. Stuiver. 1991. Late Wisconsin and early Holocene glacial history, inner Ross embayment, Antarctica. Pages 55–86 in R. A. Bindschadler, editor. *West Antarctic Ice Sheet Initiative*. NASA, Washington, D.C.

Dierssen, H. M., R. C. Smith, and M. Vernet. 2002. Glacial meltwater dynamics in coastal waters West of the Antarctic Peninsula. *Proceedings of the National Academy of Science* 99: 1790–1795.

Emslie, S. D. 1995. Age and taphonomy of abandoned penguin rookeries in the Antarctic peninsula. *Polar Record* 31: 409–418.

Emslie, S. D., W. R. Fraser, R. C. Smith, and W. O. Walker. 1998. Abandoned penguin colonies and environmental change in the Palmer Station region, Anvers Island, Antarctic Peninsula. *Antarctic Science* 10: 255–266.

Fraser, W. R., and D. L. Patterson. 1997. Human disturbance and long-term changes in Adélie penguin populations: A natural experiment at Palmer Station, Antarctic Peninsula. Pages 445–452. in B. Battaglia, J. Valencia, and D. W. H. Walton, editors. *Antarctic Communities, Species, Structure and Survival*. Cambridge University Press, New York.

Fraser, W. R., and W. Z. Trivelpiece. 1996. Factors controlling the distribution of seabirds: Winter-summer heterogeneity in the distribution of Adélie penguin populations. Pages 257–272 in R. M. Ross, E. E. Hofmann, and L. B. Quetin, editors. *Foundations for Ecological Research West of the Antarctic Peninsula*. American Geophysical Union, Washington, D.C. (Antarctic Research Series, V. 70).

Fraser, W. R., W. Z. Trivelpiece, D. G. Ainley, and S. G. Trivelpiece. 1992. Increases in Antarctic penguin populations: Reduced competition with whales or a loss of sea ice due to environmental warming? *Polar Biology* 11: 525–531.

Fraser, W. R., and E. E. Hofmann. 2003. Krill-sea ice interactions, part I: A predator's perspective on causal links between climate change, physical forcing and ecosystem response. *Marine Ecology Progress Series*, in press.

Jacka, T. H. 1990. Antarctic and Southern Ocean sea-ice and climate trends. *Annals of Glaciology* 14: 127–130.

Jacka, T. H., and W. F. Budd. 1991. Detection of temperature and sea ice extent changes in the Antarctic and Southern Ocean. Pages 63–70 in G. Weller, C. L. Wilson, and B. A. B. Severin, editors, *International Conference on the Role of the Polar Regions in Global Change*. Proceedings of a conference held June 11-15, 1990 at the University of Alaska Fairbanks. Vol. I., Geophysical Institute/Center for Global Change and Arctic System Research, Fairbanks, Alaska, University of Alaska.

King, J. C. 1994. Recent climate variability in the vicinity of the Antarctic Peninsula. *International Journal of Climatology* 14: 357–369.

King, J. C., and S. A. Harangozo. 1998. Climate change in the western Antarctic Peninsula since 1945: Observations and possible causes. *Annals of Glaciology* 27: 571–575.

Marshall, G. J., and J. C. King. 1998. Southern Hemisphere circulation anomalies associated with extreme Antarctic Peninsula winter temperatures. *Geophysical Research Letters* 25: 2437–2440.

Meehl, G. A. 1991. A reexamination of the mechanism of the semiannual oscillation in the Southern Hemisphere. *Journal of Climate* 4: 911–926.

Mo, K. C., and G. H. White. 1985. Teleconnections in the Southern Hemisphere. *Monthly Weather Review* 113: 22–37.

Quetin, L. B., R. M. Ross, T. K. Fraser, and K. L. Haberman. 1996. Factors affecting distribution and abundance of zooplankton, with an emphasis on Antarctic krill, *Euphausia superba*. Pages 357–371 in R. M. Ross, E. E. Hofmann, and L. B. Quetin, editors, *Foundations for Ecological Research west of the Antarctic Peninsula*. American Geophysical Union, Washington, D.C. (Antarctic Research Series, V. 70).

Ross, R. M., E. E. Hofmann, and L. B. Quetin, editors. 1996. *Foundations for Ecological Research West of the Antarctic Peninsula*. American Geophysical Union, Washington, D.C. (Antarctic Research Series, V. 70).

Ross, R. M., L. B. Quetin, K. S. Baker, M. Vernet, and R. C. Smith. 2000. Growth limitation in young *Euphausia superba* under field conditions. *Limnology and Oceanography* 45: 31–43.

Salihoglu, B., W. R. Fraser, and E. E. Hofmann. 2001. Factors affecting fledging weight of Adelie penguin (*Pygoscelis adeliae*) chicks: A modeling study. *Polar Biology* 24: 328–337.

Sansom, J. 1989. Antarctic surface temperature time series. *Journal of Climate* 2: 1164–1172.

Schwerdtfeger, W. 1984. *Weather and Climate of the Antarctic*. Elsevier Science Publishing Company, New York.

Skvarca, P., W. Rack, H. Rott, and T. I. Y. Donangelo. 1999. Climatic trend and the retreat and disintegration of ice shelves on the Antarctic Peninsula: An overview. *Polar Research* 18: 151–157.

Smith, R. C., K. S. Baker, W. R. Fraser, E. E. Hofmann, D. M. Karl, J. M. Klinck, L. B. Quetin, B. B. Prezelin, R. M. Ross, W. Z. Trivelpiece, and M. Vernet. 1995. The Palmer LTER: A long-term ecological research program at Palmer Station, Antarctica. *Oceanography* 8: 77–86.

Smith, R. C., S. E. Stammerjohn, and K. S. Baker. 1996. Surface air temperature variations in the western Antarctic peninsula region. Pages 105–121 in R. M. Ross, E. E. Hofmann, and L. B. Quetin, editors, *Foundations for Ecological Research West of the Antarctic Peninsula*. American Geophysical Union, Washington, D.C. (Antarctic Research Series, V. 70).

Smith, R. C., D. Ainley, K. Baker, E. Domack, S. Emslie, W. Fraser, J. Kennett, A. Leventer, E. Mosley-Thompson, S. Stammerjohn, and M. Vernet. 1999. Marine ecosystem sensitivity to climate change. *BioScience* 49: 393–404.

Smith, R. C., K. S. Baker, and S. E. Stammerjohn. 1998. Exploring sea ice indexes for polar ecosystem studies. *BioScience* 48: 83–93.

Smith, R. C., and S. E. Stammerjohn. 2001. Variations of surface air temperature and sea ice extent in the western Antarctic Peninsula (WAP) region. *Annals of Glaciology* 33: 493–500.

Stammerjohn, S. E., and R. C. Smith. 1996. Spatial and temporal variability of western Antarctic Peninsula sea ice coverage. Pages 81–104 in R. M. Ross, E. E. Hofmann, and L. B. Quetin, editors, *Foundations for Ecological Research West of the Antarctic Peninsula*. American Geophysical Union, Washington, D.C. (Antarctic Research Series, V. 70).

Stammerjohn, S. E., M. R. Drinkwater, R. C. Smith, and X. Liu. 2003. Ice-atmosphere interactions during sea-ice advance and retreat in the western Antarctic Peninsula region. *Journal of Geophysical Research,* in press.

Stark, P. 1994. Climatic warming in the central Antarctic Peninsula area. *Weather* 49: 215–220.

Taylor, R. H., P. R. Wilson, and B. W. Thomas. 1990. Status and trends of Adélie penguin populations in the Ross sea region. *Polar Record* 26: 293–304.

van den Broeke, M. R. 1998. The semi-annual oscillation and Antarctic climate, part 1: Influence on near surface temperatures (1957–79). *Antarctic Science* 10: 175–183.

van den Broeke, M. R. 2000. The semi-annual oscillation and Antarctic Climate. Part 4: A note on sea ice cover in the Amundsen and Bellingshausen Seas. *International Journal of Climatology* 20: 455–462.

van Loon, H. 1967. The half-yearly oscillations in middle and high southern latitudes and the coreless winter. *Journal of the Atmospheric Sciences* 24: 472–486.

van Loon, H., and D. J. Shea. 1985. The Southern Oscillation. Part IV: The precursors south of 15°S to the extremes of the oscillation. *Monthly Weather Review* 113: 2063–2074.

van Loon, H., and D. J. Shea. 1987. The Southern Oscillation. Part VI: Anomalies of sea level pressure on the Southern Hemisphere and of Pacific sea surface temperature during the development of a warm event. *Monthly Weather Review* 115: 370–379.

Vaughan, D. G., and C. S. M. Doake. 1996. Recent atmospheric warming and the retreat of ice shelves on the Antarctic Peninsula. *Nature* 379: 328–330.

Weatherly, J. W., J. E. Walsh, and H. J. Zwally. 1991. Antarctic sea ice variations and seasonal air temperature relationships. *Journal of Geophysical Research* 96: 15,119–15,130.

White, W. B., S.-C. Chen, and R. G. Peterson. 1998. The Antarctic Circumpolar Wave: A beta effect in ocean-atmosphere coupling over the Southern Ocean. *Journal of Physical Oceanography* 28: 2345–2361.

White, W. B., and R. G. Peterson. 1996. An Antarctic circumpolar wave in surface pressure, wind, temperature and sea-ice extent. *Nature* 380: 699–702.

Yuan, X., and D. G. Martinson. 2000. Antarctic sea ice extent variability and its global connectivity. *Journal of Climate* 13: 1697–1717.

10

Climate and Hydrologic Variations and Implications for Lake and Stream Ecological Response in the McMurdo Dry Valleys, Antarctica

Kathleen A. Welch

W. Berry Lyons

Diane M. McKnight

Peter T. Doran

Andrew G. Fountain

Diana Wall

Chris Jaros

Thomas Nylen

Clive Howard-Williams

Introduction

Because polar regions may amplify what would be considered small to moderate climate changes at lower latitudes, Weller (1998) proposed that the monitoring of high latitude regions should yield early evidence of global climate change. In addition to the climate changes themselves, the connections between the polar regions and the lower latitudes have recently become of great interest to meteorologists and paleoclimatologists alike. In the southern polar regions, the direct monitoring of important climatic variables has taken place only for the last few decades, largely because of their remoteness. This of course limits the extent to which polar records can be related to low latitude records, even at multiyear to decadal timescales. Climatologists and ecologists are faced with the problem that, even though these high latitude regions may provide important clues to global climatic change, the lengths of available records are relatively short.

The McMurdo Dry Valleys Long-Term Ecological Research (MCM LTER) program was established in 1993. This program built on the monitoring begun in the late 1960s by researchers from New Zealand, who collected records of climate, lake level, and stream discharge in the Wright Valley, Antarctica. Griffith Taylor's field party obtained the first data related to lake level in 1903 as part of Scott's Discovery expedition. Analysis of the more recent data from the New Zealand Antarctic and MCM LTER programs when compared to the 1903 datum indicates that the first half of the twentieth century was a period of steadily increasing streamflows, followed in the last half of the century by streamflows that have resulted in more slowly increasing or stable lake levels (Bomblies et al. 2001). Thus, meteorological and hydrological records generated by the MCM LTER research team, when coupled with past data and the ecological information currently being obtained, provide the first detailed attempt to understand the connection between ecosystem structure and function and climatic change in this region of Antarctica. In addition, the program helps to fill an important gap in the overall understanding of climatic variability in Antarctica.

Even though most climatic records from the Antarctic continent are relatively short and/or lack associated biological monitoring to be useful in an ecological sense, a number of investigators have demonstrated interannual variations in Antarctic climate signals and responses. For example, Cullather et al. (1996) have shown that precipitation in west Antarctica covaries with ENSO, but the sign of the correlation changed in 1990. White and Peterson (1996) have speculated that a number of circum-Antarctic climatic parameters (i.e., sea level pressure, sea surface temperature, sea ice extent, and meridional wind stress), termed the Antarctic Circumpolar Wave (ACW), show interannual variability that may be related to ENSO. Sea ice extent in the Ross Sea region of Antarctica has also been shown to vary with the higher latitude ENSO signal (Ledley and Huang 1997). Finally, a detailed ice core record from West Antarctica, extending back in time approximately 1100 years, also shows dominant periodicities in chemical concentrations that are coincident with the Southern Oscillation Index (SOI) (Kreutz et al. 2000).

The primary emphasis of this chapter is on which types of ENSO or other multiyear climate variations might be observed at the MCM site and how these variations impact the ecosystem of the dry valleys. These considerations are aided by a basic physical understanding of climate and hydrology linkages. The key climatic parameters influencing ecosystem structure and function in the McMurdo Dry Valleys are the ones that affect the physical state of water. Small interannual variations in summer temperatures, the number of days above freezing, and solar radiation can have a large impact because the availability of liquid water is such an important driver for the ecosystem (Fountain et al. 1999). Absorption of solar radiation by the surfaces and faces of the glaciers generates meltwater that either soaks into the alluvium of the streambeds or is carried by streams to the lakes in the valley floors. Meltwater generation depends on a fine balance of radiation, temperature, and the albedo of the glacier surface, which can be increased by snowfall or decreased by the input of eolian dust. Water is lost to the atmosphere from the stream and lake systems by sublimation from streambeds and ablation of the ice covers on the lakes. Because the transport and chemistry of water are the primary factors controlling habitat characteristics of the streams and lakes (Kennedy 1993), we focus our discussion on the aquatic components of the MCM ecosystem.

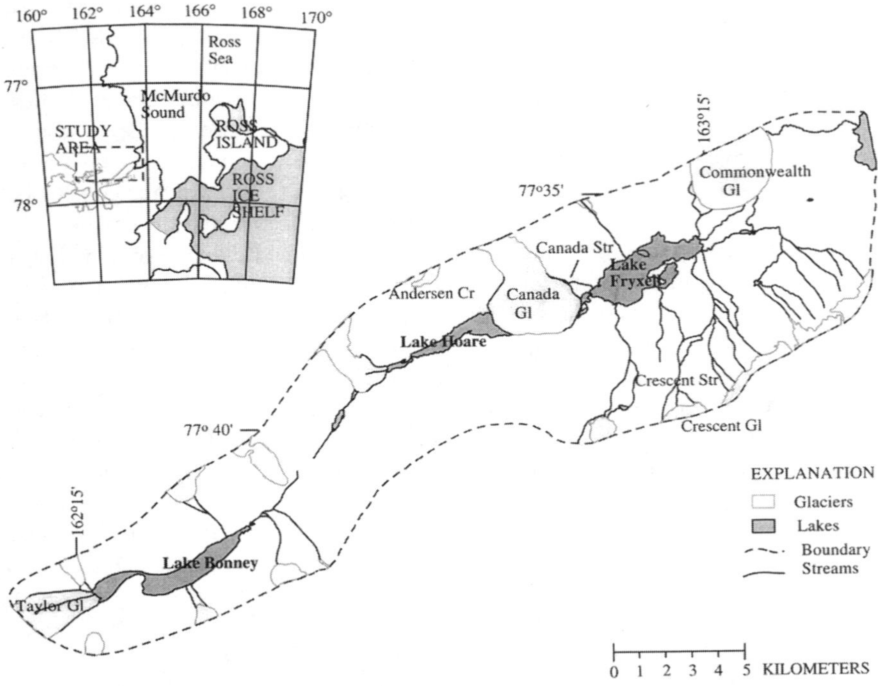

Figure 10.1 Map of Taylor Valley, Antarctica.

Site Description

The McMurdo Dry Valleys are the largest single ice-free expanse in Antarctica
(~ 4800 km²). The valleys are a mosaic of glaciers, ephemeral streams, perennially
ice-covered lakes, soils, and bedrock (Moorhead et al. 1999). They are among the
driest and coldest deserts on the planet, with annual precipitation of <10 cm yr^{-1}
and mean annual temperatures between −14.8 and −30.0°C on the valley floor at
different locations (Doran et al. 2002a). Taylor Valley has been the focal point of
MCM LTER research (figure 10.1). Taylor Valley consists of three major, closed-
basin ice-covered lakes and about 25 streams (McKnight et al. 1999). For 4–10
weeks of the austral summer, the streams are fed by glacial meltwater from the surrounding glaciers (figure 10.1). Because the distribution of liquid water greatly influences the function and biodiversity of the MCM ecosystem, investigation of the
role of climate variability has been a major emphasis of the MCM LTER (Fountain
et al. 1999). To accomplish this, the delineation of the hydrologic budget of each
subbasin within Taylor Valley (Bonney, Hoare, and Fryxell), has been undertaken,
with glacier mass balance, stream discharge, and lake level variations being closely
monitored. These, along with other data from MCM LTER, can be found at our
web site, http://huey.colorado.edu. Descriptions of the methods used to collect meteorological data are in Doran et al. (1995). The methods used to collect stream dis-

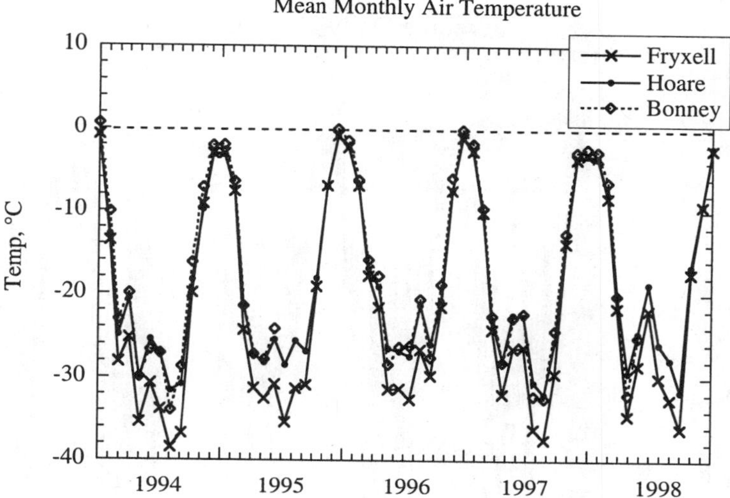

Figure 10.2 Mean monthly temperatures from meteorological stations in the Lake Bonney, Hoare, and Fryxell basins.

charge data are described in Von Guerard et al. (1995) and McKnight et al. (1994). Detailed discussions of the Taylor Valley physiochemical and biogeochemical systems have recently been published in Priscu (1998) and volume 49, no. 12 of *BioScience* and are not repeated here.

Results and Discussion

Meteorological Data

Mean monthly air temperatures for the three major lake basins in the Taylor Valley are shown in figure 10.2. Mean monthly air temperatures are at or below freezing even in the summer months. Temperatures are similar for all basins in the summer, with Lake Bonney being approximately 1°C warmer than Lake Hoare and Lake Fryxell. The occurrence of warm, dry katabatic winds flowing from the polar plateau and relatively cold easterly winds from the sea may explain the spatial differences in temperature (Clow et al. 1988). The Lake Bonney basin is the farthest inland and is most influenced by these winds from the polar plateau, whereas the influence is less pronounced nearer the coast. Mean daily air temperatures in the Lake Hoare basin (figure 10.3) have reached 2–4°C in the summer months during this period. The number of days above freezing also varies from year to year, as shown for the Lake Hoare basin. Even during the summer, there may be only a few days with temperatures above freezing (figure 10.3).

One major constraint on annual productivity of the aquatic ecosystem is the total darkness for approximately 4 months of the year in winter, from May through Au-

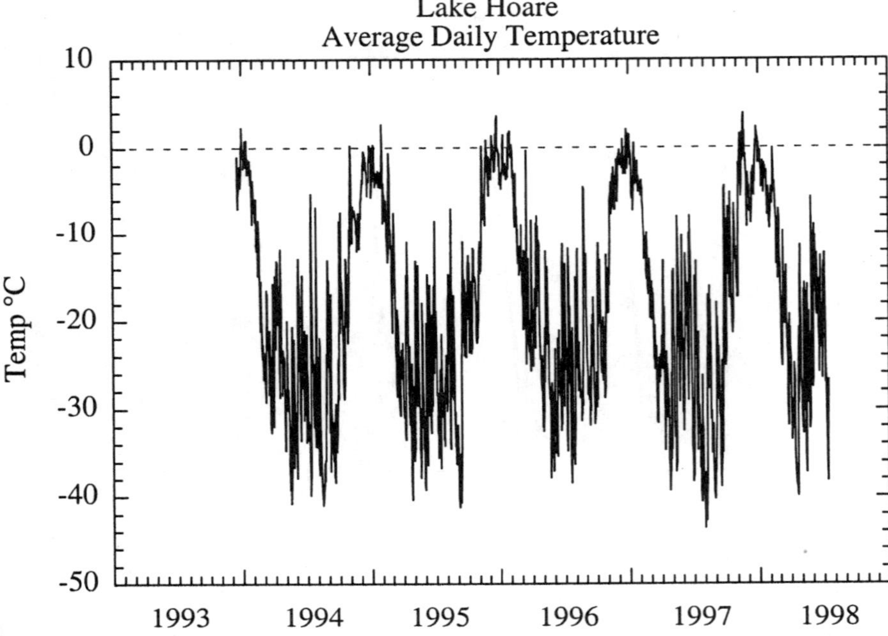

Figure 10.3 Mean daily air temperatures in the Lake Hoare basin.

gust. During the other seasons, there is interannual variability in the light regime. Solar radiation, as short wave radiation (SW), is shown for the Lake Hoare basin (figure 10.4). This parameter is important to monitor because it can influence the production of glacial melt. Water vapor and cloud cover are the primary variables responsible for the interannual variability in solar radiation (Dana et al. 1998). Because of the total darkness in winter, most of the variability in the ecosystem response will occur in the summer when there is sufficient solar radiation to drive photosynthesis and melt snow and ice. These seemingly small interannual variations in summer temperatures, the degree-days above freezing, and solar radiation can have a large impact on the availability of liquid water (Doran et al. 2002b; Fountain et al. 1999). The generation of liquid water on a glacier surface is determined by very small changes in surface temperature.

Stream Discharge

As part of the hydrologic monitoring component of the LTER, the major inflow streams in each of the three large lake basins in Taylor Valley are gauged. Stream discharge data are shown for selected streams in the Lake Bonney, Lake Hoare, and Lake Fryxell basins (figure 10.5). The inflow streams in the Lake Fryxell basin were monitored beginning in the austral summers of 1990–1991 and 1991–1992,

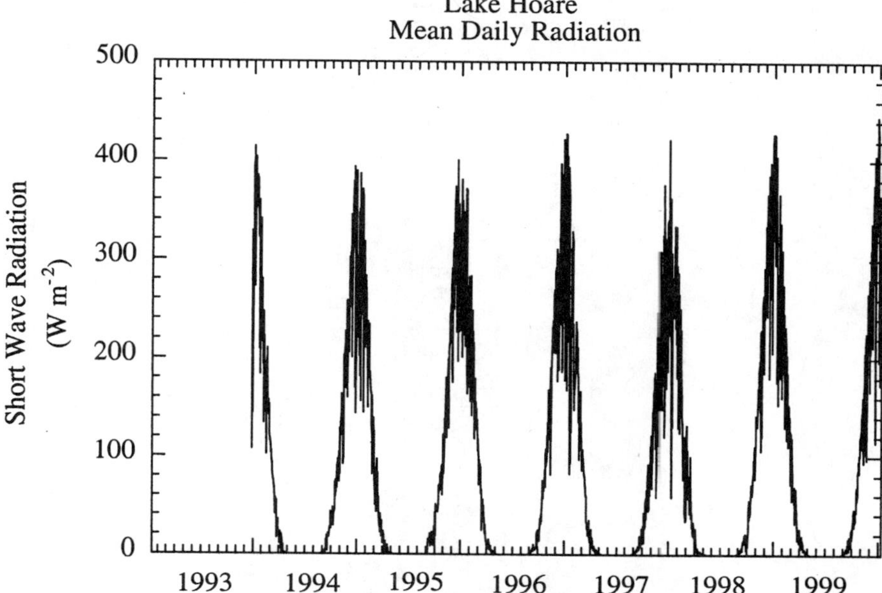

Figure 10.4 Mean daily short wave radiation for the Lake Hoare basin.

before the start of the MCM LTER in 1993. The relatively high streamflows in the Lake Fryxell basin at that time have not recurred through the 2000–2001 flow season. The stream-gauging network was expanded to include the other lake basins in the Taylor Valley in 1993, and in 1994 the LTER project assumed responsibility for the Onyx River stations in the Wright Valley that had been operated by the New Zealand Antarctic Programme since 1968. These data are also available on the MCM LTER web site.

Since 1993, the patterns of interannual variations in streamflow have differed among the lake basins, and in some cases, within basins. The different response is in part related to differences in glacier position within the valley and to stream length (Fountain et al. 1999). In addition, studies of the radiation balance of the glaciers indicate that during low flow years more of the meltwater comes from the glacier faces rather than from the subhorizontal surfaces (Fountain et al. 1998). This is supported by more pronounced diurnal variation in streamflows during low flow periods, when peak flows are associated with the time of day that solar radiation directly impacts the face of the glaciers (Conovitz et al. 1998). Storage of water in the alluvium underneath and adjacent to the stream (referred to as the hyporheic zone) is a greater control on annual discharge in low flow years than in high flow years. Storage of water in the hyporheic zone can increase with stream length. During low flows years the shorter streams in the Lake Fryxell basin account for a greater pro-

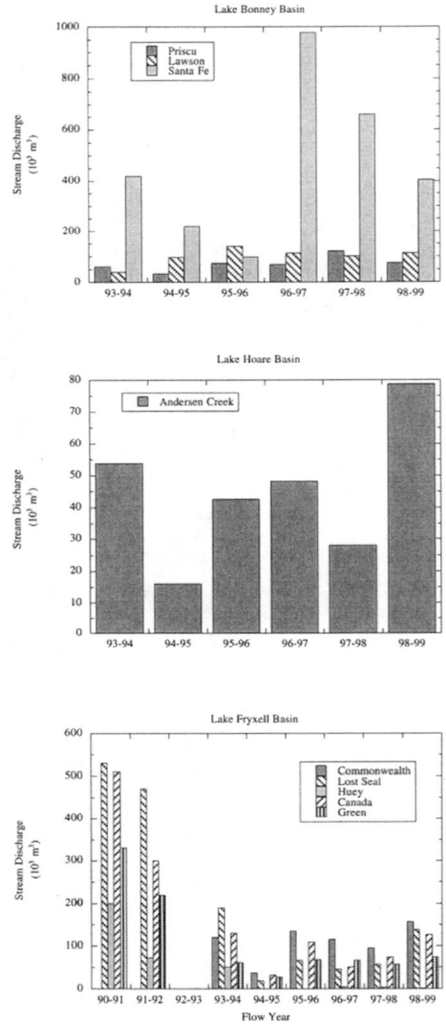

Figure 10.5 Total stream discharge data for each summer for selected streams in the Lake Bonney, Lake Hoare, and Lake Fryxell basins.

portion of total inflow to the lake than they do in the high flow years (House et al. 1995; Conovitz et al. 1998).

Table 10.1 compares variations in annual discharge for Andersen Creek in the Lake Hoare basin to available meteorological data. As shown in figure 10.1, Andersen Creek is immediately adjacent to the west side of the Canada Glacier and flows through a short (200-m) delta before entering Lake Hoare. Algal mats are sparse in this short deltaic reach, and hyporheic zone interactions do not influence flow as much as they would in the longer streams in the Lake Fryxell and Lake

Bonney basins (Conovitz et al. 1998). Annual stream discharge is positively correlated to average summer temperature, degree-days above freezing, and solar radiation (both photosynthetically active radiation, [PAR] and SW) (table 10.1). Even maximum temperatures hover around the freezing point, and hence, small variations in solar radiation can also greatly influence glacier melt and subsequent streamflows (Lewis et al. 1998; Fountain et al. 1999). During the two coldest summers (1994–1995 and 1997–1998), the Lake Hoare average summer (December and January) temperatures were only –3°C. Solar radiation was also low, and streamflows were the lowest observed during this period. The highest streamflow occurred during 1998–1999. This was not the warmest summer, but the solar radiation was relatively high (table 10.1).

Variations in annual discharge for Canada and Crescent streams in the Lake Fryxell basin are also compared in table 10.1. Canada Stream drains the east side of the Canada Glacier, and the lower 1.5 km of the stream flows in a channel that is 0.2–0.5 km east of the glacier face. Crescent Stream is 5.4 km long and flows from the Crescent Glacier on the south side of the lake. Canada Stream has very abundant algal mats and mosses as does Crescent Stream in its upper reaches near the glacier source (Alger et al. 1997; McKnight, pers. comm.). During this period, Crescent Stream had much lower annual discharges than either Andersen Creek or Canada Stream, reflecting its greater length and greater storage of meltwater in the hyporheic zone. The annual discharge of Canada Stream is positively correlated with annual discharge in Andersen Creek to a greater extent than to Crescent Stream. The annual discharge in Canada Stream is also positively correlated with solar radiation and air temperature.

The correlations with PAR and air temperature are weaker for annual discharge in Crescent Stream, but the correlation with SW radiation is stronger. The highest annual discharge in Crescent Stream occurred in 1993–1994 when SW radiation was greatest, and the lowest annual discharges occurred during the two coldest summers. The weaker relationships with meteorological data in Crescent Stream are attributable to a greater influence of hyporheic zone processes in longer streams. The meltwater stored in the hyporheic zone supports growth of mosses and cyanobacterial mats along the edges of the streams, but does not directly contribute to the lake ecosystem. Storage in the hyporheic zone may also influence total discharge for a particular stream. Relatively short streams such as Andersen Creek should be influenced less by hyporheic zone storage, and therefore, should respond rapidly to local environmental variables, making them more important streams to monitor to assess hydrologic responses to climate variability.

Streamflow Variability and ENSO

Given that there are previously documented ENSO climate effects in Antarctica, we postulate that the interannual variability of meteorological parameters, and hence streamflow in the dry valleys, might be influenced by ENSO. As previously described, temperature, barometric pressure, solar radiation, and albedo on glacier surfaces are thought to affect glacier melt and hence streamflow in the dry valleys. How might the ENSO influence these variables in the region of the dry valleys?

Table 10.1 Meteorological data from the Lake Hoare basin and total stream discharge from Andersen Creek, Canada Stream and Crescent Stream[a]

Flow year	PAR μmol s^{-1} m^{-2}	SW radiation W m^{-2}	Barometric Pressure mbar	Degree-days above 0°C	Air Temperature °C	Andersen Discharge 10^3 m^3	Canada Discharge 10^3 m^3	Crescent Discharge 10^3 m^3
93/94	595	322		8.8	−2.0	54.0	130.0	39.0
94/95	534	273	976	9.0	−3.1	15.0	32.0	4.8
95/96	562	267	978	32.3	−1.5	41.5	109.8	8.6
96/97	573	271	982	17.4	−1.9	44.8	50.9	7.9
97/98	529	254	983	28.2	−3.3	28.0	74.2	5.4
98/99	600	288	978	24.7	−1.6	88.6	125.9	10.2
99/00	609	308	971	3.8	−2.6			

	PAR	SW radiation	Barometric Pressure	Degree days above 0°C	Air Temperature	Andersen	Canada	Crescent
PAR	1							
SW RAD	0.81	1						
Pressure	−0.57	−0.85	1					
Degree-days	−0.38	−0.70	0.63					
Temperature	0.60	0.27	−0.03	0.29	1			
Andersen	0.89	0.48	−0.12	0.20	0.72	1		
Canada	0.73	0.59	−0.19	0.28	0.63	0.75	1	
Crescent	0.60	0.92	−0.18	−0.49	0.31	0.31	0.61	1

[a] Meteorological data are averages for December and January with the exception of degree-days above 0°C which is summed for the entire summer. A correlation matrix is included in the lower part of the table.

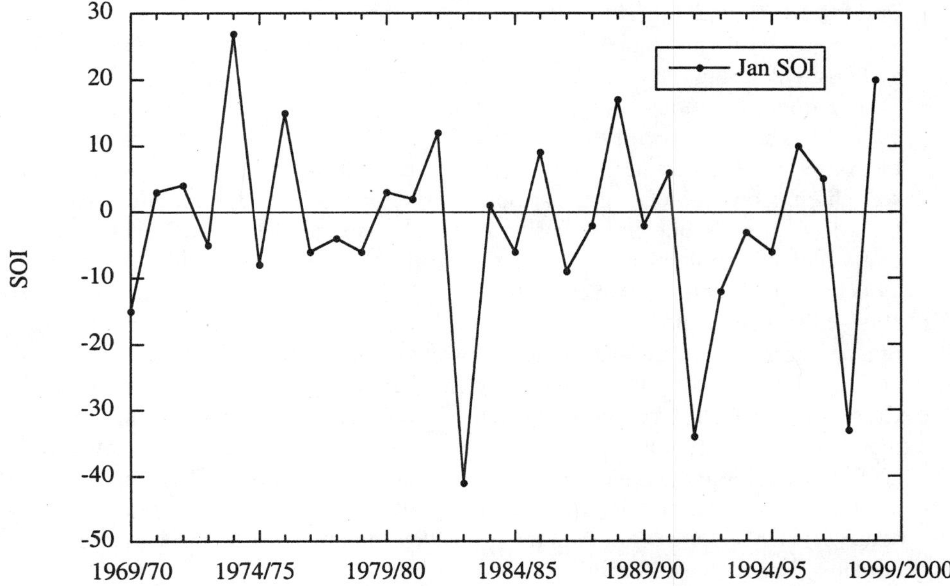

Figure 10.6 Southern Oscillation Index (SOI) for January for the period 1969–2000. Strongly negative SOI indicates the El Niño (warm) phase, whereas strongly positive SOI indicates the La Niña (cold) phase of the ENSO.

To determine the possible manifestations of ENSO in the dry valleys, we need a better understanding of the climate mechanisms related to ENSO that might influence the regional climatic and hydrologic records. As mentioned, previous links between the SOI and Antarctic climate have been observed, for example, higher sea surface temperatures and reduced sea-ice concentrations in the Ross Sea region during El Niño years (Ledley and Huang 1997). Variations in sea-ice extent may influence moisture availability, as well as the extent of cloud cover and precipitation in the dry valleys. Lower sea-ice extent and warmer sea surface temperatures during El Niño may lead to cloudier and cooler conditions in the region as well as increased precipitation. The presence of fresh snow on the glacier surfaces can increase the surface albedo and effectively limit meltwater production from the glaciers (Fountain et al. 1998).

In addition to variation in sea-ice extent and sea surface temperatures, atmospheric circulation, which influences extreme warm and cold winter temperatures in the Antarctic Peninsula region, is thought to be related to ENSO (Marshall and King 1998). It is not clear how atmospheric circulation in the Ross Sea region would be influenced by ENSO. However, variations in circulation patterns could influence such things as storm tracks and the episodes of katabatic winds in the dry valleys, which in turn could influence temperature, cloud cover, precipitation, and solar radiation.

Figure 10.6 shows the Southern Oscillation Index (SOI) for January for the period from 1969 to 2000. The SOI is calculated from normalized Tahiti and Darwin sea-level pressure anomalies. The SOI is negative during the El Niño (warm) phase and positive during the La Niña (cold) phase of the ENSO.

The high streamflows observed in the Lake Fryxell basin in 1990–1991 and 1991–1992 have not recurred through 2000–2001 (figure 10.5). The austral summer of 1991–1992 was an El Niño year, and streamflows in the Lake Fryxell basin were high. However, during the strong El Niño of 1997–1998, streamflows in the Lake Fryxell and Lake Hoare basins were relatively low (figure 10.5). The relatively high streamflows in Andersen Creek in 1995–1996, 1996–1997, and 1998–1999 coincided with a positive SOI (cold, La Niña) phase, whereas 1994–1995 and 1997–1998 were cooler summers with lower streamflows, and the January SOI was in a negative (warm, El Niño) phase. However, 1993–1994 was a year with near-neutral SOI, but with high streamflows in Andersen Creek, Canada Stream, and Crescent Stream. These records from the Taylor Valley span a short time period; therefore, it is difficult to quantify the relationship between streamflows and ENSO. In addition, the Lake Bonney, Lake Hoare, and Lake Fryxell basins, as well as other locations in the dry valleys, may exhibit a different climate response because of their relative position in the dry valleys and their slightly different microclimates (Fountain et al. 1999).

To further examine the possible connection between ENSO and climate in the dry valleys, longer time period records from this region of Antarctica are needed. The longest record of stream discharge in Antarctica is for the Onyx River, located approximately 20 km to the north of Taylor Valley in Wright Valley. The Onyx River is the longest river in Antarctica, flowing approximately 40 km from the Lower Wright Glacier inland to Lake Vanda (figure 10.7). In addition, mean summer (December and January) air temperature (figure 10.8) and mean summer (December and January) barometric pressure (figure 10.9) records are available for McMurdo Station, Antarctica, 100 km east of the dry valleys.

The annual discharge of the Onyx River varies by an order of magnitude during the period 1970–1999 (figure 10.7). The correlation matrix for the Onyx River discharge, McMurdo Station temperature and pressure, and the SOI shows that the discharge in the Onyx River is positively correlated to temperature and pressure and negatively correlated to the January SOI (table 10.2). This analysis suggests that during this time period higher discharges in the Onyx River may be associated with the El Niño phase of ENSO. None of the correlations is significant at the 95% confidence level. However, during summers when temperatures in the region are high, streamflow also tends to be high, although during some warm summers, for example, in 1974, streamflow is low. The McMurdo barometric pressure was also low in 1974, which could indicate lower incoming solar radiation because of increased cloud cover and storms in the region. As mentioned, the correlations between the SOI and McMurdo temperature and pressure are not statistically significant. However, when the SOI is low, both temperature and pressure tend to be higher.

Four of the five summers with the highest streamflows on the Onyx River (1971, 1985, 1987, and 1991) occurred when the SOI was near neutral. The SOI was low

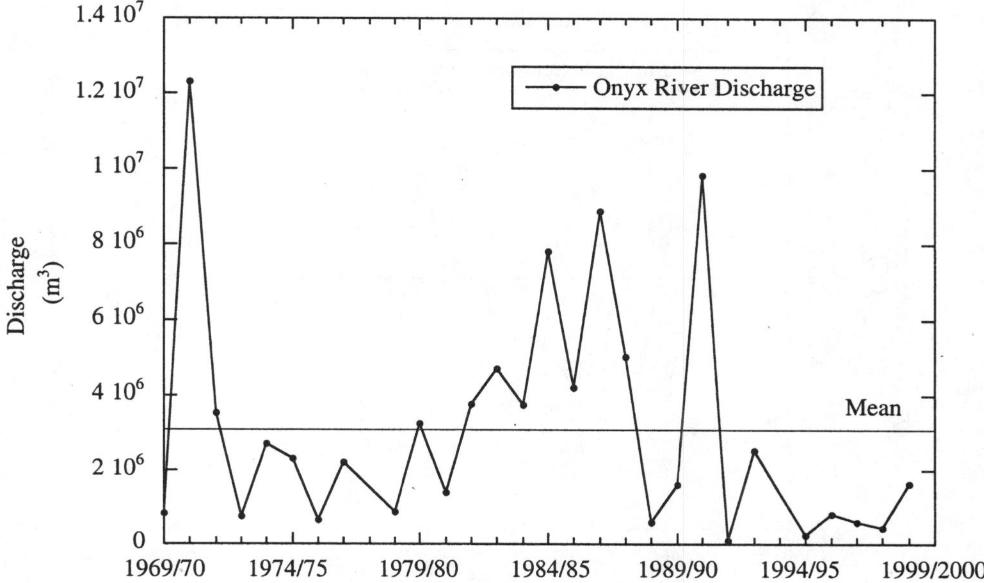

Figure 10.7 Total stream discharge for each summer from the Onyx River for the period 1969–1999.

(El Niño) in the high flow summer of 1992. During the years of lowest flow, the SOI is neutral or high (La Niña). For example, the SOI was high in the low flow summers of 1974 and 1976.

This analysis is somewhat problematic because the discharge measured on the Onyx River near Lake Vanda is also influenced by other nonclimatic factors such as potential lag between the generation of meltwater and streamflow, as well as variability of storage in the hyporheic zone. However, there is an additional record of discharge on the Onyx River near the Wright Lower Glacier, the major source of meltwater for the Onyx River. The annual discharge records between the two sites are strongly correlated (R = 0.83, n = 23) during this period. Comparing data from McMurdo Station and the Wright Valley can also be somewhat problematic because of regional climate differences. Although McMurdo Station and the dry valleys are far enough apart that they can experience different weather, it is thought that the general trends in climate of the sort that might be influenced by ENSO should be similar throughout this region. The limited (more recent) meteorological data from Lake Hoare and Lake Vanda that can be compared to the records from McMurdo Station show that, although the mean summer temperatures are different, the trends are correlated (significant at 95% confidence level). The various records from the Ross Sea region, including the ones presented here from Taylor Valley, Wright Valley, and McMurdo Station can be used to examine the influence of ENSO on the regional climate of the southern Ross Sea.

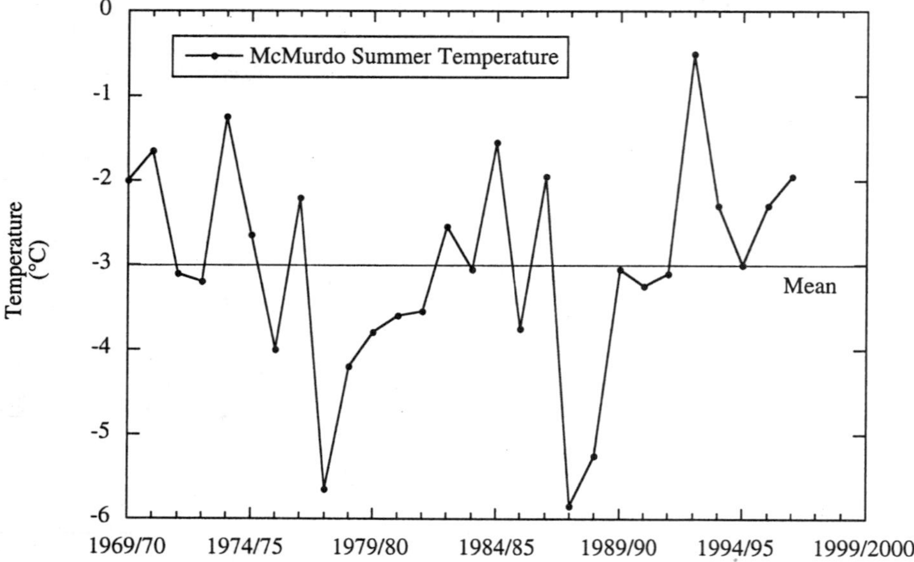

Figure 10.8 Mean summer temperature recorded at McMurdo Station, Antarctica. The record represents the average of December and January mean monthly temperatures for each summer.

During El Niño years, perhaps conditions in the Ross Sea region are such that barometric pressures are high, allowing for relatively high solar radiation, warmer air and sea surface temperatures, and high streamflows. Alternatively, lower sea-ice concentrations or sea-ice extent that might be expected during El Niño (Ledley and Huang 1997) could lead to conditions that would not favor high incoming solar radiation as a result of increased moisture availability and cloud cover. At longer timescales, high lake stands during the last glacial maximum are attributed to cool and dry conditions, which are linked to fewer clouds, less snowfall, higher radiation, and more melt (Hall et al. 2001). Clearly, additional work is needed to elucidate possible linkages among the SOI, sea-ice, solar radiation, and other climate parameters.

Climate Variability and the Hydrologic Cycle Within the McMurdo Dry Valleys

It is well documented that the closed basin lakes within the Taylor and Wright valleys have waxed and waned since the last glacial maximum. The timing and evidence for these large-scale volume changes in the lakes will not be discussed here, as they have been reviewed recently (Doran et al. 1994; Hall et al. 2001; Lyons et al. 1998b). These fluctuations in the sizes of the lakes were brought about by subtle century- to millennial-scale climatic perturbations (Fountain et al., chapter 16 this volume; Hall et al.

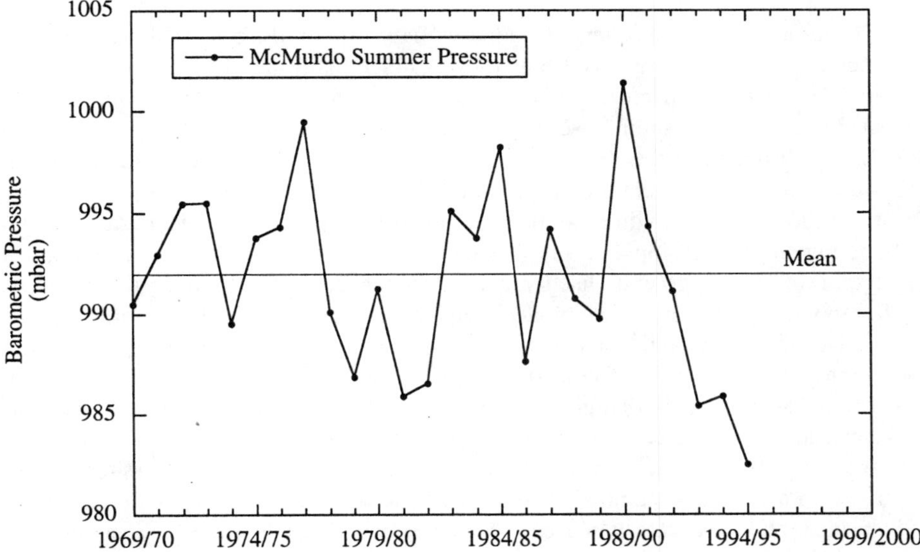

Figure 10.9 Mean summer pressure recorded at McMurdo Station, Antarctica. The record represents the average of the December and January mean monthly temperature.

2001). Similar century-scale climate variations have been observed in the Antarctic coastal marine system (Leventer et al. 1996; Smith et al. 1999).

Next, we will examine the impact short-term climatic variations might have on the McMurdo Dry Valleys ecosystem. Again, our discussion will be based on the fact that the McMurdo Dry Valleys ecosystem is sensitive to very small variations in climate because the change between solid and liquid water is delicately poised, and thus small changes in temperature and radiant energy are amplified by large nonlinear changes in the hydrologic budgets that can cascade through the system.

Table 10.2 Correlation matrix for the Onyx River discharge, McMurdo mean summer (Dec–Jan) temperature and pressure, and January SOI for the period 1970 to 1996 (N=28)

	Onyx River discharge	McMurdo mean summer temperature	McMurdo mean summer pressure	January SOI
Discharge	1			
Temperature	0.21	1		
Pressure	0.27	0.13	1	
January SOI	0.05	−0.15	−0.13	1

Impact on Stream Ecosystems: Cascade I

Glacial melt and subsequent streamflow may respond differently to several consecutive warm and sunny days than they would to many days that hover near freezing. The result of differing flow rates may lead to the same volume of water generated over an entire summer season, however, the response of the algal community in some stream habitats can be significantly different for a short-duration pulse of water versus a more continuous low flow. The volume of meltwater per unit time from the glacier has a direct impact on the stream ecosystems. In addition, glacial routing differences, rapid versus slow melting rates on the glacier, and variations in hyporheic exchange with discharge affect water quality, including nutrient distribution, hence they influence biogeochemical processes within the streams.

The algal mats in the streams are composed of cyanobacteria and persist from summer to summer in a freeze-dried state (Vincent and Howard-Williams 1986). The stream reaches with high abundance of algal mats typically have moderate gradients and a streambed armored by a stone pavement. The stream reaches with less or sparse algal mats are typically steep gradient or sandy deltaic reaches that do not have an armored streambed. These latter reaches are turbid and sediment laden during high flow events, which can scour the streambeds, limiting the accumulation of algae to form a mat (Vincent and Howard-Williams 1986; Howard-Williams et al. 1986; Alger et al. 1997; McKnight et al. 1998). Our stream algal surveys in 1994 (2 years after the high flows in 1990–1991 and 1991–1992) and in 1998 (after 6 years of lower flow) indicate that stream reaches with steep gradients or deltaic reaches with shallow gradients but unstable substrates actually accumulate phototrophic biomass as well as algal diversity under low flow conditions (McKnight et al. 1998). We attribute these increases to recovery from the effects of scouring in 1990–1992 during the current period of low and stable streamflows. Stream morphology changed as well, with a widening of stream channel and redistribution of surface flow paths. The primary phototrophs that expand their coverage of the streambed during low flow conditions are orange algal mats and mosses in some locations. However, the stream reaches that have a stone pavement and support abundant algal mats have stable mat communities and do not change much with interannual variation in streamflow.

In addition, although decreased flow may reduce the total amount of nutrients passing through the stream ecosystem, it allows for more efficient uptake of nutrients because of slower flow rates and greater contact with algal mats within the stream environment. Our modeling of nitrogen uptake in the streams of the Lake Fryxell basin indicates that uptake rates are extremely sensitive to flow regime in the streams (Moorhead and Priscu 1998), with higher uptake at lower discharges. In addition, there is the influence of diel changes in flow on nutrient concentrations. Concentrations increase at night and early morning as flows decrease. These changing concentrations must influence nutrient uptake through straightforward Michaelis Menten kinetics. All these data suggest that stream algal communities may be more viable and more efficient at removing nutrients at lower flow rates when the hydrologic connectivity (e.g., from glacier to lake) and sediment transport are minimized. This finding implies that low streamflow years lead not only to increased

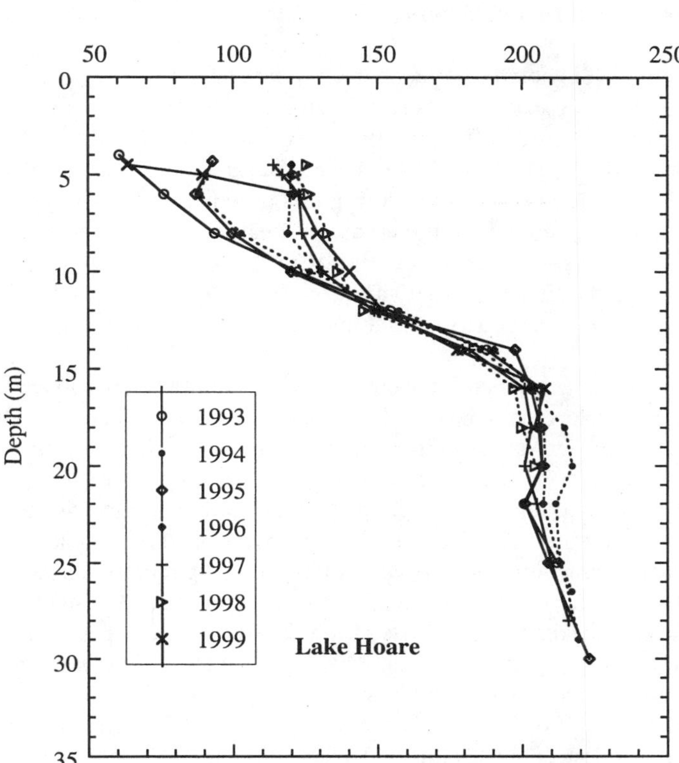

Figure 10.10 Cl versus depth at the beginning of the austral summer at Lake Hoare from 1993 to 1999.

biomass within the streams, but also to decreased nutrient and suspended sediment load concentrations to the lakes.

Impact on Lake Ecosystems: Cascade II

To evaluate the role of hydrologic and, hence, climatic variability on the lake environments, we will focus our attention on Lake Hoare, the least saline of the Taylor Valley lakes. Figure 10.10 shows a time series of Cl versus depth during early summer (i.e., prior to the initiation of streamflow) from the same location in Lake Hoare. Over the period 1993–1999, the surface water Cl (and hence total dissolved solids, TDS) has increased as a result of decreased inputs of glacial meltwater during this time period (Welch et al. 2000). As described previously, water is lost from these lakes through ablation of the perennial ice covers. During the winter, new ice is accreted onto the bottoms on the ice covers, and salts are exsolved from the ice matrix. Unless the water loss via sublimation is balanced by an input of freshwater, this input of salt will lead to an increase in salinity in the upper water column and a decrease in water column stability over time. This is clearly what we have observed in Lake Hoare from 1993 to 1999 during these relatively low flow years (figure 10.10).

Lyons et al. (1998a) noted that geochemistry data from Lake Fryxell in the McMurdo Dry Valleys indicated an overall change in the ionic composition of the lakes when data collected in the mid-1990s are compared to older, but reliable, data obtained in the early 1960s. Our recent data demonstrated an increase in $Ca:Cl$, $HCO_3:Cl$, and $SO_4:Cl$ ratios during this time period (30–35 years). This change in ratios is attributed to an increase of weathering products into the lake because of increased streamflows from the 1970s to 1991 (Lyons et al. 1998a). No matter what the cause, it is clear that variations in the hydrologic cycle within the McMurdo Dry Valleys do impart measurable geochemical changes in the streams and lakes.

In addition to changes in major ion chemistry, low streamflows also lead to a decrease in nutrient fluxes in the lakes, which are accentuated by increased algal mat uptake of fixed nitrogen species within the streams, as detailed previously. In the Taylor Valley lakes, inorganic nutrients such as nitrate, ammonium, and phosphate are low or below detection in the upper part of the water column prior to streamflow. This has an important consequence because there is little nutrient available to sustain primary production when the lakes first receive light in the spring. Another important consequence of lower streamflows is a decrease in CO_2 input into these lakes. Unlike most lakes in other parts of the world that are supersaturated with CO_2 in their surface waters (Cole et al. 1994), the Taylor Valley lakes can be extremely undersaturated in CO_2 (with respect to the atmosphere), because of their ice covers and the lack of terrestrial organic carbon input (Neumann et al. 2001). Decreased streamflow enhances this undersaturation and may lead to CO_2 limitation (Neumann et al. 2001).

Phytoplankton enumeration from 1989 to 1999 in Lake Fryxell has allowed us to evaluate the influence of streamflow and climate on the phytoplankton populations (McKnight et al. 2001). During low flow years, *Chroomonas lacustris* is the dominant algal species, whereas in high flow years, *Cryptomonas* sp. is dominant. In addition, filamentous cyanobacteria have a maximum abundance at depth during low flow years. Because phytoplankton populations persist during the winter in the upper water column, although at a lower concentration than in the summer (McKnight et al. 2000), the species distribution of the phytoplankton in early summer may be influenced by conditions during the previous summer as well as by the nutrient and light regime at that time. Although we are just beginning to digest and analyze these data, clearly the variability in the hydrologic cycle greatly influences the biological communities of these lakes (Doran et al. 2002b).

Impacts on Soil Ecosystems: Cascade III

We have primarily focused our discussion on the aquatic systems within the McMurdo Dry Valleys as they may be affected by ENSO-type climatic fluctuations, but soil environments may also be influenced by short-term climate variations. Treonis et al. (2000) have recently observed that soil nematodes are in a state of anhydrobiosis in soils with moisture contents less than 2%. Wetting soils to 12% leads to the animals coming out of anhydrobiosis within 6 hours (Treonis et al. 2000). Recent snow patch experiments by our group initiated to simulate increased snow input to the landscape have shown that soils covered by increased snowpack can have moisture contents as high as 12%, compared to nearby control soils with 0.2 to 0.7% moisture (Gooseff et al. 2000). Therefore, variations in the amount of precipitation within the McMurdo Dry Valleys brought about by both short- and long-term climate variability could have an important impact on soil moisture and hence changes in population dynamics of the dominant animals in the dry valleys. Ongoing research will continue to evaluate this concept.

Summary of Potential and Observed Cascades through the Ecosystem

The key climatic parameters influencing ecosystem structure and function in the McMurdo Dry Valleys are the ones that affect the conversion of solid to liquid water. These include, but are not limited to, temperature, solar radiation, and precipitation changes. Variations in meltwater production on the glaciers could be due to changes in solar radiation as a function of cloud cover and/or storminess, surface albedo of the glaciers, and temperature. Variations in melt will then impact stream discharge, which, in turn, can impact the distribution, species composition, and nutrient uptake rate of the algal mats within the stream channels. Fluxes of water, as well as dissolved and particulate matter, into the lakes will also be affected. This, in turn, will lead to chemical and biological changes within the lakes themselves. Thus the short-term climate fluctuations that have been observed can have multiyear impacts on the aquatic ecosystems of the dry valleys (Doran et al. 2002b).

Although it is still not clear just how ENSO and other types of climatic phenomena influence the climate of the dry valleys, it is clear that even very small changes in climate can play an important role in the ecosystem dynamics of the McMurdo Dry Valleys. This is evident from the information presented within this chapter and that of Doran et al. (2002b), as well as from our long-term records (e.g., see Fountain et al. 2003, chapter 16). Lyons et al. (2000) termed the McMurdo Dry Valleys an "unstable" system because the recovery from past climatic disturbance is quite long. Its sensitivity to small-scale (by temperate latitude measures) climate variability is an extraordinary feature of the McMurdo Dry Valley ecosystem. Although the records are not long enough to extract an ENSO signal from the climatological and hydrological data from Taylor Valley, our primary LTER site, there is little doubt that variations in local sea-ice extent, which has been related to ENSO, can have a major impact on climate and water budgets in the McMurdo Dry Valleys as well (e.g., Welch et al. 2000; Hall et al. 2001). Model simulations under a global warming scenario indicate an increasingly positive Antarctic Oscillation (AAO), or in other words, lower barometric pressures over the polar regions (Fyfe et al. 1999). Therefore, our understanding of the impact of storminess, increased snowfall, and albedo change may become more important in attempting to predict the impact of climate change on the McMurdo Dry Valley system.

Conclusion

A number of conclusions can be drawn from this research. Data collected by the MCM LTER scientists since 1993 clearly demonstrate that small variations in climate, especially in the austral summer, can have a significant multiyear impact on the ecosystem. Although we have observed interannual climate variability in the dry valleys region of Antarctica, it is unclear whether it is related to ENSO, the AAO, the ACW, or some other climatic forcing. Our record from Taylor Valley is simply too short to discern the long-term trends. The longer term climate records from the McMurdo region and Onyx River discharge may vary with the SOI; however, the correlations are not statistically significant. Nonetheless, the available records indicate that high streamflows in the Onyx River in Wright Valley occur during neutral to low SOI and that low streamflows occur during neutral to high SOI, suggesting the influence of ENSO. However, stream discharge in Andersen Creek in Taylor Valley seems to be positively correlated to SOI, with higher flows occurring during the positive (La Niña) phase of the SOI. This difference between discharge and SOI between the Onyx River and Andersen Creek probably represents differences in the physical characteristics of these systems. Much larger storage capacity in the hyporheic zone of the Onyx River mediates the flow. If this is true, the monitoring of the shorter, higher gradient streams like Andersen Creek may lead to better records of climate teleconnections. Clearly, additional work is needed to elucidate possible linkages among the SOI, sea ice, solar radiation, and other climatic parameters in this region of Antarctica, and longer climate records are needed.

The variability in climate manifested through changes in the hydrologic cycle within the McMurdo Dry Valleys cascade through the aquatic portion of the ecosystem. Variations in temperature, snowfall, and solar radiation influence meltwater production on the glaciers. The resulting streamflow influences stream microbial populations and morphology. The closed basin lakes that are fed by the streams can respond in a variety of ways, including changes in phytoplankton populations, which have been observed to be different in high and low streamflow years, possibly driven by changes in lake water chemistry. We speculate that climate variability can also cascade through the soil environment as well. Only when longer climatic records from the McMurdo Dry Valleys and from other parts of Antarctica are available will scientists be able to evaluate the overall ecological impact of ENSO climate variations in the southern polar regions.

Acknowledgments   This work was supported by NSF grants OPP-9211773 and OPP-9813061. We thank David Greenland for allowing us to participate in the LTER ASM workshop that was the original focus of this work. We thank all our MCM LTER colleagues who helped with collection and analysis of these data. We thank the New Zealand National Institute of Water and Atmospheric Research Ltd. and particularly Pete Mason and Kathy Walter for hydrological data analysis on the Onyx River. SOI data were obtained from The National Centers for Environmental Prediction (NCEP) Climate Prediction Center web site. Meteorological data for McMurdo Station were compiled from the *Antarctic Journal of the United States*. We appreciate the discussions regarding short-term climatic variations in

Antarctica with Matt Lazara, University of Wisconsin, Paul Mayewski, University of Maine, and Chris Shuman, University of Maryland. We also thank Carmen Nezat for reviewing the original manuscript. We especially thank two anonymous reviewers, whose thoughtful suggestions greatly improved the manuscript.

References

Alger, A. S., D. M. McKnight, S. A. Spaulding, C. M. Tate, G. H. Shupe, K. A. Welch, R. Edwards, E. D. Andrews, and H. R. House. 1997. Ecological processes in a cold desert ecosystem: The abundance and species distribution of algal mats in glacial meltwater streams in Taylor Valley, Antarctica. *University of Colorado, Institute of Arctic and Alpine Research, Occasional Paper*, 51. 108pp.

Bomblies, A., D. M. McKnight, and E. D. Andrews. 2001. Retrospective simulation of lake level rise in Lake Bonney based on recent 21-year record: Indication of recent climate change in the McMurdo Dry Valleys, Antarctica. *Journal of Paleolimnology* 25(4): 477–492.

Cole, J. J., N. F. Caraco, G. W. Kling, and T. K. Kratz. 1994. Carbon dioxide supersaturation in the surface waters of lakes. *Science* 265:1568–1570.

Clow, G. D., C. P. McKay, G. M. Simmons, Jr., and R. A. Wharton. 1988. Climatological observations and predicted sublimation rates at Lake Hoare, Antarctica. *Journal of Climatology* 1:715–728.

Conovitz, P. A., D. M. McKnight, L. H. MacDonald, A. G. Fountain, and H. H. House. 1998. Hydrologic processes influencing streamflow variation in Fryxell basin, Antarctica. Pages 93–108 in J. C. Priscu, editor, *Ecosystem Dynamics in a Polar Desert: The McMurdo Dry Valleys, Antarctica.* American Geophysical Union Antarctic Research Series 72, Washington, D.C.

Cullather, R. I., D. H. Bromwich, and M. L. Van Woert. 1996. Interannual variations in Antarctic precipitation related to El Niño-Southern Oscillation. *Journal of Geophysical Research* D 101:19109–19118.

Dana, G. L., R. A. Wharton, and R. Dubayah. 1998. Solar radiation in the McMurdo Dry Valleys, Antarctica. Pages 39–64 in J. C. Priscu, editor, *Ecosystem Dynamics in a Polar Desert: The McMurdo Dry Valleys, Antarctica.* American Geophysical Union Antarctic Research Series 72, Washington, D.C.

Doran, P. T., G. Dana, J. T. Hastings, and R. A. Wharton. 1995. The McMurdo LTER Automatic Weather Network (LAWN). *Antarctic Journal of the United States* 30(5):276–280.

Doran, P. T., C. P. McKay, G. D. Clow, G. L. Dana, A. G. Fountain, T. Nylen, and W. B. Lyons. 2002a. Valley floor climate observations from the McMurdo Dry Valleys, Antarctica, 1986–2000. *Journal of Geophysical Research, Atmospheres*, 10.1029/2001JD002045.

Doran, P. T., J. C. Priscu, W. B. Lyons, J. E. Walsh, A. G. Fountain, D. M. McKnight, D. L. Moorhead, R. A. Virginia, D. H. Wall, G. D. Clow, C. H. Fritsen, C. P. McKay, and A. N. Parsons. 2002b. Antarctic climate cooling and terrestrial ecosystem response. *Nature* 415:517–520.

Doran, P. T., R. A. Wharton, Jr., and W. B. Lyons. 1994. Paleolimnology of the McMurdo Dry Valleys, Antarctica. *Journal of Paleolimnology* 10:85–114.

Fountain, A. G., G. L. Dana, K. J. Lewis, B. H. Vaughn, and D. McKnight. 1998. Glaciers of the McMurdo Dry Valleys, Southern Victoria Land, Antarctica. Pages 65–75 in J. C. Priscu, editor, *Ecosystem Dynamics in a Polar Desert: The McMurdo Dry Valleys,*

Antarctica. American Geophysical Union Antarctic Research Series 72, Washington, D.C.

Fountain, A. G., W. B. Lyons, M. B. Burkins, G. L. Dana, P. T. Doran, K. J. Lewis, D.M. McKnight, D. Moorhead, A. N. Parsons, J. C. Priscu, D. H. Wall, R. A. Wharton, Jr., and R. A. Virginia. 1999. Physical controls on the Taylor Valley Ecosystem Antarctica. *BioScience* 49(12):961–971.

Fyfe, J. C., G. J. Boer, and G. M. Flato. 1999. The Arctic and Antarctic oscillations and their projected changes under global warming. *Geophysical Research Letters* 26:1601–1604.

Gooseff, M. N., J. E. Barrett, and P. Doran. 2000. Seasonal snow pack coverage and its influence on soils of a polar desert. ESA meeting, February 2001, Santa Fe, NM. Abstracts with Program.

Hall, B. L., G. H. Denton, and B. Overturf. 2001. Glacial Lake Wright, a high-level Antarctic lake during the LGM and early Holocene. *Antarctic Science* 13:53–60.

House, H. R., D. M. McKnight, and P. von Guerard. 1995. The influence of stream channel characteristics on streamflow and annual water budgets for lakes in the Taylor Valley. *Antarctic Journal of the United States* 30(5):284–287.

Howard-Williams, C., C. L. Vincent, P. A. Broady, and W. F. Vincent. 1986. Antarctic stream ecosystems: Variability in environmental properties and algae community structure. *Internationale Revue der gesamten Hydrobiologie* 71:511–544.

Kennedy, A. D. 1993. Water as a limiting factor in the Antarctic terrestrial environment: A biogeographical synthesis. *Arctic and Alpine Research* 25:308–315.

Kreutz, K. J., P. A. Mayewski, I. I. Pittalwala, L. D. Meeker, M. S. Twickler, and S. I. Whitlow. 2000. Sea-level pressure variability in the Amundsen Sea region recorded in a West Antarctic glaciochemical record. *Journal of Geophysical Research* 105(D3):4047–4059.

Ledley, T. S., and Z. Huang. 1997. A possible ENSO signal in the Ross Sea. *Geophysical Research Letters* 24:3253–3256.

Leventer, A., E. W. Domack, S. E. Ishman, S. Brachfeld, C. E. McClennen, and P. Manley. 1996. Productivity cycles of 200–300 years in the Antarctic Peninsula region: Understanding linkages among the sun, atmosphere, oceans, sea ice, and biota. *Geological Society of America Bulletin* 108:1626–1644.

Lewis, K. J., A. G. Fountain, and G. L. Dana. 1998. Surface energy balance and meltwater production for a Dry Valley glacier, Taylor Valley, Antarctica. *Annals of Glaciology* 27:603–609.

Lyons, W. B., A. Fountain, P. T. Doran, J. C. Priscu, K. Neumann, and K. A. Welch. 2000. Importance of landscape position and legacy: The evolution of the lakes in Taylor Valley, Antarctica. *Freshwater Biology* 43:355–367.

Lyons, W. B., S. W. Tyler, R. A. Wharton, Jr. D. M. McKnight, and B. H. Vaughn, 1998a, A late Holocene desiccation of Lake Hoare and Lake Fryxell, McMurdo Dry Valleys, Antarctica. *Antarctic Science* 10:247–256.

Lyons, W. B., K. A. Welch, K. Neumann, J. K. Toxey, R. McArthur, C. Williams, D. M. McKnight, and D. Moorhead. 1998b. Geochemical linkages among glaciers, streams and lakes within the Taylor Valley, Antarctica. Pages 77–92 in J. C. Priscu, editor, *Ecosystem Dynamics in a Polar Desert: The McMurdo Dry Valleys, Antarctica.* American Geophysical Union Antarctic Research Series 72, Washington, D.C.

Marshall, G. J., and J. C. King. 1998. Southern Hemisphere circulation associated with extreme Antarctic Peninsula winter temperatures. *Geophysical Research Letters* 25:2437–2440.

McKnight, D., H. House, and P. Von Guerard. 1994. McMurdo LTER: Streamflow measurements in Taylor Valley. *Antarctic Journal of the United States* 29(5):230–232.

McKnight, D. M., A. Alger, C. M. Tate, G. Shupe, and S. Spaulding. 1998. Longitudinal patterns in algal abundance and species distribution in meltwater streams in Taylor Valley, Southern Victoria Land, Antarctica. Pages 109–127 in J. C. Priscu, editor, *Ecosystem Dynamics in a Polar Desert: The McMurdo Dry Valleys, Antarctica.* American Geophysical Union Antarctic Research Series 72, Washington, D.C.

McKnight, D. M., B. L. Howes, C. D. Taylor, and D. D. Goehringer. 2000. Phytoplankton dynamics in a stably stratified Antarctic lake during winter darkness. *Journal of Phycology* 36:852–861.

McKnight, D. M., D. K. Niyogi, A. S. Alger, A. Bomblies, P. A. Conovitz, and C. M. Tate. 1999. Dry Valley streams in Antarctica: ecosystems waiting for water. *BioScience* 49(12):985–995.

McKnight, D. M., E. VonMaytre, J. Priscu, and W. B. Lyons. 2001. Response of an Antarctic lake ecosystem to climate variation: Linkages between phytoplankton species dynamics and streamflow. ASLO meeting, Albuquerque, NM, February 12–16. Abstracts with the Program.

Moorhead, D. L., P. Doran, A. G. Fountain, W. B. Lyons, D. M. McKnight, J. C. Priscu, R. A. Virginia, and D. H. Wall. 1999. Ecological legacies: Production, persistence and influence on soil and lake ecosystems of the Antarctic dry valleys. *BioScience* 49(12):1009–1019.

Moorhead, D. L., and J. C. Priscu. 1998. Modeling nitrogen transformations in Dry Valley streams, Antarctica. Pages 141–151 in J. C. Priscu, editor, *Ecosystem Dynamics in a Polar Desert: The McMurdo Dry Valleys, Antarctica.* American Geophysical Union Antarctic Research Series 72, Washington, D.C.

Neumann, K., W. B. Lyons, J. C. Priscu, and R. J. Donahoe. 2001. CO_2 concentrations in perennially ice-covered lakes of Taylor Valley, Antarctica, *Biogeochemistry* 56:27–50.

Priscu, J. C. 1998. *Ecosystem Dynamics in a Polar Desert: The McMurdo Dry Valleys, Antarctica.* American Geophysical Union, Antarctic Research Series 72, Washington, D.C.

Smith, R. C., D. Ainley, K. Baker, E. Domack, S. Emslie, B. Fraser, J. Kennett, A. Leventer, E. Mosley-Thompson, S. Stammerjohn, and M. Vernet. 1999. Marine ecosystem sensitivity to climate change. *BioScience* 49(12):393–404.

Treonis, A. M., D. H. Wall, and R. A. Virginia. 2000. The use of anhydrobiosis by soil nematodes in the Antarctic dry valleys. *Functional Ecology* 14:460–467.

Vincent W. F., and C. Howard-Williams. 1986. Antarctic stream ecosystems: Physiological ecology of a blue green algal epilithon. *Freshwater Biology* 16:219–233.

Von Guerard, P., D. M. McKnight, R. A. Harnish, J. W. Gartner, and E. D. Andrews. 1995. Streamflow, water-temperature, and specific-conductance data for selected streams draining into Lake Fryxell, Lower Taylor Valley, Victoria Land, Antarctica, 1990–92. U.S. Geological Survey Open-File Report 94-545.

Welch, K. A., K. Neumann, D. M. McKnight, A. Fountain, and W. B. Lyons. 2000. Chemistry and lake dynamics of the Taylor Valley lakes, Antarctica: The importance of long-term monitoring. Pages 282–287 in W. Davison, C. Howard-Williams, and P. Broady, editors, *Antarctic Ecosystems: Models for Wider Ecological Understanding.* New Zealand Natural Sciences, Christchurch.

Weller, G. 1998. Regional impacts of climate change in the Arctic and Antarctic, *Annals of Glaciology* 27:543–552.

White, W. B., and R. G. Peterson. 1996. An Antarctic circumpolar wave in surface pressure, wind, temperature and sea-ice extent. *Nature* 380(25):699–702.

The Quasi-Quintennial Timescale—Synthesis

Raymond C. Smith

Xiaojun Yuan

Jiping Liu

Douglas G. Martinson

Sharon E. Stammerjohn

Introduction

The El Nino–Southern Oscillation (ENSO) is one of the most important contributors to interannual variability on Earth (Diaz and Markgraf 2000). It is an aperiodic phenomenon that tends to reoccur within the range of 2 to 7 years, and it is manifest by the alternation of extreme warm (El Niño) and cold (La Niña) events. There is also evidence (Allen 2000) that the aperiodic ENSO phenomenon must be considered in conjunction with climate fluctuations at decadal to multidecadal time frames that may modulate ENSO's lower frequency variability. Numerous studies show global climatic impacts associated with the ENSO phenomenon. Further, there is considerable evidence to indicate that ENSO impacts the climate of both middle and high latitudes, and a recent analysis (figure S.1, discussed below) provides a global picture of warm versus cold ENSO conditions. Consequently, it is not surprising that many LTER sites, from the Arctic to Antarctic, show evidence of ENSO-related fluctuations in environmental variables.

The quasi-quintennial timescale of variability is second only to seasonal variability in driving worldwide weather patterns. Consequently, an important theme in part II is the worldwide influence of ENSO-related climate variability and the teleconnected spatial patterns of this variability. Also, a common theme for several ecosystems discussed in this section is their high sensitivity to small climatic changes that are subsequently amplified and cascaded through the system. For example, the narrow temperature threshold for an ice-to-water phase change may create a pronounced nonlinear ecosystem response to what is a relatively small temperature shift (as demonstrated for the McMurdo Dry Valleys). Or alternatively, this narrow temperature threshold may shift a sea ice–dominated ecosystem (Palmer

LTER) to a more oceanic marine ecosystem by reducing the seasonality and magnitude of the sea ice habitat. Such nonlinear amplifications of small climatic changes can increase the ecological response and make it more detectable within the natural background of variability. We explore these themes here.

Global Teleconnections

To illustrate the global footprint of ENSO variability, composites of yearly averaged El Niño and La Niña conditions for surface air temperature (SAT) and sea surface temperature (SST, Reynolds and Smith 1994) were generated. The SAT data were derived from the National Center for Environmental Prediction and National Center for Atmospheric Research (NCEP/NCAR) reanalysis (Kalnay et al. 1996) for the period 1980 to 2000 to avoid the lack of data in earlier years and analysis problems in the Southern Ocean (Kistler et al. 2001). In addition, using data from the last 20 years also avoids the climate regime shift of the late 1970s. Consequently, these results reflect the more recent ENSO teleconnections.

Our analysis involves two averaging steps in generating the composite maps. First, the ENSO yearly average includes data from June of the ENSO onset year to May of the following year. Then, data from all identified El Niño (La Niña) years are averaged together to produce an El Niño (La Niña) composite (see Liu et al. 2002 for details). Finally, the difference between the El Niño and La Niña composites reveals a global ENSO teleconnection pattern in these two temperature fields (figure S.1a,b). The characteristic El Niño pattern with warm anomalies in the central and eastern tropical Pacific and tropical South Indian Ocean and cold anomalies in the western tropical Pacific and subtropical Pacific in both hemispheres stands out strikingly. Associated with the 1–2.5°C warm SAT anomaly (solid contours in figure S.1a) in the tropical Pacific is a warming of the same magnitude in the eastern Ross Sea sector of the Antarctic and a cooling (dashed contours) of about 1.5°C in the Bellingshausen and Weddell Seas. As part of the ENSO footprint, this out-of-phase relationship in the Pacific and Atlantic sectors of the Antarctic represents a high-latitude climate mode named the Antarctic Dipole (Yuan and Martinson 2000, 2001). A Northern Hemisphere counterpart of the Antarctic Dipole appears in the SAT of the western Canada and northeast Canada/Greenland regions as well.

Between the polar and tropical regions, there is also a large cooling anomaly (about 1°C) in the subtropical region of the southwestern United States (in addition to the cooling anomalies in the subtropical North and South Pacific mentioned previously). This analysis clearly reveals a global ENSO footprint. Similar teleconnection patterns exist in the SST field (figure S.1b), except the warm anomaly in the tropics is twice as large as the warm anomaly in the polar regions. Moreover, the ENSO signal not only appears in sea surface temperatures but also is found below the surface. A study of vertical structure in polar oceans reveals that in the Weddell Gyre (Antarctic subpolar Atlantic and westernmost Indian Oceans) upper ocean heat content, salt budget, and water column stability are well correlated with ENSO indices (Martinson and Iannuzzi, 2003).

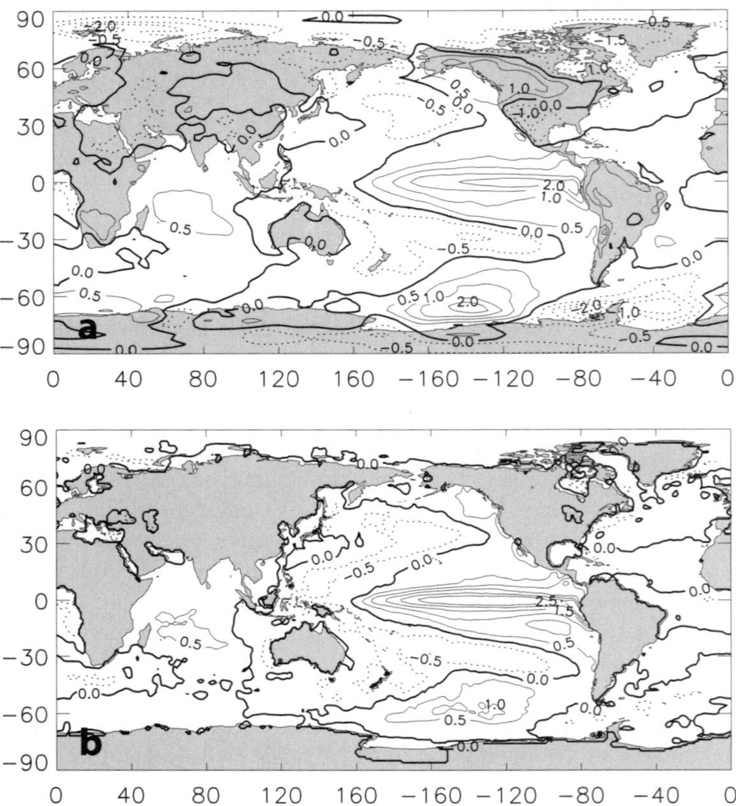

Figure S.1 (a) The composite differences of surface air temperature (SAT, °C) between El Niño (1982–1983, 1986–1987, 1987–1988, 1991–1992, 1997–1998) and La Niña (1984–1985, 1985–1986, 1988–1989, 1995–1996, 1998–1999, 1999–2000) years spanning 1979–2000 from NCEP/NCAR reanalysis. (b) Similar to (a), except for the Reynolds and Smith sea surface temperature (SST, °C) spanning 1981–2000 (Liu et al. 2002). Positive temperatures shown as solid contours, negative temperature shown as dashed contours.

This global ENSO footprint in temperature is similar to spatial patterns revealed in other climate variables. For example, Antarctic sea ice concentrations observed by satellite microwave imagery (Stammerjohn and Smith 1996; Comiso et al. 1997) also show the ENSO teleconnection pattern. Singular value decomposition (SVD), a powerful analysis tool often employed to isolate spatial patterns in two climate fields that tend to covary in time with one another (Bretherton et al. 1992; Wallace et al. 1992), was applied to global SATs and Antarctic sea ice concentrations. Figure S.2 shows the spatial patterns of the leading SVD mode in both variables. The leading spatial pattern in SAT shows a striking similarity to the ENSO footprint revealed in the composite analysis of figure S.1a. Coupled with this pattern is decreased sea ice concentration associated with the warming anomaly in the central and eastern Pacific sector of the Antarctic, and increased sea ice concentration as-

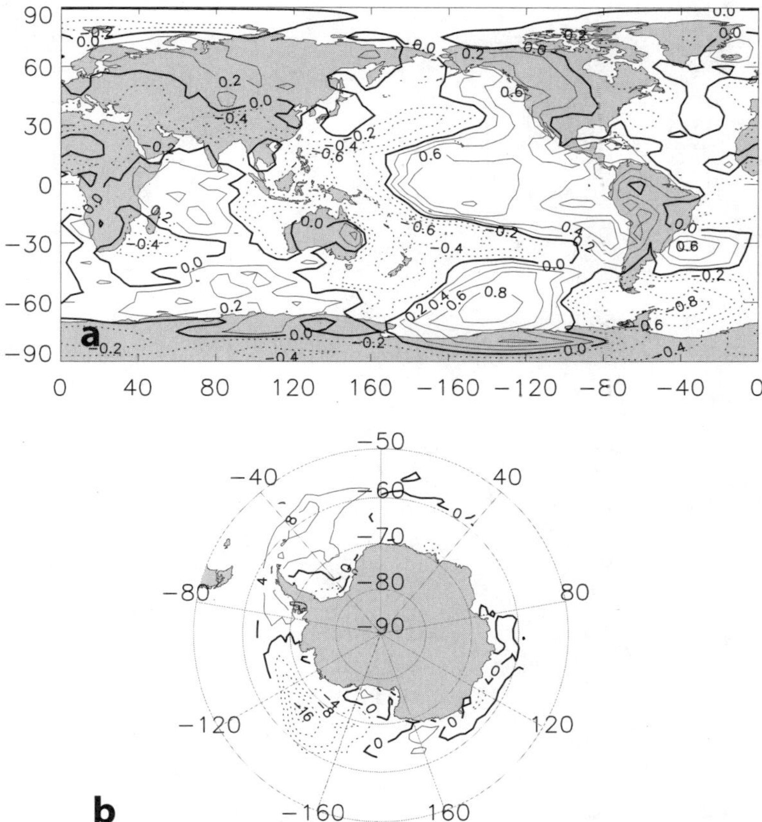

Figure S.2 The spatial patterns of the leading coupled mode in global SAT anomaly field (a) and Antarctic sea ice concentration anomaly field (b) derived by a Singular Value Decomposition (SVD) analysis. The total squared covariance explained by this mode is 15%. Sea ice concentration data were derived from satellite microwave observation from September 1978 to December 1999, whereas SAT data were taken from NCEP/NCAR reanalysis during the same period. Positive (negative) anomalies shown as solid (dashed) contours and are in arbitrary units.

sociated with the cooling anomaly in the Bellingshausen and Weddell Seas (i.e., the Antarctic Dipole in the sea ice field). This analysis indicates that the Antarctic Dipole in the sea ice field is regionally coupled with the dipole in the SAT field and remotely associated with the tropical ENSO pattern as part of the global ENSO teleconnection. The temporal correlation coefficient between the two spatial patterns in figure S.2 is 0.8, indicating that the spatial variability pattern observed in the sea ice field (as shown in figure S.2b) is most likely associated with the spatial variability pattern in the temperature field (as shown in figure S.2a). This is consistent with Yuan and Martinson (2000), who found statistically significant correlations between detrended sea ice edge anomalies in the dipole region and tropical ENSO indices.

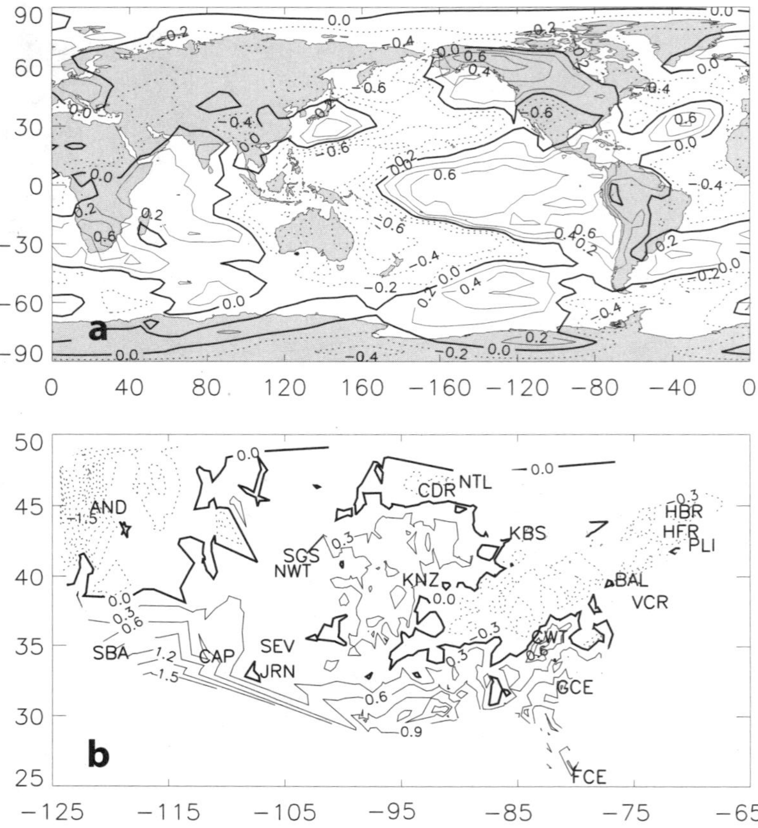

Figure S.3 The second coupled mode between monthly global SAT and U.S. precipitation derived by the SVD analysis. The total squared covariance explained by this mode is 3%. Both temperature and precipitation data span January 1975 to December 1996. The precipitation data were the weather station data provided by NCAR. Positive (negative) anomalies shown as solid (dashed) contours and are in arbitrary units. LTER site symbols as given in table 1.1.

Another example of the global ENSO teleconnection pattern is given by the SVD analysis between global SAT and precipitation in the United States (figure S.3). The monthly precipitation data are from U.S. Historical Climate Network weather stations available from the National Climate Data Center (NCDC). This data set provides high-quality and long-term meteorological data for climate studies (Easterling and Peterson 1995). The second coupled mode reveals nicely the global ENSO teleconnection pattern in the SAT field. Associated with the SAT pattern is a precipitation pattern reflecting above-normal precipitation in the southwestern, midwestern, and southeastern United States (Cayan 1996; Cayan and Peterson 1989), and below-normal precipitation in the northwest, Ohio Valley, and northeast of the United States during warm phases of ENSO (vice versa for cold phases). These precipitation patterns are consistent, for example, with the reported patterns from several LTER sites, as discussed next.

Two pathways can transport climate signals from low to high latitudes: the oceanic bridge and the atmospheric bridge. Although many researchers have studied the atmospheric pathway linking tropics and high latitudes, the underlying mechanisms remain poorly understood. In the Southern Hemisphere, earlier studies suggested that the variability of tropical convection generates Rossby waves that propagate anomalous atmospheric signals out of the tropical regions at intraseasonal timescales (Mo and Higgins 1998; Renwick and Revell 1999). These Rossby waves comprise a barotropic standing wave train of alternating anomalies in the pressure/height fields that extends southeastward from the subtropical Pacific near Australia, across the Antarctic Peninsula, and into the southwestern Atlantic, forming the Pacific–South America (PSA) pattern. The PSA is modulated at interannual timescales, with changes in the strength and phases of the Subtropical Jet, the Polar Front Jet, and the Amundsen Sea Low that comprise the PSA pattern (Mo and Ghil 1987; Karoly 1989; Mo and Higgins 1998).

In addition to the Rossby wave hypothesis, a recent study (Liu et al. 2002) suggests that changes in the regional mean meridional circulation (consisting of the Hadley, Ferrel, and Polar Cells) also drives tropical-polar teleconnections. This study shows that ENSO events change the strength of the regional Ferrel Cell. For example, poleward heat transport in the lower level of the Ferrel Cell in the South Pacific during El Niño years is larger than during La Niña years, with the opposite occurring in the Bellingshausen and Weddell Seas (i.e., the Antarctic Dipole in Ferrel Cell variability). Lui and colleagues (Lui et al. 2002) suggest that changes in the regional Ferrel Cell associated with ENSO variability influence the temperature and sea ice fields at high southern latitudes through the modulation of the mean meridional heat flux. These global teleconnections provide the climate variability linkages observed among the spatially disperse LTER sites.

LTER Site Responses

Many U.S. and international LTER sites are located within the global ENSO footprint. There is good, but not always detailed, congruence between Greenland's overview analysis (table 6.3), showing strength (as determined by correlation analysis) of the SOI for selected LTER site locations (figure 1.1), and the warm and cold locations shown here for the composite surface air temperatures (figure S.1a). There also is good congruence between the overview analysis (table 6.3) and ENSO-related precipitation (figure S.3b) across the United States. In particular, the Southwest United States (Central Arizona, CAP; Jornada Basin, JRN; Sevilleta, SEV) and the Northwest United States (Andrews, AND) show consistency between the earlier correlation analysis and the SVD derived patterns. Sites in the Northeast (HBR, HFR) and Midwest (KNZ, SGS, NWT) that showed weak correlation in Greendland's analysis are areas that show relatively low anomalies in figure S.3b. One reason for a lack of complete congruence is that the various indices (Southern Oscillation Index, SOI; El Niño–Southern Oscillation, ENSO; Multivariate ENSO

Index, MEI; etc.), although highly correlated, are not exactly the same. As Trenberth (1997) points out, both the definition of indexes, which are continually evolving, and the base period climatology of the index can make a difference in spatial coherency, temporal variability, and subsequent correlation analyses with various environmental variables. Beyond this, there are several requirements for an ENSO signal to be imprinted on an ecosystem. First, the given location (i.e., the LTER site locations shown in figure 1.1) must be within the ENSO teleconnection footprint. Figures S.1 and S.3b provide a visual indication of the possible spatial congruence of an LTER site with the ENSO-related temperature or precipitation variability, respectively. Second, the strength of an ecological response to climate variability is itself highly variable and, as noted in the overview by Greenland (chapter 6), dependent on many factors including preexisting conditions, possible cascades and threshold triggers, characteristic ecosystem time scales, and linearity or nonlinearity of the system response.

The arid and semiarid ecosystems of the southwestern United States (CAP, SEV, JRN) provide examples of systems that typically show a strong response to precipitation. These ecosystems are most strongly linked to the ENSO phenomenon during fall, winter, and spring. Some connections are strong enough that it may be possible to make seasonal forecasts on impending conditions based on key index precursors (e.g., chapters 7 and 15). Under El Niño conditions, precipitation and temperature tend to increase and decrease, respectively, whereas regional soil moisture deficit typically decreases. In these ecosystems periods of precipitation offset drought-driven impacts, which are linked to dust storms, wildfires, changes in vegetation, and reduced soil moisture, water quantity, and quality, as well as serious reductions in agriculture and livestock production. Also, as Brazel and Ellis discuss for the CAP urban site (chapter 7), human-modified urban and agricultural ecosystems may display unanticipated and amplified feedbacks with respect to climate-mediated ecosystem relationships. Consequently, efforts to understand underlying mechanisms and to improve predictive capabilities have significant economic and ecological incentives. An oscillation between wet and dry conditions would appear to be the norm for these arid and semiarid ecosystems. Processes whereby these ecosystems may be tuned to quasi-quintennial periods are less clear, particularly given the more than century-long anthropogenic impacts. Brazel and Ellis (chapter 7) provide several examples of how the CAP urban-rural ecosystem responds, typically via several complex cascades, to alternating periods of wet and dry conditions. Clearly, the ENSO signal is imprinted on many components of these arid and semiarid ecosystems that often show amplification because of strong nonlinear responses.

In sharp contrast to the ecosystems of the southwestern United States, the tropical rainforest in Puerto Rico (LUQ) has one of the highest annual rates of precipitation and shows only a weak, if any, quasi-quintennial variability with respect to precipitation. (However, it does show a strong response to temperature consistent with table 6.3 and figure S.1 for the LUQ site.) As Schaefer has noted (chapter 8), hurricanes impact Puerto Rico with an average annual interval of 9.5 years. However, most of the extreme rainfall events are not linked to hurricanes, tropical storms, or tropical depressions. Again, for this ecosystem the response to extreme rainfall events is very nonlinear, with 75% of sediment export from watersheds occurring during only 1% of the days with the greatest rainfall. Schaefer notes that surface flow of water is the minimum necessary precondition for sediment mobilization and that this occurs whenever rainfall rates exceed the hydraulic conductivity of the surface soil. Also, soil erosion and sediment deposition do not generally return to a previous state after an extreme event, and the biota may or may not persist in the remaining soil resource. Thus, an important characteristic of these tropical ecosystems is their nonlinear response to rainfall. Consequently, the variability of annual rainfall totals is of less significance than the variability of extreme events.

For the Antarctic sites (MCM, PAL), temperature is a key climatic factor influencing ecosystem structure and function, because it directly influences the phase state of water. In the McMurdo Dry Valleys (MCM), relatively small interannual variations in temperature, cloudiness, and solar radiation determine the number of days above freezing, thus the availability of liquid water. In turn, this availability of liquid water is an important driver for the function and biodiversity of the MCM ecosystem. For the Antarctic marine ecosystem (PAL), relatively small temperature changes give rise to glacier melt, ice shelf collapse, and sea ice reduction, with subsequent impacts at all trophic levels within this sea ice–dominated marine ecosystem.

For both Antarctic ecosystems the ice-to-water phase transition is, therefore, a critical temperature threshold that induces nonlinear ecological response. Consequently, both ecosystems are extremely sensitive to climate variability. The McMurdo Dry Valleys have cooled during the past several decades (Doran et al. 2002). However, the relatively short temperature and pressure records do not exhibit a significant correlation with ENSO (Welch et al. chapter 10). In contrast, temperature trends in the western Antarctic Peninsula region (Smith et al., chapter 9) show a statistically significant warming, and temperature and sea ice extent are strongly correlated with ENSO variability. Thompson and Solomon (2002) have recently interpreted climate change in the Southern Hemisphere (SH) in terms of the SH annular mode (SAM), a large-scale pattern of variability characterized by fluctuations in the strength of the circumpolar vortex. Also, Liu et al. (2002) have discussed mechanisms linking ENSO and southern high latitude climate teleconnections. These authors provide evidence that illuminates the connections between the seemingly disparate trends observed at MCM and PAL.

The western Antarctic Peninsula, the location of the Palmer LTER (PAL), is recognized as a "hot spot" in terms of global warming (IPCC 2001). Smith and colleagues (chapter 9) review and summarize statistically significant climate changes observed in this region of the Antarctic and discuss the response of the marine ecosystem to these changes. This sea ice–dominated marine ecosystem is located between relatively warm and moist maritime conditions to the north and cold and dry continental conditions to the south. Consequently, a small climate shift can be amplified directly by a latitudinal change in climate regimes and by subsequent shifts in the sensitive balance between the solid and liquid phase of water. Thus, climate variability becomes amplified, and this physical forcing appears to influence all trophic levels of the ecosystem. This influence is relatively direct, because physical forcing and ecological response are tightly coupled in marine systems (Steele

1978). Interestingly, the ecological response most clearly seen is in the population size and distribution of upper level predators. The life histories of three sympatric, congeneric penguin species are closely related but have different breeding cycles tied to their preferred habitats: ice-free or ice-covered waters. The abundance and distribution of these species have shifted in response to the timing and magnitude of seasonal sea ice extent. Although the mechanisms that control the ecological response to shifting climate conditions are not fully understood, the paleoecological record also suggests a tight coupling between climate variability and ecological response. As noted by the authors, this implies that climate-induced ecological effects cascade through the system rapidly.

The McMurdo Dry Valleys (MCM) are cold deserts composed of a mosaic of alpine glaciers, exposed bedrock, ephemeral streams, arid soils, and perennially ice-covered lakes. Welch and colleagues (chapter 10) emphasize the nonlinear and amplifying influence of the sensitive balance between solid and liquid water. The key climate parameters that influence this ecosystem are those that affect the conversion of solid to liquid water: temperature, solar radiation, and precipitation. Since liquid water remains the primary limiting condition to life in the MCM, any such climate-related change has a significant impact on the hydrologic budget with subsequent cascades through the system. In spite of this ecosystem's sensitivity to small-scale climate variability (relative to temperate latitudes), Welch and colleagues find that the temperature and pressure records do not exhibit a significant correlation with ENSO. As the authors suggest, this may be because the records are too short to reveal statistically significant trends. It may also be because the ENSO-related temperature influence is relatively less intense in this area, as suggested by figure S.1.

Summary

The global teleconnection patterns (figures S.1–S.3) discussed previously show a very broad spatiotemporal coherency. The analysis shown here is unique because it includes high-latitude areas. Thus, LTER sites from the Arctic to the Antarctic can be placed within these global patterns. Other studies (Allan 2000) and the chapters in part III of this volume present evidence to suggest that the "classical" ENSO-like patterns discussed here must be considered within a lower frequency envelope of concurrent decadal to multidecadal time periods. Also, at longer timescales, there is growing evidence that ENSO may not be spatially or temporally stable, so the patterns shown here are expected to vary in the longer term. This reinforces the idea that the various timescales used for organizing this book cannot be viewed in complete isolation and that climate forcing and ecological response will vary across a wide range of time and space scales.

Several ecosystems discussed in this section (CAP, LUQ, PAL, MCM) were distinguished by having processes whereby nonlinear ecosystem amplification of the ENSO climatic pattern gave rise to observable response. These sites were also distinguished by being on the outside edge of the cluster of sites distributed on a plot of mean annual temperature versus precipitation, as shown in figure 1.2. Ecosystems will vary in response and susceptibility to climate variability. We can speculate that possible ecosystem characteristics that make sites likely to show a more observable response to climate variability at timescales discussed in this section include a location within the ENSO teleconnection footprint of strong variability, a nonlinear ecosystem amplification, and with a climate "on the edge" with respect to mean values of temperature and/or precipitation.

References

Allan, R. J. 2000. ENSO and climatic variability in the past 150 years. Pages 3–55 in H. F. Diaz and V. Markgraf, editors, *El Niño and the Southern Oscillation*. Cambridge University Press, Cambridge.

Bretherton, C. S., C. Smith, and J. M. Wallace. 1992. An intercomparison of methods for finding coupled patterns in climate data. *Journal of Climate* 5:541–560.

Cayan, D. R. 1996. Interannual climate variability and snowpack in the western United States. *Journal of Climate* 9:928–948.

Cayan, D. R., and D. H. Peterson. 1989. The influence of the North Pacific atmospheric circulation and streamflow in the West. In *Aspects of Climate Variability in the Western Americas*, D. H. Peterson (ed.), Geophysical. Monographs. 55, American Geophysical Union, Washington, D.C., 375–397.

Comiso, J. C., D. J. Cavalieri, C. L. Parakinson, and P. Gloersen. 1997. Passive microwave algorithms for sea ice concentration: A comparison of two techniques. *Remote Sensing Environment* 60:357–387.

Diaz, H. F., and V. Markgraf. 2000. *El Niño and the Southern Oscillation*. Cambridge University Press, Cambridge.

Doran, P. T., J. C. Priscu, W. B. Lyons, J. E. Walsh, A. G. Fountain, D. M. McKnight, D. L. Moorhear, R. A. Virginia, D. H. Wall, G. D. Clow, C. H. Fritsen, C. P. McKay, and A. N. Parsons. 2002. Antarctic climate cooling and terrestrial ecosystem response. *Nature* 415:517–520.

Easterling, D. R., and T. C. Peterson. 1995. A new method of detecting undocumented discontinuities in climatological time series. *International Journal of Climatology* 15:369–377.

IPCC. 2001. *Climate Change 2001, Synthesis Report*. A contribution of Working Groups I, II and III to the Third Assessment Report of the Intergovernmental Panel on Climate Change [R. T. Watson and the Core Writing Team, editors]. Cambridge University Press, Cambridge.

Kalnay E., M. Kanamitsu, R. Kistler, W. Collins, D. Deaven, L. Gandin, M. Iredell, S. Saha, G. White, J. Woollen, Y. Zhu, M. Chelliah, W. Ebisuzaki, W. Higgins, J. Janowiak, K. C. Mo, C. Ropelewski, J. Wang, A. Lettma, R. Reynolds, R. Jenne, and D. Joseph. 1996. The NCEP/NCAR 40-year reanalysis project. *Bulletin of the American Meteorological Society* 77:437–471.

Karoly, D. J. 1989. Southern hemisphere circulation features associated with El Niño–Southern Oscillation events. *Journal of Climate* 2:1239–1252.

Kistler, R., E. Kalnay, W. Collins, S. Saha, G. White, J. Woollen, M. Cheliah, W. Ebisuzaki, M. Kanamitsu, V. Kousky, H. van den Dool, R. Jenne, and M. Fiorino. 2001. The NCEP-NCAR 50-year reanalysis: Monthly means CD-ROM and documentation. *Bulletin of the American Meteorological Society* 82:247–267.

Liu, J., X. Yuan, D. Rind, and D. G. Martinson. 2002. Mechanism study of the ENSO and southern high latitude climate teleconnections. *Geophysical Research Letters*, 29(14), 24–1:24–4.

Martinson, D. G. and R. A. Iannuzzi. 2003. Spatial/temporal patterns in Weddell Gyre characteristics and their relationship to global climate. *Journal of Geophysical Research Oceans*, in press.

Mo, K. C., and M. Ghil. 1987. Statistics and dynamics of persistent anomalies. *Journal of Atmospheric Science* 44:877–901.

Mo, K. C., and R. W. Higgins. 1998. The Pacific-South American modes and tropical convection during the Southern Hemisphere winter. *Monthly Weather Review* 126:1581–1596.

Renwick, J. A., and M. J. Revell. 1999. Blocking over the South Pacific and Rossby wave propagation. *Monthly Weather Review* 127:2233–2247.

Reynolds, R. W., and T. M. Smith. 1994. Improved global sea surface temperature analyses using optimum interpolation. *Journal of Climate* 7:929–948.

Stammerjohn, S. E., and R. C. Smith. 1996. Spatial and temporal variability of western Antarctic Peninsula sea ice coverage. Pages 81–104 in R. M. Ross, E. E. Hofmann, and L. B. Quetin, editors, *Foundations for Ecological Research West of the Antarctic Peninsula*, American Geophysical Union, Washington, D.C., Antarctic Research Series 70.

Steele, J. H., ed. 1978. *Spatial pattern in plankton communities*. Plenum, New York.

Trenberth, K. E. 1997. The definition of El Niño. *Bulletin of the American Meteorological Society* 78:2271–2777.

Thompson, D. W. J., and S. Solomon. 2002. Interpretation of recent southern hemisphere climate change. *Science* 296:895–899.

Wallace, J. M., and C. Smith, and C. S. Bretherton. 1992. Singular Value Decomposition of Wintertime Sea Surface Temperature and 500-mb Height Anomalies. *Journal of Climate* 5:561–576.

Yuan, X., and D. G., Martinson. 2000. Antarctic sea ice variability and its global connectivity. *Journal of Climate* 13:1697–1717.

Yuan, X., and D. G. Martinson. 2001. The Antarctic dipole and its predictability. *Geophysical Research Letters* 28:3609–3612.

Part III

The Interdecadal Timescale

Introductory Overview

Douglas G. Goodin

Timescale is the organizing framework of this volume. In various sections, we consider the effects of climate variability on ecosystems at timescales ranging from weeks or months to centuries. In part III, we turn our attention to interdecadal-scale events. The timescales we consider are not absolutely defined, but for our purposes we define the interdecadal scale to encompass effects occurring with recurring cycles generally ranging from 10 to 50 years. A recurring theme in many of the chapters in this section is the effect on ecosystem response of teleconnection patterns associated with recognized quasi-periodic atmospheric circulation modes. These circulation modes include the well-known El Niño–Southern Oscillation (ENSO) phenomenon, which is generally thought to recur at shorter, interdecadal timescales but also includes some longer-term periodicities. Several other climate variability modes, including the Pacific North American index (PNA), North Atlantic Oscillation (NAO), Pacific Decadal Oscillation (PDO), and North Pacific index (NP) also show strong interdecadal scale signatures and figure prominently in the chapters of part III.

McHugh and Goodin begin the section by examining the climate record at several North American LTER sites for evidence of interdecadal-scale fluctuation. They note that interdecadal-scale contributions to climate variability can best be described in terms of two types of variation: (1) discontinuities in mean value, and (2) the presence of trends in the data. Evaluation of interdecadal periodicities in LTER data is complicated by the relatively short time series of observations available. McHugh and Goodin approach the problem mainly through the use of power spectrum analysis, a widely used tool for evaluating the periodicity in a time series of data. Principal components analysis is used to decompose the time series of growing-season climate data for each of the LTER sites into their principal modes of variability. These modes are then subjected to power spectrum analysis to evaluate the proportions of the variance in the data occurring at various timescales. McHugh and Goodin's results suggest that significant effects on precipitation and temperature at interdecadal timescales are uncommon in these data, although significant periodicities at both shorter and longer frequencies do emerge from the data (a finding of relevance to other sections of this volume). McHugh and Goodin further note that these results are for all LTER sites considered together and that significant interdecadal effects may emerge at specific sites, an observation relevant to the cross-timescale analysis appearing elsewhere in this volume. These results also suggest something about the geography of climate variation and how data from LTER sites may contribute to an understanding of the effects of climate variability in a number of ecosystem types.

In chapter 12, Glenn Juday and colleagues turn their attention toward a more direct observation of the interaction of climate and ecosystem response. Juday et al. use dendrochronological techniques to characterize 200 years of climate variability at the Bonanza Creek LTER, near Fairbanks, Alaska. Their recent reconstruction of nineteenth century temperatures in this area of Alaska indicates a pattern of quasi-decadal-scale variability, possibly associated with the PDO teleconnection pattern. Juday et al. compare this proxy temperature record to tree-ring width records for other parts of Alaska. They then use temperature records as part of a model of seed production in white spruce trees. Their results show a surprisingly warm climate over most of Alaska during the past two centuries. Temperature and precipitation patterns are coordinated in an inverse relationship, indicating a tendency for climatic conditions to remain either hot and dry or cool and wet. Juday and colleagues show that climate variability is closely tied to the reproductive cycle of the white spruce. They note that white spruce cone and seed production occurs episodically and in response to environmental cues. They suggest climate variability as an obvious candidate to provide these environmental cues. Examination of a 39-year record of white spruce reproduction at and around the BNZ LTER shows that seed drops are episodic, with major seed production events generally occurring at intervals of 12–17 years. Juday et al. note that major seed production events appear to be initiated during El Niño years. These results are an excellent example of the close relationship between climate variability and an important ecological response.

A somewhat different but equaling compelling link between climate and ecological response is made in Greenland's chapter (13) on climate variation and Coho salmon catch. Greenland uses the example of Coho salmon catches to illustrate how climate observations made at one LTER site (H. J. Andrews Forest, Oregon, AND) were generalized to a regional scale (the coastal Pacific Northwest) and then linked to a biological observation (Coho salmon population as represented by catch) that is important both ecologically and economically to the region of interest. In so doing, Greenland also provides an illustration of the fundamental framework questions underlying this volume, and a suggestion for further research guided by these framework questions. Greenland's investigation of salmon catch grew out of his analysis of climate at the AND LTER; he found salmon catch to be well related to temperature in western Oregon and to the broader region of the coastal Pacific Northwest. Comparisons of these temperatures with salmon catch data showed a strong inverse relationship, apparently at a cycle of approximately 20 years. Greenland links these two cycles through a five-level cascade model relating the PDO to nutrient availability, and thus to salmon population. In Greenland's cascade model, air temperature effects are a by-product of pressure changes (thus having no direct effect on salmon ecology), but are nevertheless strongly linked statistically to salmon catch. Greenland's results well illustrate the indirect and sometimes complex interaction between climate and ecosystem response.

In chapter 14 Hayden and Hayden examine potential changes in storm frequency at LTER sites in North America. They use data from the National Weather Service to map the presence and path of extratropical cyclonic storms during their mature (i.e., close isobar) stage. For their analysis, they selected 19 LTER sites, all located in the middle latitudes. They divided these sites into five geographically defined groups representing the interior west, west coast, Midwest, Appalachian region, and east coast. They examined storm data for these regions in terms of three questions: (1) Has storm frequency changed over the past century, and have these changes been geographically consistent? (2) Are changes in storminess consistent with model forecasts of climate change? (3) Is there a characteristic pattern associated with ENSO effects? Hayden and Hayden approach this problem using a gridded database of storms. They selected the grid cells associated with each LTER site and then extracted the index of storm counts from each LTER grid cell. Their results indicate that storm frequencies over the continental United States have changed during the past century, but the pattern of change can only be resolved at spatial scales below continental. Large changes in storm frequency are noted at most LTER sites; however, when averaged at a continental scale, no changes are apparent. The ENSO phenomenon showed little influence over storm frequency, and changes in storminess were not consistent with model predictions. As in McHugh and Goodin's chapter, Hayden and Hayden's results provide insight into the geographic effects of climate variability, as well as advancing important conclusions about the effects of scale in the observation of climate change.

In the final chapter of part III, Milne et al. use dendrochronology to analyze drought cycles at the Sevilleta LTER in New Mexico. They converted annual estimates of precipitation into anomalies by subtraction of the long-term mean, then used a probabilistic technique based on cumulated precipitation anomalies to evaluate the data time series. Their analysis revealed repeated, persistent wet and dry periods, most notably seven lengthy strings of dry years constituting persistent drought conditions. Milne et al.'s analysis tentatively suggests a 55- to 62-year recurrence interval for drought conditions, an interval coinciding with recent major droughts in the 1890s and 1950s. Based on this recurrence interval, they predict a prolonged period of lower precipitation in this region beginning in 2001 and persisting well into the second decade of the twenty-first century. Along with their climatic analysis, Milne et al. consider the effect that persistent drought might have on the transitional biome found at the SEV LTER. They suggest that long-term drought conditions could lead to changes in species composition, particularly between shrubland and woodland. Based on past interaction between two dominant species *Larrea tridentata* (a desert shrub) and *Juniper monosperma* (a woodland species) during the drought of the 1950s, Milne et al. hypothesize that drought conditions should favor establishment of *L. tridentata,* but thus far a simple model of species interaction under drought conditions remains elusive. Results of Milne et al. suggest the complex links between climate and ecosystem response and the variety of mechanisms by which these links may be expressed.

The five chapters in this section represent a cross section of climate and ecosystem effects at interdecadal scales. The variety of climatic processes considered and the frequent indication of periodicities not strictly defined at an interdecadal scale clearly indicate the difficulty of isolating one time period or one climatic process for intensive study. These chapters reinforce the across-the-scale complexity of the climate-biosphere system.

11

Interdecadal–Scale Variability: An Assessment of LTER Climate Data

Maurice J. McHugh

Douglas G. Goodin

Interdecadal-scale climate variability must be considered when interpreting climatic trends at local, regional, or global scales. Significant amounts of variance are found at interdecadal timescales in many climate parameters of both "direct" data (e.g., precipitation and sea surface temperatures at specific locations) and "indirect" data through which the climate system operates (e.g., circulation indices such as the Pacific North American index [PNA] or the North Atlantic Oscillation index [NAO]). The aim of this study is to evaluate LTER climate data for evidence of interdecadal-scale variability, which may in turn be associated with interdecadal-scale fluctuations evident in ecological or biophysical data measured throughout the LTER site network.

In their conceptualization of climatic variability, Marcus and Brazel (1984) describe four types of interannual climate variations: (1) Periodic variations around a stationary mean are well known to occur at short timescales, such as diurnal temperature changes or the annual cycle, but are difficult to resolve at decadal or longer timescales. (2) Discontinuities generated by sudden changes in the overall state of the climate system can reveal nonstationarity in the mean about which data vary in a periodic or quasi-periodic manner. These sudden alterations can result in periods perhaps characterized by prolonged drought or colder than normal temperatures. (3) The climate system may undergo trends such as periods of slowly increasing or decreasing precipitation or of warming or cooling until some new mean "steady" state is reached. (4) Climate data may exhibit increasing or decreasing variability about a specific mean value or steady state. Interdecadal contributions to climate variability can be described in terms of types 2 and 3 of Marcus and Brazel's conceptual classification—discontinuities in the mean and trends in the data. Records of the Northern Hemisphere's average land surface temperature show discontinuities in the mean state of the hemispheric temperature record in conjunction with obvious trends. Conceptually, it is hard to distinguish between these aspects of climate variability. Trends are an essential component of an alteration in the mean state of the temperature series, as they serve as a temporal linkage between the different mean states.

Climate Variability: Complexity and Chaos

Evaluating the causes of climatic variability tends to be difficult for a variety of reasons. Data, even when accurately measured during a lengthy period, tend to be significantly affected by local factors such as topography, urbanization, or location relative to large water bodies. Additionally, significant amounts of variability can often be attributed to larger scale associations with climate variability in distant locations. Climatologists, therefore, evaluate climate data in terms of in situ trends and variations but also in terms of the processes causing or associated with those variations, such as ocean-atmosphere circulation anomalies in far-distant locations, which are referred to as "teleconnections"—literally meaning "remote connection." Therefore, climate data must be understood to be influenced by both local- and large-scale climatic forcing. One of the inherent problems in evaluating teleconnections is that the apparent association between measures of atmospheric circulation or climate variability and far-flung locations may result from indirect causal pathways. For example, forcing by variable C may produce the apparently significant correlation between variables A and B.

The NAO, for example, is a measure of the sea level pressure gradient between Iceland and the Azores, and indirectly of the westerly wind speed across the North Atlantic Ocean, and it is long known to have statistically significant associations with rainfall over portions of India or China. Whether the NAO actually causes rainfall variability in India or China or whether those phenomena are correlated because they are both being driven by some external phenomenon remains to be determined.

Further complicating the matter, the climate system is inherently nonlinear in nature because of the large number of feedback mechanisms of which we are aware. Dynamic and thermodynamic thresholds within the climate system ensure that analysis and evaluation of climate variability and change is complicated. A myriad of small changes can result in the atmosphere changing from a stable state to an unstable state, from a laminar to a turbulent flow regime, and from ice melting to remaining solid. For example, alterations in atmospheric circulation can affect surface temperatures and increase snowmelt. Reduced snowcover results in decreased albedo, with energies that otherwise would be reflected now available to cascade through various climate system components. Each of these components may further enhance those energies, producing a positive feedback or eliminating the energy from the climate system altogether in a negative feedback. The often large and important differences in atmospheric state produced by subtle changes in climatic variables attest to the existence and power of these thresholds and feedback mechanisms. These thresholds and feedbacks can produce, or magnify, the apparently chaotic nature of the climate system.

Traditionally, statistical analysis of climate data relies heavily on linear methods, such as linear regression or simple correlations. Nonlinear statistical methods using techniques based on Bayesian analysis and neural networks are beginning to slowly permeate the atmospheric sciences. In some circumstances, neural networks are merely being used to produce results comparable to those of a linear regression, but in other circumstances the method is being used in truly nonlinear applications such as nonlinear principal components analysis (Hsieh and Tang 1998; Hsieh 2001).

Analysis and evaluation of the causal nature of the interrelations among climate trends, periodicities, and remote linkages is therefore extremely complex, with atmospheric scientists, including climatologists, well aware of the analytical problems they face. Given these issues, many atmospheric scientists believe that the only correct and appropriate manner in which causal analysis can be performed is through the use of sophisticated numerical models of the entire climate system, including coupling between the atmosphere and components such as the oceans, cryosphere, and, importantly, the biosphere.

Recurrent Variability in the Climate System

Recurrent or quasi-periodic variability intrinsic in the climate system can be found at timescales of 1 day, 365 days, several years, and possibly several decades. Our records of the climate system have been gathered on a near-global scale only since the end of the 1940s, but they have been truly global only since the advent of the modern satellite in the 1970s (Kalnay et al. 1996). This relatively short timescale prevents truly global analyses of intermediate and long-term timescales of variability. However, individual stations have been recording data since the 1600s in Europe (Vose et al. 1992). Climatologists have used various records of climate change and variability from before the 1600s, such as records of French grape harvest dates (e.g., Lamb 1977), documentary evidence of historical climate changes (e.g., Lamb 1977; van Loon and Rogers 1978; Mantua et al. 1997), dendrochronology (e.g., Minobe 1997), ice core data (e.g., Thompson et al. 1998), and sedimentary records from the ocean bottoms (e.g., Hays et al. 1976).

Causes of interdecadal-scale variability in the climate system are not very well understood and are thought to include solar variability, air-sea interaction, and certain coupled modes of atmospheric variability that exhibit quasi-periodic behavior. Sunspot numbers have a pronounced periodicity at 11 years (Labitzke and van Loon 1988; Mitra et al. 1991; Currie and O'Brien 1992) and a lesser known periodicity at 22 years (Vines 1986). Some well-known modes of atmospheric variability are also associated with interdecadal-scale variability. In particular, the Pacific Decadal Oscillation (PDO) in sea level pressure over the northern Pacific Ocean (Trenberth and Hurrell 1994) is known to play an important role in modulating sea surface temperatures (SSTs) across the North Pacific, which in turn affect the Pacific Salmon catch (e.g., Mantua et al. 1997; chapter 13 of this volume). This is only one example of the well-documented and extensive changes in the climate over the North Pacific that occurred in the late 1970s (e.g., Namias 1978; Trenberth 1990;

Graham 1994; Trenberth and Hurrell 1994)—changes that appear strongly coupled to concurrent ecological changes (e.g., Venrick et al. 1987; Ebbesmeyer et al. 1991; Francis and Sibley 1991; Brodeur and Ware 1992; Hollowed and Wooster 1992; Beamish and Boullion 1993; Francis and Hare 1994; Mantua et al. 1997).

Given the varied nature of these interdecadal-scale signals in the climate system, their detection in instrumental climate data proves to be quite difficult for a variety of reasons. Quasi-periodic signals at a variety of timescales can be envisaged reverberating throughout the climate system simultaneously with a multitude of other nonperiodic signals. Because signals tend to oscillate at certain periodicities, occasionally the combined amplitude of the signals may overlap and reinforce each other, or cancel each other out. Climate signals may have direct effects on climate data (e.g., through modulation of precipitation or temperature), and thus indirectly on ecological data, whereas other climatic signals may have direct effects on ecological data (e.g., solar variability). In addition to quasi-periodic signals in the climate system, nonoscillatory factors such as humankind's alteration of Earth's surface and atmospheric chemistry, as well as changes in the energy and carbon content of oceanic bottom waters, may induce additional degrees of variance into the system, making it more difficult to isolate and describe one specific periodic signal.

Data and Methodology

Additional problems arise in the statistical description and evaluation of the existence and significance of these quasi-periodic signals. One technique commonly used to evaluate periodic components of data variance is spectrum analysis, which relies on its ability to estimate proportions of variance at specific frequencies (periods/year). It is known that some proportion of the variance at one frequency can "leak" into adjacent frequencies and contaminate the spectral estimates by spreading the information too widely across several frequencies. This is referred to as spectral leakage (Mitchell et al. 1966). Other biases may occur if trends in the data are not removed (e.g., Warner 1998).

LTER climate data used include monthly mean, maximum, and minimum temperatures and precipitation where available at LTER stations. Data at Palmer Station in Antarctica were analyzed in addition to North American data. Data were obtained from the LTER CLIMDES database, available on the internet (http://intranet.lternet.edu/archives/documents/Publications/climdes/index.html) (figure 11.1). To relate significant findings to ecologically relevant research, only average growing-season (March to September) data are used.

There are two potential approaches to the efficient and accurate identification of interdecadal-scale variability in the LTER climate data. One approach is to examine the relationship between LTER data and indexes of climatic phenomena known to have a substantial portion of their variance occurring at interdecadal timescales, for example, sunspot numbers or the PDO. The second approach is to use power spectrum analysis to evaluate the proportions of variance in the climate data occurring at interdecadal timescales. Both of these methods were used in this study,

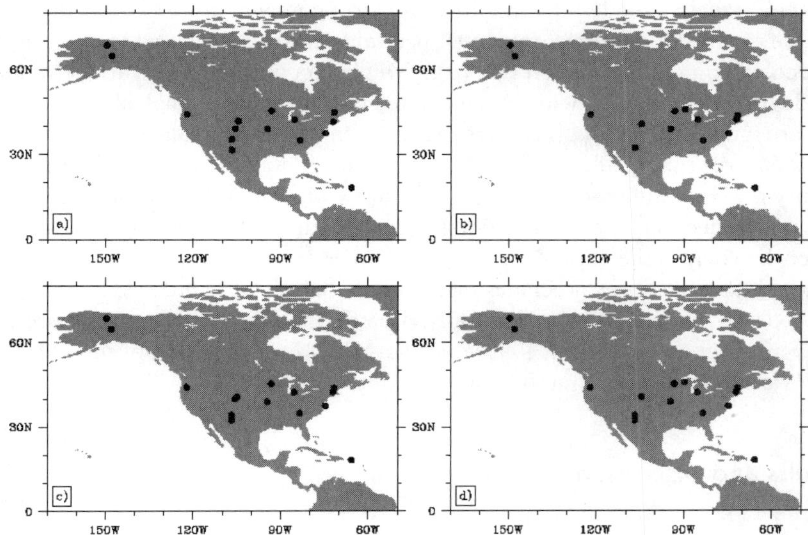

Figure 11.1 Locations of LTER stations recording (a) mean temperature, (b) maximum temperature, (c) minimum temperature, and (d) precipitation data used in this study. Palmer Station, Antarctica, is included in the mean temperature data set but is not plotted here.

but only the latter approach is described in detail. This is because most of the well-known climate phenomena exhibiting significant interdecadal-scale variability are most readily apparent and are at their strongest during winter. This may be due to the nature and origin of the climate phenomena and/or to the relatively low variance found in winter climate data that allows easier signal detection.

For example, interdecadal-scale variability is dominant in Trenberth and Hurrell's (1994) index of North Pacific sea level pressure between November and March. Other times of the year show little, if any, interdecadal-scale variability over the North Pacific. This merely represents a shift from polar domination of midlatitude climate variability during the winter months to a regime increasingly influenced by tropical variability during the nonwinter months. Correlations performed between Trenberth and Hurrell's (1994) North Pacific sea level pressure index and LTER growing-season data do produce statistically significant relationships, but it is possible that large portions of the covariance between the series do not occur at interdecadal timescales, but at interannual timescales during the growing season. The timescales at which the covariance occurs may be assessed by cross-spectrum analysis (Warner 1998). However, Goodin et al. (chapter 20 in this volume) show that some interdecadal-scale climate signals are found to correlate significantly with growing-season (April–September) temperature, precipitation, and annual net primary productivity at the Konza Prairie LTER site near Manhattan, Kansas.

To allow efficient analysis of LTER climate data, principal component analysis (PCA) is used to decompose time series of the mean growing season climate data into their principal modes of variability, principal components (PCs). The first four PCs are retained for varimax rotation. Spectral analysis is then used to evaluate the variance spectra of the rotated PCs (e.g., Mitchell et al. 1966). PCs are detrended, and deviations from the mean are calculated prior to spectral analysis; all missing data are ignored. The relatively large proportion of missing data prior to the mid-1950s ensures that results from the LTER climate data set for the first half of the century is biased in favor of those few stations with data. It is possible that the results may reflect the interdecadal components of variability relevant to those sites, rather than across the data set as a whole. However, elimination of those stations without long time series precludes meaningful analysis of interdecadal-scale variability at most LTER sites. Additionally, use of LTER climate data over more limited periods (e.g., Greenland 1999) does not allow for evaluation of interdecadal-scale variability in a meaningful manner.

Results and Discussion

Mean Temperature

The first four rotated PCs of growing-season mean temperature account for almost 62% of its variance. Spectral analysis performed on each of the rotated PCs (figure 11.2) depicts statistically significant periodicities in PCs 1, 2, and 4. Although mean temperature data are dominated by low-frequency variability, statistically significant peaks are found between 3.0 and 3.5 years in the first two components, and a significant peak occurs at 2.88 years in PC 4, indicating the importance of quasi-triennial periodicities in the LTER growing season mean temperature data. These quasi-triennial periods appear similar to those found in the El Niño–Southern Oscillation (ENSO) signal that has variance concentrated between 2 and 3 years and also between 4 and 6 years (Rogers 1984; Keppene and Ghil 1992; Mann and Park 1994; McHugh 1999; Greenland 1999). Significant quasi-quintennial periodicities are observed in PC 4, and a prominent but insignificant peak in PC 1 occurs in this latter band of ENSO variance.

Statistically significant low-frequency variability between 50 and 100 years is evident in the spectra of each PC, except PC 3. A similar 50- to 70-year spectral peak was found in North American tree ring data (Minobe 1997), whereas the 100-year periodicity may be related to solar forcing (Friis-Christensen and Laasen 1991) and/or anthropogenic factors (IPCC 1995). However, PC 3 exhibits a significant concentration of spectral variance at interdecadal periodicities, but the relatively small amount of variance explained by PC 3 indicates that this interdecadal signal is of fairly small importance to the overall data set. More prominent, but statistically insignificant, signals at and near this periodicity are also evident in the spectrum of PC 1. The spatial distribution of PC loadings is rendered relatively meaningless because of the minuscule number of sites across North America, the lack of any coherent trend or distribution, and the juxtaposition of nearby sites with strong positive and negative loadings. Loading maps are therefore not displayed.

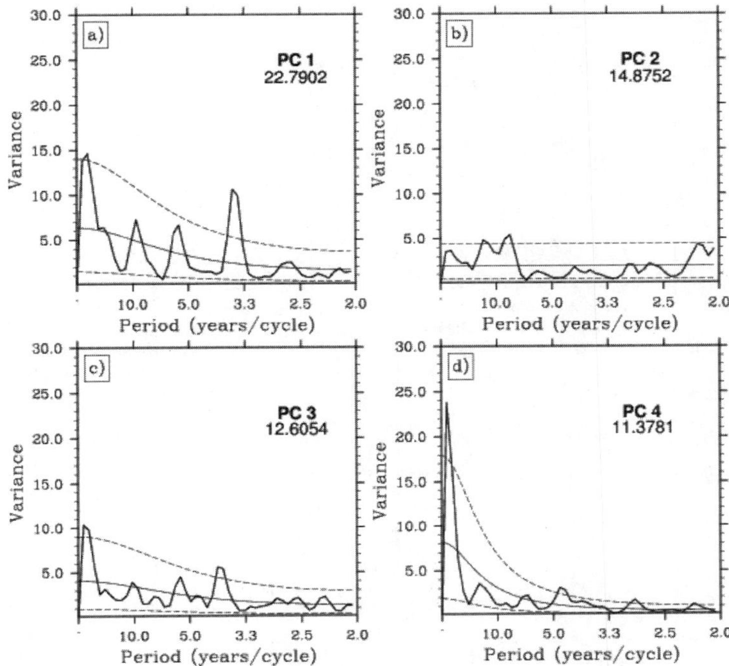

Figure 11.2 Spectral analysis of the principal components of mean temperature (a) PC 1, (b) PC 2, (c) PC 3, and (d) PC 4. The percentage of the data set's variance accounted for by each PC is measured along the y-axis. The thick solid line depicts the spectral estimates, which are unitless; period lengths are indicated along the x-axis in years/cycle. The red noise spectral power estimates represented by the uppermost dashed line represent the 95% confidence limit; the lower dashed line represents the red noise spectral power estimates at the 5% confidence limit. The middle (solid) line represents the 50% confidence level.

Maximum Temperature

In comparison to the mean temperature spectra, low-frequency variability dominates the spectra of PC 1, PC 3, and PC 4 (figure 11.3); significant periodicities at about 50 years are evident in all three spectra. Significant interdecadal-scale variability is evident in PC 3, with smaller and insignificant peaks also observed at this periodicity in PC 1. The spectrum of PC 1 closely resembles that of the first PC of mean temperature, depicting a significant quasi-triennial periodicity in addition to a smaller nonsignificant periodicity between 5 and 6 years. These periodicities are similar to those associated with ENSO, and suggest an interrelationship such as the one described by Greenland (1999). Statistically significant quasi-biennial variability is evident in PC 3 and PC 4, in addition to the aforementioned low-frequency signals.

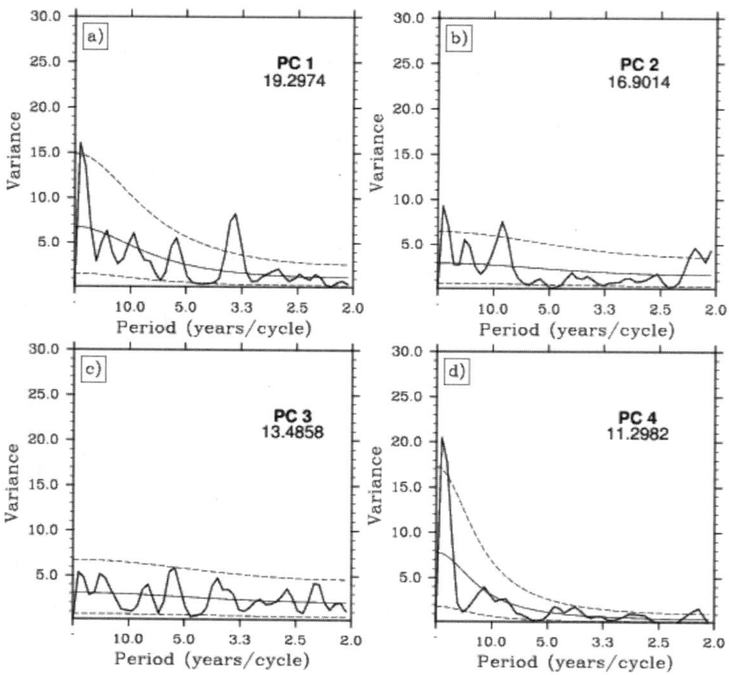

Figure 11.3 Spectral analysis of the principal components of maximum temperature (a) PC 1, (b) PC 2, (c) PC 3, and (d) PC 4. The percentage of the data set's variance accounted for by each PC is measured along the y-axis. The thick solid line depicts the spectral estimates, which are unitless; period lengths are indicated along the x-axis in years/cycle. The red noise spectral power estimates represented by the uppermost dashed line represent the 95% confidence limit; the lower dashed line represents the red noise spectral power estimates at the 5% confidence limit. The middle (solid) line represents the 50% confidence level.

Minimum Temperature

Spectral characteristics of growing-season minimum temperatures (figure 11.4) exhibit significant variability at low frequencies, but more high-frequency signals are observed in comparison to the spectra of mean or maximum temperatures. PC 1, PC 3, and PC 4 all have large proportions of their variance at low frequencies, but they also have statistically significant signals at quasi-quintennial periodicities. All PCs depict significant or prominent spikes at quasi-biennial or shorter periodicities. None of the PCs exhibit much power at interdecadal periodicities.

Precipitation

Precipitation spectra for the four retained PCs are depicted in figure 11.5. These spectra show little quasi-periodic variability across the LTER sites, and few significant periodicities are found. Prominent signals at quasi-biennial periods are observed in PC 2, PC 3, and PC 4, with marginally significant variance observed at

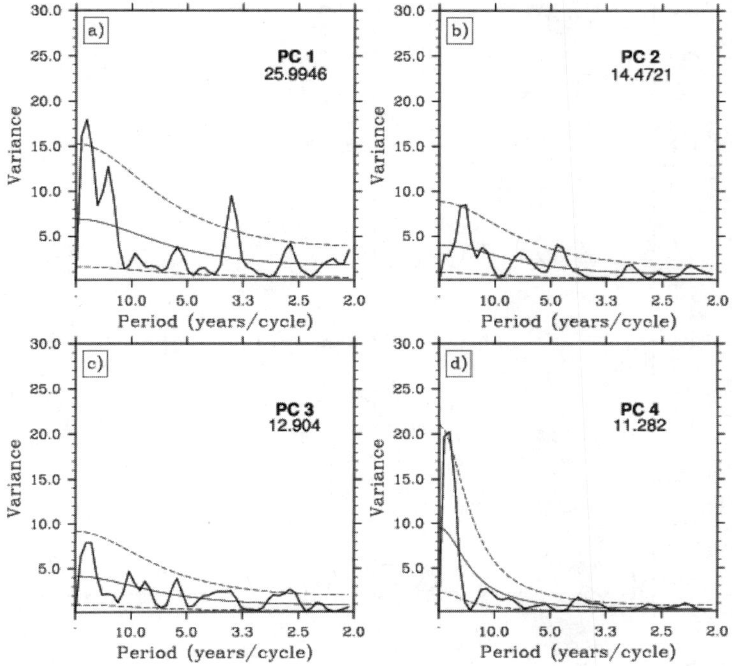

Figure 11.4 Spectral analysis of the principal components of minimum temperature (a) PC 1, (b) PC 2, (c) PC 3, and (d) PC 4. The percentage of the data set's variance accounted for by each PC is measured along the y-axis. The thick solid line depicts the spectral estimates, which are unitless; period lengths are indicated along the x-axis in years/cycle. The red noise spectral power estimates represented by the uppermost dashed line represent the 95% confidence limit; the lower dashed line represents the red noise spectral power estimates at the 5% confidence limit. The middle (solid) line represents the 50% confidence level.

low frequencies in PC 2 and PC 3. There is little, if any, spectral variance observed in any of the PCs at interdecadal timescales.

Discussion and Conclusions

Prominent or significant spectral powers at interdecadal timescales appear uncommon in the LTER climate data on mean growing season (March–September) used in this study. High-frequency variability, especially that observed at quasi-biennial and quasi-quintennial periodicities, appear as important as the variability occurring at the interdecadal timescale. Although some significant quasi-periodic components of variability are observed, few, if any, are consistently observed across the four variables examined. Significant periodicities appear relatively consistently between the temperature variables at timescales associated with ENSO, but some subtle differences are noted, in particular in the high-frequency spectral domain. Much of

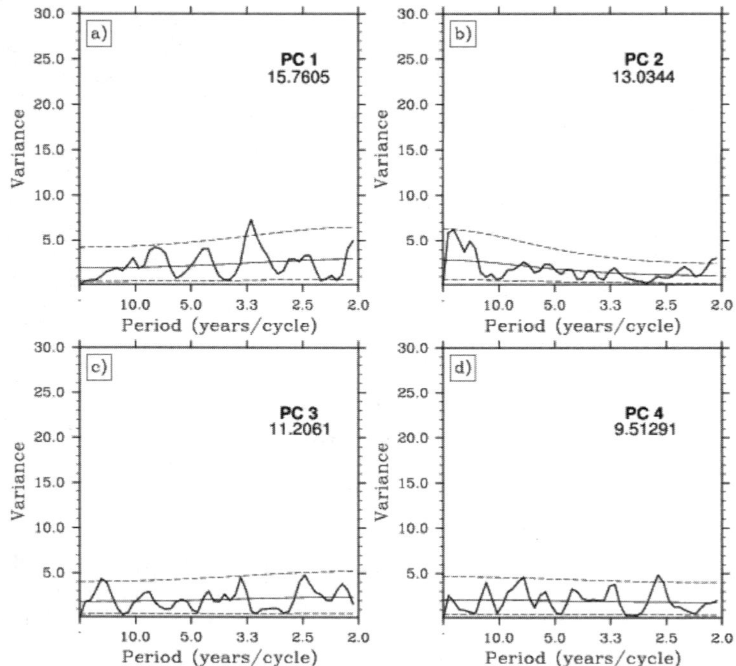

Figure 11.5 Spectral analysis of the principal components of precipitation (a) PC 1, (b) PC 2, (c) PC 3, and (d) PC 4. The percentage of the data set's variance accounted for by each PC is measured along the y-axis. The thick solid line depicts the spectral estimates, which are unitless; period lengths are indicated along the x-axis in years/cycle. The red noise spectral power estimates represented by the uppermost dashed line represent the 95% confidence limit; the lower dashed line represents the red noise spectral power estimates at the 5% confidence limit. The middle (solid) line represents the 50% confidence level.

the variance in temperature data is found to contain significant, or prominent, spectral powers at periodicities at 50 years or longer.

However, the interpretation of these oscillations that have such extremely long periodicities must be questioned in a data set containing, at best, 104 years of data. Given such a short time period, only two complete 50-year cycles could possibly occur. Statistically, these signals may be significant according to the chi-square test or other criteria, but it is doubtful that their significance can be accurately assessed given that there are potentially only two complete cycles. Realistically, this oscillation would be unlikely to have completed these two cycles during the past century. Even if this were so, it is not likely that the oscillation would take exactly 50 years to complete: some variance around an approximate 50-year period would likely have occurred because of probable interactions between this phenomenon and other climate system components. Such signal variability would result in decreased spectral power and observed significance at the 50-year periodicity, making accurate estimates of this cycle's statistical significance unlikely.

We must also remember that these results do not necessarily reflect the individual power spectra of the climate data at each LTER site. Rather, they reflect the spectral characteristics of the whole LTER network as a result of the decomposition of data into principal modes of variability across the whole data set. Actual periodicities at individual sites *may* be somewhat stronger or weaker relative to those found across the entire network.

References

Beamish, R.J., and D.R. Boullion, 1993. Pacific salmon production trends in relation to climate. *Canadian Journal of Fisheries and Aquatic Science.* 50:1002–1016.

Brodeur, R.D., and D.M. Ware, 1992. Interannual and interdecadal changes in zooplankton biomass in the subarctic Pacific Ocean. *Fisheries Oceanography.* 1:32–38.

Currie, R.G., and D.P. O'Brien, 1992. Deterministic signals in USA precipitation records: II. *International Journal of Climatology.* 12: 281–304.

Ebbesmeyer, C.C., D.R. Cayan, D.R. McLain, F.H. Nichols, D.H. Peterson, and K.T. Redmond, 1991. 1976 step in the Pacific climate: Forty environmental changes between 1968–1975 and 1977–1985, Proceedings Seventh Annual Pacific Climate Workshop, Asilomar, CA, California Department of Water Research, 115–126.

Francis, R.C., and S.R. Hare, 1994. Decadal-scale regime shifts in the large marine ecosystems of the north-east Pacific: A case for historical science. *Fisheries Oceanography.* 3:279–291.

Francis, R.C., and T.H. Sibley, 1991. Climate change and fisheries: What are the real issues? *North West Environmental Journal.* 7:295–307.

Friis-Christensen, E., and K. Laasen, 1991. Length of the solar cycle: An indicator of solar activity closely associated with climate. *Science.* 254:698–700.

Graham, N.E., 1994. Decadal-scale climate variability in the 1970s and 1980s: Observations and model results. *Climate Dynamics.* 10:135–159.

Greenland, D., 1999. ENSO-related phenomena at long-term ecological research sites. *Physical Geography.* 20:491–507.

Hays, J.D., J. Imbrie, and N.J. Shackleton, 1976. Variations in the Earth's orbit: Pacemaker of the ice ages. *Science.* 194:1121–1131.

Hollowed, A.B., and W.S. Wooster, 1992. Variability of winter ocean conditions and strong year classes of Northeast Pacific groundfish, *ICES Marine Science Symposium.* 195:433–444.

Hsieh, W.W., 2001. Nonlinear canonical correlation analysis of the tropical Pacific climate variability using a neural network approach. *Journal of Climate.* 14:2528–2539.

Hsieh, W.W., and B. Tang, 1998. Applying neural network models to prediction and data analysis in meteorology and oceanography. *Bulletin of the American Meteorological Society.* 79:1855–1870.

IPCC, 1995. *Climate Change 1995—The Science of Climate Change: The Second Assessment Report of the Inter-Governmental Panel on Climate Change.* J.T. Houghton, L.G.Meira Filho, B.A. Callander, N. Harris, A. Kattenberg, and K. Maskell (eds.). Cambridge University Press, New York.

Kalnay, E., M. Kanimatsu, R. Kistler, W. Collins, D. Deaven, L. Gandin, M. Iredell, S. Saha, G. White, J. Wollen, Y. Zhu, M. Chelliah, W. Ebisuazaki, W. Higgins, J. Janowiak, K.C. Mo, C. Roplewski, J. Wang, A. Leetmaa, R. Reynolds, R. Jenne, and D. Joseph, 1996. The NCEP/NCAR 40-year reanalysis project. *Bulletin of the American Meteorological Society.* 77:437–471 plus CD-ROM.

Keppene C.L., and M. Ghil, 1992. Adaptive filtering and prediction of the Southern Oscillation index. *Journal of Geophysical Research.* 97:20,449–20,454.

Kodera, K., H. Koide, and H. Yoshimura, 1999. Northern Hemisphere winter circulatiuon associated with the North Atlantic Oscillation and stratospheric polar-night jet, *Geophysical Research Letters.* 26:443–446.

Labitzke, K., and H. van Loon, 1988. Associations between the 11-year solar cycle, the QBO and the atmosphere, Part I. The troposphere and the stratosphere in the Northern Hemisphere in winter. *Journal of Atmospheric and Terrestrial Physics.* 50:197–206.

Lamb, H.H., 1977. *Climatic history and the future.* Princeton, N.J.: Princeton University Press. 711 pp.

Mann, M. E., and J. Park, 1994. Global-scale modes of surface temperature variability on interannual to century timescales. *Journal of Geophysical Research.* 25:819–25,833.

Mantua, N.J., S.R. Hare, Y. Zhang, J.M. Wallace, and R.C. Francis, 1997. A Pacific interdecadal climate oscillation with impacts on Salmon production. *Bulletin of the American Meteorological Society.* 78:1069–1079.

Marcus, M.G., and S.W. Brazel, 1984. Climate changes in Arizona's future. Arizona State Climate Publication No. 1, Office of the State Climatologist, Arizona State University, Tempe, Arizona.

McHugh, M.J., 1999. *Precipitation over southern Africa and global-scale atmospheric circulation during boreal winter.* Columbus: Ohio State University, Ph.D. Dissertation, 221 pp.

Minobe, S., 1997. A 50–70 year climatic oscillation over the North Atlantic and North America. *Geophysical Research Letters.* 24:683–686.

Mitchell, J.M., B. Dzerdzeevskii, H. Flohn, W.L. Hofmeyr, H.H. Lamb, K.N. Rao, and C.C. Wallén. 1966. *Climatic Change.* WMO Tech. Note No. 79, WMO No. 195TP100, 79 pp. World Meteorological Organization, Geneva, Switzerland.

Mitra, K., S. Mukherji, and S.N. Dutta. 1991. Some indications of 18.6 year luni-solar and 10–11 year solar cycles in rainfall in northwest India, the plains of Uttar Pradesh, and north-central India. *International Journal of Climatology* 11:645–652.

Namias, J., 1978. Multiple causes of the North American abnormal winter of 1976–77. *Monthly Weather Review.* 106:279–295.

Rogers, J.C., 1984. The association between the North Atlantic Oscillation and the Southern Oscillation in the Northern Hemisphere. *Monthly Weather Review.* 112: 1999–2015.

Thompson, L.G., M.E. Davis, E.M. Thompson, T.A. Sowers, K.A. Henderson, V.S. Zagorodnov, P.N. Lin, V.N. Mikhalenko, R.K. Campen, J.F. Bolzan, J. Cole-Dai, and B. Francou, 1998. A 25,000 year tropical climate history from Bolivian ice cores. *Science.* 282:1858–1864.

Trenberth, K.E., 1990. Recent observed interdecadal climate changes in the Northern Hemisphere. *Bulletin of the American Meteorological Society.* 71:988–993.

Trenberth, K.E., and J.W. Hurrell, 1994. Decadal atmosphere-ocean variability in the Pacific. *Climate Dynamics.* 9:303–319.

van Loon, H., and J. C. Rogers. 1978. The seesaw in winter temperatures between Greenland and Northwestern Europe, I, General description. *Monthly Weather Review.* 106:296–310.

Venrick, E.L., J.A. McGowan, D.R. Cayan, and T.L. Hayward, 1987. Climate and chlorophyll: Long term trends in central North Pacific Ocean. *Science.* 238:70–72.

Vines, R.G., 1986. Rainfall patterns in India. *Journal of Climatology.* 6:135–138.

Vose, R.S, R.L. Schmoyer, P.M. Steurer, T.C. Peterson, R. Heim, T.R. Karl, and J.K.

Eischeid, 1992. *The global historical climatology network: Long-term monthly temperature, precipitation, sea-level pressure and station pressure data.* ORNL/CDIAC-53, NDP-041. Carbon Dioxide Information Analysis Center, Oak Ridge National laboratory, Oak Ridge, Tennessee. 315 pp.

Warner, R.M. 1998. *Spectral Analysis of Time Series Data.* New York. Guilford Press. 190 pp.

12

A 200-Year Perspective of Climate Variability and the Response of White Spruce in Interior Alaska

Glenn Patrick Juday

Valerie Barber

Scott Rupp

John Zasada

Martin Wilmking

Introduction

The two most important life functions that organisms carry out to persist in the environment are reproduction and growth. In this chapter we examine the role of climate and climate variability as controlling factors in the growth of one of the most important and productive of the North American boreal forest tree species, white spruce (*Picea glauca* [Moench] Voss). Because the relationship between climate and tree growth is so close, tree-ring properties have been used successfully for many years as a proxy to reconstruct past climates. Our recent reconstruction of nineteenth-century summer temperatures at Fairbanks based on white spruce tree-ring characteristics (Barber et al. in press) reveals a fundamental pattern of quasi-decadal climate variability. The values in this reconstruction of nineteenth-century Fairbanks summer temperatures are surprisingly warm compared to values in much of the published paleoclimatic literature for boreal North America. In this chapter we compare our temperature reconstructions with ring-width records in northern and south-central Alaska to see whether tree-growth signals in the nineteenth century in those regions are consistent with tree-ring characteristics in and near Bonanza Creek (BNZ) LTER (25 km southwest of Fairbanks) that suggest warm temperatures during the mid-nineteenth century. We also present a conceptual model of key limiting events in white spruce reproduction and compare it to a 39-year record of seed fall at BNZ. Finally, we derive a radial growth pattern index from white spruce at nine stands across Interior Alaska that matches recent major seed crop events in the BNZ monitoring period, and we identify dates after 1800 when major seed crops of white spruce, which are infrequent, may have been produced.

Climate and Climate Variability in the Boreal Region

Climate of the Boreal Region

The boreal region is characterized by a broad zone of forest with a continuous distribution across Eurasia and North America, amounting to about 17% of the earth's land surface area (Bonan et al. 1992). The boreal region is often conceived of as a zone of relatively homogenous climate, but in fact a surprising diversity of climates are present. During the long days of summer, continental interior locations under persistent high-pressure systems experience hot weather that can promote extensive forest fires frequently exceeding 100 kilohectares (K ha). Summer daily maximum temperatures are cooled to a considerable degree in maritime portions of the boreal region affected by air masses that originate over the North Atlantic, North Pacific, or Arctic Oceans. Precipitation is abundant in the boreal zone of most of the Nordic countries and western Russia, whereas east-central Siberia experiences low winter snow depths because the strong Siberian High suppresses precipitation. The boreal landscapes of far eastern Siberia and western North America are mountainous, whereas the topography of most of central and western Siberia and eastern Canada is less mountainous. The mountainous boreal regions are characterized by local climates (Pojar 1996), aspect-controlled differences in forest types (Viereck et al. 1986), and a much more irregular boundary between forest and tundra (figure 12.1). In the topographically complex landscapes of Alaska, northwestern Canada, and central and northeastern Siberia, precipitation limits forest growth, so natural grasslands are part of the landscape. Precipitation at BNZ is influenced by the Pacific Monsoon and reaches a distinct maximum in the late summer. In other parts of the boreal forest region, precipitation is more evenly distributed throughout the year or even reaches a winter maximum. All this regional climatic variation must be taken into account as a fundamental backdrop when considering temporal variability and ecological response in boreal forests.

On both continents, neither the northern nor the southern boundaries of the boreal zone are aligned at the same latitude east to west. The Icelandic Low and Aleutian Low deflect storm tracks and advect relatively mild air masses northward as they approach the western margin of Eurasia and North America, respectively. As a result, the boreal forest belt is located considerably farther north in both the Nordic countries and western North America than in the center of the continents. By contrast, cold polar air flowing southward follows a persistent path along the eastern portion of both continents, and, consequently, the boreal forest belt reaches its southernmost limits there. Essentially, all of the boreal forest in Alaska is north of 60° N, and practically all the boreal forest of eastern Canada is south of 60° N (figure 12.1). The boreal forest region is particularly prone to climatic variability because minor variations in these key features of atmospheric circulation can either intensify the advection of warm air into this naturally cold region or enhance the distribution of cold air southward through the region.

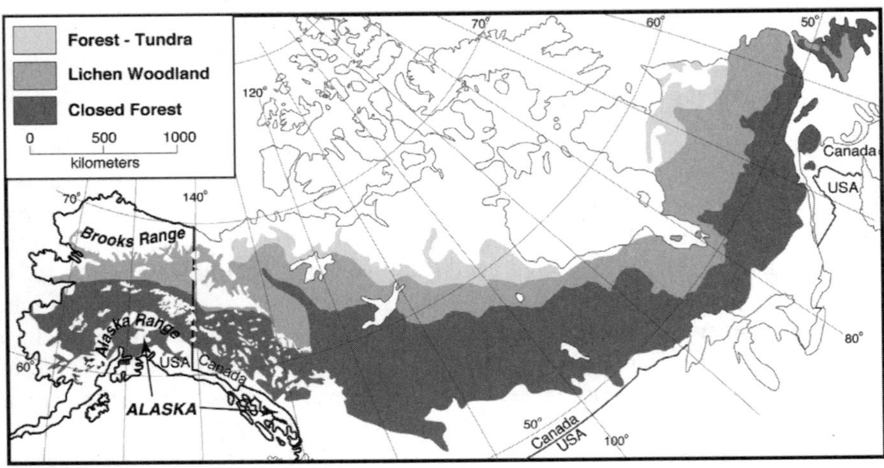

Figure 12.1 Distribution of the boreal forest of North America. Note the southward depression of both the northern and southern extent of the forest in eastern Canada, and the northward extension of boundaries in Alaska. Mountain topography in western North America is associated with complex and irregular forest boundaries.

Climate of BNZ

BNZ LTER represents the Interior Alaska boreal forest (figure 12.2A) and is made up of two sites, Caribou-Poker Creeks Research Watershed and Bonanza Creek Experimental Forest. All subsequent references to the BNZ in this chapter refer to the Bonanza Creek site (figure 12.2B,C). Interior Alaska is a well-defined region of complex physiography, delimited in the north by the Brooks Range and in the south by the Alaska Range. The region essentially covers the area between 63 and 67° N. Interior Alaska extends from the Yukon Territory at 141° W westward to the Bering Sea climatic boundary at about 155° W, where precipitation exceeds 400 mm (Edwards et al. 2001). The region is made up of two large, low-lying tectonic basins, the Tanana Valley and Yukon Flats, separated by uplands 500–1000 m in elevation. The regional climate is cold continental with January means −20°C or colder and July means 15–20°C, depending on elevation and location within the region.

The Brooks and Alaska Ranges act as topographic barriers to moisture-laden air from surrounding oceans. Consequently, Interior Alaska is semiarid, with annual precipitation ranging from 400 to <200 mm (Patric and Black 1968). Precipitation generally declines to the east and is strongly influenced by topography (Edwards et al. 2001). About 60% of the annual precipitation falls as summer rain. Because annual precipitation in central Interior Alaska is low and summers are often warm, a precipitation deficit (excess of potential evapotranspiration over precipitation) develops. Calculated precipitation deficit values in Alaska include 9.5 cm at Bettles in the central Brooks Range, 18.8 cm at Fairbanks, and 28.9 cm at Fort Yukon (Slaughter and Viereck 1986).

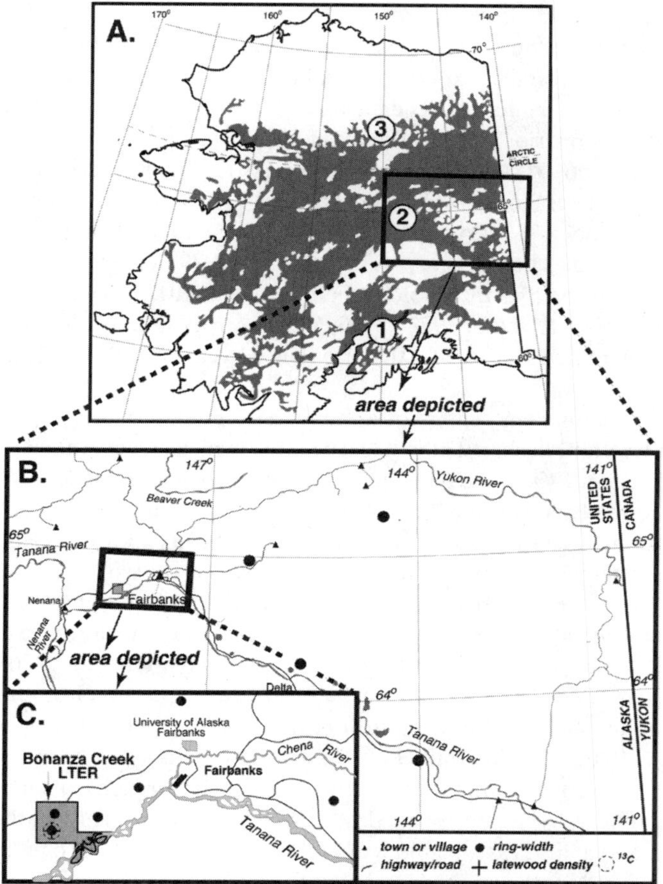

Figure 12.2 (A) Distribution of boreal forest in Alaska. Location 1 = south-central Alaska (Fort Richardson) tree-ring sample site. Location 2 = BNZ. Location 3 = Brooks Range tree-line sample site. (B) Location of tree-ring sample sites and BNZ in east-central Alaska. (C) Detail of tree-ring sample sites in the vicinity of Fairbanks and BNZ.

Because of its high-latitude location just south of the Arctic Circle, BNZ experiences nearly continuous daylight at the summer solstice, and short daylight hours (minimum 3 hours 42 minutes) with only a few degrees of sun elevation in the winter. BNZ is located in the Tanana Valley, a broad low-elevation (133 m above sea level) lowland situated immediately north of the Alaska Range and south of the Yukon-Tanana Uplands of central Alaska (figure 12.2C). BNZ extends from islands and the north bank of the Tanana River up to the crest of the first range of hills or low mountains (440 m asl) northward. The high-elevation barriers of the Brooks Range and Alaska Range mountains surrounding BNZ deflect or subdue direct clashes of warm and cold air masses, resulting in moderate winds and periods with little air mass mixing.

Climate records in Alaska are relatively sparse; most date from no earlier than the mid-twentieth century. The oldest continuous record from Interior Alaska is a combination of University Experiment Station (UES) and Fairbanks Airport data (Juday 1984). Trends in Fairbanks data are highly representative of the boreal forest region across Interior Alaska (Barber et al. 2000). During the period of instrument records (1906–2000) of the combined UES/Fairbanks (hereafter Fairbanks) station, mean annual temperature averaged –3.1°C. Average July temperature for Fairbanks was about 16.3°C during the period of record, and annual precipitation averaged 282 mm. The 30-year average (1950–1979) of growing degree-days (GDD) based on the 5°C threshold at Fairbanks was 1075 (Viereck et al. 1986). During the 24-year period of 1973–1997, GDD at the Fairbanks International Airport station increased 10.9% over the previous 24 years (1949–1972) (Juday et al. 1998).

Three semipermanent centers of atmospheric circulation are important to the weather in Interior Alaska: (1) the Aleutian Low, which sits over the North Pacific Ocean and occurs 25% of the time, making it the dominant influence on the Gulf of Alaska (Overland and Heister 1980); (2) the East Pacific high-pressure system that is present throughout the year off the coasts of California and Baja California; and (3) the Siberian high-pressure system located over eastern Asia. The position of the Aleutian Low moves southeastward from the Bering Sea into the Gulf of Alaska between August and December. In January, the Aleutian Low moves to the western Aleutians, where it slowly dissipates through July (Favorite et al. 1976). The East Pacific High reaches maximum intensity and northward position from June through August. It dominates most of the North Pacific, including the Gulf of Alaska (Favorite et al. 1976). The Siberian High is associated with a huge pool of very cold winter air over eastern Asia and northern Alaska, which reaches its maximum intensity in January. When entrenched, it causes a southward shift in the Aleutian storm track and an increase in cold winds blowing from the north over the western Gulf of Alaska (Wilson and Overland 1986).

Interior Alaska has two distinct summer circulation patterns apparent in records of interannual climate. During the twentieth century, the summer climate in Interior Alaska has alternated between periods of colder/wetter and warmer/drier weather than the long-term mean (Edwards et al. 2001; Mock et al. 1998). The synoptic pattern for colder/wetter conditions is produced by an eastward shift of the East Asian trough, an upper level jet-type feature centered over eastern Asia, and a stronger than normal Pacific subtropical high. Both the eastward shift and the intensification of the subtropical high increase the frequency of storms containing moisture-laden air entering the interior basin from the southwest, the only direction free of a major topographical obstacle. A high-pressure ridge located north to northeast of Alaska produces the synoptic pattern for warmer/drier conditions. The high center brings clear skies and warm dry continental air from the east at the season of maximum surface heating from the long daylight hours near the Arctic Circle. Negative surface-pressure height anomalies over the Yukon Territory of Canada and over northern Siberia represent a northwest shift of the average pressure system. The Pacific Subtropical High located south of Alaska is weak under this regime, resulting in reduced flow of moist air from the west. Persistent blocking ridge conditions are directly correlated with periods of extensive wildfires across the western North American boreal forest (Johnson 1992).

Mean summer (May–August) temperature at Fairbanks is highly variable in several respects (figure 12.3A). The smoothed mean summer temperature record displays an apparent quasi-decadal cycle through the twentieth century in which highs and lows of cycles alternate regularly every 7 to 8 years (figure 12.3A). Results of spectral analysis of the longer (1800–1996) and highly correlated reconstructed summer temperature series are discussed later. The coefficient of variation (unbiased) for summer temperatures during the period 1906–2000 is 7.99, indicating high variability. The 1976–1977 regime shift, evident throughout Alaska and the North Pacific region of North America (Ebbesmeyer et al. 1990), represents the most important long-term change during the period of record. After the regime shift, sustained levels of warmth as measured by mean annual temperature (MAT), winter mean temperatures, and summer mean temperatures ramped up to a new level not seen previously in the twentieth century in Alaska (Juday 1984; Juday et al. 1998). Mean summer temperature in the 20 years following the shift (1977–1996) was 14.15°C, an increase of 0.52°C compared to the 20 years prior. The 1977–1996 mean summer temperature after the shift was 1.46°C warmer than the coldest 20-year interval (1922–1941). The warmest mean summer temperature (15.83°C) occurred in 1990 after the shift. The coldest mean summer temperature (10.40°C) occurred in 1922. The 5.43°C difference between the warmest and coldest summers is so large that it represents a considerable challenge for the dominant plants to have a set of adaptive traits able to cope with such divergent conditions.

The smoothed Fairbanks growth year precipitation record (October–September) also displays a marked quasi-decadal cyclicity (figure 12.3A). The tendency of the peaks of the temperature cycle to occur at the troughs of the precipitation cycle is an expression of the strong tendency for climate to remain in either the cool or cold/wet mode versus the warm or hot/dry mode. However, the long-term trend of Fairbanks precipitation is less variable than temperature. A few years of heavy snowfall occurred in the early 1990s, but otherwise annual precipitation has decreased slightly in the late twentieth century compared to the early and middle part of the century. Other Interior Alaska stations such as Bettles and McGrath also show little change, although some coastal northern Alaska stations have experienced an increase in precipitation.

Increased temperature without a concurrent increase in precipitation means that effective moisture is lowered. From about 1912 to 1968, the predominant mode of climate was cool and wet with a relative moisture surplus, broken only briefly by warmer and drier intervals (figure 12.3A). Cool/wet periods lasted on average for 9–10 years (min. 6 years, max. 14 years). During a transition period between 1951 and 1970, precipitation and temperature averaged near the long-term mean, although muted cycles were present (figure 12.3A). After the mid-1970s, climate switched to a hot and dry mode characterized by a strong relative moisture deficit (figure 12.3A) (Barber et al. 2000).

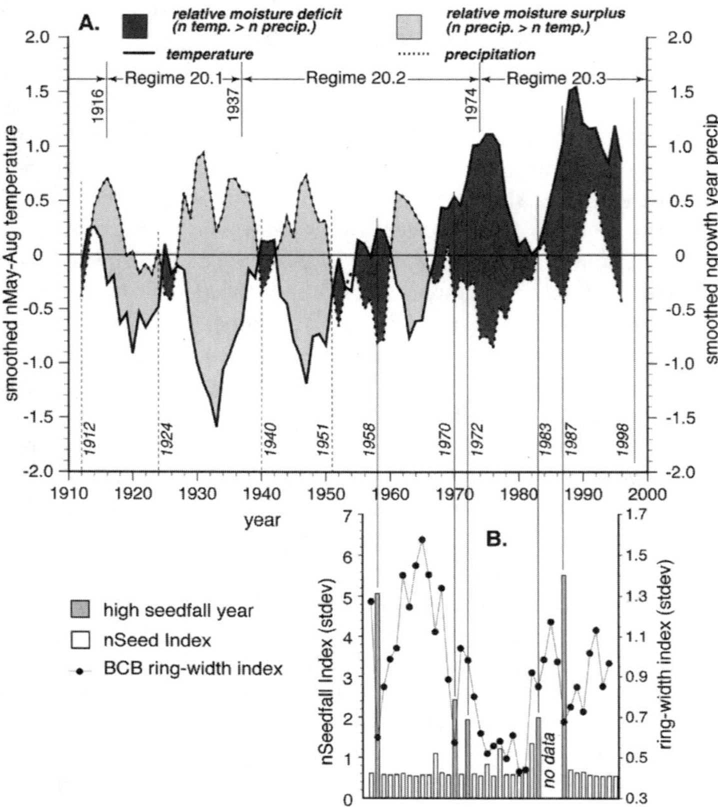

Figure 12.3 (A) Normalized May–August mean temperature and growth year (October–September) precipitation during the period of record at Fairbanks, Alaska. Values are smoothed with a 5-year running mean. Horizontal line represents long-term mean of both temperature and precipitation. Periods of above long-term mean precipitation and below mean summer temperature (relative moisture surplus) are indicated in light shading. Periods of above long-term mean summer temperature and below mean precipitation (relative moisture deficit) are indicated by dark shading. Summer temperature regimes (see figure 12.5 for definition) are indicated. Years of known (solid) or inferred (dashed) high white spruce seed crops are indicated by vertical lines. (B) Seed fall index compared to radial growth at Bonanza Creek Bluff (BCB), located near the seed fall collection site. Seed fall index is equal to 1 plus the normalized (subtraction method) raw seed counts from seed traps. The UP1A record began in 1957 and ended in 1982; the stand was killed in the 1983 Rosie Creek Fire. The UP3A seed fall record began in 1968 and continued until 1983, then again from 1987 to the present. During the period of overlap, the mean of seed counts at both sites is used. The five major seed crop years are highlighted. Solid line with circles represents ring-width index (normalized and detrended radial growth). Note the steep 1-year declines in radial growth during years of major seed crops.

Climate Variability and Growth of White Spruce

Ecology of White Spruce

White spruce is the predominant forest cover, occupying 12.1 million ha in Alaska or about 26% of boreal forest (Labau and van Hees 1990). Nearly 2.8 million ha of the white spruce forest (about 23% of white spruce–dominated stands) meets the USDA Forest Service standard of commercially productive forest. The two main types of white spruce–dominated ecosystems are floodplain and upland (Viereck et al. 1986). White spruce–dominated floodplain stands are particularly extensive along the broad and shifting floodplains of the large aggrading rivers (rivers with rising relative elevation because of sediment buildup in the riverbed) fed by glacial meltwater such as the Yukon, Matanuska, and Tanana (Van Cleve et al. 1993). Much of the focus of the BNZ research program historically has consisted of direct observations, measurements, and experiments on primary successional surfaces of the Tanana River floodplain, including late successional white spruce (e.g., Van Cleve and Viereck 1981; Yarie et al. 1998). However, radial growth of glacial floodplain white spruce generally does not correlate at a statistically significant level with Fairbanks air temperature or precipitation; thus the focus of this chapter is on the very different upland white spruce stands of BNZ and similar sites across Interior Alaska (figure 12.2).

Upland stands dominated by white spruce are broadly distributed across valley, hillside, and mountain landscapes. White spruce is usually present in at least minor amounts in most upland forest community types, even on sites where it is not dominant, with the exception of permafrost-dominated sites (Viereck et al. 1992). White spruce is an important, if not the dominant, tree-line species in much of Interior Alaska in the Brooks Range, Alaska Range, and highlands between them (Juday et al. 1999; Viereck 1979). To a large extent the distribution of dominant tree species in most forest communities in Interior Alaska has been shaped by fire (Dyrness et al. 1986). The Interior Alaska boreal forest appears to be adapted to a stand-replacement disturbance system. Typically, succession on burned white spruce sites in Interior Alaska leads to reestablished white spruce dominance in 150 to 300 years (Foote 1983).

Climate Reconstruction and Tree-Ring Characteristics

Tree rings contain a wealth of information about the conditions affecting the growth and health of trees (Fritts 1976). Although ring-width measurements are familiar, new technology allows measurements of other tree-ring properties, such as x-ray densitometry (Jacoby et al. 1988) and stable isotope analysis (Leavitt and Danzer 1992) that have been explored only recently for the information they contain. Tree rings are a reliable indicator of the state of a tree's health or vigor, as well as a direct measure of the rate of product accumulation in stands managed for forest products (Nyland 1996). Larger tree rings are possible only under optimum growing conditions, although smaller tree rings result from a variety of potentially stressful conditions.

In northern conifers maximum latewood density increases when moisture stress during the mid- to late growing season is great (D'Arrigo et al. 1992). Maximum latewood density of boreal conifers may be an appropriate index for canopy growth as indicated by satellite-sensed Normalized Difference Vegetation Index (NDVI) values where productivity is temperature related (D'Arrigo et al. 2000). Carbon 13 (^{13}C) isotope content is generally measured as discrimination or change in the amount of isotope in sampled plant tissue compared to a reference standard. Less ^{13}C discrimination (greater ^{13}C) indicates production of the subject plant tissue under a condition of restricted stomatal exchange, generally as a result of moisture stress (Livingston and Spittlehouse 1996). Maximum latewood density, ^{13}C isotope discrimination, and ring-width of upland white spruce stands in BNZ and similar sites are generally well-correlated with each other, although the first two ring properties are less autoregressive than ring-width, and so contain information generally specific to the climatic conditions of the year of ring formation (Barber et al. 2000). In this chapter summer temperature reconstruction is based on the maximum latewood density and ^{13}C isotope discrimination, and white spruce growth responses are based on ring-width (Barber et al. in press).

Classic dendroclimatological technique starts with the selection of widely spaced trees assumed to be free of crown-to-crown canopy competition. Trees growing in marginal environments are preferred where one clear factor of the environment is assumed to limit growth. The goal is to extract a pure climate signal from the measured tree-ring properties. Tree lines are considered to be a particularly suitable environment for dendroclimatological sampling because trees found there are expected to be limited by lack of growing season warmth. Northern Alaska tree lines, both elevational and latitudinal, have been a popular subject of dendrochronological investigations (Cropper 1982; D'Arrigo et al. 1992; D'Arrigo and Jacoby Jr. 1992; Jacoby and D'Arrigo 1995; Jacoby et al. 1999; Jacoby et al. 1996; Overpeck et al. 1997). In virtually all these studies, the radial growth response of trees at tree line is positively correlated to temperature.

Conversely, samples of white spruce on upland sites in BNZ display a strong negative sensitivity to summer temperature (Juday et al. 1998; Juday et al. 1999). The negative relationship of radial growth to summer temperature in upland white spruce is consistent throughout the twentieth century, occurring in 20 stands (including 2 stands in BNZ) representative of dominant and codominant trees in mature and old stands throughout Interior Alaska, including trees across a broad range of diameters (Barber et al. 2000). During the twentieth century, smoothed radial growth of these trees displays a 15-year cyclicity in which opposite trends in radial growth occur regularly over periods of 7 to 8 years (Barber et al. 2000). Properties of tree rings, including ring-width, ^{13}C isotope discrimination, and maximum latewood density, are consistent with growth limitation because of temperature-induced drought stress (Barber et al. 2000). Ten of the 20 stands with ring-width records extended back far enough in time to cover most of the nineteenth century (figure 12.4A,B). Unprocessed ring-width data (1806–1996) subjected to spectrum analysis contains intermediate peaks at 8 and 16.7 years. However, preprocessing (differenced, 64-year segments sampled at 32-year intervals) treatment of the 191-year regional ring-width sample displays no significant periods at the 95% confidence level.

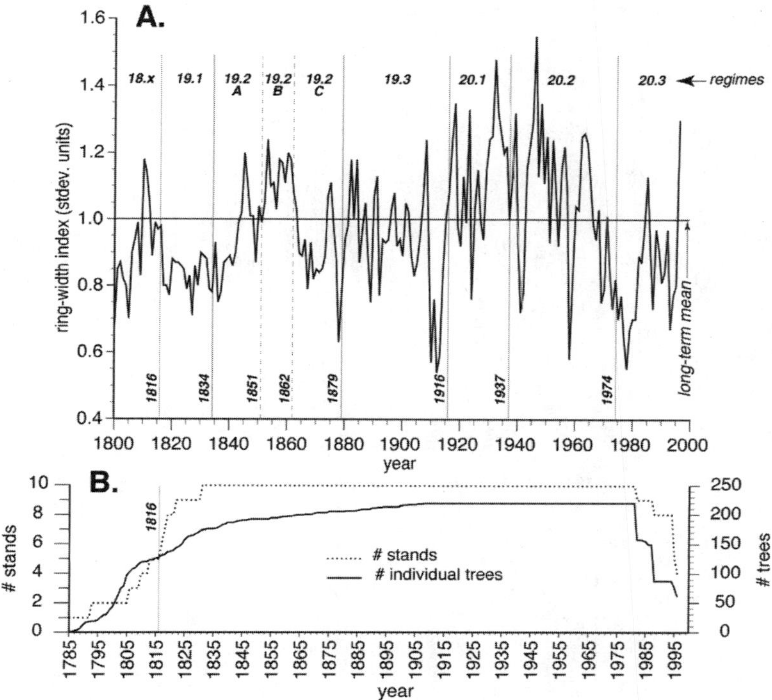

Figure 12.4 (A) Mean ring-width index (normalized and detrended radial growth) for 10 white spruce stands depicted in figure 12.2. Summer temperature regimes (see figure 12.5) are identified across the top; vertical lines represent boundary dates of regimes. (B) Sample depth of ring-width sample, represented as both number of stands and number of individual trees contributing. Decline in sample size makes values prior to 1816 less reliable.

We have recently extended the measurement series of ^{13}C discrimination and maximum latewood density from the Reserve West stand in BNZ back to 1800 (Barber et al. in press). These two tree-ring properties passed a test of significance in a multiple regression/principal components-based reconstruction model to serve as proxies for a reconstruction of May through August temperatures at Fairbanks (Barber et al. in press) (figure 12.5A). Decadal-scale shifts appear in our reconstruction prior to the instrument record (figure 12.5A). Reconstructions of the annual Pacific Decadal Oscillation (PDO) index based on western North American tree-ring records, which account for up to 53% of the instrumental variance and extend as far back as A.D. 1700, indicate that decadal-scale climatic shifts have occurred in the northeast Pacific region prior to the period of instrumental records (D'Arrigo et al. 2001). Spectral analysis of raw (unprocessed) reconstructed Fairbanks summer temperature values display peaks at 9 and 18 years. However, again, peaks of preprocessed (differenced) reconstructed summer temperature (1800–1996) may not be significant at the 95% level. High short-term (1- to 3-year) variability combined with rapid shifts between regimes tends to mask the quasi-decadal cyclicity.

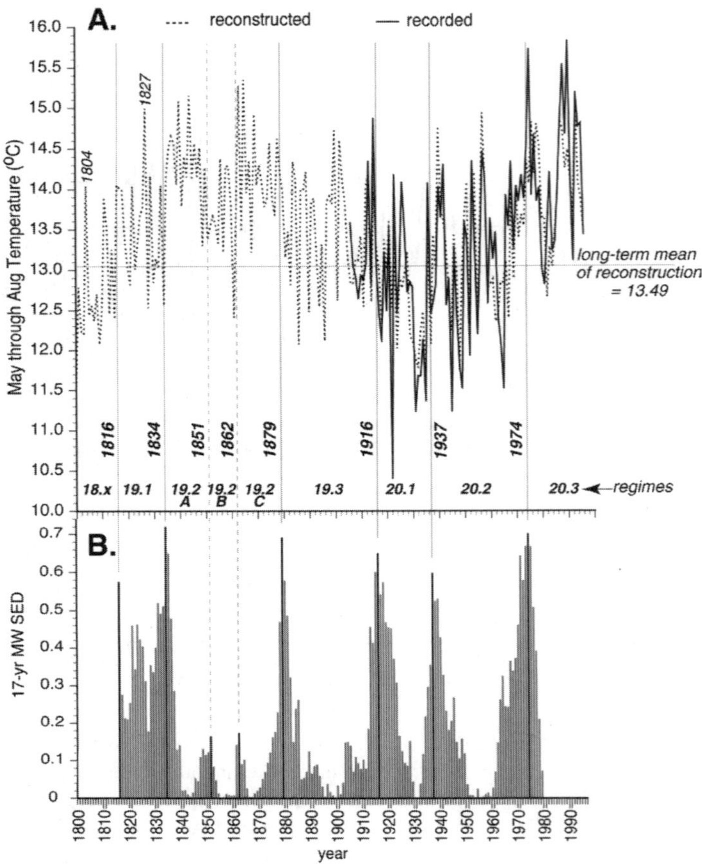

Figure 12.5 (A) Recorded and reconstructed summer (May through August) temperature (annual data) at Fairbanks, Alaska. Reconstruction is based on a two-proxy model composed of ^{13}C discrimination and maximum latewood density in white spruce collected at the Reserve West site (figure 12.2C) in BNZ (Barber et al. in press). Regimes of summer temperature are identified along the horizontal axis. Vertical lines represent dates of regime change. The regime numbering system consists of century identifier (left of decimal) and sequential numeral within the century (right of decimal). (B) Nodes of change identified using a 17-year moving split window squared Euclidean distance (MW SED) metric.

To identify nodes of change in the May–August temperature reconstruction, we used a moving split window analysis with squared Euclidean distance metrics (MW SED) (Johnson et al. 1992; Turner et al. 1991) (figure 12.5B). We used a moving window period (17 years) approximately as long as the putative cycle length in the reconstructed record in order to maximize change index values at transitions. The MW SED result shows peaks or spikes where the greatest change occurred, and we used these spikes to define the borders of climate regimes (figure 12.5B). Our proposed regimes, then, represent multidecadal periods of characteristic summer temperatures extending between periods of rapid climate change (figure 12.5A). For convenience, we number summer temperature regimes for the century in which they began, and assign a decimal of 1 for the first regime initiated in the century with increasing decimal numbers for successive regimes of that century. Thus, the first regime initiated in the twentieth century is labeled 20.1. We have further divided one regime (19.2) into three subregimes (A, B, and C) of lesser magnitude change.

Based on these criteria, the twentieth century contains Regimes 20.1, 20.2, and 20.3, and the nineteenth century is divided into three regimes as well. We tentatively identify a rapid climate change at about 1816 based on an apparent major change in variables compared to Regime 19.1, although we do not have MW SED values calculated before that year. These early years of the nineteenth century appear to have been part of a regime that began in the eighteenth century, and since we are unsure of the number of regimes contained in that century, we label that regime as 18.x.

Regime 20.3 began in the mid-1970s and continues today. The most notable feature of this regime is the elevated summer temperature without a concurrent increase in precipitation, making this period the hottest and driest in the instrument record (figure 12.3). Spruce tree-ring properties respond in a pattern consistent with both of these features, especially because ^{13}C discrimination is very low and maximum density is very high (Barber et al. 2000) (figure 12.5A). Relative tree growth, in contrast, has become increasingly lower (figure 12.4A), an effect we attribute to drought stress (Barber et al. 2000).

The previous regime (20.2) began around 1937 and was characterized by cool summer temperatures and favorable moisture. Regime 20.1 (1916–1937) was the coldest in our 200-year series and had the highest growth year precipitation in the twentieth century. These climatic characteristics are consistent with an intensified summer maritime influence caused by a strong southwesterly flow bringing moist air through a gap in the mountain barriers that surround central Alaska. Measured tree-ring properties strongly reflect this climate signal, displaying high ^{13}C discrimination, low maximum density values, and extremely large ring-width during both these regimes (figure 12.4A) (Barber et al. 2000).

Regime 19.3, which extends from 1879–1916, was developed largely from the reconstructed proxy record. It was an unusually quiescent period notable for its lack of extremes (figure 12.5A). All three proxies were relatively flat, indicating no large swings in climate for a period of about 35 years. We conclude from the proxy data that Regime 19.3 was a period of moderate summer temperatures. As a result, white spruce radial growth was neither particularly high nor low (figure 12.4A). Consistent with our findings, tree-ring series from the northeast Pacific as well as PDO reconstructions indicate a shift toward less pronounced interdecadal variability after about the middle 1800s (D'Arrigo et al. 2001). Tree-ring proxies as well as the recorded Fairbanks data indicate a rapid shift (1–3 years) from Regime 19.3 to Regime 20.1.

We divided Regime 19.2 into three subregimes based on distinct periods of change but of a much lesser magnitude than a complete regime shift (figure 4A,B). Overall, Regime 19.2 indicates relatively warm to hot summer temperatures defined by rapid climate changes in 1834 and 1879 with decadal-scale cycles throughout. Although this regime is broken up by a somewhat cooler period between 1851 and 1862 (Regime 19.2B), the overriding signal from the proxies in Regime 19.2 is extreme summer warmth. The ring characteristics during this period most closely resemble (but not completely) those in Regime 20.3, with low overall ^{13}C discrimination and high maximum density and small ring-widths, signals we attribute to extreme moisture stress during the twentieth century (Barber et al. 2000). Densities during Regime 19.2 are higher than the mean for the entire record but lower than during Regime 20.3 (warm summer temperature proxy). If the proxies maintained fidelity with environmental conditions observed in the twentieth century, then Regime 19.2 should be interpreted as a period of warm to hot/dry summer conditions.

We tentatively identify a rapid climate change at about 1816, based on an apparent major change in ^{13}C discrimination and radial growth compared to Regime 19.1. The pre-1816 values of ring-width and density must be viewed with caution because of the low sample depth between 1800 and 1816 (figure 12.4B).

The reconstructed summer warmth of Regime 19.2 is surprising and not consistent with earlier dendroclimatological reconstructions in Interior Alaska (Barber et al. in press; Garfinkle and Brubaker 1980; Jacoby and D'Arrigo 1989; Jacoby and Ulan 1983). However, the ^{13}C discrimination technique had not been applied previously to Alaskan white spruce, and it has some superior properties compared to the ring-width technique, such as lower autoregression and independence from some factors (biotic) that can affect radial growth for nonclimatic reasons. So it would be useful to determine whether long-term records of white spruce radial growth display responses about the time of Regime 19.2 that are consistent with strong summer warmth in Alaska.

Ring-width records from a location near the tree line (upper North Fork of the Koyukuk River) exhibit both positive and negative sensitivity to reconstructed Fairbanks summer temperature, but both sensitivities show a response consistent with summer warmth during Regime 19.2, the period 1834–1878 (figure 12.6A,B). Similarly, a population of spruce (probably hybrids of white and Sitka spruce) at Fort Richardson on the south-central coast of Alaska near Anchorage includes some trees with positive and some with negative correlation of radial growth to Fairbanks summer temperatures (figure 12.6C,D). Again, both south-central coast populations show a radial growth pattern consistent with summer warmth during the period of Regime 19.2. The integration of these results and known features of the Interior Alaska climate system and upland white spruce ecosystem presents a coherent picture.

Climate variation and reproduction of white spruce

White spruce cone and seed production occurs only episodically in northern latitude forests, including BNZ (Zasada et al. 1992; Zasada and Viereck 1970). Given the prolific reproductive ability of boreal trees that compete with white spruce (Zasada et al. 1992), infrequent reproduction should be a real limitation to its success

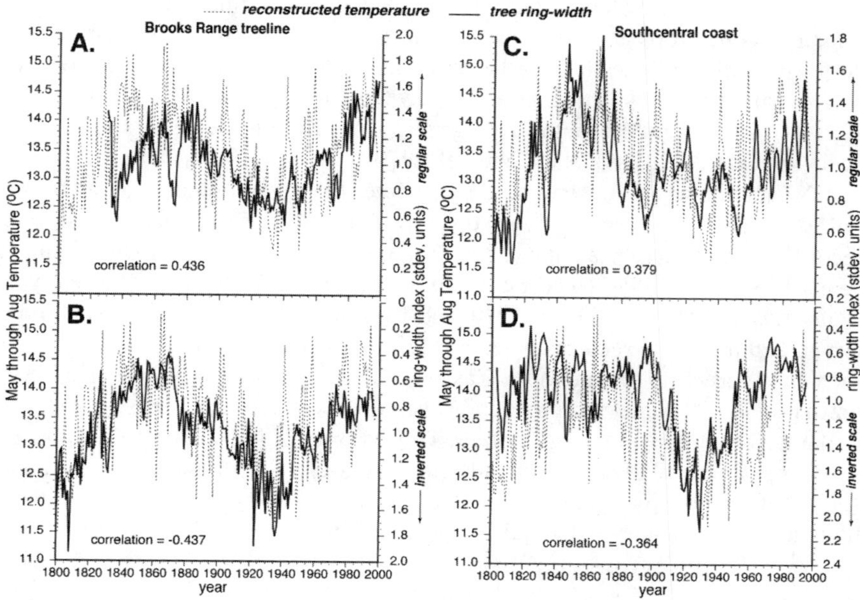

Figure 12.6 Relationship of white spruce radial growth (ring-width index) to reconstructed summer temperature at Fairbanks. Note the inverted radial growth scale of the lower two graphs. Correlation scores are between radial growth and reconstructed summer temperature in the previous year, which has been shown to maximize correlation (Barber et al. 2000). (A) Trees (n = 8) at a site near the tree line in the central Brooks Range (location 3, figure 12.2A) with positive sensitivity to summer temperature. (B) Trees (n = 4) from the same site in the Brooks Range with negative sensitivity to summer temperature. (C) Trees (n = 12) from Fort Richardson in Anchorage (location 1, figure 12.2A) with positive sensitivity to Fairbanks summer temperature. (D) Trees (n = 5) from Fort Richardson with negative sensitivity to Fairbanks summer temperature.

or even persistence in the environment. In fact, white spruce is the only species for which a significant artificial regeneration effort is made on managed forest land in Interior Alaska, and relying on natural regeneration of this species is recognized as risky (Fox et al. 1984). The fact that the consequences of failing to reproduce are so serious for the survival prospects of a species suggests that episodic reproduction is actually a strategy to match reproduction to conditions that are especially favorable. It then follows that environmental stimuli are needed to cue the plant to the time when conditions are most suitable for reproduction. Climate variability, as the main feature of year-to-year environmental variability, is the obvious candidate to examine as a regulator of white spruce reproduction.

Different investigators have reported a variety of observations on the magnitude, timing, and periodicity of white spruce cone and seed crops (Alden 1985; Dobbs 1976; Waldron 1965; Zasada 1972; Zasada 1986; Zasada 1995; Zasada et al. 1992; Zasada and Viereck 1970). The timing of white spruce cone production varies somewhat among trees, sites, and years (Waldron 1965; Zasada et al. 1992; Zasada and Viereck 1970), with both latitude and elevation affecting the periodicity. The timing of cone crops is quite consistent in the area of white spruce sampling reported here (figure 12.2). In British Columbia, the average frequency of good crops was 7 years for low-elevation sites and 12 years for high-elevation sites (Coates et al. 1994). In Interior Alaska, the average frequency for low-elevation sites is 10–12 years (Zasada 1972; Zasada and Viereck 1970).

BNZ maintains a long-term record of white spruce reproduction based on seed fall as measured by seed traps on the forest floor (figure 12.3B). The seed-fall record presented here is a composite of the UP1A and UP3A sites. During the 39-year period, 1957–1995, seed-fall index values stand out for five different years (figure 12.3B). The years 1958, 1970, and 1987 had the major seed crops of the interval, and moderate seed crops were measured in 1972 and 1983. The 1958 (Zasada and Viereck 1970) and 1970 (Zasada et al. 1978) seed crops have been recognized previously as major reproduction events for white spruce. The 1987 seed crop in the BNZ record is the largest of all. The majority of years functionally have no seed crop (figure 12.3B). Seed counts after 1996 were not completed for this analysis, but it was obvious that 1998 was a major seed-fall year, comparable to, although probably somewhat less than, 1958 and 1987.

A Model of White Spruce Seed Production at BNZ

Figure 12.7 is a model of white spruce reproduction based on the literature and observations and data sets developed at BNZ. The model is based on key events or "gateways" that must occur to allow the reproduction process to proceed. Some of the gateways are categorical: If the event does not pass a threshold then the reproduction process does not proceed no matter how favorable the conditions for the gateways that preceded or will follow it. Some of the gateways operate in a scalar fashion: Better conditions will allow a bigger crop, whereas less favorable conditions will reduce the crop.

The prerequisite (figure 12.7, step 1) of the process appears to be a sufficient level of growth reserves as indicated by recent favorable climate and radial growth. Several years in the period of seed-fall monitoring at BNZ have nearly all of the crop-promoting factors, but the trees start from such a chronically stressed condition that few or no cones develop, indicating that this is a categorical gateway. In the chronically warm and dry conditions of Regime 20.3, white spruce experienced threshold conditions to pass the second gateway consistently (figure 12.3A). The limiting factor became exceeding the threshold for growth reserves in the first gateway (figure 12.7). Under these circumstances, cool/moist conditions were limiting. Specifically, the 1983 crop of moderate size came at the first period of significantly improved climate favorability for radial growth following the longest drought stress period of the century to that point (figure 12.3A).

The next gateway (figure 12.7, step 2) is associated with a drought stress signal at the time of the formation of the bud primordia, which occurs at the end of vegetative shoot elongation. The period of bud differentiation is known to be a critical control over white spruce reproductive success, and reproduction can be induced artificially by stressors such as root pruning (Owens and Molder 1979; Owens and

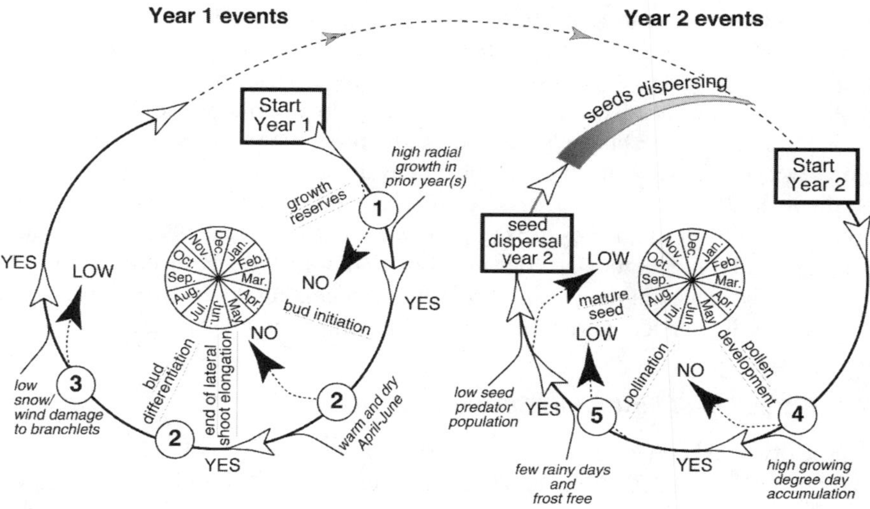

Figure 12.7 Conceptual model of critical gateways (numbered circles) leading to the production of white spruce seed crops. For detailed explanation of gateways, see text. Events begin at the top of the left-hand circle (start year 1) and continue clockwise, then proceed to the top of the right-hand circle and proceed clockwise to seed dispersal at the end of year 2. Timing of gateways during the year is indicated by position on yearly circle with months indicated. Open arrows indicate critical thresholds being met by conditions identified on the outer perimeter of circle and seed-crop production proceeds through the gateway; filled arrows (pointed inward on circle) indicate either condition not met or marginal conditions likely to result in a low seed crop.

Molder 1977). The literature has anecdotally suggested a relationship between warm, dry weather in the current year (inducing the crop) and following year (maturing the crop) and the production of cone crops (Alden 1985; Viereck 1973; Zasada et al. 1978; Zasada et al. 1992). Drought stress appears to be the factor that causes the preferential differentiation of the bud primordia into reproductive staminate and ovulate buds rather than vegetative buds. During the generally cool temperatures and favorable moisture conditions of Regime 20.2, warm/dry stressful weather was confined to the intersection of peaking temperature cycles and precipitation cycles (figure 12.5) and appeared to be so infrequent as to be limiting. For example, the major cone crop of 1958 followed the warm and exceptionally dry summer of 1957 (figure 12.3A, B).

The third gateway (figure 12.7, step 3) is a scalar process and involves the lack of severe pruning of reproductive shoots from wind and canopy snow loading in the fall and winter of the first year. The shoots are delicate and can be broken easily, especially when they have swollen and begun to elongate and then are subjected to extra loading from heavy rain or wet snow. The fourth gateway (figure 12.7, step 4) takes place in the late spring and early summer of the second year. High growing degree-day heat sums promote the maturation of the pollen and cone buds in time for the remainder of the steps to be completed before frost ends the growing season. Lack of sufficient heat will terminate the incipient crop. The final gateway (figure 12.7, step 5) involves both a categorical risk, the survival of pollen and cone buds in early stages (e.g., lack of killing frost), and a scalar process, heavy pollen flight (e.g., lack of persistent rains) to ensure high levels of cross pollination.

A physiological response, produced by the previous year's white spruce seed crop, has been shown to affect the number of available sites for differentiation of reproductive buds, therefore successive crops are not possible, and excellent years are followed by crop failures or very poor years (Zasada et al. 1992). The moderately large and moderate-size cone crops of 1970 and 1972, respectively, are the most closely spaced in the observation period, indicating that a 2-year spacing of crops is possible under appropriate, but probably rare, circumstances.

Reconstruction of Past White Spruce Seed Crops

The "typical" cone/seed crops (those that did not follow within 2 years of one another) of 1958, 1970, 1983, and 1987 are marked with a specific radial growth signal. The year of cone/seed maturation is marked by a deep reduction in radial growth compared to the trend of the year prior and year following (figure 12.3B), even in trees entirely free from competition (Youngblood 1991). Thus the year of a major white spruce seed crop is marked by a steep V-shaped radial growth pattern. Single-year radial growth reduction during the year of production of a major cone crop is also known for douglas-fir, grand fir, and western white pine (Eis et al. 1965). A numerical index that reflects the degree of this signal is the mean of radial growth in the preceding and following years divided by radial growth in the contemporary year (figure 12.8).

"Suspect" years that have a growth signal consistent with a major cone/seed crop match the BNZ seed-fall monitoring record, especially 1958 and 1987 (figure 12.8). The year 1969 began with weather in Interior Alaska that induced cone crop development, but an early season frost destroyed it (Zasada 1971), setting the stage for a somewhat reduced crop the following year. In northern British Columbia, 1970 was also a major cone crop year in white spruce (Eis and Inkster 1972). Prior to the period of seed-crop monitoring at BNZ, the Fairbanks climate record indicates strongly favorable cone crop–inducing weather (poor conditions for radial growth) in 1910, 1912, 1924, and 1940 (figure 12.3A,B, figure 12.8). For example, the earliest date (20 April) of ice breakup on the Tanana River at Nenana, 35 km downstream from BNZ, occurred in 1940 (Juday et al. 1998) because of a very early and warm spring that continued through May and early June. The record warm late spring and early summer of 1940 was ideal for the second gateway (figure 12.7, step 2) of the reproductive process. The second earliest breakup time occurred on 20 April 1998, the year of the maturation of another major cone crop in which warm early summer weather was ideal for the fourth gateway (figure 12.7, step 4). The induction of the 1998 cone crop came in 1997, a year of a very strong El Niño effect in Interior Alaska.

The frequency of viable white spruce seed crops at the tree line is even more limited than in lower elevation forests. Populations of vigorous white spruce seedlings dating to about 1940 have been noted at several locations in northern and In-

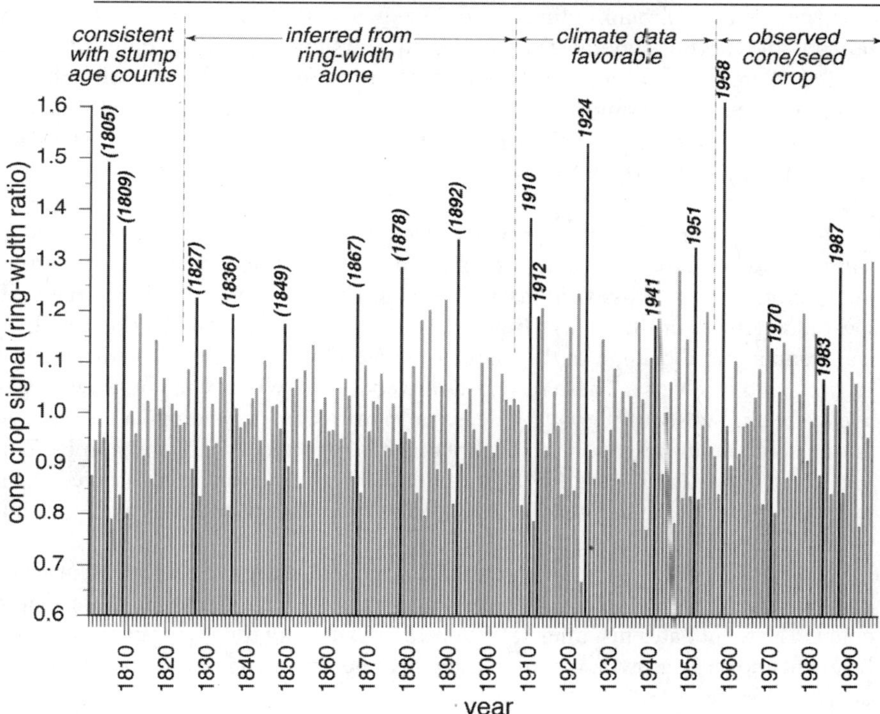

Figure 12.8 V index of white spruce radial growth from nine stands across Interior Alaska. Sample includes all stands depicted in figure 12.2B,C, with the exception of the Reserve West stand that burned in 1983. Index is the mean of radial growth in the year preceding and following the test year divided by growth in the test year. Values are high when 1-year radial growth reduction is great. Dates highlighted include years of major seed crops monitored at BNZ (post-1957), years in which weather at Fairbanks was conducive to seed-crop initiation or production (1906–1956), years with no other information concerning seed crops (1827–1905), and years in which stump age counts suggest cohorts of seedlings (1800–1826).

terior Alaska (Viereck 1979). The age structure of white spruce populations at the Rock Creek Long-Term Ecological Monitoring (LTEM) site in Denali National Park at tree-line in the central Alaska Range suggests that the major periods of recruitment to that population during the twentieth century took place as the result of seed crops whose timing matches the V signal identified here. The greatest recruitment to the Rock Creek LTEM population took place in the early 1940s, with additional peaks of recruitment at about 1910 and in the mid-1920s (Juday et al. 1999). Seedling white spruce with inferred dates of origin in the second decade of the twentieth century, suggesting a match with the 1910 and 1912 dates, were the second most abundant seedling age-class at Rock Creek.

Little information is available to identify potential major white spruce cone/seed crops before the Fairbanks climate record began in 1906. The ring-width V index serves as a working hypothesis about which years may have been dates of major cone/seed crops, a sort of paleocrop proxy. It is interesting to observe that the intervals between major ring-width V signals from 1827 to 1892 are 9, 13, 18, 11, and 14 years (figure 12.8), intervals that are very consistent with those between major seed-fall years during the BNZ monitoring period of 1957 to the present (13, 17, 11 years).

Of course, steep single-year growth reductions may occur across a population of trees for reasons other than climatic signals or allocation of photosynthates to cone maturation. In stem cross sections at several locations in Interior Alaska, the very low radial growth in 1878–1879 (figure 12.4) is also marked by indications of physical trauma and a change in the status of some trees from slow growing to fast growing and vice versa. These responses are consistent with a wind or snow breakage event and subsequent growth release or suppression, depending on the damage to the stem and its neighbors. Snow breakage is reported as happening in BNZ in 1967 (Van Cleve and Zasada 1970), and a major breakage event was observed again in 1990–1991. So, the V index for 1878 may not represent a reproductive event but a growth reduction from mechanical injury.

Several older white spruce stands in and near BNZ that have been clearcut allow large numbers of low stump surfaces to be ring counted. Many of these stump counts cluster at dates just after 1805 and 1810. Allowing for a few years for establishment and early growth, we can reasonably infer that major seed crops were produced in 1805 and 1809.

Conclusions

The ecology of white spruce growth and reproduction at BNZ is an excellent example of important ecological responses related to climate variability. The long time series available allows the identification of repeated outcomes of a system that includes as its interacting parameters climate (indicated by recorded data and reconstructed climate), tree growth (indicated by ring-width), seed crop timing and abundance, and stand age cohorts. The consistency of the response of the system across multiple cycles ("realizations" in time series analysis) is strong evidence in favor of an underlying causative mechanism rather than random patterns.

Several of the years of key events described in this chapter happened during El Niño years. For example, the 1941, 1958, 1983, 1987, and 1998 seed crops were initiated and/or matured during strong El Niños. Since 1940, 15 out of the 17 years with the greatest area burned in Alaska occurred during moderate to strong El Niños (Hess et al. 2001). The contribution of the El Niño signal to the climate–growth–reproduction system described here should be further examined. Strong and moderate El Niños produce positive temperature anomalies in Interior Alaska but generally below normal precipitation, particularly in the winter (Hess et al. 2001; Ropelewski and Halpert 1986). El Niños represent a deepening and an expansion of the Aleutian Low, so that El Niño conditions are simply an amplification of the system we have already described. However, the El Niño effect is not completely consistent in Interior Alaska, which may relate to the path the Aleutian Low takes as it approaches Alaska.

Changing climate sensitivity in the boreal forest is potentially a big issue for the future. Across the Northern Hemisphere, conifer trees at tree line have experienced a loss in climatic sensitivity at some sites (Briffa et al. 1998), causing reduced growth during warmer years as compared to the past. At least some tree-line populations of white spruce in Alaska have lost sensitivity altogether (Jacoby et al. 1999) and no longer respond to increased warmth. A recent review of the climate-related growth dynamics of tree-line forests in Alaska (Juday et al. 1999) has demonstrated that Alaska tree lines are responding in different ways to recent climate warming, depending on the location and environmental setting of the tree line. If climatic warming typical of that experienced in Regime 20.3 continues, changing sensitivity could become a widespread phenomenon with deep implications for resource management and the ecology of the Alaskan boreal forest. It also means that tree-line climate reconstructions should be viewed with caution.

The original logical inference about the strategy of periodic reproduction was that it is designed to time infrequent reproductive events to particularly favorable periods and to minimize reproductive costs otherwise. So, what would be particularly favorable for white spruce reproduction in the pattern of timing described in this chapter? The key cue of environmental variability for white spruce is a critical period of warm and dry early summer weather in successive years (figure 12.7). The plant has an internal, hormonally driven system to detect and respond with maximum reproductive effort to the relatively infrequent intervals when strong stress is generated (e.g., by drought) (Owens and Molder 1979; Owens and Molder 1977). This same dry early spring weather and clear weather during the long days near the summer solstice represent fire weather (Johnson et al. 1992). It appears that the described reproductive timing of white spruce maximizes the odds that seeds will be released into a landscape in which fires have occurred recently.

The thick organic mat of the forest floor of the boreal forest is a significant obstacle to white spruce seedling germination, survival, and growth (Zasada 1968). The reduction or removal of coarse or refractory organic material by fire or other methods can significantly improve white spruce seedling germination, growth, and survival (Zasada et al. 1992). The area of boreal forest burned annually in Interior Alaska reaches a distinct peak about every 10 years (Juday et al. 1998). Total area burned in Alaska is relatively well correlated with the May through August temperature in Fairbanks (Juday et al. 1998). The two years with the highest total area burned in Alaska since records began in 1955 are 1957 and 1969 (Juday et al. 1998)—years that are especially well matched to the large cone crops of 1958 and 1970. Surviving trees around the fire margin in both years, 1958 and 1970, released large seed crops onto well-prepared and receptive seedbeds.

Clearly, a disproportionate share of the living young white spruce trees less than 50 years old in Interior Alaska are the result of reproduction in only four years (1958, 1970, 1987, 1998), based simply on probability. If an interval of 12 years is maintained between large white spruce seed crops over the life of a typical BNZ 200-year-old mature white spruce stand and the trees do not become reproductive for their first 30 to 40 years, then on average only about 14 major reproductive events occur during the life of a stand.

Presumably, the entire Interior Alaskan boreal ecosystem is responding to the variability and alternating state of the climatic system (figure 12.3A) analyzed in this chapter. Our description of the periodic white spruce reproduction system and its apparent environmental controls is presented here for the first time, so corresponding responses for other plants and animals within the system have not yet been identified. However, jt is not difficult to imagine potential subjects for investigation, ranging from regional populations of browsing animals dependent on early successional postfire vegetation to white spruce seed predators. The temporal association of tree reproduction with a particular sequence of variability in the climate system is itself a "cascade" through the system with respect to the flow of energy and information.

If only the instrument-based record at Fairbanks (figure 12.3A) is examined, a substantial summer warming trend in the twentieth century is obvious. The 1977–2001 postshift mean summer temperature was 0.97°C warmer than the instrument record prior to that time (1906–1976). However, the longer view provided by the near 200-year reconstruction (figure 12.5A) suggests at least the possibility of a cyclic process in which the warming at the end of twentieth century is roughly matched by a warm period in the mid-1800s. As previously noted, peak mid-nineteenth century warmth in Interior Alaska was not reflected in Northern Hemisphere trends at the time, yet the warming is clearly indicated by Alaska tree-ring properties. By contrast, an annual temperature reconstruction based on 20 tree-ring width records for the past three centuries at tree-line sites in northern North America, Scandinavia, Siberia, and Mongolia is well correlated with a global energy-balance model that uses solar, volcanic, and anthropogenic trace gas and aerosol variations as three primary forcing functions (D'Ariggo et al. 1999). It appears that Interior Alaska summer temperatures were opposite the global trends during much of the nineteenth century and early twentieth century (Mann et al. 1998). However, during Regime 20.3 trends in Interior Alaska summer temperatures and northern hemisphere warm season mean temperatures are in agreement (Mann et al. 1998), and both sets of temperature values have reached some of the highest levels of the last two or more centuries. The consequences of these sustained high summer temperatures in Interior Alaska during Regime 20.3 include prolonged climatic stress that has resulted in decreased radial growth of upland white spruce, and large, frequent fires, decreasing carbon uptake and storage in this ecosystem.

References

Alden, J. 1985. Biology and management of white spruce seed crops for reforestation in Subarctic Taiga Forests. *Agricultural and Forestry Experimental Station, School of Agriculture and Land Resources Management*, Bulletin 69, 51.

Barber, V. A., Juday, G. P., and Finney, B. P. 2000. Reduced growth of Alaskan white spruce in the twentieth century from temperature-induced drought stress. *Nature*. 405:668–673.

Barber, V. A., Juday, G. P., Finney, B. P., and Wilmking, M. In press. Reconstruction of summer temperatures in Interior Alaska from tree-ring proxies: Evidence for changing synoptic climate regimes. *Climatic Change.*

Bonan, G. B., Pollard, D., and Thompson, S. L. 1992. Effects of boreal forest vegetation on global climate. *Nature.* 359:716–718.

Briffa, K. R., Schweingruber, F. H., Jones, P. D., Osborn, T. J., Shiyatov, S. G., and Vagnov, E. A. 1998. Reduced sensitivity of recent tree-growth to temperature at high northern latitudes. *Nature.* 391:678–682.

Coates, D. C., Haeussler, S., Lindeburgh, S., Pojar, R., and Stock, A. J. 1994. Ecology and Silviculture of Interior Spruce in British Columbia. Forest Resource Development Report. 182 pp.

Cropper, J. P. 1982. Climate reconstructions (1801 to 1938) inferred from three-ring width chronologies of the North American Arctic. *Arctic and Alpine Research.* 14:223–241.

D'Arrigo, R. D., and Jacoby Jr., G. C. 1992. Dendroclimatic evidence from northern North America. Pp. 296–311 in R. S. Bradley and P. D. Jones, editors. *Climate Since A.D. 1500.* Routledge, London.

D'Arrigo, R. D., Jacoby, G. C., and Free, R. M. 1992. Tree-ring width and maximum latewood density at the North American tree line: Parameters of climate change. *Canadian Journal of Forest Research.* 22:1290–1296.

D'Arrigo, R., Jacoby, G., Free, M., and Robock, A. 1999. Northern hemisphere temperature variability for the past three centuries: Tree-ring and model estimates. *Climatic Change* 42:663–675.

D'Arrigo, R. D., Malmstrom, C. M., Jacoby, G. C., Los, S. O., and Bunker, D. E. 2000. Correlation between maximum latewood density of annual tree rings and NDVI based estimates of forest productivity. *International Journal Of Remote Sensing.* 21:2329–2336.

D'Arrigo, R., Villalba, R., and Wiles, G. 2001. Tree-ring estimates of Pacific decadal climate variability. *Climate Dynamics.* 18 (3-4):219–224.

Dobbs, R. C. 1976. White spruce seed dispersal in central British Columbia. *Forestry Chronicle.* 52:225–228.

Dyrness, C. T., Viereck, L. A., and Van Cleve, K. 1986. Fire in taiga communities of Interior Alaska. Pp. 74–86 in F. S. Chapin III, P. W. Flanagan, L. A. Viereck, and K. Van Cleve, eds. *Forest ecosystems in the Alaskan taiga: A synthesis of structure and function.* Springer-Verlag, New York.

Ebbesmeyer, C. C., Cayan, D. R., McLain, D. R., Nichols, F. H., Peterson, D. H., and Redmond, K. T. 1990. 1976 step in the Pacific climate: Forty environmental changes between 1968–1975 and 1977–1984. *Proceedings of the Seventh Annual Pacific Climate (PACLIM) Workshop.* California Department of Water Resources. Sacramento.

Edwards, M. E., Mock, C. L., Finney, B. P., Barber, V. A., and Bartlein, P. J. 2001. Potential analogues for paleoclimatic variations in eastern interior Alaska during the past 14,000 yr: Atmospheric-circulation controls of regional temperature and moisture responses. *Quaternary Sciences Reviews.* 20:189–202.

Eis, S., Garman, E. H., and Ebell, L. F. 1965. Relation between cone production and diameter increment of Douglas-fir (*Pseudotsuga menziesii* (Mirb) Franco), grand fir (*Abies grandis* (Dougl.) Lindl.), and Western white pine (*Pinus monticola* Dougl.). *Canadian Journal of Botany.* 43:1553–1559.

Eis, S., and Inkster, J. 1972. White spruce cone production and prediction of cone crops. *Canadian Journal of Forest Research.* 2:460–466.

Favorite, F., Dodimead, A. J., and Nasu, K. 1976. Oceanography of the subarctic Pacific region, 1960–71. International North Pacific Fisheries Commission Bulletin No. 33.

Foote, M. J. 1983. Classification, description, and dynamics of plant communities after fire in the taiga of interior Alaska. *Research Paper PNW-307.* USDA Forest Service, Pacific Northwest Forest and Range Experiment Station. Portland, Oregon.

Fox, J. D., Zasada, J. C., Gasbarro, A. F., and Van Veldhuizen, R. 1984. Monte Carlo simulation of white spruce regeneration after logging in interior Alaska. *Canadian Journal of Forest Research.* 14:617–622.

Fritts, H. C. 1976. *Tree-rings and climate.* Academic Press, New York.

Garfinkle, H. L., and Brubaker, L. B. 1980. Modern climate-tree-growth relationships and climatic reconstruction in sub-Arctic Alaska. *Nature.* 286:872–874.

Hess, J. C., Scott, C. A., Hufford, G. L., and Fleming, M. D. 2001. El Niño and its impact on fire weather conditions in Alaska. *International Journal of Wildland Fire.* 10:1–13.

Jacoby, G. C., and D'Arrigo, R. 1989. Reconstructed northern hemisphere annual temperature since 1671 based on high-latitude tree-ring data from North America. *Climatic Change.* 14:39–59.

Jacoby, G. C., and D'Arrigo, R. D. 1995. Tree ring width and density evidence of climatic and potential forest change in Alaska. *Global Biogeochemical Cycles.* 9:227–234.

Jacoby, G. C., D'Arrigo, R. D., and Juday, G. P. 1999. Tree-ring indicators of climatic change at northern latitudes. *World Resource Review.* 11:21–29.

Jacoby, G. C., Ivanciu, I. S., and Ulan, L. D. 1988. A 263-year record of summer temperature for northern Quebec reconstructed from tree-ring data and evidence of a major climatic shift in the early 1800's. *Palaeogeography, Palaeoclimatology, Palaecology.* 64: 69–78.

Jacoby, G. C., and Ulan, L. D. 1983. Tree ring indications of uplift at Icy Cape, Alaska, related to the 1899 earthquake. *Journal of Geophysical Research.* 88(B11):9305–9313.

Jacoby, G. C., Wiles, G., and D'Arrigo, R. D. 1996. Alaskan dendroclimatic variations for the past 300 years along a north-south transect. Pages 235–248 in Tree Rings, Environment and Humanity, J. S. Dean, D. M. Meko, and T. W. Swetnam, eds., Tucson, Arizona.

Johnson, C. A., Pastor, J., and Pinay, G. 1992. Quantitative methods for studying landscape boundaries. pp. 107–128. In *Landscape Boundaries: Consequences for Biotic Diversity and Ecological Flows.* A. J. Hansen and F. di Castri, eds. Springer-Verlag, Berlin.

Johnson, E. A. 1992. *Fire and Vegetation Dynamics; Studies from North American Boreal Forest.* Cambridge University Press. London.

Juday, G. P. 1984. Temperature Trends in the Alaska Climate Record. *Proceedings of the Conference on the Potential Effects of Carbon Dioxide-Induced Climatic Changes in Alaska.* 7–8 April, 1982 Fairbanks, Alaska, pp. 76–88.

Juday, G. P., Barber, V. A., Berg, E., and Valentine, D. 1999. Recent dynamics of white spruce treeline forests across Alaska in relation to climate. *Sustainable Development in the Northern Timberline Forests. Proceedings of the Timberline Workshop, May 10-11, 1998,* Whitehorse, Canada. pp. 165–187.

Juday, G. P., Ott, R. A., Valentine, D. W., and Barber, V. A. 1998. Forests, Climate Stress, Insects and Fire. *Proceedings of The Conference on Implications of Global Climate Change in Alaska and the Bering Sea Region.* 3–6 June 1997 Fairbanks, Alaska. pp. 23–50.

Labau, V. J., and W. S. van Hess. 1990. An inventory of Alaska's boreal forests: Their extent, condition, and potential use. In *Boreal Forest: Condition, dynamics, anthropological influences.* Proceedings of the International Symposium, 16–26 July, 1990. Archangelsk, Russia. State Forestry Committee of the USSR. Part 6. Pages 30–39.

Leavitt, S. W., and Danzer, S. R. 1992. Method for batch processing small wood samples to holocellulose for stable-carbon isotope analysis. *Analytical Chemistry.* 65:87–89.

Livingston, N. J., and Spittlehouse, D. L. 1996. Carbon isotope fractionation in tree-rings early and late wood in relation to intra-growing season water balance. *Plant Cell Environment*. 19:768–774.

Mann, M. E., R. S. Bradley, and M. K. Hughes. 1998. Global-scale temperature patterns and climate forcing over the past six centuries. *Nature* 392:698–702.

Mock, C. J., Bartlein, P. J., and Anderson, P. M. 1998. Atmospheric circulation patterns and spatial climatic variations in Beringia. *International Journal of Climatology*. 18:1085–1104.

Nyland, R. D. 1996. *Silviculture Concepts and Applications*. McGraw-Hill, New York.

Overland, J. E., and Heister, T. R. 1980. Development of a synoptic climatology for the northeast Gulf of Alaska. *Journal of Applied Meteorology*. 19:1–14.

Overpeck, J., Hughen, K., Hardy, D., Bradley, R., Case, R., Douglas, M., Finney, B., Gajewski, K., Jacoby, G., Jennings, A., Lamoureux, S., Lasca, A., MacDonald, G., Moore, J., Retelle, M., Smith, S., Wolfe, A., and Zielinski, G. 1997. Arctic environmental change of the last four centuries. *Science*. 278:1251–1256.

Owens, J. N., and Molder, M. 1977. Bud development in *Picea glauca*. II. Cone differentiation and early development. *Canadian Journal of Botany*. 55:2746–2760.

Owens, J. N., and Molder, M. 1979. Sexual reproduction in white spruce (*Picea glauca*). *Canadian Journal of Botany*. 57:152–169.

Patric, J. H., and Black, P. E. 1968. Potential evapotranspiration and climate in Alaska by Thornthwaite's classification. *PNW-71*, U.S.D.A. Forest Service, Juneau, Alaska.

Pojar, J. 1996. Environment and biogeography of the western boreal forest. *Forestry Chronicle*. 72:51–58.

Ropelewski, C. F., and Halpert, M. S. 1986. North American precipitation and temperature patterns associated with the El Niño/Southern Oscillation (ENSO). *Monthly Weather Review*. 114:2352–2362.

Slaughter, C. W., and Viereck, L. A. 1986. Climatic characteristics of the taiga in Interior Alaska. Pages 9–21 in K. Van Cleve, F. S. I. Chapin, L. A. Viereck, and C. T. Dyrness, eds. *Forest ecosystems in the Alaska taiga. A synthesis of structure and function*. Springer-Verlag, New York.

Turner, S. J., O'Neill, R. V., Conley, M. R., and Humphries, H. C. 1991. Pattern and scale: Statistics for landscape ecology. Pages 17–50 in M. G. Turner and R. H. Gardner, eds. *Ecological Studies: Quantitative Methods in Landscape Ecology: the Analysis and Interpretation of Landscape Heterogeneity*. Springer-Verlag, New York.

Van Cleve, K., Dyrness, C. T., Marion, G. M., and Erickson, R. 1993. Control of soil development on the Tanana River floodplain of interior Alaska. *Canadian Journal of Forest Research*. 23:941–955.

Van Cleve, K., and Viereck, L. A. 1981. Forest succession in relation to nutrient cycling in the boreal forest of Alaska. Pp. 185–212 in D. C. West, H. Shuggart, and D. B. Botkin, eds. *Forest succession concepts and application*. Springer-Verlag, New York.

Van Cleve, K., and Zasada, J. 1970. Snow breakage in black and white spruce stands in interior Alaska. *Journal of Forestry*. 68:82–83.

Viereck, L. A. 1973. Wildfire in the taiga of Alaska. *Quaternary Research*. 3:465–495.

Vierek, L. A. 1979. Characteristics of treeline plant communities in Alaska. *Holoarctic Ecology*. 2:228–238.

Viereck, L. A., Cleve, K. V., and Dyrness, C. T. 1986. Forest ecosystem distribution in the taiga environment. Pp. 22–43 in K. Van Cleve, F. S. Chapin III, P. W. Flanagan, L. A. Viereck, and C. T. Dyrness, eds. *Forest ecosystems in the Alaskan taiga: A synthesis of structure and function*. Springer-Verlag, New York.

Viereck, L. A., Dyrness, C. T., Batten, A. R., and Wenzlik, K. J. 1992. *The Alaska vegetation classification system*. U.S. Department of Agriculture, Forest Service, Pacific Northwest Research Station., Portland, Oregon.

Waldron, R. M. 1965. Cone production and seedfall in a mature white spruce stand. *Forestry Chronicle.* 41:314–329.

Wilson, J. G., and Overland, J. E. 1986. Meteorology. Pages 31–54 in D. W. Hood and S. T. Zimmerman, eds. *The Gulf of Alaska; Physical Environment and Biological Resources.* Alaska Office. Ocean Assessments Division, NOAA. U.S. Department of Commerce. Mineral Management Publication MMS 86-0095, Springfield, Virginia.

Yarie, J., Viereck, L., Van Cleve, K., and Adams, P. 1998. Flooding and ecosystem dynamics along the Tanana River. *Bioscience.* 48:690–695.

Youngblood, A. P. 1991. Radial growth after a shelterwood seed cut in a mature stand of white spruce in interior Alaska. *Canadian Journal of Forest Research.* 21:410–413.

Zasada, J. C. 1971. *Frost damage to white spruce cone in interior Alaska.* USDA Forest Service, Pacific Northwest Forest and Range Experiment Station, Portland, Oregon.

Zasada, J. C. 1968. Natural regeneration of interior Alaska forests—seed, seedbed, and vegetative reproduction considerations. In *Proceedings, Fire in the northern environment, a symposium.* Portland, Oregon, pp. 231–246.

Zasada, J. C. 1972. *Guidelines for obtaining natural regeneration of white spruce in Alaska.* USDA Forest Service, Pacific Northwest Forest and Range Experiment Station, Portland, Oregon.

Zasada, J. C. 1986. Natural regeneration of trees and tall shrubs on forest sites in interior Alaska. Pages 44–73 in F. S. Chapin III, P. W. Flanagan, L. A. Viereck, and K. Van Cleve, eds. *Forest ecosystems in the Alaskan taiga: A synthesis of structure and function.* Springer-Verlag, New York.

Zasada, J. C. 1995. Natural regeneration of white spruce—Information needs and experience from the Alaskan boreal forest. In *Innovative silviculture systems in boreal forests, a proceedings,* Edmonton, Alberta, pp. 40–46.

Zasada, J. C., Foote, J. M., Deneke, F. J., and Parkerson, R. H. 1978. Case history of an excellent white spruce cone and seed crop in Interior Alaska: Cone and seed production, germination, and seedling survival. *General Technical Report PNW-65,* USDA Forest Service, Pacific Northwest Forest and Range Experiment Station, Portland, Oregon.

Zasada, J. C., Sharik, T. L., and Nygren, M. 1992. The reproductive process in boreal forest trees. Pp. 85–125 in H. H. Shugart, R. Leemans, and G. Bonan, eds. *A Systems Analysis of the Global Boreal Forest.* Cambridge University Press, Cambridge.

Zasada, J. C., and Viereck, L. A. 1970. White spruce cone and seed production in interior Alaska, 1957–1968. *USDA Forest Service Research Note PNW-129,* Pacific Northwest Forest and Range Experiment Station, Portland, Oregon.

13

Decadal Climate Variation and Coho Salmon Catch

David Greenland

When temporally smoothed data are used for the period 1925 to 1985 there is a close inverse statistical relationship acting at an interdecadal timescale between the Pacific Northwest (PNW) air temperatures and Coho salmon catch off the coast of Washington and Oregon. This relationship is now well known, although not fully explained, but at the time of its discovery in 1994 it was part of advances being made by several research groups on interdecadal-scale climate/ecological changes in the PNW (Greenland 1995). The discovery and later, related findings may be usefully examined within the context of the framework questions of this book (see chapter 1) because it provides a very interesting example of climate variability and ecosystem response found, in part, by Long-Term Ecological Research (LTER) investigators. The logical progression for this chapter is first to review a little of the relationship between Coho salmon and climate and then to explain how a study at one LTER site led to a finding with regional implications. An update of the findings at interdecadal-scale climate/ecological changes in the PNW is then appropriate, followed by a discussion of the topic with the framework questions of this book. The PNW is defined, for the purposes of this chapter, as the area of Washington and Oregon west of the crest of the Cascade Range. The term *decadal* is used loosely in this chapter to refer to changes that focus on time periods of about 10 to 30 years in length.

Salmon and Climate

Salmon live part of their lives in terrestrial, freshwater environments and part in marine, saltwater environments. The salmon life history starts with fertilized eggs remaining in gravel in freshwater stream beds and hatching after 1–3 months. One to five months later, fry emerge in the spring or summer. Juvenile fish are in freshwater from a few days to 4 years, depending on species and locality. After the juveniles change to smolts, they can migrate to the ocean, usually in spring or early summer, often taking advantage of streamflows driven by snowmelt. The fish spend 1–4 years in the ocean and then return to their freshwater home stream to spawn and die. More specifically, the typical life cycle for Oregon Coho spans 3 years (18 months in freshwater and 18 months in the ocean). Climatic factors affect the fish at all stages of their fascinating life history, and Greenland (1995) has reviewed some of these factors. However, it is believed that salmon are most vulnerable to climate variations when they are at the migrating smolt stage (Pearcy 1992). This stage is also where decadal-scale climate variability plays a role.

There are many other factors besides climate that control salmon survival rates. Indeed the "cyclic" variability described here is played out against a background of a century-long decrease in the salmon population in the PNW—a decrease resulting mainly from a suite of direct and indirect human-caused factors. The following account relates only to Coho salmon, although somewhat similar or inverse patterns have also been found with Sockeye, Pink, and Chinook salmon in, and north of, the PNW (Mantua et al. 1997). Where the term *salmon* is used herein without a qualifier, the reference is implicitly to Coho salmon, although the point being made may well apply to other species as well.

We must first examine how research results at an individual LTER site give rise to further discoveries at a regional scale.

From the Andrews Forest to the Pacific Ocean

An important part of the LTER program is its interdisciplinary nature and how research findings in one subdiscipline stimulate discoveries in another. This connectivity was particularly important in the results reported in this chapter, and the steps of the connection are worth recording as well as the relationship to which they led.

In the early 1990s, I completed an introductory analysis of the climate of the H. J. Andrews LTER site (HJA) that is located in the foothills of the Cascade Range in Oregon (Greenland 1994; see also chapter 19). Several times I presented the results of the climate analysis to the HJA LTER research group. One of the results of my climate study indicated the relationship of the El Niño–Southern Oscillation (ENSO) phenomenon to the Andrews climate. One graphic showed that the value of the Southern Oscillation Index (SOI) for 1982–1983 was markedly displaced from an otherwise clear linear relationship between SOI and winter water year precipitation. This graphic reminded Dr. Stanley Gregory, a stream ecologist, of an analogous graphic of the relationship between the number of previous year Coho jacks (early maturers) versus an overall Oregon Production Index (Pearcy 1992, fig. 5.5). The 1982–1983 data point was also an outlier in the Coho jack relationship. Outliers in both graphs are a function of the very intense El Niño event of that year.

I was then asked to present papers at Fishery Science meetings on the relationship between climate and salmon within the context of the effect of climate on, and the linkages between, the terrestrial and oceanic part of the salmon habitat. In preparation for these presentations, I found an important paper by Frances and Sibley (1991) on decadal variation of Coho salmon catch in the northeastern Pacific Ocean. Frances and Sibley (1991) had previously reported a close relationship between winter (November to March) air temperatures at Sitka, Alaska, and surface water temperature at Langara Spit, Queen Charlotte Island, British Columbia, and the catch of Pink salmon in the Gulf of Alaska. They had also reported an inverse relationship between the catch of Pink salmon in the Gulf of Alaska and the catch of Coho salmon off the coast of Washington and Oregon. Both relationships covered the period 1925 to 1985 and were found when the data were normalized and subjected to a 7-year weighted filter (R. C. Frances, pers. comm., 1994). Given these relationships I reasoned that it would be likely that there would be an inverse relationship between air temperatures in the PNW and the catch of Coho salmon off the coast of Washington and Oregon.

I used a 5-year unweighted filter of the annual mean air temperatures at the H. J. Andrews Long-Term Ecological Research site. The filter was applied to values of Andrews' temperatures that were normalized to the long-term mean for the 1925–1985 period. It had been shown elsewhere that the Andrews' temperatures are well related to those of western Oregon in general (Greenland 1994). To a large extent the temperatures of western Oregon are also related to those of the broader PNW region. Coho Salmon catch data were extracted from the graphs of Frances and Sibley and a close inverse relationship was indeed found (figure 13.1). This relationship and time series suggest an approximate 20 year "cycle" in both air temperatures in the PNW and the catch of Coho salmon off the coast of Washington and Oregon. The term "cycle" is placed in quotes because the data sets are not of sufficient length to determine the existence of true cycles. However, recent tree-ring studies by Biondi et al. (2001) for the southwestern United States and Gedalof and Smith (2001) for northwestern North America have extended the length of the time series to about 400 years. A similar decadal-scale variability shows up throughout this 400-year record. The interdecadal variation in the tree rings of Interior Alaska (chapter 12) are also related to PDO changes.

In summary, when temporally smoothed data are used for the period 1925 to 1985, there is a close inverse statistical relationship between the PNW air temperatures and Coho salmon catch off the coast of Washington and Oregon. To attempt to explain this inverse relationship, we must explore the large spatial and long-term aspects of atmospheric and ocean currents as well as further thinking of Frances and Sibley and their colleagues. Although there are biophysical variations of a variety of time and space scales that affect salmon catch, we are mostly interested here in the decadal-scale changes that are inevitably linked with Pacific basinwide spatial variations in the atmosphere and ocean.

A Possible Cascade

A cascading system is generally regarded as one that exhibits flow of material, energy, or information. The following cascade is proposed to partially explain the re-

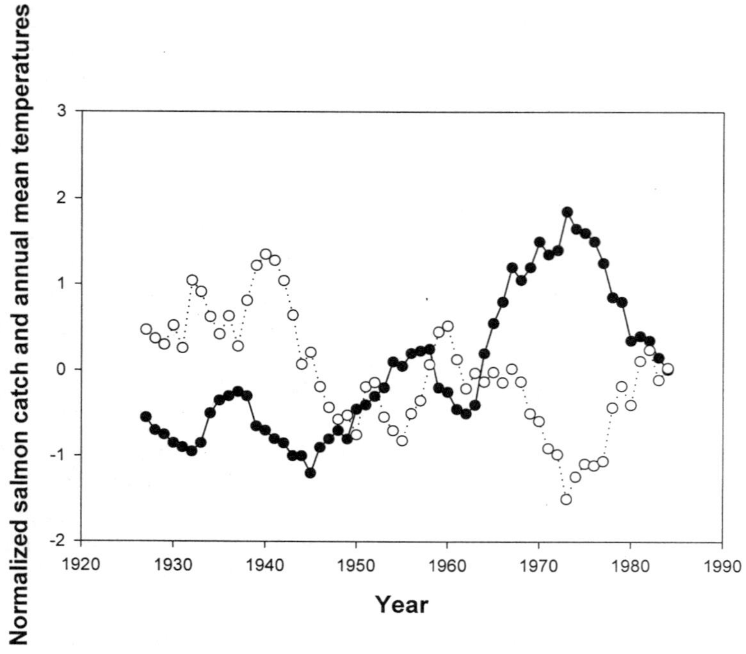

Figure 13.1 Five-year moving average of annual mean temperature at the H. J. Andrews LTER site (open circles) and seven-year moving average of Coho salmon catch off the coast of Washington and Oregon (filled circles). With permission from the California Department of Water Resources. (Salmon data from Frances and Sibley, 1991.)

lationship between the PNW air temperatures and Coho salmon catch off the coast of Washington and Oregon and the resulting variability at the interdecadal timescale. A five-level cascade actually starts with a coupled ocean/atmosphere variability now called the Pacific Decadal Oscillation (PDO) (Level 1). The two major modes of the PDO affect the air circulation over the northeastern Pacific Ocean area, and this leads to changes in the direction of the air currents and the intensity of the ocean currents (Level 2). These changes in the direction of the air currents give rise to the changes in the air temperatures of the PNW. The air temperatures are a by-product in this cascade and have little to do with salmon catch. It is hypothesized that the operation of the ocean current leads to the provision of greater or lesser nutrients (Level 3) that in turn work their way up the food chain and eventually result in a variable abundance of salmon (Level 4) and subsequent salmon catch (Level 5). There can be several more levels in this cascade, depending on the degree of detail at which the food chain is resolved.

Attempts to specify the system cascade in such a way have the advantage, among other things, of identifying topics that require further investigation. In this cascade the atmospheric changes are known in some detail. The related provision, or lack of provision, of nutrient-rich ocean water under different atmospheric conditions also seems understandable, although it is known in less detail. However, there are major uncertainties about the linkages between the lower and upper parts of the food chain, as well as aspects of salmon ecology such as the comparative role of wild and hatchery-bred salmon. Besides uncertainties in food chain linkage, there are other possibilities for the "cascade." For example, an alternative explanation is that changes in the physical habitat, such as temperature changes, alter the distribution and effectiveness of predators as suggested by Fisher and Pearcy (1988). The establishment and further development of this cascade provides us with a fertile research agenda.

Advances in Atmospheric and Fisheries Science

Since the link between the climate at HJA and Coho fish catch off the coast of the PNW was discovered, there have been many advances in our knowledge of the climate and fisheries of the region.

At the outset it must be explained that the basic geography of surface ocean currents in the northeastern Pacific Ocean consists first of the subarctic current moving water eastward across the North Pacific Ocean at about 45–50° N, the approximate latitude of the cities of Seattle and Vancouver. As the water nears the North American continent, it bifurcates. Part of it flows northward near the Canadian and Alaskan coastlines and into the Gulf of Alaska to help form the Alaskan gyre. Another part of the water from the subarctic current flows southward, forming the Californian current.

Even at the time that Frances and Sibley were reviewing the inverse relation between Gulf of Alaska and PNW salmon catch, Frances (1993) quoted the ideas of Hollowed and Wooster (1991) to explain the oceanographic and atmospheric differences in the two modes of the (then unnamed) PDO. It was argued that the inverse relationship in catch in the two ocean areas might be explained by the north or south movement of the divergence, or bifurcation, zone between the Alaskan and California currents. The suggested model is bimodal and postulates two states or modes of operation of the ocean currents (figure 13.2). When the bifurcation zone is more to the north (PDO cold phase), the Aleutian low-pressure zone in the atmosphere is weak and more cold subarctic current water is taken into the Californian current and the upwelling of nutrient-rich water off the Oregon and Washington coasts is enhanced. When the bifurcation zone is farther to the south (PDO warm phase), the Aleutian low-pressure zone is deep, and it swings winds, rain-bearing storms, and more cold subarctic current water into the Alaskan current, leaving the water off the Oregon and Washington coasts relatively warmer. The reality is more complex than simply changing water and air temperatures by altering water transport. The Aleutian Low pressure zone forces changes the upper ocean temperatures in the northeastern Pacific by a combination of surface heat fluxes, vertical mixing, and Ekman transports (Miller et al. 1994).

The Hollowed and Wooster model may be placed in a larger atmospheric context by noting its relationship to the synoptic climatology and teleconnective index values (figure 13.2). What later became known as the PDO cold phase is associated with a weak Aleutian low pressure with winds coming more directly from the west

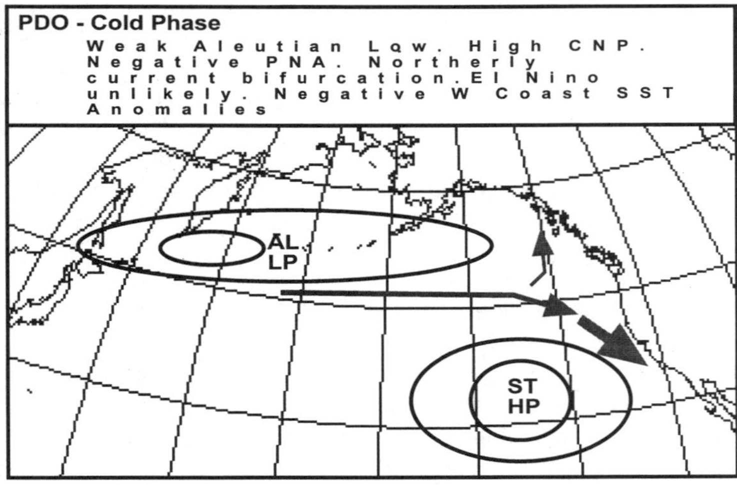

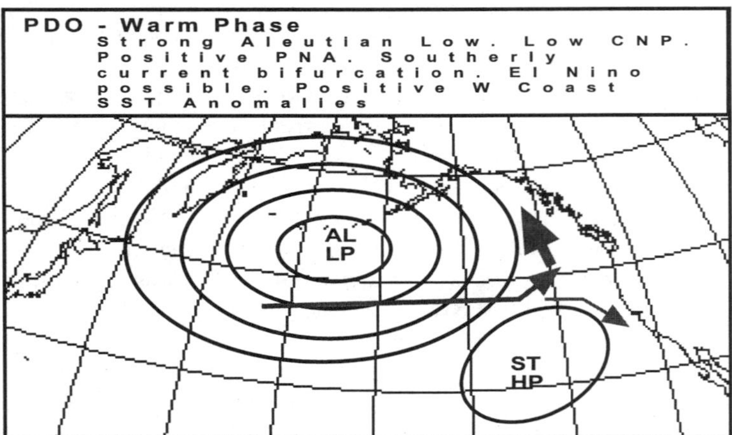

Figure 13.2 A schematic of the surface atmospheric pressure distribution accompanying the two main modes of the Pacific Decadal Oscillation. Solid black lines represent isobars. Gray arrows represent hypothesized accompanying bifurcation of the subarctic ocean current into the Alaskan current and the Californian current.

across the Pacific at the latitudes of Washington and Oregon. The more northerly bifurcation of the subarctic current pushes more water into the Californian current and gives rise to negative sea surface temperature (SST) anomalies. These circumstances are also associated with surface offshore wind flow and ocean upwelling that brings food sources to the marine food chain near the coast. These conditions are not generally consistent with intense El Niño conditions. At the opposite mode of the suggested model, the PDO warm phase is associated with a strong Aleutian low pressure and enhanced southwesterly winds in the northeast Pacific. The more southerly subarctic current bifurcation enhances northward ocean flow into the Alaskan current, giving rise to positive SST anomalies in the eastern part of the northeastern Pacific. These conditions are consistent with the results of El Niño events. The relative values of other atmospheric indices of pressure distribution and implied atmospheric motion such as the Central North Pacific (CNP) index and the Pacific North American (PNA) index are also shown in figure 13.2.

The most important discovery was the codification of the interdecadal-scale climate regime shifts in the northeastern Pacific Ocean and the northwestern North American continent now known to be related to changes in the ocean and named the Pacific Decadal Oscillation (PDO). University of Washington researchers noted that the PDO was in a cool phase from about 1900 to 1925 and from 1945 to 1977 (Mantua et al. 1997). The PDO was in a warm phase from 1925 to 1945 and after 1977. Another phase change may have occurred in the mid-1990s. Additional work has attempted to identify the terminal points of the regime shifts. Overland et al. (1999) found that since the turn of the century, 37% of the winter interannual variance of the Aleutian low is at timescales greater than 5 years. An objective algorithm detected zero crossings of Aleutian low central pressure anomalies in 1925, 1931, 1939, 1947, 1959, 1968, 1976, and 1989. Ware and Thomson (2000) noted that the climate of the northeastern Pacific Ocean has oscillated at three dominant timescales over the last 400 years: the well-known 2- to 8-year El Niño–Southern Oscillation (ENSO) timescale, a 20- to 40-year interdecadal timescale, and a 60- to 80-year multidecadal timescale. The latter oscillation has been the dominant mode of air temperature variability along the west coast of North America during the last 400 years. During this period, there have been conspicuous temporal modulations of the ENSO and the interdecadal signals. Low-frequency temperature oscillations at periods greater than 10 years in the northeast Pacific have been significantly coherent and in-phase from southern California to British Columbia. However, with the exception of the ENSO signal, higher frequency variability has been weakly coherent along the west coast. Biondi et al. (2001) found the PDO is closely matched by the dominant mode of tree-ring variability that provides a preliminary view of multiannual climate fluctuations spanning the past four centuries. A reconstructed PDO index features a prominent bidecadal oscillation, whose amplitude weakened in the late 1700s to mid-1800s. A comparison with proxy records of ENSO suggests that the greatest decadal-scale oscillations in Pacific climate between 1706 and 1977 occurred around 1750, 1905, and 1947.

Undoubtedly, more investigation of the various timescales of variability will be made, but the focus in this account remains on the decadal-scale variability. Studies at this scale have extended our knowledge further. Climatic factors influence the type, distribution, and abundance of predators of young salmon. During warm coastal ocean years (PDO and ENSO warm phase), migrating predatory fish such as Pacific hake and Pacific mackerel arrive in the coastal ocean of the PNW earlier in the year and are closer to the shore (JISAO CIG 1999). Large changes in abundance of predators and prey, as well as changes in species composition of the zooplankton community off Newport, Oregon, and Vancouver Island, have been reported in recent years (Peterson and Mackas 2001). Stratification of ocean water has also been noted as a factor. The lower abundance of salmon in the PNW during warm phases of the PDO is believed to have occurred in more stratified coastal ocean water that is less nutrient-rich from upwelling during this phase. University of Washington scientists have also suggested why the decadal-scale relationship between climate and salmon abundance is stronger than the year-to-year ENSO relation (Hare et al. 1999). First, because not all fish spend the same number of years in the ocean, the year classes that may have been affected differently by interannual climate variation are smeared together. Second, an individual fish may feel beneficial effects of ENSO one year and detrimental effects during the opposite ENSO phase, but the PDO tends to have year-to-year persistence.

Overall, the more we learn about these decadal-scale sets of events, the more fascinating they become.

The Framework Questions and Future Research

I now address some of the framework questions of the book (see chapter 1). We have already seen in this chapter how the climate phenomenon of the PDO gives rise to a cascade of ecological events, resulting in greater or lesser fish abundance. The cascading principle is thus directly applicable in this example. However, there are other events taking place outside the principal coastal ocean cascade that may play a role in the overall system. For example, the changing availability of nutrients in the coastal ocean currents that is affected by changes from one PDO mode to the other is only one of the manifestations of the PDO modal shift. Another important resulting change is that, during the PDO cold phase, terrestrial snowpacks and later meltwater are relatively high, causing bountiful stream water flow conditions in the PNW for salmon during their downstream migration. The opposite is true during the PDO warm phase. Thus, there is also a terrestrial component acting outside the coastal ocean cascade that is the focus of this account. A more complete cascade/model would include both the oceanic and the terrestrial part and their potential interaction in the final determination of the size of the salmon populations. An even more complete consideration would also include the deep, as well as the coastal, ocean. The subarctic current itself is relatively nutrient rich, so another research topic is to determine the relative size of the nutrient pool provided by upwelling in the coastal ocean compared to that provided by the import at the surface from the subarctic current. In the example of this chapter, a more profound consideration of the principal cascade leads to the recognition of the importance of its consideration as an open system.

The issue of preexisting conditions has at least two major implications for salmon survival in the PNW. These relate to cyclic changes superimposed on longer term trends. First, if the human-related insults to salmon populations such as water pollution and habitat destruction are not reversed or removed, the populations will disappear regardless of any natural decadal-scale cycles that may exist. Long-term trends provide a new set of preexisting conditions for the start of each decadal-scale "cycle." Second, for the future, climate models suggest increased winter flooding and decreased spring and summer streamflows along with elevated stream and estuary temperatures that, singly and together, will be detrimental to salmon stocks even before they enter the ocean. If such climate changes are indeed realized, the prospect for PNW salmon stock has been described as "bleak" (JISAO CIG 1999).

As far as the decadal-scale "cycle" goes, and ignoring longer term trends, the ecological effects, operating through the cascade, are mostly complete by the time the next climate cycle begins. This is because the timescale of an Oregon Coho salmon cohort, of about 3 years, is much shorter than the approximate 15- to 20-year decadal climate regime shift.

The example of decadal-scale change in PNW salmon abundance does have at least an implied upper and lower limit. These limits are identified by the two major modes of climate variability (figure 13.2) and the values of climatic and other variables that result from these modes. As a working hypothesis we may suggest that, in the absence of other factors operating at other timescales, the cascading effects of the climatic variation will stop at these extreme points and not proceed in the sense of a continuing positive feedback. In addition, in this example the climatic event or episode does appear to reverse to some "original" or at least repeated state because of its cyclic nature. The observed historic record does not show evidence of the climate state going beyond an original position, but tree-ring records have been useful in better defining the size and intensity of the climate regime shifts during a 400-year period (Biondi et al. 2001; Gedalof and Smith 2001). We may also hypothesize that the cascade in this example does reverse in a more or less direct manner that does not display some kind of hysteresis. This reversal occurs in the sense that once the physical climatic conditions that give rise to a lack of oceanic water nutrients and to low fish abundance have changed to the other extreme, if all other factors remain the same, nutrient availability and fish abundance returns. The phytoplankton at the lower end of the food chain have not been irreversibly destroyed by adverse climatic conditions in the warm phase of the PDO—they have simply been made unavailable to the coastal oceanic salmon. The broad timing of these decadal-scale events is fairly well known, so as far as this is concerned, a simulation model of the system could be initiated. However, there is a lack of information on the smaller timescale events such as the operation of the food chain within any given spring or summer.

The "if all other factors remain the same" in the previous discussion is very important and has not been realized since human activities have been carried on intensively—at least during the twentieth century. A reviewer of this chapter points out that "while it seems likely that the impacts of decadal climate cycles are "reversible" in many ways, the potential for extinction of wild Coho is very real. Losing locally adapted breeding populations may not be reversible, and this is a critical issue that has caused a great deal of concern in restoration circles. Clearly, PDO climate cycles haven't been the sole cause of Coho extinctions or threats to extinction, but in cases where populations are severely reduced in number by habitat loss and degradation, overfishing, and/or poor management practices, a string of years with unfavorable climate conditions may be the last straw."

We might speculate that further insights into the decadal-scale changes of climate and salmon abundance might be obtained by examining the system as if it were a chaotic one. Certainly there are many nonlinear changes in the various components of the system. Even more intriguingly, the two modal positions of the climate regimes (figure 13.2) give the appearance of being attractors for the oceanic/climatic part of the system. Further specification of the nature of the chaos, if it exists, will await the quantification of the system.

Acknowledgment   This chapter is based on work performed in association with the H. J. Andrews LTER group that is funded in part by grants from the National Science Foundation Division of Environmental Biology, Long-Term Studies. I am very grateful for extremely helpful comments and contributions provided by Dr. Nathan Mantua of the University of Washington.

References

Biondi, F., A. Gershunov, and D.R. Cayan. 2001. North Pacific decadal climate variability since 1661. *Journal of Climate*. 14:5–10.

Fisher, J., and W. Pearcy. 1988. Growth of juvenile Coho salmon off Oregon and Washington, USA, in years of differing coastal upwelling. *Canadian Journal of Fisheries and Aquatic Science*. 45:1036–1044.

Francis, R.C. 1993. Climate change and salmonid production in the North Pacific Ocean. Pages 33–42 in *Proceedings of the Ninth Annual Pacific Climate (PACLIM) Workshop*. April 21–24, 1992. Asilomar, California. California Department of Water Resources. Sacramento.

Francis, R.C., and T.H. Sibley. 1991. Climate change and fisheries: What are the real issues? *The Northwest Environmental Journal*. 7:295–307.

Gedalof, Z., and D. Smith, 2001. Interdecadal climate variability and regime-scale shifts in Pacific North America. *Geophysical Research Letters*. Vol. 28:1515–1518.

Greenland, D. 1994. The Pacific Northwest regional context of the climate of the H. J. Andrews Experimental Forest Long-Term Ecological Research Site. *Northwest Science* 69:81–96.

Greenland, D. 1995. Salmon Populations and Large Scale Atmospheric Events. *Proceedings of American Fisheries Society 1994 Northeast Pacific Chinook and Coho Salmon Workshop. Salmon Ecosystem Restoration: Myth and Reality*. Nov. 7–10. Pages 103–114. Republished by request under the title "Offshore Coho Salmon Populations near the Pacific Northwest and Large Scale Atmospheric Events" in *Proceedings of Twelfth Pacific Climate (PACLIM) Workshop*, May 2–5, 1995. C.M. Isaacs and V.L. Tharp (eds). 1996. California Department of Water Resources. Interagency Ecological Studies Program, Technical Report 46, pp. 109–120.

Hare, S.R., N.J. Mantua, and R.C. Francis. 1999. Inverse production regimes: Alaskan and West Coast Pacific Salmon. *Fisheries*. Vol. 21, No. 1:6–14.

Hollowed, A.B., and W.S. Wooster. 1991. Year class variability of marine fishes and winter ocean conditions in the Northeast Pacific Ocean. *ICES Variability Symposium*. June 5–7, 1991. Mariehamn, Finland. Paper No. 33. Session 3.

JISAO CIG. 1999. *Impacts of climate variability and change in the Pacific Northwest. Pacific Northwest Contribution the National Assessment in the Potential Consequences of Climate Change for the United States*. The Joint Institute for the Study of Atmosphere and Ocean Climate Impacts Group, Box 354235. University of Washington, Seattle, Washington. JISAO contribution #715. 109 pp.

Mantua, N.J., S.R. Hare, Y. Zhang, J.M. Wallace, and R.C. Francis. 1997. A Pacific inter-decadal climate oscillation with impacts on salmon production. *Bulletin of the American Meteorological Society.* 78:1069–1079.

Miller, A.J., D.R. Cayan, T.P. Barnett, N.E. Graham, and J.M. Oberhuber. 1994. The 1976–7 climate shift of the Pacific Ocean. *Oceanography.* 7:21–26.

Overland, J.E., J. Miletta Adams, and N.A. Bond. 1999. Decadal Variability of the Aleutian Low and Its Relation to High-Latitude Circulation. *Journal of Climate.* 12:1542–1548.

Pearcy, W.G. 1992. *Ocean Ecology of North Pacific Salmonids.* University of Washinsgton Press, Seattle.

Peterson, W., and D. Mackas, 2001: Shifts in zooplankton abundance and species composition off central Oregon and southwestern British Columbia. *PICES Press*, North Pacific Marine Science Organization. Vol. 9:28–31.

Ware, D.M., and R.E. Thomson. 2000. Interannual to Multidecadal Timescale Climate Variations in the Northeast Pacific. *Journal of Climatology.* 13: 3209–3220.

14

Decadal and Century-Long Changes in Storminess at Long-Term Ecological Research Sites

Bruce P. Hayden

Nils R. Hayden

E cological disturbances at Long-Term Ecological Research (LTER) sites are often the result of extreme meteorological events. Among the events of significance are tropical storms, including hurricanes, and extratropical cyclones. Extratropical storms are low-pressure systems of the middle and high latitudes with their attendant cold and warm fronts. These fronts are associated with strong, horizontal thermal gradients in surface temperatures, strong winds, and a vigorous jet stream aloft. These storms and their attendant fronts generate most of the annual precipitation in the continental United States and provide the lifting mechanisms for thunderstorms that, on occasion, spawn tornadoes. Off the United States West and East Coasts, extratropical storms generate winds, wind waves, wind tides, and long-shore currents that rework coastal sediments, alter landscape morphology, and change the regional patterns of coastal erosion and accretion (Dolan et al. 1988). Although extratropical storms do not match hurricanes in either precipitation intensity or in the strength of the winds generated, they are much larger in size and have a more extensive geographic impact. On occasion, extratropical storms will intensify at an extraordinary rate of 1 millibar (mb) per hour for 24 hours or more. Such storms are classed as "bomb" and are comparable to hurricanes. Extratropical storms occur in all months of the year but are most frequent and more intense in winter when the north-south temperature contrast is large and dynamic support for storm intensification from the stronger jet stream aloft is great. In this chapter, we will explore the history of storminess for those LTER sites in the continental United States at which more than a century of data on storms and their storm tracks are readily available. Specifically, we will look at the record of changes in storminess at both the regional and national scales.

During the 1990s, significant storms along the U.S. West Coast and droughts and fires in Florida in an El Niño year led to a hypothesis that El Niño and La Niña conditions were associated with a modulation in the frequency of storms. In addition, it has been suggested that the frequency of El Niño and La Niña events and, by inference, storminess, has increased during the past century. Several LTER sites have recorded changes in climatic and ecological states that are coherent with El Niño and La Niña cycles. In this chapter we will examine the record of storminess averaged over all of the El Niño and La Niña events since the mid-1880s relative to these ideas. Concerns about global warming arising from elevated atmospheric carbon dioxide concentrations have led to interpretations of model studies of a world with more storms and with more intense storms resulting in more frequent flooding. Many LTER sites have established research capabilities designed to detect global warming signals in ecosystem dynamics. We will also look at the history of storms relative to the projections of storminess from General Circulation Models (GCMs) in response to elevated atmospheric carbon dioxide. GCM sensitivity in the level of storminess relative to the projected atmospheric carbon dioxide increases will be explored through a study of the historical record of change.

Background

Davis et al. (2000) characterized extratropical cyclones and their nature.

> An extratropical cyclone is a synoptic-scale, migratory atmospheric disturbance that forms and evolves entirely outside of tropical latitudes. Its horizontal dimensions are on the order of 10^3 km. The most generally recognized manifestation of the extratropical cyclone occurs near the earth's surface in the form of a large region of surface low pressure in which air spirals cyclonically (counter-clockwise in the Northern Hemisphere, reverse in the Southern Hemisphere) and inward to try to fill the depressed region. Large horizontal temperature gradients provide the primary energy source for the disturbance as the potential energy created by these gradients is converted to kinetic energy.

> The extratropical cyclone is always associated with an upper trough. In a well-developed extratropical cyclone, there is a three dimensional balance, such that air rushing into the system at lower-levels is forced upward to high altitudes, where it diverges out and away from the cyclone's center. The rising motion often gives rise to the inclement weather typical of these storms.

Most extratropical storms of the middle latitudes move generally from west to east and may cross the continental United States in 4 to 6 days. The paths taken by these centers of low pressure may be followed from hour to hour and day to day by inspection of surface weather maps. Derivative maps of these tracks can be produced from a sequence of consecutive individual weather maps. The National Weather Service has been publishing such derivative maps of storm tracks on a monthly basis since the early 1880s. In this chapter compilations of monthly storm tracks published by the National Weather Service are summarized by month for 180 latitude and longitude grid cells from southern Canada, the continental United States, and the western North Atlantic Ocean. Archives of maps of monthly storm tracks have been used for more than a century to delineate the climatology of storms. We will focus on the history of storms at LTER sites. Bigelow's study of 1897, Klein's studies of 1951, 1958, and 1965, Hayden's 1975 and 1981 studies, Reitan's work in 1974, and Key and Chan's 1999 report are but a few of the significant compendial studies of monthly storm frequency and storm track climatologies.

Hosler and Gamage (1956) studied U.S. extratropical cyclone frequencies in 5° latitude and longitude grid cells for the 50 years ending in 1955. They found no evidence in the annual means of periodicities, trends, or shifts in regions of maximum frequency. Reitan (1974) updated Hosler and Gamage's work through the 1960s and reported a general but modest decline in cyclone frequency. Other regional studies have come to opposite conclusions. Hayden (1975), for the period 1942–1975, showed an increased frequency of storms off the U.S. Atlantic Coast that generate deepwater waves of 1.3 m or greater height. Hayden concluded that the number, severity, and duration of Atlantic Coast storms had increased between 1942 and 1974. Mather et al. (1964) and Resio and Hayden (1975) reported an increase in Atlantic Coast storm severity from the 1920s to the 1960s. Resio and Hayden (1975) attributed the increase in Atlantic Coast storm frequency and magnitude to anticyclonic blocking in the Northern Hemisphere. Dickinson and Namias (1976) related the offshore displacement of the Atlantic Coast storm track, detected in the 1950s, as arising from a general cooling in the U.S. Southeast during the twentieth century and to an expansion of the Northern Hemisphere polar vortex. Hayden (1981) explored the long-term (1885–1978) history of storminess for North America east of the 100th meridian and found large century-long changes in storm frequency characterized by increased storminess off the U.S. Atlantic Coast from 1885–1894 to 1925–1934. The opposite trend was found over the Great Plains. Because these storms eventually track into the high latitudes of the North Atlantic, it is likely that patterns of storms, especially those in the eastern half of North America are related to the North Atlantic Oscillation (NAO). Although the connections to hemispheric atmospheric circulation patterns are not the subject of this chapter, it is clear that the patterns of storm climate change can be put into a larger context.

Extratropical storms provide the essential rain and snowfalls that sustain ecological systems and human activities in the middle and high latitudes. The storms also give rise to extreme, damaging levels of precipitation (Uccellini et al. 1995), wind waves (Resio and Hayden 1975), and short-term sea level changes that may result in coastal erosion and sedimentation (Dolan and Hayden 1981), ecosystem changes (Hayden et al. 1995), and the loss of lives and property (Mather et al. 1964; Hayden 1988; Davis and Dolan 1993). These storms, and their associated fronts, provide most of the excess precipitation over evaporation and evapotranspiration during the recharge part of the water year. They are responsible for the snowfalls that build the snow pack and water storage in mountainous areas, and are the main source of water for winter and spring flooding. Hayden (1988) detailed the relationship between flooding regimes and the general circulation of the atmosphere and classified flood regions of the world according to the types of precipitation systems that caused the flooding.

Extratropical storms are also responsible for most of the major East Coast snowstorms (Knappenberger and Michaels 1993; Kocin and Uccellini 1990). In general, extratropical storms are central to the problems of and stresses on water resource systems. In addition, on theoretical grounds, if climate change occurs because of increases in carbon dioxide, changes in spatial and temporal patterns in storminess are to be expected. In coastal regions of the United States, these same storms are the proximate cause of beach erosion and the flooding of wetlands with saline waters as well as the maintenance of coastal ecosystems (Dolan and Hayden 1981). Property damage along the U.S. coast from storms at sea occurs each year. It is generally recognized among coastal scientists that with coastal development the risk of property damage from storms has increased yearly (Davis and Dolan 1993). In addition, the current eustatic sea level rise of about 11 mm/yr (Emery and Aubrey 1991) brings the water's edge to a more landward position each year, thus increasing the hazard of storms and the risk of property damage and loss of lives. Two-thirds of the nation's shorelines are eroding (Fenster and Dolan 1994), resulting in a progressively greater exposure of coastal infrastructure to storm wave and storm surge damage.

The fundamental aspects of climate change in response to an enrichment of carbon dioxide in the atmosphere have been known since the end of the last century (Arrhenius 1896) and have been confirmed by GCMs in the last two decades. Arrhenius found that the high latitudes warm more than the low latitudes, the continental areas warm more than the oceanic areas, the Northern Hemisphere warms more than the Southern Hemisphere, the winter half of the year warms more than the summer half of the year, and the planetary atmosphere becomes generally more moist. In addition, Arrhenius estimated that the bottom 15 km of the atmosphere would warm while cooling occurred at higher altitudes. General Circulation Models show the same pattern that Arrhenius determined in 1896, but GCMs also provide information about pressure and wind fields. The model projections of carbon dioxide–forced climate change indicate warming in the high latitudes more than in the low latitudes. The expected result is a reduced north-south temperature gradient, a smaller circumpolar vortex, and a weaker, more northerly latitudinal position of both the zone of the maximum north-south thermal gradient and the jet stream (Hayden and Dolan 1977; Davis and Dolan 1993). Storms in such an altered climate should, on average, be fewer in number, weaker in intensity, and displaced northward in geographic position.

The questions that motivate the present study are, Have there been changes at LTER sites in storminess over the past century? Are the changes consistent or the same nationwide? Is the change in storminess over the past century consistent with GCM projections of climate change based on the increases in carbon dioxide to date? Is there a characteristic pattern of storminess associated with El Niño and La Niña cycles?

Data

The histories of extratropical storms studied here are not based on human visual observations of storms but rather are revealed on weather maps on which pressure fields are analyzed and the centers of low pressure are surrounded by one or more

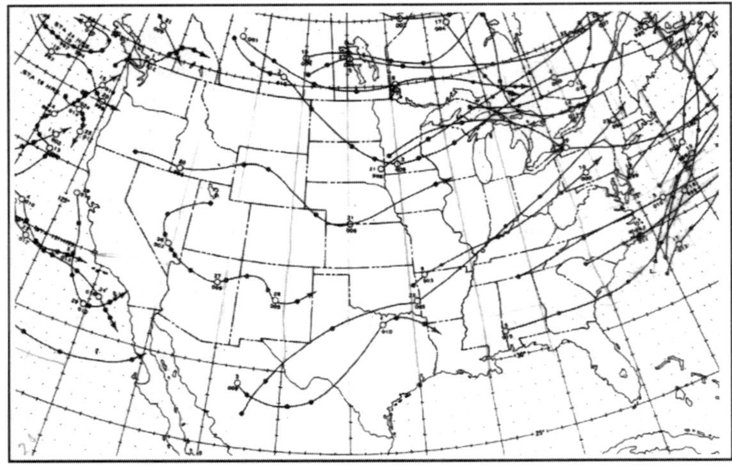

Figure 14.1 Tracks of the centers of cyclones at sea level, January 1957. Open circles are the positions of storm centers at 7:30 A.M. EST and filled circles are storm positions at 6-hour increments.

closed isobars. The storms are accompanied by counterclockwise circulation (in the Northern Hemisphere) and are associated with convection, cloudiness, and precipitation, which are also indicated on the weather maps. This definition of a storm excludes from study storms during their period of cyclogenesis and prior to the circular closure of the isobaric field when central pressures are less than 4 mb lower than the surrounding area. Also excluded by the National Weather Service in their analyses are low-pressure centers that last less than 6 hours. Such incipient or weak storms are, however, not insignificant in the lower midlatitudes, where they may produce significant rainfall and associated thunderstorms. Because rainfall arises from such "prestorm" conditions, especially in the lower latitudes, we must be aware of this problem in the operational definition. Through the study of sequences of weather maps, using analyzed records of barometric pressure at weather stations, the centers of such low-pressure systems are recorded and mapped as storm tracks (figure 14.1). Maps of such storm tracks compiled by month by the National Weather Service are available for the years 1885–1996. Storm frequency (storminess) was tabulated from monthly charts of the "Tracks of the Centers of Cyclones at Sea Level," published by the Monthly Weather Review, Mariners Weather Log, and in recent years provided directly to the authors by the National Weather Service.

In the present study of storminess at LTER sites, storms for each year (1885–1996) are tabulated for the 180 2.5°-latitude by 5.0°-longitude grid cells in the study area (25° N to 55° N and 60° W to 125° W). Unfortunately, the National Weather Service can no longer afford to prepare these important charts. The annual totals of cyclones passing through each grid cell were summed from the monthly totals following the procedures of Hayden (1981).

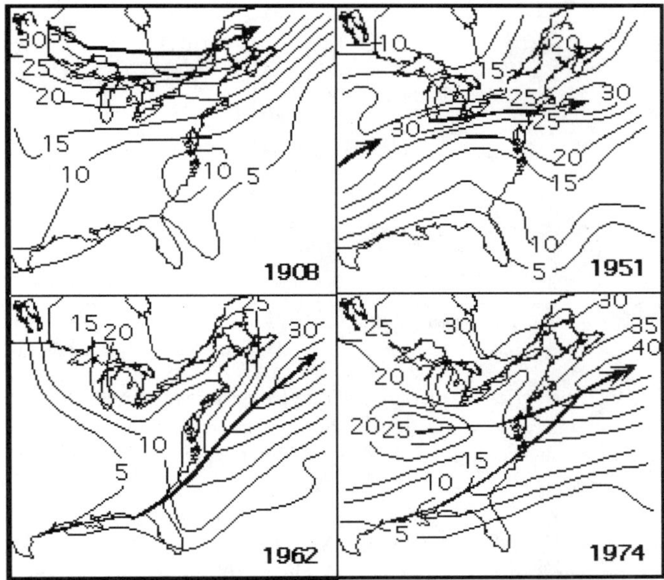

Figure 14.2 Annual composites of storm frequencies for four representative years: 1908, 1951, 1962, and 1974. Heavy lines with arrows indicate the dominant storm tracks for the year.

Year-to-Year Variations in Storm Tracks

Figure 14.2 shows the pattern of storminess over eastern North America in four example years: 1908, 1951, 1962 and 1974. These years were selected because they illustrate the range of storm track patterns that may occur. In the year 1908, the majority of storms tracked eastward along the northern margin of the Great Lakes. Storm frequencies declined abruptly from the Great Lakes southward. In 1951, the dominant storm track was out of Colorado, extending eastward south of the Great Lakes and exiting the Atlantic Coast near Long Island, New York. In this year there were few storms to the north of the Great lakes in southern Canada (Angel and Isard 1998). The Atlantic Coast storm track dominated in 1962 with very few storms over the interior of the continent. This was one of the stormiest years on record for Atlantic Coast nor'easters. In 1974, the Atlantic Coast storm track was displaced northward and westward of the 1962 position, and a second track of storm is observable along the Ohio Valley, exiting the Atlantic Coast in central New Jersey.

National Trends in Storminess: 1885–1996

The number of storms recorded in each of the 180 cells in the study region for each year was summed to provide a count index of the frequency of North American

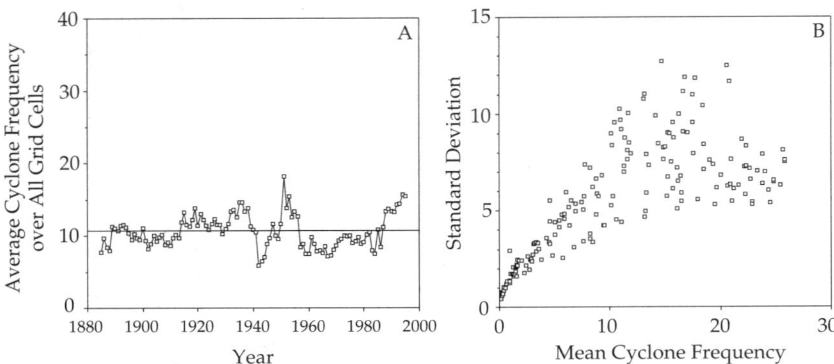

Figure 14.3 (A) Time series of average cyclone frequencies (number of storms per year) over the entire study area shown in figure 14.4 and (B) scattergram of mean cyclone frequencies and the standard deviations of mean cyclone frequencies for the study area.

storminess from 1885–1996 (figure 14.3A). There is no indication of any overall North American trend in storminess. There is neither a systematic increase in the number of storms since 1885 nor a decrease in storminess for that period. The slope of the regression line is zero. Figure 14.3B shows the relationship between the standard deviation of storm counts in each year and the mean cyclone frequency. Clearly, the variability in storm climate across the years of record is strongly correlated with the number of storms that occurred. For this attribute of climate, there has been no nationwide change in the average occurrence of storms since 1885. We must also conclude from figure 14.3B that the nationwide variability has not changed either.

Figure 14.4A shows the spatial variation in the mean number of storms for the entire period of record (1885–1996), and figure 14.4B shows the spatial field of standard deviations about the long-term means. The three-letter acronym of each LTER site identifies the locations of the 19 LTER sites included in this study. The long-term mean frequency of cyclones (figure 14.4A) shows three major storm tracks: (1) the Alberta storm track arising out of southern Alberta and continuing eastward across the northern edge of the Great Lakes and out the St. Lawrence Seaway; (2) the Colorado storm track extending into the Great Lakes and merging with the Alberta storm track in the Great Lakes region; and (3) the Atlantic Coast storm track paralleling the Gulf and East Coasts just a bit offshore. Storms along the Alberta and Colorado storm tracks often decay over the mountains as they come from the west and may reform in the lee of the Rocky Mountains. The formation of storms, cyclogenesis, is common on the lee side of the mountains at the locations of the Alberta and Colorado storm tracks. Along the Atlantic Coast, such cyclogenesis often occurs just to the east of the coast and along the usual storm tracks.

It is clear from figures 14.4A and 14.4B that the positive correlation between temporal variability and mean storm frequency in time (figure 14.3B) also applies spatially as well. Areas of high storm frequency are also areas of high interannual variability. The Alberta and Colorado storm tracks are also axes of maximum vari-

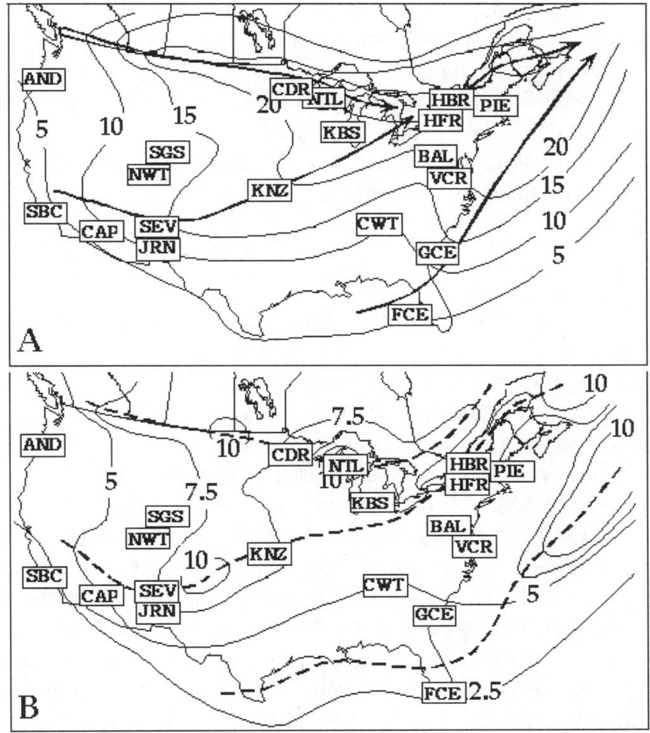

Figure 14.4 Long-term (1885–1996) mean storm frequencies (A) and standard deviations (B). The units are storms per year. Arrows in panel A indicate the long-term mean storm tracks. The dashed lines in panel B show the locus of maximum storm frequency variability. The locations of LTER sites are shown and indicated by the three-letter acronym for the sites.

ability in annual storm frequency. The axis of maximum variability along the Atlantic Coast is displaced southward and eastward (seaward) of the axis of the storm track itself. This displacement undoubtedly accounts for some of the scatter of points in figure 14.3B. Caution is in order regarding the three storm tracks shown in figure 14.4A. The senior author, in a study of the patterns of storm tracks in each of the 112 years of record, found few years in which all three storm tracks were present in the annual mean statistics for individual years. In particular, years when both the Colorado and Atlantic storm tracks are frequented by storms are rare. Thus, figure 14.4A must be viewed as a time-integrated composite that is not often realized in individual years.

Storminess at LTER Sites

Nineteen of the 24 LTER sites lie within the middle latitudes, where extratropical storms are the dominant meteorological and ecosystem disturbance events. Although there is no apparent national trend in storm frequency or variability during the twentieth century, it does not follow that regional trends do not exist. However, nationally, the sum of all such regional trends must balance out. For the purpose of detailing the histories of storminess at LTER sites, the sites have been grouped into regions with common storm tracks and where common temporal patterns of storminess are found. Five regions are defined: the West Coast, the Interior West, the Midwest, the Appalachians, and the East Coast.

The West Coast

The West Coast LTER sites are at Santa Barbara, California, and Andrews Experimental Forest, Oregon. Figure 14.5 shows the time history of storm frequencies in the latitude by longitude grid cells that contain each West Coast LTER. In addition, the time series of storms for the Puget Sound area is shown because storms (indicated by a center of low pressure) coming ashore at this location and to the north produce wintertime rainfall as far south as San Diego, California. In the Puget Sound area (figure 14.5A), storm frequencies were highest near the turn of the century and declined somewhat until the middle of the century. Missing data for the 1960s, however, limits our confidence in this observation. At the H. J. Andrews Experimental Forest LTER site (figure 14.5B), the average annual number of storms is about five, and there are no secular trends in storminess between 1885 and 1996. Although five storms for this stormy coast seems to be very low, we must remember that the West Coast is significantly impacted by storms that come ashore far to the north along the British Columbia and southern Alaska coasts as the trailing fronts. Such fronts from these storms produce rainfall all along the West Coast. The centers of these eastward-moving storms, however, occur infrequently in the southern sections of the West Coast, and then mostly in the winter months. Santa Barbara Coastal Ecosystems LTER site (figure 14.5C) exhibits a pattern much like areas to the north, with an annual average number of storms around four per year. Like points to the north, there is little if any indication of secular change in storm frequency in southern California during the twentieth century. Although these data do not contain information on storm intensity, there is generally a log-linear relationship between storm intensity and storm frequency. Years with many storms are more likely to have intense storms. We would not expect to find a historical trend in storm intensity along the U.S. West Coast if such data were available because there is no trend in the total number of occurrences.

The Interior West

Although the LTER sites on the West Coast show no appreciable long-term trends in storm frequency, the Interior West LTER sites have experienced significant storm climate changes (figure 14.6). There are five LTER sites in this region: Phoenix (CAP), the Sevilleta (SEV), Jornada (JRN), Niwot Ridge (NWT), and Short Grass Steppe (SGS). Two of the sites, Sevilleta and Jornada, fall within the same grid cell. All four storm count grid cells exhibit a rise in storm frequencies from the start of the records in 1885 to the late 1920s and a decline thereafter, with

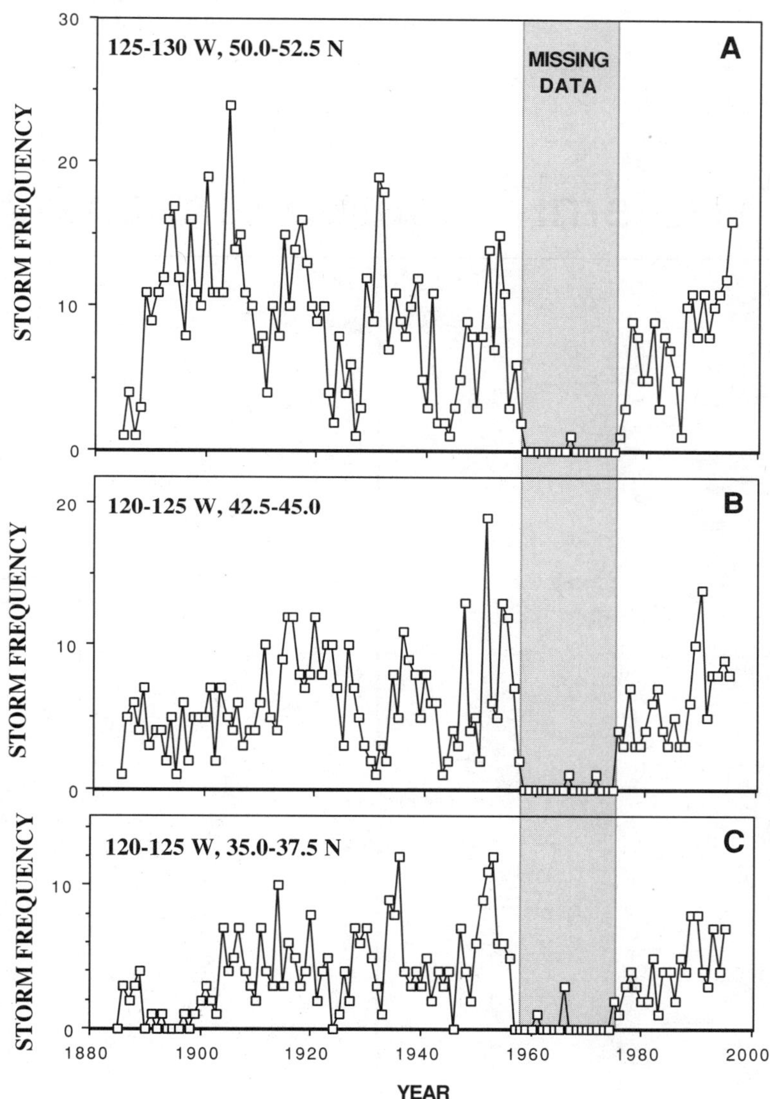

Figure 14.5 The number of storms per year in the Puget Sound area (A), the Andrews LTER site (B), and the Santa Barbara LTER site (C). Storm frequencies are given in storms per year. The latitude and longitude grid cell used to tabulate storm frequencies is shown in each panel.

a rise in storm numbers in the last two decades of the period of record. At Niwot Ridge LTER site (figure 14.6A), storm frequencies rose from 20 storms per year to 35 storms per year early in the record and fell to around 15 storms per year in the middle part of the period of record. Because the majority of these storms fall in the winter half of the year, such large climate changes should be manifested as signif-

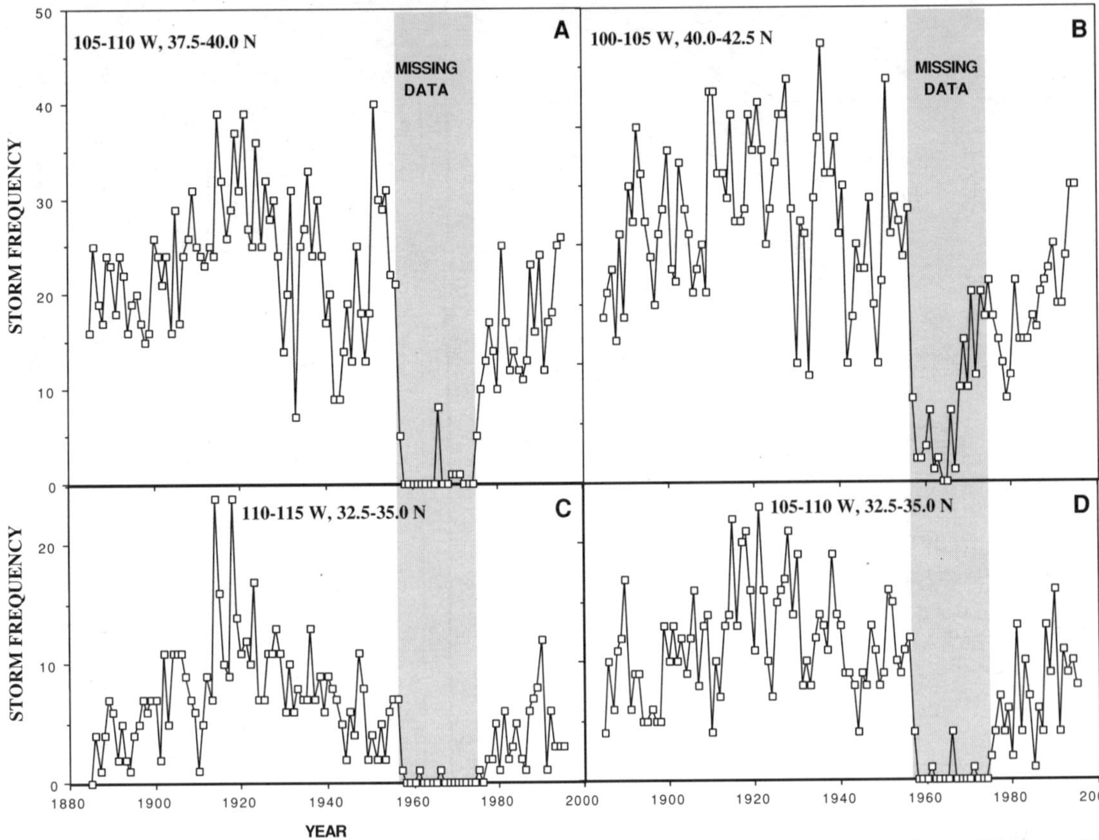

Figure 14.6 The number of storms per year at the Niwot Ridge LTER site (A), the Short Grass Steppe LTER site (B), the Central Arizona Phoenix LTER site (C), and the Jornada and Sevilleta LTER sites (D). Storm frequencies are given in number of storms per year. The latitude and longitude of the grid cell used to tabulate storm frequencies is shown in each panel.

icant changes in the snowpack in the mountains of the area. A similar pattern is seen at the Short Grass Steppe (figure 14.6B). Storm frequencies doubled between 1885 and 1930 and declined by that much or more in the decades that followed. In the case of the Phoenix LTER site (figure 14.6C), the rise in storm frequencies from 1885 to 1930 amounted to an increase of nearly 10 storms per year to be followed by a decline of about the same to the middle of the century. This change amounts to more than twice the year-to-year variation in storm frequency, a large climate change by any standard. At the Sevellita and Jornada LTER sites, storm frequencies increased from around 5 per year in the 1880 to around 15 storms per year by the 1920s and declined thereafter. Because the number of cyclones is low in the lower latitudes and high in the higher latitudes, the 10-storms-per-year change in New Mexico is a very large increase—300%. In contrast, the 20-storms-per-year increase at the Niwot LTER site is "only" a 100% increase. In any case, these are very large climate changes.

Each of these five LTER sites have seen significantly increased storminess over the past two decades. In addition, the form of these four time series of storm frequencies (figure 14.6) resembles that of the record of Northern Hemisphere global temperature. This history implies that when the Northern Hemisphere warms the interior of the United States, the west experiences increased storminess. If the recent history of warming is the result of human enrichment of the atmosphere with carbon dioxide giving rise to the northward displacement of the zone of greatest thermal contrast (as projected by climate models), then we would expect fewer rather than more storms during periods of warming. At this point, we cannot make a good case for a causal link between storminess at these western LTER sites and the observed history of Northern Hemisphere warming.

The Midwest

Cedar Creek, North Temperate Lakes, Kellogg Biological Station, and Konza Prairie LTER sites fall within this region. These sites are grouped together because they exhibit similar histories of storminess. There have been significant secular changes in storminess across this region during the twentieth century (figure 14.7). The trends at these locations resemble those for the interior west LTER sites with relatively high storm frequencies in the early decades of the past century, a minimum in storm frequency in the middle decades, and a return to higher frequencies in recent decades. At the Cedar Creek LTER (figure 14.7A), the amplitude of the century-long changes in storm frequency amounts to about 15 storms per year. A similar amplitude is seen at the North Temperate Lake LTER site (figure 14.7B). At the Konza Prairie LTER, the amplitude of change from high- to low-frequency decades is nearly 20 storms per year. Finally, at the Kellogg Biological Station LTER (figure 14.7D), a similar trend is seen but with a more modest, yet still large, climate change of about 10 storms per year. In each of these four cases, the long-period amplitude is much larger than the year-to-year variability and, as such, clearly indicates the large magnitude of climate change present in the frequency of storms. Because the storm intensity versus storm frequency relationship is log-linear, we would expect more intense storms in years with many storms. In addition,

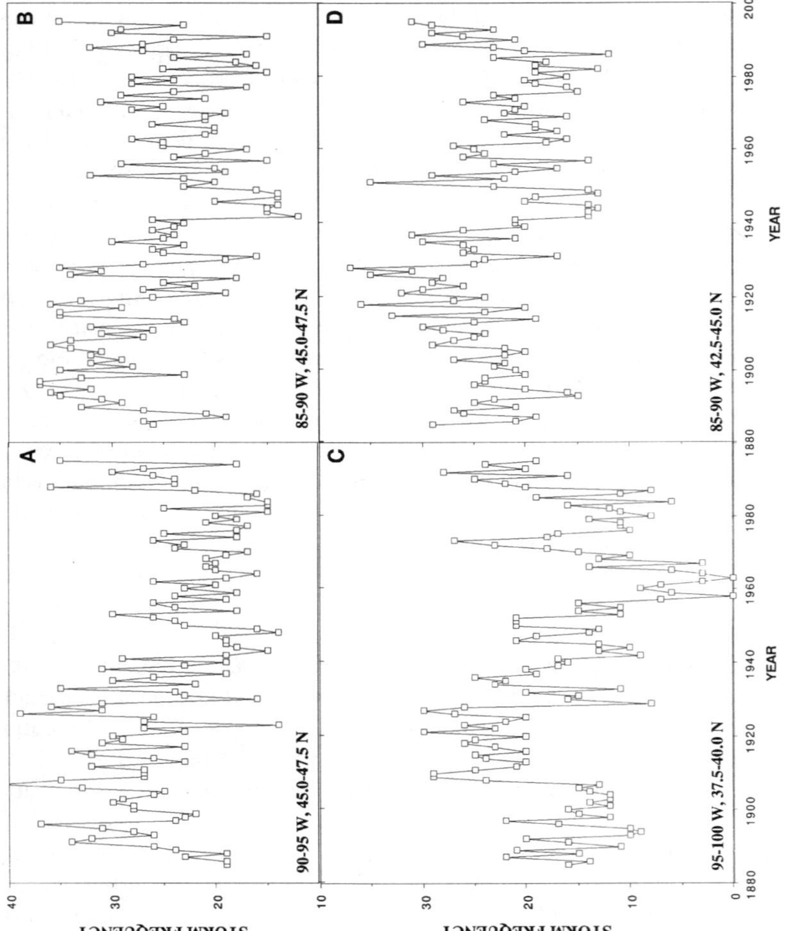

Figure 14.7 The number of storms per year at the Cedar Creek LTER site (A), North Temperate Lakes LTER site (B), the Konza Prairie LTER site (C), and Kellogg Biological Station LTER site (D). Storm frequencies are given in storms per year. The latitude and longitude of the grid cell used to tabulate storm frequencies are shown in each panel.

decades with frequent storms should exhibit large year-to-year variability in storm frequency.

The Appalachians

Hubbard Brook, Harvard Forest, Baltimore, and Coweeta LTER sites are grouped into the Appalachian region. These LTER sites lie between the Colorado and Atlantic Coast storm tracks. Hubbard Brook (figure 14.8A) and Coweeta (figure 14.8C) exhibit little if any systematic climate change. Indeed, the year-to-year variations at these sites greatly exceeds any century-long variation in storm frequencies that might be present. At the Harvard Forest (figure 14.8B) and Baltimore (figure 14.8D) LTERs, there are clear indications of an upward trend in storm frequencies from the beginning of record to the 1990s. In the case of Harvard Forest, the numbers of storms has increased from around 15 per year to 25 per year. At the Baltimore LTER site, an increasing trend of about 10 storms per year is evident. This upward trend in storminess becomes more pronounced as we move east of the Appalachians toward the coastal LTER sites and the marine environment.

The East Coast

There are four LTER sites along the U.S. Atlantic Coast: Plum Island, Virginia Coast Reserve, Georgia Coast, and the Florida Everglades. Each of these sites is influenced by storms along the Atlantic storm track. The time series of storminess for these LTER sites for the past century are shown in figure 14.9. Changes in century-long storm frequency are very pronounced. At Plum Island LTER (figure 14.9A), the amplitude of change amounts to more than 25 storms per year, with a low of 10 storms per year around the turn of the century to 30 or 35 storms per year in recent decades. A strong climate change in storminess is clearly evident. At the Virginia Coast Reserve LTER (figure 14.9B), the change in storm numbers amounted to 15 storms per year from 1885 to 1990. The magnitude of storm change since 1885 declines as we move southward. At the Georgia Coast LTER (figure 14.9C), the amplitude of change is about 8 storms per year. Although this change is modest, it has been systematic over the past century with a nearly linear trend. Little if any trend in storm numbers is evident at the Florida Everglades LTER (figure 14.9D). However, much of the freshwater passing through the Everglades of southern Florida, arises in northern Florida and much of the winter rain across Florida comes from storm centers passing even farther to the north. Also, rainfall from storms that do not pass the threshold used in this study often produce a large amount of rainfall at these low latitudes.

It should be clear that, when viewed from the national perspective, no overall trend in storminess or variability in storminess has been detected (figure 14.3A). It should be equally clear that, at the regional scale, there have been substantial changes in storm climate. In places where there has been an increase in storminess, we should expect to see more frequent larger storms and higher storm frequency variability as well. Conversely, it is true that, with declines in storm frequency,

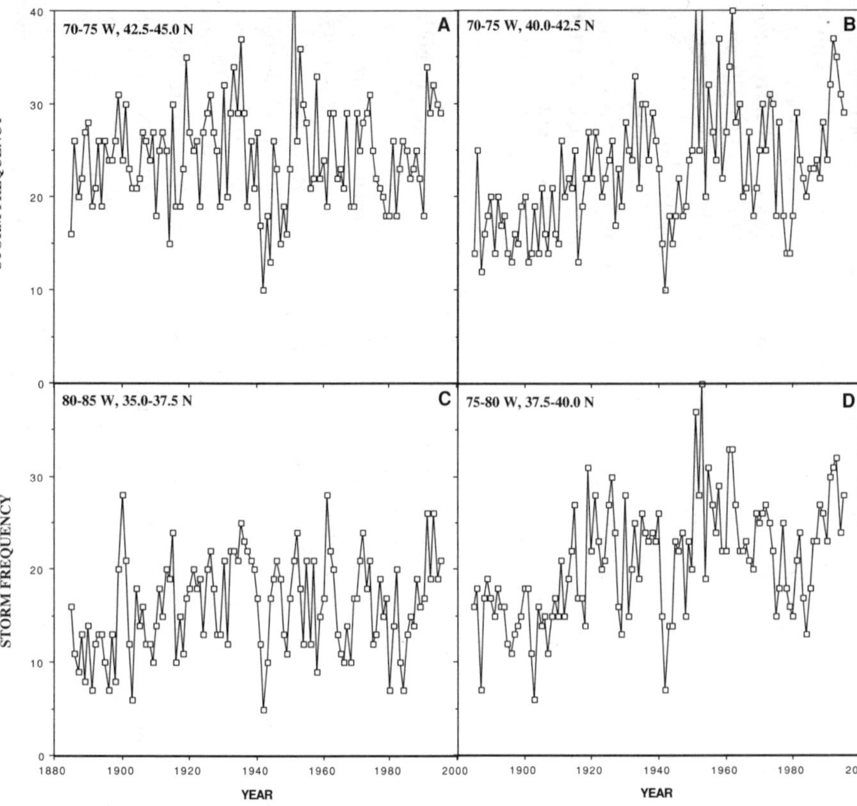

Figure 14.8 The number of storms per year at the Hubbard Brook LTER site (A), Harvard Forest LTER site (B), the Coweeta LTER site (C), and the Baltimore LTER site (D). Storm frequencies are given in storms per year. The latitude and longitude of the grid cell used to tabulate storm frequencies is shown in each panel.

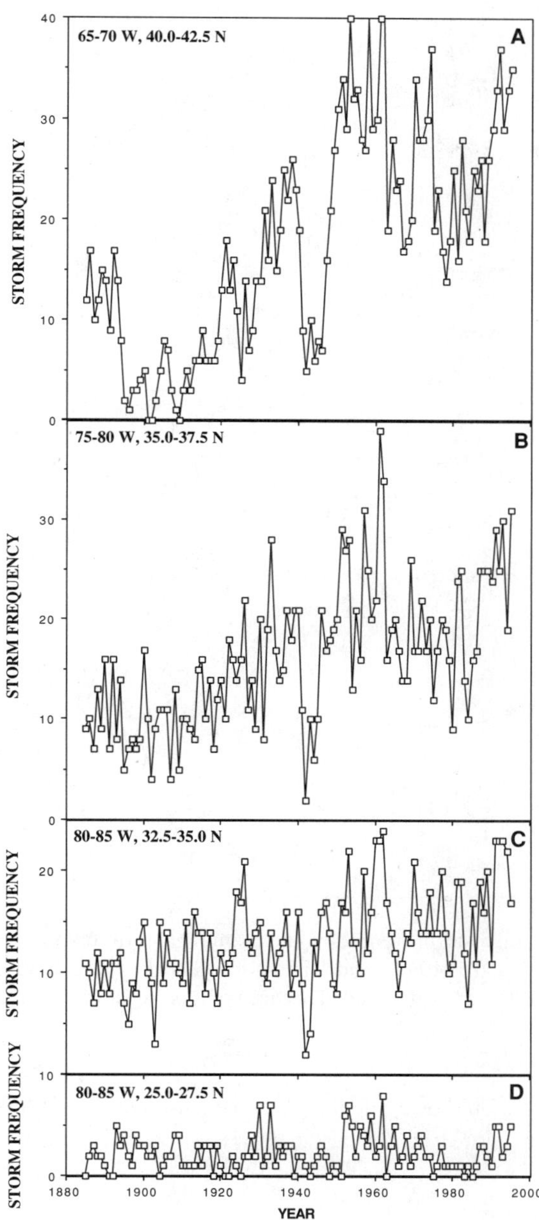

Figure 14.9 The number of storms per year at the Plum Island LTER site (A), Virginia Coast Reserve LTER site (B), the Georgia Coastal Ecosystem LTER site (C), and the Florida Everglades Ecosystem LTER site (D). Storm frequencies are given in storms per year. The latitude and longitude of the grid cell used to tabulate storm frequencies is shown in each panel.

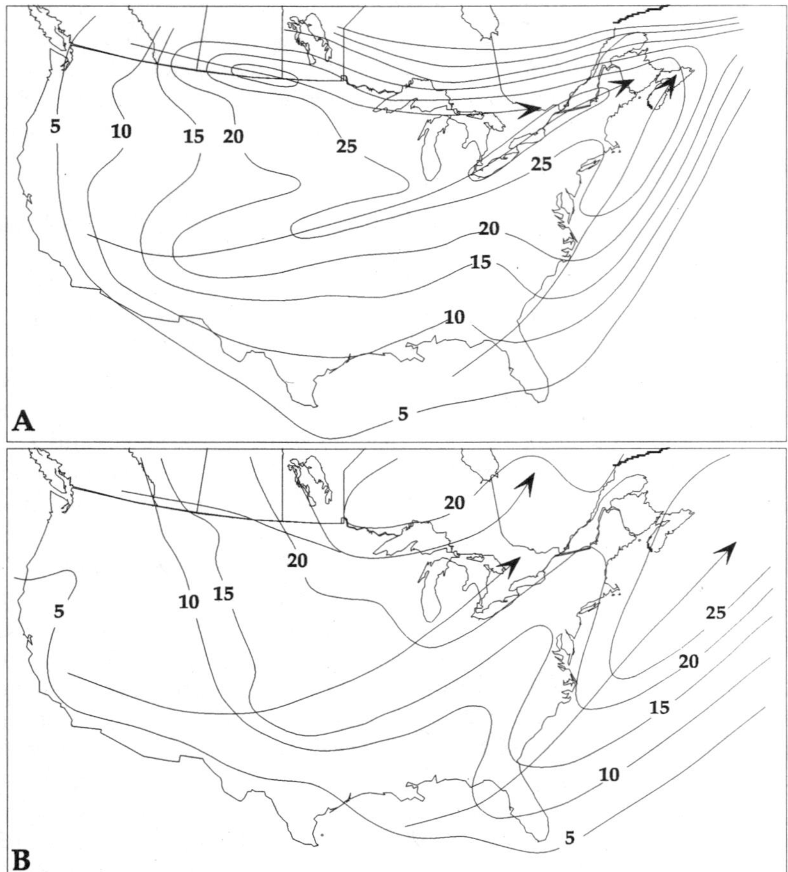

Figure 14.10 Average storm frequency and dominant storm tracks (arrows) for El Niño years between 1885 and 1940 (A) and for El Niño years between 1940 and 1996. El Niño events during the 1960s are not included because of missing storm track data.

fewer large storm should be evident and the year-to-year variability should decline as well.

Extratropical Storms, El Niño, and La Niña

Because there exist large storm climate changes at the time scale of the length of the record of data used in this study and because there are suggestions that the El Niño and La Niña events are becoming more frequent, the analysis of storminess during El Niño and La Niña periods is divided into two periods: before 1940 and after 1940. The two panels in figure 14.10 compare El Niño storminess patterns in the early period of record to those in more recent decades. The El Niño and La Niña events of the 1960s were not included in this study because of missing data for the western United States during the 1960s. Figure 14.10A shows the spatial field of average annual extratropical cyclone frequencies for six El Niño episodes before 1940 (1888–1889, 1905–1906, 1911–1912, 1914–1915, 1919–1920, and 1925–1926). Figure 14.10B shows the spatial field of average annual storm counts for seven El Niño events after 1940 (1940–1941, 1951–1953, 1977–1978, 1982–1983, 1986–1987, 1991–1992, 1993–1994). The Atlantic Coast storm track is displaced seaward in the latter period compared to the pre-1940 portion of the record. The differences in the two periods, however, are not the result of instability or a fundamental signature of storminess in El Niño periods. Rather, there is a confounding in the data due to the very large centennial-scale changes in storm climate that have occurred over the course of the past century. The storm climate of earlier decades of the period of record is fundamentally different from those of more recent decades. A comparison of figure 14.10A,B with figure 14.4A leads to the conclusion that pattern of storminess over North America during El Niño years is not different from the pattern of storms averaged over the period of record (112 years). The case for increased El Niño storminess along the California coast as a signature of El Niño events cannot be made from data on storm counts taken from weather maps by the National Weather Service.

Figure 14.11 provides an analysis of storminess patterns during La Niña episodes. The spatial field of average annual extratropical cyclone frequencies for six La Niña episodes before 1940 (1889–1890, 1989–1899, 1909–1910, 1921–1922, 1928–1929, and 1937–1938) are shown in figure 14.11A. Figure 14.11B shows the spatial field of annual average storm counts for five La Niña episodes after the 1940 period of the record (1943–1944, 1949–1950, 1954–1955, 1975–1976, and 1988–1989). With the exception of a displacement of storms offshore along the Atlantic Coast in the post-1940 period, no differences can be seen between La Niña events before and after 1940. As was the case for the analysis of El Niño, we find no substantive difference between storminess during La Niña years and an average year for the entire period of record (figure 14.4A).

Comparison of figures 14.10 and 14.11 provides an opportunity to examine differences during storminess in El Niño years with storminess during La Niña years. Storminess patterns for the 112 years of record are no different in El Niño or La Niña years. It is probable that news-media reports of El Niño and La Niña storm patterns are based on the vivid memory of the recent events and thus conclusions are drawn based on a modest sample size. This, however, does not mean that there is no El Niño/La Niña signal in rainfall statistics. Closed isobar storms along the southern states from Southern California to Florida are not common, but rainfall during the period of cyclogenesis leading up to a definable cyclone is often substantial. In addition, rainfall arising from frontal systems associated with cyclones may exhibit an El Niño/La Niña signal. The present study indicates only that developed cyclones over the continental United States do not exhibit a distinct El Niño/La Niña signal.

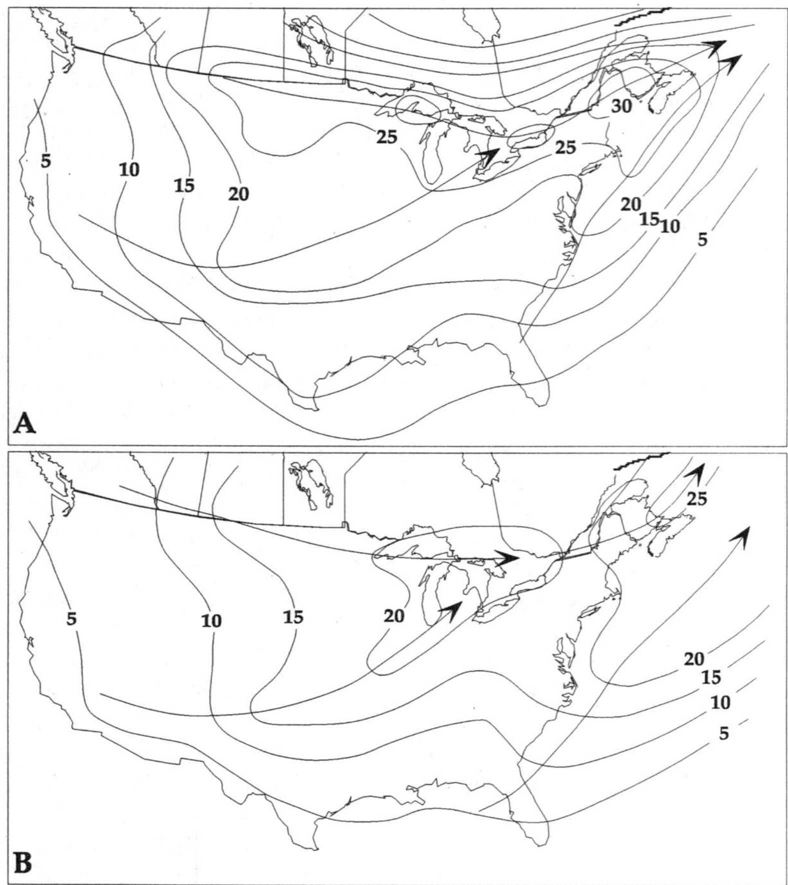

Figure 14.11 Average storm frequency and dominant storm tracks (arrows) for La Niña years between 1885 and 1940 (A) and for La Niña years between 1940 and 1996 (B). La Niña events in the 1960s are not included because of missing storm track data.

Storminess and Global Warming

All models, from the most sophisticated General Circulation Models, such as the greenhouse gas–forced models of Manabe et al. (1991) or models that combine the cooling effects of sulfate aerosols with the warming from greenhouse gases (Taylor and Penner 1994), indicate a relaxation of the meridional (tropics-to-equator) temperature gradient in winter. This reduced thermal gradient forces a general weakening of the midlatitude westerlies. The more "summerlike" (strictly, less "winterlike") characteristic of the greenhouse-enhanced atmosphere clearly projects a general retreat of the jet stream to a more northerly location (Davis and Benkovic

1992, 1994). Extratropical cyclones under such a regime should be fewer in number and displaced northward on average. Figure 14.3A clearly indicates that, to date, there has been no trend in the number of cyclones over North America during the twentieth century. In addition, there is no evidence for a change in the variability of storminess over North America at the continental scale. Davis et al. (2000) suggested that an increase in Ohio Valley storms and a decrease in Atlantic Coast storms, as seen in figures 14.7 and 14.9, is consistent with a preferential warming of the continent and a reduced thermal contrast across the U.S. Atlantic Coast.

Storm data from the Hadley Center General Circulation Model (HADCM2) have been used to build scenarios of the nature of storm climate in a carbon dioxide–enriched world (Johns et al. 1997; Carnell et al. 1996). Thirty-year average HADCM2 indexes of storminess centered on the years A.D. 2005, A.D. 2020, and A.D. 2085 prepared for the U.S. National Assessment of Climate Change are used here. The spatial fields of the 30-year average storm counts for the three times are plotted in figure 14.12. Average HADCM2 storm frequency index values for the 30 years centered on A.D. 2005 are thought to be representative of current atmospheric conditions; however, figure 14.12A differs from the observed record in fundamental ways. In the A.D. 2005 HADCM2 model output, the Alberta Storm track is not revealed at all. The Colorado storm track is delineated well, but the strong, bullseye maximum in Colorado is not found in the observational record of storm frequencies derived from weather maps. This maximum is probably fixed in this location because dynamic/orographic interactions gave rise to excessive, downstream cyclogensis in the lee of the Rocky Mountains. The storms tracking out of northern Mexico in the model ouptut data are not a feature found in the observational record. The Atlantic Coast storm track is found in the model output but is displaced much farther offshore than observed in the historical record. The current version of the model used in this assessment (HADCM2) does not produce a geography of storm tracks that is congruent with the observed pattern of storminess over North America.

The storm track output in the HADCM2 model for the 30 years centered on 2020 and 2085 (figures 14.12B,C) is essentially the same as that projected for the current conditions (model year 2005). It is clear from figure 14.12 that there is no model storm sensitivity to elevated carbon dioxide. This does not mean that a greenhouse-warmed world will have the same storm patterns we see today, only that such changes are not resolved in the current version of the Hadley model. Changes in storm frequency, magnitude, and specific geography should, on theoretical grounds, arise from the altered thermal gradient that is the signature of greenhouse global warming. The Hadley model with its current spatial resolution and physics is unable to resolve storms to a degree adequate for regional studies of storminess or changes in storm frequency and magnitude in a carbon dioxide–enriched world. With advances in computational capacity and in spatial resolution, GCMs should markedly improve the capacity to resolve frontal storms in the model outputs. At that time, the sensitivity of storminess arising from changes in atmospheric carbon dioxide should be revisited. With respect to storm frequency and storm track, the lack of detectable sensitivity by the model is disturbing. If the models cannot make adequate projections about these fundamental synoptic features of

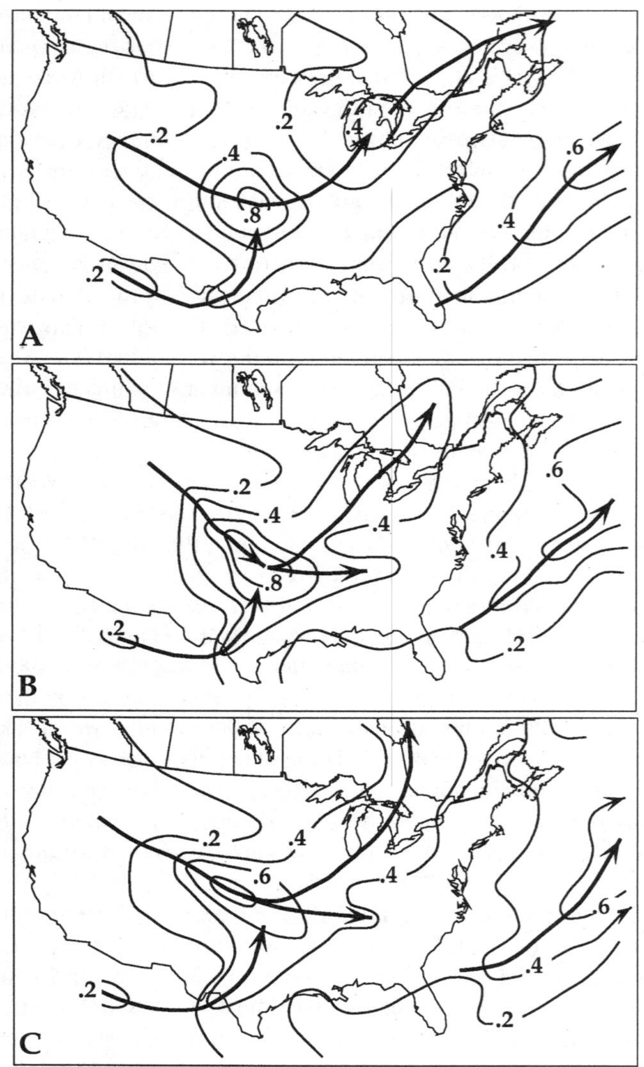

Figure 14.12 Charts of an index of storm frequency and storm tracks (arrows) derived from the Hadley Center's General Circulation Model (HADCM2) for the average 30-year periods centered on (A) 2005, (B) 2020, and (C) 2085.

the general circulation of the atmosphere, how then are parameterized output variables of the model such as precipitation and related hydrological processes to be estimated with confidence?

Discussion and Conclusions

Storm frequencies have changed over much of the continental United States since 1885. Very large changes in storm frequency are found at most LTER sites. However, when averaged over the entire continental United States, no net change in storminess is detected. We also conclude that there has been no change in the variability of storminess on a national basis. The storm climate changes observed at the 19 LTER sites are regional in nature and are very large in magnitude. Although we can be confident that storm climate has changed, we are less able to detail the consequences such as flooding or drought, erosion or deposition, and diminished or enriched water supply. These consequences require detailed site-specific study.

El Niño/La Niña variations in storminess were not detected either nationally or regionally. This conclusion is based on the analysis of all El Niño and La Niña events since 1885, with the exception of the 1960s for which data are missing. We conclude that indications of patterns of El Niño and La Niña storminess arose from a very limited sample size.

The patterns of change in storminess at the 19 LTER sites were not consistent with theoretical expectations for a world with greater warming in the high latitudes than in the low latitudes as expected from an atmosphere enriched with carbon dioxide. We also conclude that the current generation of atmospheric General Circulation Models do not generate any changes in storminess patterns for the next 100 years.

References

Angel, J. R., and S. A. Isard. 1998. The frequency and intensity of Great Lakes cyclones. *Journal of Climate* 11:1861–1871.

Arrhenius, S. 1896. On the influence of carbonic acid in the air upon the temperature of the ground. *The Philosophical Magazine* and *Journal of Science* 41(251):237–276.

Bigelow, F. H. 1897. Storms, storm tracks and weather forecasting. *Weather Bureau Bulletin* 20(114):5–87.

Carnell, R. E., C. A. Seniro, and J. F. B. Mitchell. 1996. An assessment of measures of storminess: Simulated changes in Northern Hemisphere winter due to increasing CO_2. *Climate Dynamics* 12:467–476.

Davis, R. E., and S. R. Benkovic. 1992. Climatological variations in the Northern Hemisphere circumpolar vortex in January. *Theoretical and Applied Climatology* 46:63–73.

Davis, R. E., and S. Benkovic 1994. Spatial and temporal variations of the January circumpolar vortex over the Northern Hemisphere. *International Journal of Climatology* 14:415–428.

Davis, R. E., and R. Dolan. 1993. Nor'easters. *American Scientist* 81:428–439.

Davis, R. E., P. J. Michaels, and B. P. Hayden. 2000. Overview of extratropical cyclones. Pages 401–426 in R. A. Pielke, Sr. and R. A. Pielke, Jr., editors. *Storms*. Routledge, New York.

Dickinson, R., and J. Namias. 1976. North American influences on the circulation and climate of the North American sector. *Monthly Weather Review* 104:1255–1265.

Dolan, R., and B. P. Hayden. 1981. Templates of change: Storms and shoreline hazards. *Oceanus* 23(4):32–37.

Dolan, R., H. Lins, and B. P. Hayden. 1988. Mid-Atlantic coastal storms. *Journal of Coastal Research* 4(3):417–433.

Emery, K. O., and D. G. Aubrey. 1991. *Sea levels, land levels and tide gauges*. Springer-Verlag, New York.

Fenster, M. S., and R. Dolan. 1994. Large-scale reversals in shoreline trends along the U.S. mid-Atlantic coast. *Geology* 22(6):543–546.

Hayden, B. 1975. Storm wave climates at Cape Hatteras, North Carolina: Recent secular variations. *Science* 190:981–983.

Hayden, B. P. 1981. Secular variations in Atlantic coast extratropical cyclones. *Monthly Weather Review* 100(1):159–167.

Hayden, B. P. 1988. Flood climates. Pages 13–27 in V. R. Baker, R. C. Kochel, and P. C. Patton, editors. *Flood geomorphology*. Wiley, New York.

Hayden, B. P., and R. Dolan. 1977. Seasonal changes in the planetary wind system and their relationship to the most severe coastal storms. *Geoscience and Man* 18:113–119.

Hayden, B.P., M. C. F. V. Santos, G. Shao, and R. C. Kochel. 1995. Geomorphological controls on coastal vegetation at the Virginia Coast Reserve. *Geomorphology* 13:283–300.

Hosler, C. L., and L. A. Gamage. 1956. Cyclone frequencies in the United States for the period 1905–1954. *Monthly Weather Review* 84:388–390.

Johns, T. C., R. E. Carnell, J. F. Crossley, J. M. Gregory, J. F. B. Mitchell, C. A. Seniro, S. F. B. Tett, and R. A. Wood. 1997. The second Hadley center coupled ocean-atmosphere GCM: Model description, spinup and validation. *Climate Dynamics* 13:103–134.

Key, J. R., and A. C. K. Chan. 1999. Multidecadal global and regional trends in 1000 mb and 500 mb cyclone frequencies. *Geophysical Research Letters* 26(14):2053–2056.

Klein, W. H. 1951. A hemispheric study of daily pressure variability at sea level and aloft. *Journal of Meteorology* 8:332–346.

Klein, W. H. 1958. The frequency of cyclones and anticyclones in relation to the mean circulation. *Journal of Meteorology* 15:98–102.

Klein, W. H. 1965. *Application of synoptic climatology and short-range numerical prediction to five-day forecasting*. Weather Bureau Research Paper No. 46, 109 pp.

Knappenberger, P. C., and P. J. Michaels. 1993. Cyclone tracks and wintertime climate in the mid-Atlantic region of the USA. *International Journal of Climatology* 13:509–531.

Kocin, P. J., and L. W. Uccellini. 1900. Snowstorms along the northeastern coast of the United States 1955–1985. *Meteorological Monographs* 22(44), 280 pp.

Manabe, S., Stouffer, R. J., Spelman, M. J., and K. Bryan. 1991. Transient Responses of a Coupled Ocean-atmosphere Model to Gradual Changes of Atmospheric CO_2. *Journal of Climate* 4:785–818.

Mather, J., H. Adams, and G. Yoshioka. 1964. Coastal storms of the eastern United States. *Journal of Applied Meteorology* 3:693–706.

Reitan, C. H. 1974. Frequencies of cyclone activity over North America 1951–1970. *Monthly Weather Review* 102:961–868.

Resio, D., and B. Hayden. 1975. Recent secular variations in mid-Atlantic extratropical storm climate. *Journal of Applied Meteorology* 14:1223–1234.

Taylor, K. E., and J. E. Penner. 1994. Response of the Climate System to Atmospheric Aerosols and Greenhouse Gases. *Nature* 369:734–737.

Uccelini, L. W., P. J. Kocin, R. S. Schneider, P. M. Stolkols, and R. A. Dorr. 1995. Forecasting the 12–14 March 1993 superstorm. *Bulletin of the American Meteorological Society* 7(2):183–199.

15

Multidecadal Drought Cycles in South-Central New Mexico: Patterns and Consequences

Bruce T. Milne
Douglas I. Moore
Julio L. Betancourt
James A. Parks
Thomas W. Swetnam
Robert R. Parmenter
William T. Pockman

Extreme, regional droughts are the most common form of disturbance in semiarid ecosystems typified by relatively slow recovery rates. Drought-driven impacts can include regionally synchronized insect outbreaks, wildfires, and tree mortality (Swetnam and Betancourt 1990), as well as disastrous failures of agriculture, silviculture, and livestock production (Mainguet 1994). Drought conditions, accompanied by anthropogenic land mismanagement, have led to subsequent invasions of grasslands and farmlands by woody shrubs and nonnative forbs and grasses, contributing to the modern "desertification" process manifested in many parts of the world (Archer et al. 1988).

In the American Southwest, the drought of the 1950s was one of the most severe climate events of the past millennium because of wide ramifications for the region's ecology (Herbel et al. 1972; Swetnam and Betancourt 1998), water resources (Thomas 1963), and economy (Regensberg 1996). As human population and resource needs increase in the Southwest, so will the economic sensitivity to large-scale drought. A clear understanding of extreme droughts is necessary not only to understand long-term ecosystem dynamics, but also to mitigate socioeconomic impacts.

The goals of this chapter are to use the Sevilleta LTER site in central New Mexico to (1) quantify the decadal variability in precipitation inferred from a 394-year record of tree rings, (2) relate the repeated decadal fluctuations in precipitation to major droughts of the 1890s and 1950s, (3) assess the ecological responses associated with droughts of the last century, and (4) elucidate the biotic-atmospheric feedbacks that may influence future responses. We assess the magnitude, timing, and consequences of decadal fluctuations in annual precipitation.

Sevilleta LTER Site Description

The Sevilleta LTER research site is located at the Sevilleta National Wildlife Refuge (NWR), Socorro County, New Mexico (34°20' N, 106°50' W). The Sevilleta NWR comprises 100,000 ha of grassland, desert, and woodland bordered by two mountain ranges and the Rio Grande Valley in between. Elevations range from 1,350 m at the Rio Grande to 2,797 m at Ladrón Peak in the northwestern portion of the refuge. Topography, geology, soils, and hydrology, interacting with major air mass dynamics, provide a spatial and temporal template that makes the region a transition zone between several biomes. The region contains communities that both represent and intersect Great Plains Grassland, Great Basin Shrub-steppe, Chihuahuan Desert, Interior Chaparral, and Montane Coniferous Forest (Brown 1982).

Sevilleta Climate Description

The Sevilleta LTER study region straddles the boundary between major seasonal air masses (e.g., winter arctic frontal systems descend southward across the Great Plains and influence Sevilleta's eastern edge; Great Basin polar air masses extend to Sevilleta's northern edge; the Bermuda High generates summer convective storms over the mountains, which track northeast across Sevilleta's lowlands). Superimposed on these spatial patterns are the temporal dynamics of the El Niño–Southern Oscillation (ENSO) phenomenon. These climate phenomena are further translated by the orography of the southern Rocky Mountains.

The annual climate of the Sevilleta area includes two different patterns of seasonal storms. During the late fall, winter, and spring, storms develop in both the northeastern and tropical Pacific Ocean. These storms are steered into the region by both the polar and subtropical jet streams, and their annual frequencies and total precipitation over central New Mexico are in part modulated by both interannual (ENSO) and decadal-scale (Pacific Decadal Oscillation) variability in Pacific climate. In general, central New Mexico tends to have wet falls, winters, and springs during El Niño events, or times when the Pacific Decadal Oscillation (PDO) is positive, and the opposite conditions during La Niña (for ENSO, see Andrade and Sellers 1988, Molles and Dahm 1990, Redmond and Koch 1991, Cayan and Webb 1992, Kahya and Dracup 1993; for PDO, see Cayan et al. 1998, Mantua and Hare 2002). At Socorro, just south of the Sevilleta, precipitation during the period from October through May increased by 53% during El Niño years and decreased by slightly more than half during La Niña events when compared to medial years over the past 80 years (figure 15.1, table 15.1). Although this fall-winter-spring precipitation increase may appear small, the colder conditions and dormant vegetation favor soil moisture recharge, thereby increasing soil water availability for the growing season. Also, most of these cool-season frontal storms have broad tracks

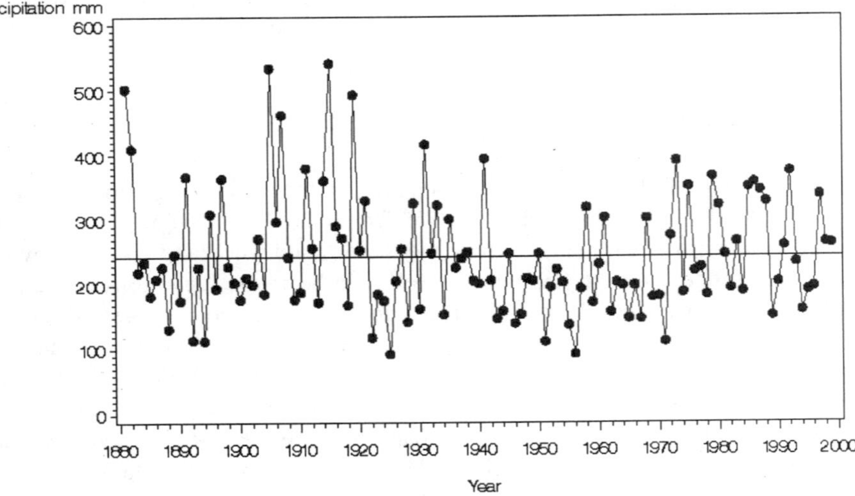

Figure 15.1 Precipitation record for Socorro, New Mexico (1880–2000). Note periods of low precipitation from the late 1890s to early 1900s and during the period 1950–1957.

(>100 km) and tend to synchronize ecosystem response to precipitation events across the region.

Summertime precipitation is driven by the North American Monsoon. During the summer, the jet stream retreats to more northern latitudes, thereby reducing the influence of Pacific storm systems on the region. Normally, the period of greatest precipitation in much of the Southwest occurs during the months of July, August, and September, and is associated with convective thunderstorms during this North American Monsoon. Monsoon precipitation originates from moist air masses over the Gulf of Mexico and is pushed into the Southwest by the Bermuda High (Mitchell 1976; Neilson 1986). The resulting precipitation is distributed heterogeneously on the landscape by thunderstorms that originate over the mountains and move over the lowlands. Although temporal variability is low, spatial variability is

Table 15.1 Mean precipitation amounts by ENSO class from Socorro, New Mexico, 1914–1993[a].

ENSO class	N	Precipitation (mm)		
		Annual	Oct–May	Jun–Sep
El Niño	15	275.8 a	156.2 a	119.6 a
Medial	56	239.4 a	102.3 b	137.1 a
La Niña	9	162.5 b	49.9 c	112.5 a

[a] ENSO classes with different letters are significantly different ($P < 0.05$).

high. The linkage between the ENSO phenomenon and summer precipitation in New Mexico is weak (Andrade and Sellers 1988; Molles et al. 1992; Harrington et al. 1992). There seems to be some predictability of the onset and/or magnitude of the monsoon moisture (Higgins and Shi 2000), but the predictability is much poorer than for the winter and spring moisture associated with extremes of the ENSO cycle. Although the monsoon moisture of July, August, and September generally accounts for well over half of the annual precipitation in the Sevilleta region, high evapotranspiration prevents summer moisture from recharging soil moisture to depth. Thus, vegetation responses to summer moisture are ephemeral and patchy.

Therefore, we regard the fall-winter-spring precipitation as separate from that of the summer (June–September) precipitation. Continuous monitoring of soil moisture in the Sevilleta shows that the increased temperatures and plant growth of May and June deplete surface soil moisture from the nonmonsoon period before the monsoons begin in July. Monsoon moisture is generally lost quickly (2–3 days) between storms through evaporation, so that the transition from summer to winter is often a period of dry soils as well.

Multidecadal Precipitation Patterns

Direct measurements of precipitation at the Sevilleta LTER began in 1988, and records for Socorro County date only to the 1870s. Consequently, we elucidated persistent interdecadal weather patterns from a record of annual precipitation obtained from tree rings. A 394-year precipitation history (1598–1991) was reconstructed by Parks et al. (in press) by combining piñon pine (*Pinus edulis*) tree-ring chronologies from live and dead trees (killed by the 1950s drought) in Arroyo de Milagro in the Los Pinos Mountains, in the eastern sector of the Sevilleta NWR (figures 15.2, 15.3a). The original reconstructed precipitation depths are presented as deviations from the long-term mean (286 mm, s.d., = 63.5) to emphasize anomalous wet and dry years (figure 15.3a). Details of this particular reconstruction are explained in Betancourt et al. (1993) and Parks et al. (in press). The Sevilleta tree-ring index chronology was statistically compared to actual precipitation measured in Socorro for the period of overlap (1892–1991). We compared several seasons, but the highest correlation (r = 0.74) was for a 13-month interval covering the prior August through the current August of the growth year. For southern New Mexico conifers, Grissino-Mayer et al. (1997) found a similar conditioning of cambial growth by rainfall during the previous summer and fall and the current spring and summer, with winter rainfall (January through March) having little effect. Linear regression was used to estimate August-to-August precipitation from tree-ring indices.

We used multitaper method spectral analysis (MTM; Dettinger et al. 1995) to measure the temporal variability in the reconstructed precipitation record. MTM spectral analysis uses three orthogonal waveforms to measure the variance in a time series at a variety of frequencies, or inversely, periods. Confidence intervals (95 and 99%) were defined under the geophysically appropriate null hypothesis

Figure 15.2 Piñon-juniper woodlands in the Los Pinos Mountains, Sevilleta National Wildlife Refuge, New Mexico. Piñon pines in this mountain range provided samples for reconstruction of the long-term precipitation record.

that the time series is red noise, that is, more correlated than white noise and less correlated than a random walk. Thus, peaks in the MTM spectrum that exceed the confidence limits indicated persistent oscillations (figure 15.3b). At the decadal scale, the Sevilleta tree-ring reconstruction evidenced significant (99% confidence interval) variability for periods of 72.9, 68.5, 64.1, 60.2, 56.8, 53.8, 51.3, 48.8, 44.4, 42.7, 40.9, 39.4, and 36.6 years (mean = 51.9, s.d. = 11.3). Thus, on average, droughts and wet periods have occurred about every 52 years in Arroyo de Milagro. The data must be evaluated with caution because detrending during the construction of the original tree-ring record may have led to an overestimate of variation in the low frequencies (D. Meko, pers. comm., 2002).

Droughts present plants with conflicting demographic challenges. A seed is in a race between (1) the appearance of suitable conditions for germination and (2) the ever-diminishing chance of surviving or escaping predation. Droughts retard the former and may accelerate the latter. Thus, an ecologically relevant analysis of annual precipitation records should focus on the deviations from the mean precipitation because the deviations measure the uniqueness of any given year. Moreover, a string of wet years favors germination and establishment of species that require anomalously high amounts of water, such as the blue grama grass, *Bouteloua gracilis* (Neilson 1986). A string of dry years should exhaust seed banks through attrition and thereby color plant community dynamics for years to come. Thus, we recast the annual estimates of precipitation as deviations from the mean, or "normal" (figure 15.3a), and examined the strings of successive deviations to find epochs of persistent wet and dry conditions.

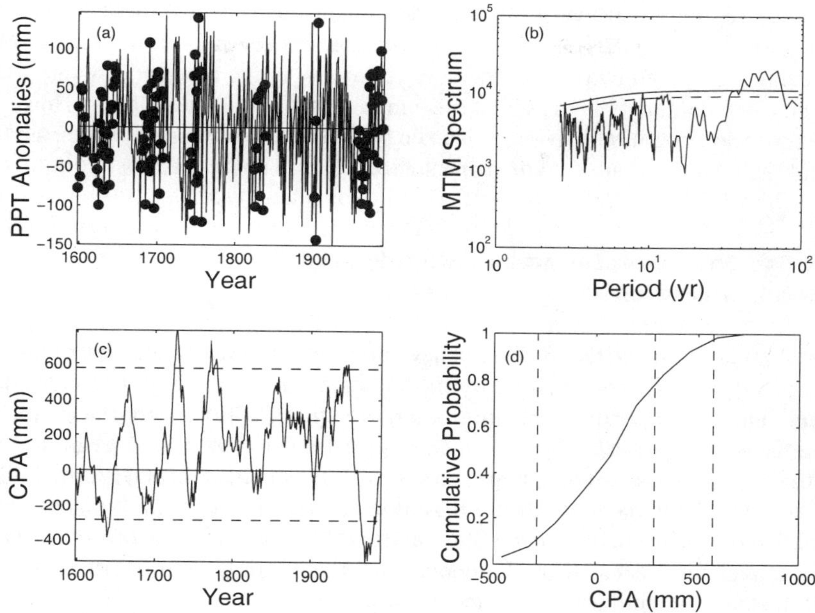

Figure 15.3 Patterns of long-term variation of precipitation based on the Arroyo de Mila-gro tree-ring record. (a) Annual precipitation depths minus the long-term mean. Dots indi-cate years that occurred in dry epochs. (b) Multitaper method spectrum of the raw precipi-tation depths. Horizontal lines indicate 95% (dashed) and 99% (solid) confidence intervals. (c) Running sum of the precipitation anomalies, that is, cumulative precipitation anomalies, CPA. (d) Cumulative probability distribution of the CPAs.

Following Feller (1968) and the technique of "mass curves" in hydrology (e.g., Peng and Buras 2000), we revealed persistent, repeated wet and dry periods by forming the running sum of the anomalies, that is, cumulated precipitation anom-alies or CPAs (figure 15.3c). Peaks and troughs correspond to high and low net ac-cumulations, respectively. The maximum CPA (787 mm) occurred in 1727, and the low of −523 mm occurred in 1971. Plants living in 1727 or 1971 would have been at the ends of extremely wet or dry epochs, respectively. The cumulative probability density of the CPA indicated that only 10% of the years were in major droughts like the early 1600s and 1950s (figure 15.3d). In contrast, the skewed distribution of anomalies produced wet epochs that exceeded one standard deviation of the CPA 20% of the time. Individual years were somewhat more likely to be in the midst of a wet epoch than a dry one.

There were 7 lengthy strings of dry years that constituted persistent droughts. Dividing the record length by 7 suggested 56 years as an approximate drought re-currence interval, which is in agreement with the time series analysis (figure 15.3b). The interval coincided with recent major droughts of the 1890s and 1950s that were about 60 years apart.

We concluded that long-term fluctuations in precipitation have occurred over the

Sevilleta region about every 52 years, although intervals of 41 to 63 years are within one standard deviation of the expected 52 years. Attempts to predict future droughts are limited by the considerable variability in the recurrence interval and the lack of a definitive theory about the climatic origin of the oscillation. However, this limitation does not deny the existence of decade-long droughts in the past that could recur with major ecological and socioeconomic consequences.

Possible Sources of Decadal to Multidecadal Climate Oscillations

The ultimate source of the low-frequency variations in the Sevilleta tree-ring series is presently unresolved. Oscillatory modes at decadal to century timescales have been identified in annual temperature and precipitation series from instrumental records (e.g., Cayan et al. 1998; McCabe and Dettinger 1999; Dettinger et al. 2001) and tree rings across western North America (e.g., Biondi et al. 2001; Gedalof and Smith 2001; Villalba et al. 2001). It is assumed that most of this low-frequency variation originates in the Pacific Basin and involves interaction of the ENSO mode with longer term decadal to centennial fluctuations in climate (Dettinger et al. 2001). Note that similar low-frequency variations have also been identified for the Atlantic Ocean (e.g., Delworth et al. 1993; Enfield and Mestas-Nuñez 1999) and could produce similar periodicities in temperature and precipitation, particularly in summer. In the North Pacific, much of the sea surface temperature variance occurs in a mode with decadal (~20–30 years) timescales and is accompanied by variability in the strength and position of the Aleutian Low in winter. This has been called the North Pacific Oscillation (NPO; Gershnov and Barnett 1998) or Pacific Decadal Oscillation (PDO; Mantua and Hare 2002).

Among the leading explanations for low-frequency variations such as the PDO are stochastic atmospheric forcing, atmospheric teleconnections, midlatitude ocean-atmosphere interactions, tropical-extratropical interactions, oceanic teleconnections, and intrinsic ocean variability (see summaries in Hare et al. 2000; Dettinger et al. 2001; Mantua and Hare 2002). There is considerable discussion about the steady state versus chaotic behavior of decadal- to century-scale variability, and thus about its predictability. An optimistic view is that knowledge about the present phase of the long-term mode (e.g., PDO) can be used to forecast climate several years ahead (Latif and Barnett 1996; Dettinger et al. 2000; Schneider and Miller 2001).

The twentieth century was marked by two full PDO cycles. The "cool" or negative PDO (more La Niña-like) regime prevailed from 1890 to 1924 and 1947 to 1976, whereas the "warm" or positive PDO regime prevailed from 1925 to 1946 and from 1977 to 1994. Using wavelet analysis, Minobe (1997, 1999) found that fluctuations in North Pacific sea surface temperatures, sea level pressures, and temperature reconstructions based on North American tree rings were most energetic at periodicities of 15–25 years (for boreal winter) and 50–70 years (for boreal winter and spring). According to Minobe (1999), the two periodicities synchronize with a relative period of three and produce a "regime shift" in North Pacific climate in the 1920s, 1940s, and 1970s when they reverse phase. There is a slight indication for such a regime shift starting in 1998 when the PDO index turned sharply negative ("cool" mode), the tropical Pacific began to cool after the prolonged post-1976 warming, and the North Atlantic started to warm after prolonged cooling since the 1960s. Such were the conditions during the regime shift of the 1940s that led into the 1950s drought, an episode that, according to Sevilleta tree rings, tends to recur every 41–63 years. We resist the temptation to make any long-term forecast, but we suggest instead that New Mexico politicians, resource managers, and ranchers alike have little cause for optimism. And as Swetnam and Betancourt (1998) have pointed out, given the possibility for another extreme drought, local biologists should be poised to take advantage of such "natural experiments."

Reliability of Arroyo de Milagro Reconstruction for Regional Assessment

Our analyses of the tree-ring precipitation chronology indicated that major droughts have occurred at Arroyo de Milagro in the Sevilleta every 41–63 years (figure 15.3b,c). Overlap between the tree-ring record and modern meteorological measurements enabled us to evaluate the extent to which the Sevilleta record indicates conditions over the broader Middle Rio Grande Basin.

Measured total annual precipitation can be a poor predictor of ecosystem productivity because the seasonal timing of precipitation is critical to its biotic effectiveness. Timing and other factors such as soil moisture holding capacity can influence the effect of the precipitation on growth (Valentine and Norris 1964). The Palmer Drought Severity Index (PDSI; Palmer 1965) can identify ecologically effective wet and dry conditions. The PDSI uses precipitation, temperature, and soil moisture to give a measure of moisture availability. The index is standardized to the local climate, so that relative dryness and wetness can be compared across the entire United States. The PDSI is extrapolated over climate regions within states.

The PDSI record for the Sevilleta region covers the entire period from 1895 to the present (figure 15.4; source http://lwf.ncdc.noaa.gov/oa/climate/onlineprod/drought/main.html). Periods when the index is above 0 are considered wet periods, whereas those below zero denote dry conditions. Drought classifications begin at −2.0 and increase in severity with decreasing index values. For the relatively short period of the Sevilleta LTER (1989–present), conditions have oscillated from the dry La Niña of 1989 to quite wet during the 1992–1993 El Niño. The index then stayed mostly dry until the wet monsoon of 1996, which carried on through the 1997–1998 El Niño and then back to mostly dry for the extended La Niña of 1998 to the present (2002). The wet to dry oscillation is obviously the norm.

The most noticeable exceptions to this pattern are during two extended periods during the 1890s to early 1900s and from 1950 to 1957. Then, the index stayed below zero for the entire time. The 1890s to early 1900s drought was responsible

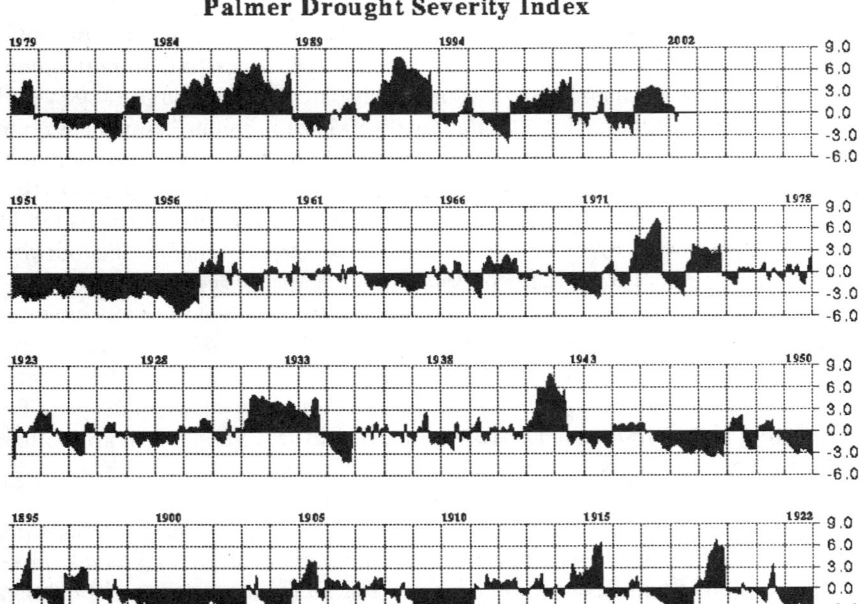

Figure 15.4 Palmer Drought Severity Index for central New Mexico (1895–2002). Note the periods of low (dry) indices during the periods of 1898–1905 and 1950–1957.

for massive die-offs of livestock and shrub encroachment on southwestern range-lands in the 1890s. The period 1950–1957 is known as the "50's drought." In the Southwest, this was much more severe than the "Dust Bowl" drought of the 1930s in the panhandle region of Oklahoma and Texas. Table 15.2 lists the years with mean annual PDSI values less than minus 2.0. These two dry periods indicated by the PDSI are consistent with the tree-ring chronology and reflect the low rainfall periods in the Socorro precipitation record.

The PDSI values are significantly related to the reconstructed precipitation depths. A regression analysis using reconstructed precipitation from the tree-rings (Betancourt et al. 1993; Parks et al. in press) and the mean annual PDSI values from Climate Region 5 (Middle Rio Grande Valley) produced an $R^2 = 0.46$ ($p < 0.0001$, $n = 95$; figure 15.5). Among months, May and June exhibited the strongest relations ($R^2 = 0.497$, $p < 0.0001$, $n = 95$ and $R^2 = 0.469$, $p < 0.001$, $n = 95$), respectively. Thus, the reconstructed precipitation depths were statistically related to regional moisture conditions. However, the considerable unexplained variation, possibly due to soil, hydraulic, and orographic conditions at the Arroyo de Milagro site, cautions against unbridled application of the record to the region.

Table 15.2 Years from 1895 to 2000 with mean annual Palmer Drought Severity Index (PDSI) of minus 2.0 or less. Note multiyear periods of drought in early 1900s and 1950s.

Year	PDSI	Year	PDSI	Year	PDSI	Year	PDSI
1900	−3.12	1918	−2.59	1946	−2.50	1971	−2.25
1901	−2.44	1922	−2.41	1947	−3.01	1982	−2.27
1902	−2.65	1934	−2.56	1950	−2.09		
1904	−3.29			1951	−3.38		
1909	−2.39			1952	−3.02		
1910	−3.19			1953	−2.30		
				1954	−3.31		
				1955	−3.11		
				1956	−3.78		
				1957	−3.23		

Influence of Climate on Sevilleta's Biome Transition Zone

The Sevilleta LTER site is at the interface of several major biomes, including short grass prairie, Chihuahan Desert shrubland, and piñon-juniper woodland. Combined effects of potential evapotranspiration and available soil moisture appear to regulate the distributions of the various biomes. For example, where uniform soils occur over locally rough terrain, nuances of slope and aspect regulate the surface energy supply and produce repeating vegetation patterns (figure 15.6). Desert shrubs such as *Larrea tridentata* dominate on south-facing slopes with high moisture deficits. A few meters away, *Juniperus monosperma* dominates at the same elevation on neighboring north-facing slopes. Inspections of aerial photographs taken before and after the 1950s drought indicate that junipers found at intermediate east and west facing slopes died where conditions became marginal during the drought. Today, the two dominant species meet in such vicinities. Repeat photography indicates that *L. tridentata* became more widely established after the 1950s drought (figure 15.7), apparently reflecting the massive expansion of desert shrubland in New Mexico (Grover and Musick 1990).

We used preliminary measurements of the relative growth rates of *L. tridentata* and *J. monosperma* as functions of annual precipitation to examine the hypothesis that the species were capable of growing in the distant past at Arroyo de Milagro. The periodic wet and dry epochs over the last 400 years (figure 15.3c) should have created opportunities for species such as *L. tridentata* to establish. However, satisfactory resource availability would not overcome dispersal limitations or competition that can also affect establishment. Ongoing field experiments address the competitive interactions of these species.

Longitudinal studies of 10 marked twigs from each of 20 shrubs of each species at 7 sites ($200 \leq PPT \leq 340$ mm) were used to calibrate second-order polynomial regressions of mass specific growth rate as a function of annual precipitation (B. T. Milne, C. Restrepo, and W. Pockman, unpubl. data, 2001). The mean relative

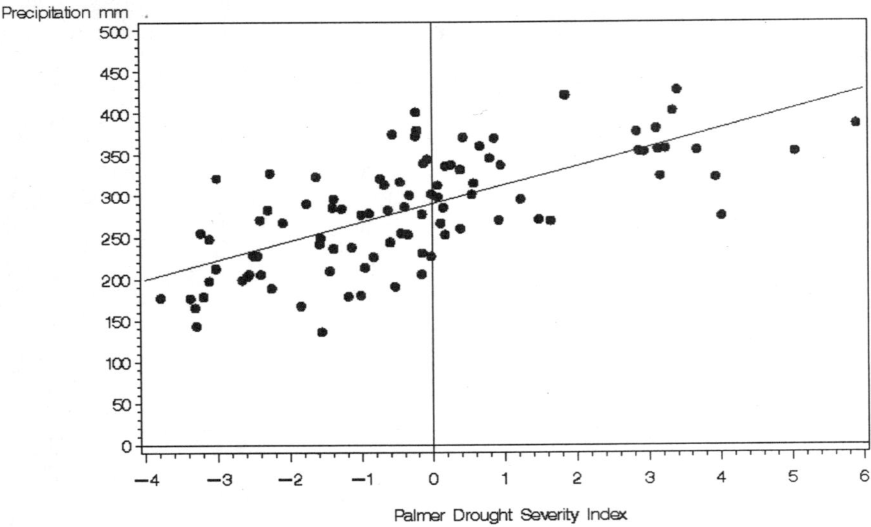

Figure 15.5 Regression of the Palmer Drought Severity Index (PDSI) values and the precipitation record reconstructed from the Arroyo de Milagro tree-ring data.

change in *L. tridentata* biomass (B) was $dB/Bdt = -137.2 + 1.28\,PPT - 0.0029\,PPT^2$, where *PPT* is annual precipitation (mm) in 2000. For *J. monosperma*, $dB/Bdt = -6.83 + 0.05\,PPT - 0.0001\,PPT^2$. We simulated annual net primary productivity for the years 1598–1991, assuming that plants began with 1 kg/m² of biomass (i.e., were not limited by dispersal) and had the potential to grow logistically to a carrying capacity of 50 kg/m². We assumed no competition. Thus, the simulation estimated the potential net primary productivity for each species given the series of annual precipitation depths estimated from the tree-ring record.

The relatively high mean annual precipitation at Arroyo de Milagro favored the growth of *J. monosperma* more than *L. tridentata*. Several decades showed persistent epochs of high juniper productivity that coincided with lethal conditions for *L. tridentata* (figure 15.8). Suitable conditions for *L. tridentata* appeared throughout the record but were interrupted repeatedly by periods of no growth. We concluded that climatic variability, dispersal limitation, and competition are possible explanations for the lack of *L. tridentata* in the area until recent times.

Of course, variability across the broader landscape could maintain source populations for either species. Source populations would ameliorate the effects of climate variability at a given location by providing an infusion of propagules. Thus, a comprehensive view would include a metapopulation approach to understand the role of temporal variation through space (Keymer et al. 2000). Climate fluctuations can shift the edges of populations as species advance beyond their previous distribution or are eliminated from occupied areas. Similarly, at the edge of a species distribution, variation in microclimate may create locally patchy distributions (Holt and Keitt 2000).

Figure 15.6 View of a small watershed at the Sevilleta LTER site, showing juniper savanna vegetation on north-facing slope (left) contiguous with Chihuahuan Desert vegetation on the south-facing slope (right).

One example of a physiological limit that is particularly relevant in arid ecosystems derives from the plant's requirement for water to sustain transpiration. Water transport through the xylem is subject to interruption by cavitation caused either by high xylem tensions associated with drought or by cycles of freezing and thawing in the xylem (Tyree and Sperry 1989). The drought or freezing conditions required to cause cavitation vary considerably among species and are determined by structural features of the xylem, pit membrane pore diameter in the case of drought (Sperry and Tyree 1988) and xylem conduit volume in the case of freezing (Davis et al. 1999; Sperry et al. 1994). This quantifiable link to xylem structure makes the definition of a physical limit to plant function more straightforward than for some physiological parameters for which the functional limit is difficult to predict. Although some species can repair the effect of cavitation (Holbrook and Zwieniecki 1999), hydraulic limits are largely fixed in an individual, making the maintenance of water transport in the individuals dependent on the drought and freezing conditions they experience while active.

Drought Effects

Highly variable precipitation reduces the chance that soil water conditions that favor germination and establishment will occur in the necessary sequence (Neilson 1986). Although dominant woody species at Sevilleta LTER (e.g., *Pinus edulis, J. monosperma, L. tridentata*) are highly drought tolerant (Linton et al. 1998; Pock-

Figure 15.7 Photographs of Palo Duro Canyon area, Sevilleta LTER site. Top: Cattle herd near windmill drinking area, circa 1928. Bottom: Same view in 1998. Note the invasion by creosotebush (*Larrea tridentata*) and grasses (*Sporobolus* spp.).

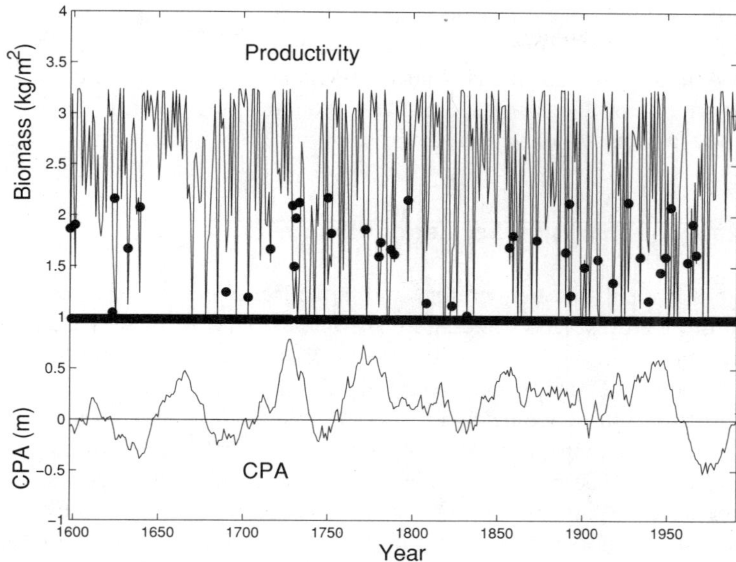

Figure 15.8 Simulated net primary productivity of *Juniperus monosperma* (solid thin line) and *Larrea tridentata* (filled dots) with respect to cumulative precipitation over the 394-year surrogate precipitation history from Arroyo de Milagro, Sevilleta LTER.

man and Sperry 2000), the limited extent of seedling root systems exposes them to extremely dry shallow soil. Under such conditions, extensive cavitation leads to mortality (Williams et al. 1997). Once established, the dominant shrub species approach the point of complete cavitation only rarely (Pockman and Sperry 2000). Conditions during extended drought periods such as the 1950s drought in New Mexico (Allen and Breshears 1998) are likely to approach the physiological limits of these species to drought induced cavitation.

Freezing Effects

At Sevilleta, the transition from shrubland, dominated by *L. tridentata*, to grassland in the north represents the northern limit of the continuous distribution of *L. tridentata*. Although its evergreen habit enables *L. tridentata* to exploit favorable conditions at any time, it must also maintain water transport during freezing conditions to support its evergreen foliage. Historical reports of *L. tridentata* at its northern limit indicate that extreme freezing events lead to heavy dieback of aboveground growth (Cottam 1937). Stems of mature *L. tridentata* in southern Arizona exhibit no embolism following freezing between 0° and –10° C, after which embolism increases linearly with decreasing temperature until embolism is complete at –16° to –20° C (Pockman and Sperry 1997). A coarse-scale analysis of long-term climate data showed that this critical temperature range corresponds to the northern limit of the species in the Sonoran and Mojave Deserts but fails to account for the northern extension of the range in the Rio Grande valley to its limit at Sevilleta. Recent data suggest that *L. tridentata* at Sevilleta are more resistant to freezing-induced cavitation than those in Arizona (Martinez-Vilalta and Pockman, 2002). Biogeographic variation provides an opportunity to examine the role of this physiological limit in determining the distribution of the species.

Ecosystem Feedbacks on Sevilleta's Climate

Although it is clear that temperature and moisture influence the distributions of plants at the Sevilleta LTER site, there is some evidence that the vegetation patterns have feedbacks on the local climate as well. One example of this occurs in an area of the Sevilleta that is dominated by *L. tridentata* (figure 15.7). It is widely hypothesized that anthropogenic disturbances of arid lands in Mexico and the American Southwest, in concert with drought cycles (e.g., the great droughts of the 1890s and 1950s; Hastings and Turner 1965), have facilitated the recent acceleration of the range expansion of woody shrubs. Rangeland overgrazing by excessive cattle, sheep and horses in the late nineteenth century, coupled with extended droughts, favored desert shrub (*L. tridentata* and mesquite, *Prosopis* spp.). Bray (1901, p. 289) noted the speed of the invasions at the turn of the century in west Texas: "Regarding the establishment of woody vegetation, it is the unanimous testimony of men of long observation that most of the chaparral [=*L. tridentata*] and mesquite covered country was formerly open grass prairie." Encroachment of desert shrubs northward and into grasslands continued during the twentieth century, particularly in New Mexico and west Texas (Gardner 1951; Branscomb 1958; Humphrey 1958; Buffington and Herbel 1965; York and Dick-Peddie 1969; Hennessy et al. 1983; Humphrey 1987; McPherson et al. 1988; Grover and Musick 1990).

Invasions of shrub species into grasslands not only have considerable consequences for ecosystem dynamics (Schlesinger et al. 1990) and human activities (Mainguet 1994), but also for potential alterations of mesoscale climatic conditions. Hayden (1998) reviewed the existing literature and analyzed large-scale temperature patterns, relating the observed minimum temperatures in deserts of the American Southwest to the predicted dew-point temperatures of the same sites. Winter minimum temperatures ranged up to 8°C warmer than predicted (Hayden 1998); at the Sevilleta LTER site, winter minimum temperatures could be as much as 4°C warmer than predicted from dew-point temperatures. The reason for this difference was attributed to nonmethane hydrocarbons released from desert vegetation (particularly terpenes from desert shrubs like *L. tridentata*); these hydrocarbons functioned as "greenhouse gasses" and decreased the emissivity of the local atmosphere above the vegetation, preventing heat loss and raising nighttime minima.

At the Sevilleta LTER site, two meteorological stations are located within 5 km of each other at the same elevation on McKenzie Flats; one is in a community dominated by *L. tridentata*, whereas the other is surrounded by grassland. The station in creosotebush vegetation typically registers higher nighttime minimum tempera-

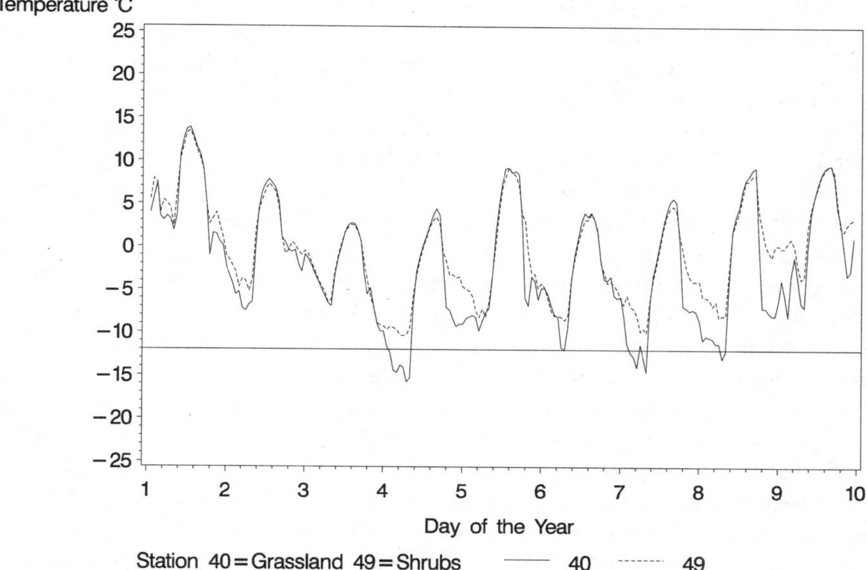

Figure 15.9 Air temperatures above desert shrub vegetation and grassland vegetation at the Sevilleta LTER site, January 1–10, 2000; weather stations are 4.4 km apart. Note warmer minimum temperatures above desert shrub vegetation.

tures than the station in the grassland (figure 15.9), particularly during calm periods (high winds mix the atmosphere over both sites). Atmospheric warming reduces the probability of killing-frosts, thereby increasing the likelihood of survival and population increase of these desert shrub species. Thus, it may be that desertification of grasslands by woody shrub invasions is accelerated by the invading shrubs via changes in atmospheric chemistry that ultimately alter the temperature regime, allowing greater survivorship and expansion of the shrub populations by frost reduction. Future research on this hypothesis will be needed to determine whether, and by how much, vegetation changes can influence the regional climates of arid and semiarid ecosystems.

Conclusions

Ecologists have been slow to measure or model biological responses to longer periodicities in climate. Most gap models, for example, simulate climate variability as randomized variance around an annual mean. Inserting decadal climatic oscillations in gap models, however, has been shown to either induce periodicity or shift the mean in biomass (Yeakley et al. 1994). A general characterization of the Sevilleta climate includes seasonal, interannual, and decadal scales of variation.

The 41- to 63-year periodicity in annual precipitation constitutes a slowly changing climatic context within which higher frequency ecological responses occur. Alterations of wild populations by management, disease, or predation might result in different responses, depending on whether the oscillation is on an upswing or a downswing. Efforts at reforestation or land reclamation would probably be most successful in the wetting phase of the cycle, during which a 20- to 30-year period of relatively wet weather would benefit the efforts. Attempts to maintain marginally productive rangeland operations during the ensuing dry phase of the oscillation are at risk of failure. Indeed, it may be no accident that the Sevilleta National Wildlife Refuge was created from a working ranch only 10 years after the end of the 1950s drought when the CPA was at an all-time low (figure 15.3c).

Policy makers and financial institutions might anticipate that economic collapses follow droughts that stimulate abrupt changes in land ownership. The Sevilleta tree-ring record, limited to only 7 multiyear droughts of different severities and durations, suggests a 52-year recurrence interval for major droughts (figure 15.3b), with the last drought period persisting between 1942 and 1972. We presently lack the necessary time depth, regional coverage, and understanding of the global-scale ocean-atmosphere interactions that underlie this apparent periodicity. We also recognize that these periodicities may be inherently unstable and could be modulated by various phenomena, including anthropogenic forcing of climate. Hence, we exercise caution in making any prediction about imminent drought in south-central New Mexico. Nevertheless, we point out that extreme drought in the early part of the twenty-first century is within the realm of expectation and that human population growth and demand for water and other resources in south-central New Mexico has amplified sensitivity to drought.

Land managers might reflect on the experiences of the 1890s. The introduction of railroads created an incentive to stock the range because of easier access to markets. Simultaneously, introduction of windmills enabled ranchers to supply drinking water for livestock during the drought. However, without precipitation, forage plants were unable to withstand high stocking rates, leading to long-lasting depletion of soil organic matter and seed banks. In future droughts, apparent technological solutions such as precision farming methods, optimized irrigation strategies, and genetically engineered crops or livestock may actually increase the risk to various components of the landscape or ecosystem, just as wells and trains did in the past. The ecology of semiarid lands is subject to environmental variation on the order of a human career and thus should probably be managed at timescales of several human generations.

Summary

Variation in precipitation occurs at seasonal, interannual, and decadal timescales. Ecological and economic consequences related to land use and resource management drive the search for repeatable patterns ascertained from direct and surrogate climate records. We studied a 394-year record (1598–1991) of precipitation derived from the annual rings of piñon pine (*Pinus edulis*) at the Sevilleta National Wildlife

Refuge and Long-Term Ecological Research site in central New Mexico, United States. A significant 52-year periodicity (standard deviation 11.3 years) of precipitation coincided with the major regional droughts of the 1890s and 1950s. Long-term ecological consequences of decade-long droughts pertain to the establishment of novel species, physiological stress, feedbacks between plants and the atmosphere, and economic repercussions related to land use. Simulated net primary productivity of *Juniperus monosperma* (one-seeded juniper) and *Larrea tridentata* (creosotebush) indicate that suitable conditions for *L. tridentata* growth occurred intermittently during the last four centuries. Assessments of the droughts of the 1890s and 1950s suggest that future technological attempts to ameliorate the effects of drought should minimize unforeseen consequences for various components of the landscape. Occasional decadal oscillations in annual precipitation are a major ecological factor in the region.

Acknowledgments   Dave Meko, Mike Dettinger, and two anonymous reviewers guided the final outcome of this effort. Support was provided by NSF grants DEB 9910123 to BTM and DEB 0080529. Sevilleta LTER publication no. 263.

References

Allen, C., and D. Breshears. 1998. Drought-induced shift of a forest-woodland ecotone: Rapid landscape response to climate variation. *Proceedings of the National Academy USA* 95:14839–14842.

Andrade, E.R., and W.D. Sellers. 1988. El Niño and its effect on precipitation in Arizona. *Journal of Climatology* 8:403–410.

Archer, S., C. Scrifres, C.R. Bassham, and R. Maggio. 1988. Autogenic succession in a subtropical savanna: Conversion of grassland to thorn woodland. *Ecological Monographs* 58:111–127.

Betancourt, J.L., E.A. Pierson, K.A. Rylander, J.A. Parks, and J.S. Dean. 1993. Influence of History and Climate on New Mexico Pinyon-Juniper Woodlands. Pages 53–62 in *Managing Juniper Ecosystems for Sustainablity and Social Needs*. USDA Forest Service General Technical Report RM-236. Rocky Mountain Forest and Range Experiment Station. U.S. Department of Agriculture. Fort Collins, Colorado.

Biondi, F., A. Gershunov, and D.R. Cayan. 2001. North Pacific decadal climate variability since 1661. *Journal of Climate* 14:5–10.

Branscomb, B. L. 1958. Shrub invasion of a southern New Mexico desert grassland range. *Journal of Range Management* 11:129–132.

Bray, W.L. 1901. The ecological relations of the vegetation of western Texas. *Botanical Gazette* 32:99–123, 195–217, 262–291.

Brown, D.E. 1982. Biotic Communities of the American Southwest—United States and Mexico. *Desert Plants* 4:1–342.

Buffington, L.C., and C.H. Herbel. 1965. Vegetational changes on a semidesert grassland range from 1858 to 1963. *Ecological Monographs* 35:139–164.

Cayan, D.R., and R.H. Webb. 1992. El Niño/Southern Oscillation and streamflow in the Western United States. Pages 29–68 in H.F. Diaz and V. Markgraf, editors, *El Niño Historical and Paleoclimatic Aspects of the Southern Oscillation*. Cambridge University Press, Cambridge.

Cayan, D.R., M.D. Dettinger, H.F. Diaz, and N.E. Graham. 1998. Decadal variability of precipitation over western North America. *Journal of Climate* 11:3148–3166.

Cottam, W.P. 1937. Has Utah lost claim to the lower Sonoran zone? *Science* 85:563–564.

Davis, S., J. Sperry, and U. Hacke. 1999. The relationship between xylem conduit diameter and cavitation caused by freezing. *American Journal of Botany* 86:1367–1372.

Delworth, T., S. Manabe, and R.J. Stouffer. 1993. Interdecadal variations of the thermohaline circulation in a coupled ocean-atmosphere model. *Journal of Climate* 6:1993–2011.

Dettinger, M.D., M. Ghil, C.M. Strong, W. Weibel, and P. Yiou. 1995. Software expedites singular-spectrum analysis of noisy time series, *Eos, Trans. American Geophysical Union* 76(2):12, 14, 21.

Dettinger, M.D., D.R. Cayan, G.J. McCabe, Jr. and K.T. Redmond. 2000. Winter-spring 2001 United States streamflow probabilities based on anticipated neutral ENSO conditions and recent NPO status. *Experimental Long-Lead Forecast Bulletin* 9:55–60.

Dettinger, M.D., D.S. Battisti, R.D. Garreaud, G.J. McCabe, and C.M. Bitz. 2001. Inter-hemispheric effects of interannual and decadal ENSO-like climate variations on the Americas. Pages 1–16 in V. Markgraf, editor, *Present and Past Interhemispheric Climate Linkages in the Americas and their Societal Effects.* Academic Press, San Diego, California.

Enfield, D.B., and A.M. Mestas-Nuñez. 1999. Multiscale variabilities in global sea surface temperatures and their relationships with tropospheric climate patters. *Journal of Climate* 12:2719–2733.

Feller, W. 1968. *An introduction to probability theory and its applications.* Vol. 1. Wiley, New York.

Gardner, J.L. 1951. Vegetation of the creosotebush area of the Rio Grande valley in New Mexico. *Ecological Monographs* 21:379–403.

Gedalof, Z., and D.J. Smith. 2001. Interdecadal climate variability and regime-scale shifts in Pacific North America. *Geophysical Research Letters* 28:1515–1518.

Gerschnov, A., and T.P. Barnett. 1998. Interdecadal modulation of ENSO teleconnections. *Bulletin of the American Meteorological Society* 79:2715–2725.

Grissino-Mayer, H.D., C.H. Baisan, and T.W. Swetnam. 1997. *A 1,374-year reconstruction of annual precipitation for the southern Rio Grande Basin.* Final Report, Department of Defense, Legacy Program, Ft. Bliss, Texas.

Grover, H.D., and H.B. Musick. 1990. Shrubland encroachment in southern New Mexico, U.S.A.: An analysis of desertification processes in the American Southwest. *Climatic Change* 17:305–330.

Hare, S.R., S. Minobe, W.S. Wooster, and S. McKinnell. 2000. An introduction to the PICES symposium on the nature and impacts of North Pacific climate regime shifts. *Progress in Oceanography* 47(2-4):99–102.

Harrington, J.A., R.S. Cerveny, and R.C. Balling, Jr. 1992. Impact of the Southern Oscillation on the North American southwest monsoon. *Physical Geography* 13:318–330.

Hastings, J.R., and R.M. Turner. 1965. *The changing mile. An ecological study of vegetation change with time in the lower mile of an arid and semiarid region.* University of Arizona Press, Tucson.

Hayden, B.P. 1998. Ecosystem feedbacks on climate at the landscape scale. *Philosophical Transactions of the Royal Society London* B, 353:5–18.

Hennessy, J.T., R.P. Gibbens, J.M. Tromble, and M. Cardenas. 1983. Vegetation changes from 1935 to 1980 in mesquite dunelands and former grasslands of southern New Mexico. *Journal of Range Management* 36:370–374.

Herbel, C.H., F.N. Ares, and R.A. Wright. 1972. Drought effects on a grassland range. *Ecology* 53:1084–1094.

Higgins, R.W., and W. Shi. 2000. Dominant factors responsible for interannual variability of the summer monsoon in the southwestern United States. *Journal of Climate* 13(4):759–776.

Holbrook, N. M., and M.A. Zwieniecki. 1999. Embolism repair and xylem tension: Do we need a miracle? *Plant Physiology* 120:7–10.

Holt, R.D., and T.H. Keitt. 2000. Alternative causes for range limits: A metapopulation perspective. *Ecology Letters* 3:41–47.

Humphrey, R.R. 1958. The desert grassland. A history of vegetational change and an analysis of causes. *Botanical Review* 24:193–251.

Humphrey, R.R. 1987. *90 years and 535 miles: Vegetation changes along the Mexican border*. University of New Mexico Press, Albuquerque.

Kahya, E., and J.A. Dracup. 1993. U.S. streamflow patterns in relation to the El Niño/Southern Oscillation. *Water Resources Research* 29:2491–2503.

Keymer, J. E., P.A. Marquet, J.X. Velasco-Hernandez, and S.A. Levin. 2000. Extinction thresholds and metapopulation persistence in dynamic landscapes. *American Naturalist* 156:478–494.

Latif, M., and T.P. Barnett. 1996. Decadal climate variability over the North Pacific and North America: Dynamics and predictability. *Journal of Climate* 9:2407–2423.

Linton, M.J., J.S. Sperry, and D.G. Williams. 1998. Limits to water transport in *Juniperus osteosperma* and *Pinus edulis*: Implications for drought tolerance and regulation of transpiration. *Functional Ecology* 12:906–911.

Mainguet, M. 1994. *Desertification: Natural background and human mismanagement*. Springer-Verlag, New York. 314 pp.

Mantua, N.J., and S.R. Hare. 2002. The Pacific Decadal Oscillation. *Journal of Oceanography* 58:35–44.

Martínez-Vilalta, J., and W.T. Pockman. 2002. The vulnerability to freezing-induced xylem cavitation of *Larrea tridentate* (Zygophyllaceae) in the Chihuahuan desert. *American Journal of Botany* 89:1916–1924.

McCabe, G.J., and M.D. Detinger. 1999. Decadal variations in the strength of ENSO teleconnections with precipitation in the western United States. *International Journal of Climatology* 19:1069–1079.

McPherson, G.R., H.A. Wright, and D.B. Wester. 1988. Patterns of shrub invasion in semiarid Texas grasslands. *American Midland Naturalist* 120:391–397.

Minobe, S. 1997. A 50-70 year climatic oscillation over the North Pacific and North America. *Geophysical Research Letters* 24:683–686.

Minobe, S. 1999. Resonance in bidecadal and pentadecadal climate oscillations over the North Pacific: Role in climatic regime shifts. *Geophysical Research Letters* 26:855–858.

Mitchell, V.L. 1976. The regionalization of climate in the western United States. *Journal of Applied Meteorology* 15:920–927.

Molles, M.C., Jr., and C.N. Dahm. 1990. A perspective on El Niño and La Niña: Global implications for stream ecology. *Journal of the North American Benthological Society* 9:68–76.

Molles, M.C., Jr., C.N. Dahm, and M.T. Crocker. 1992. Climatic variability and streams and rivers in semi-arid regions. Pages 197–202 in R.D. Robarts and M.L. Bothwell, editors, *Aquatic ecosystems in semi-arid regions: Implications for resource management*. Environment Canada, Saskatoon, Canada.

Neilson, R.P. 1986. High resolution climatic analysis and southwest biogeography. *Science* 232:27–34.

Palmer W.C. 1965 *Meteorological drought*. Research Paper Number 45, U.S. Department of Commerce Weather Bureau. Washington, D.C.

Parks, J.A., J.S. Dean, and J.L. Betancourt, in press, Tree rings, drought and the Pueblo Abandonment of south-central New Mexico in the 1670s. In D.E. Doyel and J.S. Dean, editors, *Environmental Change and Human Adaptation in the Ancient Southwest*. University of Utah Press, Salt Lake City, Utah.

Peng, C.S., and N. Buras. 2000. Practical estimation of inflows into multireservoir system. *Journal of Water Resources Planning and Management-ASCE* 126:331–334.

Pockman, W.T., and J.S. Sperry. 1997. Freezing-induced xylem cavitation and the northern limit of *Larrea tridentata*. *Oecologia* 109:19–27.

Pockman, W.T., and J.S. Sperry. 2000. Vulnerability to cavitation and the distribution of Sonoran Desert vegetation. *American Journal of Botany* 87:1287–1299.

Redmond, K.T., and R.W. Koch. 1991. Surface climate and streamflow variability in the western United States and their relationship to large-scale circulation indices. *Water Resources Research* 27:2381–2399.

Regensberg, A. 1996. General dynamics of drought, ranching and politics in New Mexico, 1953–1961. *New Mexico Historical Review,* January:25–49.

Schlesinger, W.H., J.F. Reynolds, G.L. Cunningham, L.F. Huenneke, W.M. Jarrell, R.A. Virginia, and W.G. Whitford. 1990. Biological feedbacks in global desertification *Science* 247:1043–1048.

Schneider, N., and A.J. Miller. 2001. Predicting western North Pacific Ocean climate. *Journal of Climate* 14:3997–4002.

Sperry, J.S., and M.T. Tyree. 1988. Mechanism of water stress-induced xylem embolism. *Plant Physiology* 88:581–587.

Sperry, J.S., K.L. Nichols, J.E.M. Sullivan, and S.E. Eastlack. 1994. Xylem embolism in ring-porous, diffuse-porous and coniferous trees of northern Utah and interior Alaska. *Ecology* 75:1736–1752.

Swetnam, T.W., and J.L. Betancourt. 1990. Fire–Southern Oscillation relations in the southwestern United States. *Science* 249:1017–1020.

Swetnam, T.W., and J.L. Betancourt. 1998. Mesoscale disturbance and ecological response to decadal climatic variability in the American Southwest. *Journal of Climate* 11:3128–3147.

Thomas, H.E. 1963. General summary of the effects of the drought in the Southwest: Drought in the Southwest, 1942–1956. *U.S. Geological Survey Professional Paper* 372-H.

Tyree, M.T., and J.S. Sperry. 1989. Vulnerability of xylem to cavitation and embolism. *Annual Review of Plant Physiology and Molecular Biology* 40:19–38.

Valentine, K.A., and J.J. Norris. 1964. A comparative study of soils of selected creosotebush sites in southern New Mexico. *Journal of Range Management* 17:23–32.

Villalba, R., R.D. D'Arrigo, E.R. Cook, G.C. Jacoby, and G. Wiles. 2001. Decadal-scale climate variability along the extratropical western coast of the Americas: Evidence from tree-ring records. Pages 155–172 in Vera Markgraf, editor, *Interhemispheric Climate Linkages*. Academic Press, San Diego.

Williams, J.E., S.D. Davis, and K. Portwood. 1997. Xylem embolism in seedlings and resprouts of *Adenostoma fasciculatum* after fire. *Australian Journal of Botany* 45:291–300.

Yeakley, J.A., R.A. Moen, D.D. Breshears, and M.K. Nungesser. 1994. Response of North

American ecosystem models to multi-annual periodicities in temperature and precipitation *Landscape Ecology* 9:249–260.

York, J.C., and W.A. Dick-Peddie. 1969. Vegetation changes in southern New Mexico during the past hundred years. Pages 155–166 in W.G. McGinnies and B.J. Goldman, editors, *Arid Lands in Perspective*. University of Arizona Press, Tucson.

The Interdecadal Timescale—Synthesis

Douglas G. Goodin
Maurice J. McHugh

The five chapters of part III provide a broad overview of decadal-scale climate processes and their ecological effect in a variety of ecosystems. Written by authors with disciplinary backgrounds that encompass climatology, biometeorology, and ecology, the chapters range from cross-site climate analysis with little direct attention to ecosystem effects (e.g., McHugh and Goodin, chapter 11; Hayden and Hayden, chapter 14) to more intensive studies of direct climate/ecological interaction at single sites or over more defined geographical areas (e.g., Greenland, chapter 13; Juday et al., chapter 12; Milne et al., chapter 15). Separately, each of these chapters contributes to understanding some aspect of the interaction of climate and ecology. As an integrated whole, they encapsulate many of the cross-disciplinary problems confronted by LTER scientists as they explore the interaction of climate and ecology. Despite the widely varying topics addressed and the disparate backgrounds of the contributors, similar themes emerge in each of the chapters. Here, we elucidate these themes and place them within the framework questions that have guided this volume.

Climatic Themes in Decadal-Scale Ecosystem Variability

Climatologists have long recognized the existence of cyclical or quasi-cyclical modes or patterns in the global circulation system. Typically, these patterns are characterized by variation in the strength or position of semipermanent pressure centers within the global circulation system. These variations occur at timescales ranging from seasonal to decadal, and such variability is frequently invoked as a causal mechanism for climatic trends or fluctuation at these various timescales. A variety of indexes have been constructed to characterize these pressure patterns and the teleconnections that result from them (see van Loon and Rogers 1978, Rogers 1984, and Trenberth and Hurrell 1994 for in-depth discussion of the derivation and interrelationships of atmospheric circulation indices). Evidence of some of these patterns recurs throughout each of the chapters, suggesting their importance in decadal-scale climate/ecology interactions at LTER sites.

Although the chapters in this section concentrate on interdecadal variability, climate variability is a multiscale phenomenon in both space and time. Several authors acknowledge this, notably Milne et al. (chapter 15), McHugh and Goodin (chapter 11), and Greenland (chapter 13). Each of these chapters notes the importance of nondecadal variations, particularly the El Niño–Southern Oscillation (ENSO) phenomenon. While acknowledging the presence and interaction of circulation patterns at multiple timescales, in this Synthesis we will concentrate on those that operate at the decadal timescale (i.e., cycles of 10–30 years) on which this section is organized.

Two circulation indices, the North Atlantic Oscillation (NAO) and Pacific Decadal Oscillation (PDO) emerge as principal modes of interdecadal-scale climate variability. Each of the indexes is defined in terms of the seasonal behavior of large Northern Hemisphere semipermanent pressure centers. The NAO indexes pressure changes associated with the Icelandic high-pressure system, whereas the PDO indexes the Aleutian low-pressure system. Both the NAO and PDO affect atmospheric and ocean circulation in the Northern Hemisphere, manifest as trends in dominance of meridional or zonal atmospheric flow with associated tendencies for variation in moisture and energy advection. Thus NAO and PDO phenomena affect temperature and precipitation, often at great distances from the centers of pressure themselves.

Effects of the Pacific Decadal Oscillation are apparent in nearly all of the interdecadal patterns discussed in these chapters. Greenland (chapter 13) notes that over a 60-year period from 1925 to 1985, temperature at the HJA LTER and salmon catch in the ocean off the Pacific Northwest are inversely correlated. He also notes that the time series of both of these variables suggests a cycle of approximately 20 years (noted by the author with caution, since the relative brevity of the data time series precludes strong conclusions in the absence of true cycles). Both of these cycles appear related to the PDO, although the links among PDO, temperature, and salmon harvest are indirect. Greenland attributes this close link between climate and ecosystem response to a five-level cascade coupling ocean and atmosphere variability to salmon harvest via nutrient availability.

Juday et al. (chapter 12) also note a strong relationship between PDO effects and ecosystem response, particularly the response of white spruce in the Alaskan interior. Using dendrochronological surrogates for temperature and precipitation, Juday et al. showed that quasi-decadal cycles in tree growth occurred throughout the nineteenth and twentieth centuries. The timing and periodicity of these fluctuations strongly suggest a link to the PDO, although the authors point out that, like the PDO, the North Atlantic Oscillation affects boreal zone climate through alteration of storm tracks and advection patterns. An interesting aspect of Juday et al.'s analysis is that the use of biological data (tree rings, in this case) as surrogates for climatic conditions implicitly assumes the closeness of the climate and ecology link. Juday et al. use this relationship to link reconstructed temperature at one site (Fairbanks, Alaska) to tree-ring pattern at the BNZ LTER. The coherency of results between the two sites allow these authors to integrate the effect of known features of the Alaskan interior climate system with dynamics of the upland white spruce ecosystem.

Juday et al.'s results in chapter 12 also illustrate the interaction of climate processes at multiple timescales. They note that favorable years for seed cone drop often correspond to poor radial growth years and that these key events are frequently associated with ENSO effects. ENSO, a shorter term (quasi-quintennial) phenomenon, is linked to PDO, such that ENSO represents a deepening and expansion of the Aleutian low-pressure center whose behavior is indexed by the PDO (Ropelewski and Halpert 1986).

Milne et al. (chapter 15) also used dendrochronological records to infer climate variability at the Sevilleta LTER site in the southwestern United States. Using a 393-year (1598–1991) record, they note an interdecadal periodicity of about 59–62 years, corresponding to drought cycles in the U.S. Southwest. As is the case in the Alaskan interior, climatic and ecological response in the arid southwest is linked to ENSO effect, although Milne et al. note that this regional linkage has a strong seasonal component—summer precipitation is not strongly influenced by ENSO, but fall, winter, and spring rainfall are. Although not explicitly examined by the authors of chapter 15, it is quite likely that other Pacific teleconnection processes influence the Sevilleta site. They note that fall, winter, and spring storms originate in the Pacific and are guided by zonal jet stream flows. Thus, processes influencing the balance of zonal/meridional flow, such as the strength and position of north Pacific pressure centers are likely to influence precipitation at this site. A link to indexes related to Pacific flow such as PDO and the Pacific North American index (PNA, an index of zonal to meridional flow over North America) can be hypothesized.

McHugh and Goodin's spectral analysis of temperature and precipitation at several LTER sites (chapter 11) also shows the influence of several well-known circulation patterns. These authors used Principal Components Analysis to decompose time series of mean growing season maximum/minimum temperature and precipitation into their principal modes of variability, then evaluated the proportion of variance occurring at interdecadal timescales using power spectrum analysis. Therefore, McHugh and Goodin's results reflect the temporal behavior across the entire LTER network, allowing individual sites to be placed within the context of the network. This approach also permits analysis of the geography of the network.

As in other chapters in this section, McHugh and Goodin found evidence of periodic behavior at timescales other than decadal. Quasi-quintennial variability resembling the ENSO signal occur in two of the mean temperature principal components, as well as in the maximum temperature component. Significant spectral power is also found at periods of 50 years (resembling the timescale of precipitation variability observed by Milne et al. (chapter 15), but there was little evidence of significant cycles at decadal timescales. This contrasts with results from individual sites (e.g., chapters 12, 13, 15), where decadal-scale cycles are present. Examination of periodic patterns at individual sites in the context of the whole network could reveal important conclusions about the geography of climate variability across the LTER network. The analysis of storm frequency by Hayden and Hayden (chapter 14) also suggests the importance of network geography in cross-site analysis. In addition, Hayden and Hayden's analysis showed little evidence of a link between storm frequency and ENSO across the network as a whole, nor did significant links with decadal-scale teleconnections emerge. Their results did show geographic patterns of storm frequency change, with the greatest change occurring in the western and central LTER sites. Like the findings of McHugh and Goodin, these results provide a geographic framework in which to consider the spatial aspects of climate variability and climatic change at individual LTER sites.

Discussion of the Framework Questions

The chapters in part III provide some insight into the framework questions identified at the outset of the book. The concluding chapter of this volume (chapter 21) will provide a comprehensive summary of each chapter in terms of the framework questions; in this section we will consider a few observations related to the chapters dealing with the interdecadal scale.

Most of the chapters were successful in identifying some type of climate variability, usually of a periodic or quasi-periodic type. Greenland (chapter 13) found that Coho salmon catch and temperature in the Pacific Northwest are related and that their variation seems to be linked to periodic pressure changes in the Pacific Ocean. Juday et al. (chapter 12) found evidence of similar climate variation in their analysis of white spruce in the Alaskan interior. They also attributed this variation to periodic change in Pacific pressure systems. Milne et al. (chapter 15) and McHugh and Goodin (chapter 11) found longer term variability in climatic signals: Milne et al. at a single site (Sevilleta LTER), and McHugh and Goodin across the entire network. Although all report some type of variability, of the five chapters in part III, only Hayden and Hayden (chapter 14) report any evidence of a trend.

Scale dependence is a theme that emerges from these chapters. Although nominally devoted to decadal-scale climate dynamics, each chapter includes some discussion of variability at scales other than decadal. Given the complexity of the climate system and the number of variables influencing climate, it is perhaps not surprising that strict adherence to a given time framework is not practical. In addition to this temporal-scale dependence, it is also noteworthy that in those chapters in which multiple sites were considered (i.e., chapters 11 and 14), a spatial-scale dependence could be noted. Hayden and Hayden (chapter 14) were able to geographically stratify their sites based on storm frequency parameters. McHugh and Goodin (chapter 11) noted the strength of association between individual sites and climate dynamics across the entire network, thus establishing geographic patterns of site variability.

In the chapters where the interactions between climate and ecosystems are directly considered, questions concerning preexisting conditions and cascade effects prove relevant. In chapter 13, Greenland provides a clear example of system cascade, outlining a five-step model linking atmospheric circulation pressure features in the northern Pacific to salmon harvest off the coast of Washington and Oregon. Greenland's cascade model clearly links an important circulation pattern (the PDO) to an ecosystem response (salmon harvest). His model also shows that the observed inverse correlation between temperature and salmon harvest, which motivated the investigation, is not direct, but rather both are responses to a remote driving force.

Juday et al.'s analysis of spruce tree dynamics in Alaska (chapter 12) shows the importance of both system cascades and preexisting conditions. Juday et al. note that warm, dry years are necessary to "condition" the white spruce reproductive system for extensive cone production in years of favorable climate—a dependence on short-term preexisting conditions. Juday et al. also present a type of cascade model in which a series of key events or "gateways" must be passed before the reproduction process can begin. The thresholds implied by these gateways represent a type of system cascade in which flows of material or energy are controlled by "stocks" and "regulators." These systems are noted for their ability to often produce unexpected transient behavior (Chorley and Kennedy 1971).

Milne et al. (chapter 15) also demonstrate a system cascade in their analysis of drought and its ecosystem effects in New Mexico. This cascade may be distinguished by the presence of significant local effects along with larger scale climate variability—an example of the scale effect in climate/ecosystem interaction. The local feedback mechanisms are in the form of terpenes released by desert vegetation, which establishes localized heating via a small-scale "greenhouse effect." This localized heating provides a positive feedback mechanism within the cascade and reinforces ecosystems changes associated with climate variability.

Conclusion

The chapters in part III provide an overview of climate research issues at individual LTER sites and over larger geographic areas. Several themes emerge, including scale dependence, nested effects, geographical effect, complexity of interaction, cascade effects, and persistence effects. These topics lie within the theme of this book, and they illustrate the various modes of research needed to evaluate climate/ecosystem interaction. These chapters also suggest avenues of further research.

Clearly, more investigation is needed to refine our understanding of the mechanisms by which climate and ecosystems interact. Greenland (chapter 13) notes that a full understanding of the system cascade by which climate and salmon harvest are related will require a more comprehensive model than the five-level process outline in chapter 13. Milne et al.'s analysis of shrub dynamics in central New Mexico (chapter 15) and Juday et al.'s analysis of spruce reproduction (chapter 12) also provide examples of system cascades where further understanding of climate/ecosystem interaction will refine explanatory models.

Another recurring theme in this synthesis is the interaction among climate processes at multiple time and space scales. Investigations of the interaction of one climate process (e.g., PDO, ENSO) with one or a few ecosystem parameters have been conducted, but systematic evaluation of how the processes quantified by the various indexes (e.g., NAO, PDO, PNA, etc.) interact is less commonly undertaken. Systematic investigations of the effect of multiple, nested climate processes on ecosystem response are needed, perhaps using indexes derived from existing time series of climate indicators.

References

Chorley, R.J., and Kennedy, B.A. 1971. *Physical Geography: A Systems Approach.* London: Prentice-Hall.

Rogers, J.C. 1984. The association between the North Atlantic Oscillation and the Southern Oscillation in the Northern Hemisphere. *Monthly Weather Review* 112:1999–2015.

Ropelewski, C.F., and Halpert, M.S. 1986. North American precipitation and temperature patterns associated with the El Niño/Southern Oscillation (ENSO). *Monthly Weather Review* 114:2352–2362.

Trenberth, K.E., and Hurrell, J.W. 1994. Decadal atmosphere-ocean variability in the Pacific. *Climate Dynamics* 9:303–309.

van Loon, H., and Rogers, J.C. 1978. The seesaw in winter temperatures between Greenland and Northwestern Europe. I, General description. *Monthly Weather Review* 106:296–310.

Part IV

Century to Millennial Timescale

Introductory Overview

Raymond C. Smith
Douglas G. Goodin

Elias argues (chapter 18, p. 370) that ecosystems are shaped by environmental changes that have occurred over thousands of years so that the century to millennial timescale is of particular significance because "it is on these timescales that ecosystems form, break apart, and reform in new configurations." Within this context, the authors for the three chapters in part IV evaluate evidence for climate variability since the Last Glacial Maximum (LGM) to the present. They evaluate the biological responses to these longer term changes and highlight the importance of past climatic conditions on current ecosystem function. If we view, as Elias does, glacial climate as a filter through which ecosystems have passed, then variability since the LGM comprises a significant fraction of the biotic history that shaped current ecosystems. This is an overriding theme for this section.

Fountain and Lyons (chapter 16), examining a dry valley ecosystem in Antarctica (MCM), evaluate various proxy records to establish the historic context of their landscape. They argue that this historical context is important for a full understanding of ecosystems and that it is especially important for the MCM ecosystem. Providing an excellent example of legacy, the effect of past imprints on current ecosystem function, they present evidence that past climatic variations truly dictate current ecosystem status. During the LGM, ice blocked the current Taylor Valley, forming a lake that contained phytoplankton and algal mats. Subsequent warming eliminated the blockage, drained the large lake, forming several smaller ones, and established the current landscape. The former large lake supplied nutrients to the soil and current lakes. Fountain and Lyons (p. 334) state that "the vital importance of climatic legacy in the dry valleys is due to its extreme environment, low biodiversity, and short food chains." They also observe a "polar amplification," whereby the sharp solid/liquid phase transition of water allows small changes in climate to produce relatively large variations in ecosystem response.

The Jornada Long-Term Ecological Research site (JRN) is representative of the desert shrubland and desert grassland ecosystems of the southwestern United States. Monger (chapter 17) makes use of a range of biotic (packrat middens, fossil pollen), abiotic (chronological data on lake levels, position of alpine glaciers and rock glaciers) and soil-geomorphic evidence to create a working hypothesis of the bioclimatic changes during the last 20,000 years. There is a remarkable consistency in these proxy estimates given their diversity. This proxy evidence is merged with historical and measured climate variability to address the question, How has this ecosystem responded to climate variability? Evidence points to both climate and human land use as having important, and interacting, impacts on this ecosystem.

Elias (chapter 18) discusses millennial and century climate changes in the Colorado Alpine (NWT). Using remains of beetles as indirect evidence for past environmental conditions, Elias constructs a temperature history of the Colorado Alpine since the LGM. Based on the assumption that the present climatic tolerance range of a species can be applied to its Quaternary fossil record, he builds a technique such that a fossil occurrence of a given species is used to imply a paleotemperature within the same tolerance range. This evidence provides a temperature reconstruction for this area during the past 24,000 years. Elias notes that there are conflicting interpretations of insect, pollen, and archeological data during the mid-Holocene interval, and he suggests the need for additional regional studies to clarify reconstruction of this period.

The three chapters in this section, although they represent a diverse set of ecosystems, provide climatic reconstructions since the Last Glacial Maximum. They thereby describe regional biotic and landscape history that leads to the postglacial environments of the Holocene and our current ecosystem conditions.

16

Century- to Millennial-Scale Climate Change and Ecosystem Response in Taylor Valley, Antarctica

Andrew G. Fountain

W. Berry Lyons

Introduction

The view of climate change during the Pleistocene and the Holocene was very much different a mere decade ago. With the collection and detailed analyses of ice core records from both Greenland and Antarctica in the early and mid-1990s, respectively, the collective view of climate variability during this time period has changed dramatically. During the Pleistocene, at least as far back as 450,000 years B.P., abrupt and severe temperature fluctuations were a regular occurrence rather than the exception (Mayewski et al. 1996, 1998; Petit et al. 1999). During the Pleistocene, these rapid and large climatic fluctuations, initially identified in the ice core records, have been verified in both marine and lacustrine sediments as well (Bond et al. 1993; Grimm et al. 1993), suggesting large-scale (hemispheric to global) climate restructuring over very short periods of time (Mayewski et al. 1997). Similar types of climatic fluctuations, but with smaller amplitudes, have also occurred during the Holocene (O'Brien et al. 1995; Bond et al. 1997; Arz et al. 2001). What were the biological responses to these changes in temperature, precipitation, and atmospheric chemistry? We must answer this question if we are to understand the century- to millennial-scale influence of climate on the structure and function of ecosystems.

Because the polar regions are thought to be amplifiers of global climate change, these regions are ideal for investigating the response of ecological systems to, what in temperate regions might be considered, small-scale climatic variation. Our knowledge of past climatic variations in Antarctica comes from different types of proxy records, including ice core, geologic, and marine (Lyons et al. 1997). It is clear, however, that coastal Antarctica may respond to oceanic, atmospheric, and

Figure 16.1 Landsat image of the McMurdo Dry Valleys. Both the east Antarctic Ice sheet in the bottom left of the photo and the Ross Sea ice in the top right of the photo are covered by snow. The darker shades identify the bare soil in the Dry Valleys.

ice sheet–based climatic "drivers," and therefore ice-free regions, such as the McMurdo Dry Valleys, may respond to climate change in a much more complex manner than previously thought (R. Poreda, unpubl. data 2001). Since the initiation of the McMurdo Dry Valleys Long-Term Ecological Research program (MCM) in 1993, there has been a keen interest not only in the dynamics of the present day ecosystem, but also in the legacies produced via past climatic variation on the ecosystem. In this chapter we examine the current structure and function of the dry valleys ecosystem from the perspective of our work centered in Taylor Valley. From this understanding we examine the changes in the ecosystem in response to climatic changes for the past 27,000 years and highlight the importance of past climatic conditions on current ecosystem functioning.

Site Description

The McMurdo Dry Valleys Long-Term Ecological Research (MCM) site is located on the edge of the Antarctic continent (77.5° S, 163° E) in a much larger region known as the Southern Victoria Land (figure 16.1). Only 2% of the Antarctic is ice free (Drewery et al. 1982) and the McMurdo Dry Valleys is the largest ice-free region on the continent with about 2000 km^2 of snow-free area (Chinn 1988). The valleys owe their existence to the Transantarctic Mountains that block the ice flow from the East Antarctic Ice Sheet (figure 16.1). A few lobes of the ice sheet penetrate the mountains and reach the valleys. In addition, numerous small alpine gla-

Figure 16.2 Western Taylor Valley looking west. Lake Bonney is at the lower left. Note the ice-free margin (moat) at the lower end of the lake. Central in the background is the Hughes Glacier, flowing off the Kukri Hills. To the extreme left is the Sollas Glacier, and the shadow on the right of the glacier is a cinder cone, which was used by Wilch et al. (1993) to date the glacier activity. (Photograph by Thomas Nylen)

ciers form in the mountains and flow to the valley floor. The valleys are characterized by a rocky-sandy soil devoid of vascular vegetation, thus yielding a stark landscape (figure 16.2). Perennially ice-covered lakes, which are fed from ephemeral streams originating as glacial meltwater, are present in nearly all the valleys (Fountain et al. 1999). The polar climate of the region experiences continual darkness in midwinter and continual sunlight in midsummer. Air temperatures average about −17°C, with the winter minimum about -40°C and the summer maximum a few degrees above freezing (Clow et al. 1988).

Precipitation occurs as snow and can fall at any time of the year. Annual values of snowfall are about 10 cm water equivalent (Keys 1980), most of which is lost to sublimation (Chinn 1993). The biology of the dry valleys exhibits spectacularly low biodiversity and short food chains (Priscu et al. 1999; Virginia and Wall 1999). The soil communities are limited to a few phyla, including rotifers, tardigrades, nematodes, protozoans, fungi, and bacteria (Freckman and Virginia 1998). Nematodes represent the largest predator in the valleys. The streams support communities of cyanobacteria, eukaryotic algae, and mosses (McKnight et al. 1999). The lakes host only microorganisms and benthic microbial mats (Wharton et al. 1993, Priscu et al. 1999). Small microbial communities inhabit natural melt holes (cryoconite holes) in the glaciers (Wharton et al. 1985) and in the lake ice (Priscu et al. 1998). Although

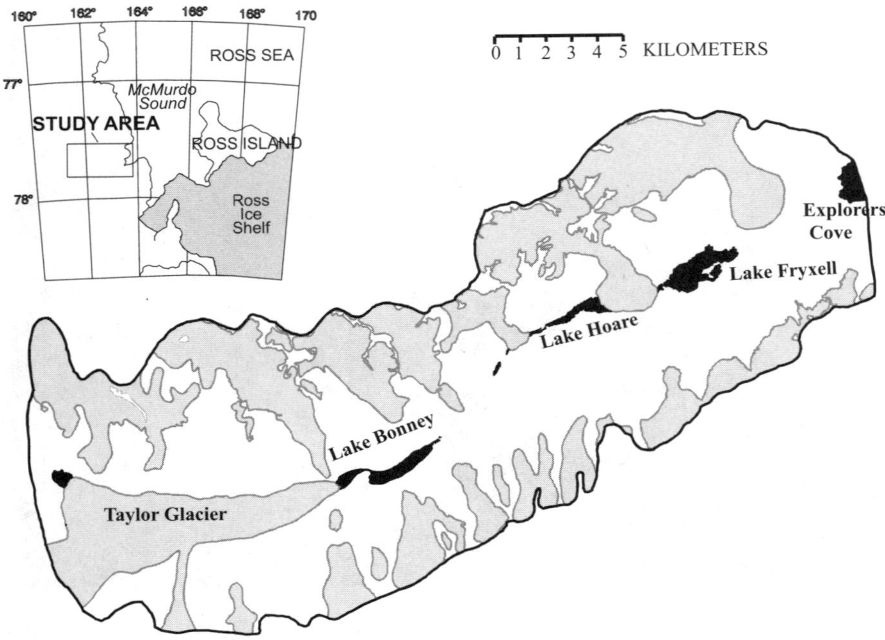

Figure 16.3 Map of Taylor Valley

life exists at higher latitudes in Antarctica, the dry valleys host one of the last functioning terrestrial ecosystems where streams, lakes, and soils are interconnected both physically and biologically.

The MCM is primarily focused on Taylor Valley, an east- to west-trending valley about 34 km long and 12 km wide (figure 16.3). To the east, Taylor Valley is open to McMurdo Sound, and to the west it is blocked by Taylor Glacier, which flows from the East Antarctic Ice Sheet (figure 16.1). Elevations range from sea level to 2000 m in the mountains that form the northern and southern boundaries of the valley. Elevations of the valley floor range from sea level to about 60 m. The MCM was initiated in 1993 and research compilations can be found in Priscu (1998) and in BioScience (1999). To understand how climatic change affects the Taylor Valley ecosystem, a more complete physical description of the valley, ecosystem, and their interaction is required.

Physical Setting

Structurally, the dry valley region has acted as a rigid block since the early Tertiary, 50 million years ago (Fitzgerald et al. 1986). Geotectonic uplift in Taylor Valley (Wilch et al. 1993), which is important to ecological processes in the valley, has been relatively slow for the past 2.57 million years. Throughout the Pliocene, small cinder cones erupted in the intermediate regions of middle Taylor Valley and apparently produced little or no ash deposits. Because the cones are located high on the valley sides, generally above 300 m in elevation, and well away from the valley floor, their role in valley ecology is probably small and is not considered here. Dry valley uplift has received close attention because of the possible ramifications for changing regional climate (Behrendt and Cooper 1991). More recent examination by Wilch et al. (1993) argues that, based on their textural characteristics, the cinder cones could not have erupted under water. Therefore, the age and elevation of the cones provide limiting values for the uplift. If uplift occurred at all over the past 2.6 million years, it could not have exceeded 300 m.

A geomorphological model divides the dry valleys into coastal, intermediate, and interior regions (Marchant and Denton 1996). These divisions roughly correspond, at least since mid-Pliocene times (Wilch et al. 1993), to low, intermediate, and high elevations. The low-elevation coastal areas exhibit modern soil movement (e.g., solifluction, ice wedge polygonal patterning), and the depth to ice-cemented sediment is less than 50 cm. The landforms and soils are less than 12,000 years old because they are formed from late-Wisconsin glacial deposits (addressed next). Midelevation intermediate zones exhibit comparatively low soil moisture. Glacier meltwater is infrequent in this zone, and consequently streamflow is uncommon except during rare extremes of warm air temperatures. Water-induced slope movement (solifluction) is limited to areas with a potential water source such as near glaciers or annually forming snow patches. The landscape of intermediate zones is not dynamic, and climatic conditions favor preservation of desert pavements and sand-wedge polygons. Based on potassium-argon dating of in situ ash deposits, slope movement has been minimal in central Taylor Valley during the past 7.1 million years (Marchant and Denton 1996). The high-elevation interior zone, found above 800 m, has virtually no soil moisture, and meltwater is entirely absent. There is no active layer over permafrost because soils lack ice. Dated ash deposits, trapped and buried in thermal contraction cracks in the soil or preserved under desert pavements, exhibit ages of up to 10–15 million years (Marchant and Denton 1996). The sedimentary structure of the ash layer indicates no slope movement or contact with water since the time of deposition. Taylor Valley is completely encompassed by zones 1 and 2 because of its relative proximity to the coast compared to other dry valleys. Nonetheless, the age of the landscape surfaces is quite old compared to its temperate counterparts: It ranges in age from 12,000 years at the valley bottom to 7.1 million years on the upper valley walls.

Since the start of the Pleistocene, significant surface modification in the lower elevations (zone 1) of Taylor Valley, like all of the dry valleys, has been by glacial activity, including water runoff from glaciers. The soils of zone 1, the valley lowlands, formed from glacial deposits or by events directly resulting from glacier activity. Rates of eolian modification seem to be small. Marchant and Denton (1996) date near-surface ash deposits to 10–15 million years in the windy high-elevation zones, indicating that desert pavements greatly inhibit significant wind erosion or deposition. The soils of zone 1 contain marine and lacustrine organic matter at the lower and intermediate levels, and endolithic sources dominate the organic matter in the higher elevations (Burkins et al. 2000).

Establishing the historic context of landscapes is crucial to understanding all ecosystems (Swanson et al. 1988), but this is particularly the case in the MCM where past climatic variations truly dictate current ecosystem status. Because of its polar location, the primary disturbances in the MCM ecosystem have been climatic, and the landscape pattern has been primarily dictated by climatic, not biotic, processes.

Climate

The dry valleys are considered a polar desert. As in much of Antarctica, precipitation is very low. Measurements in Wright Valley, adjacent to Taylor Valley, show that the annual snowfall ranged from 0.6 to 10 cm water equivalent (Bromley 1985). Precipitation is greatest nearest the coast and decreases inland (Keys 1980; Fountain et al. 1999). Based on snow depths measured along the floor of Taylor Valley after a summer snowstorm and on snow depths measured on Taylor Valley glaciers at 200–300 m in elevation (Fountain et al. 1999), precipitation accumulation decreased at a gradient of −0.06 cm km^{-1} (water equivalent) with distance from the coast. Typically, snow sublimates before melting, making little contribution to the hydrology. Only in places where snow accumulates to significant depths and is protected from the winds, such as swales, stream channels, or along glacier margins, does it make a contribution, if transient, to the hydrology of the valley.

In Taylor Valley, temperatures range from a few degrees above freezing in late December or January to about −45°C in winter (Clow et al. 1988, Doran et al. 2002b). Average annual temperature in the valley ranges from −16°C to −20°C. One of the important factors that control temperature is the katabatic winds that flow off the East Antarctic Ice Sheet. These foehn-type winds adiabatically warm as the flow from higher elevations and can raise air temperatures by 10 degrees within 15–20 minutes. Thus, winter temperatures are also partly controlled by the frequency of katabatic winds (Doran et al. 2002b). In addition to warming, these winds dramatically reduce the humidity and significantly increase sublimation from the ice surfaces (Clow et al. 1988). Aside from katabatic effects, the valleys are typically windy. Monthly average wind speeds range from 2 to 4 m s^{-1} in Taylor Valley (Clow et al. 1988).

Glaciers

Almost all of the glaciers in the dry valleys are relatively small alpine glaciers, a few km^2 in area. In Taylor Valley, the glaciers originate from the Asgard Range on the north side and from the Kukri Hills on the south side (figure 16.1), where snow accumulation exceeds loss by sublimation and wind erosion. Taylor Glacier is the largest glacier in the valley, defining its western boundary, and flows from the East Antarctic Ice Sheet. Roughly one-third of Taylor Valley, as defined from the ridge divide of each mountain range, is ice covered (Fountain et al. 1998). The glaciers are polar glaciers with bases frozen to substrate. The larger glaciers terminate in vertical ice cliffs about 20 m high. Unlike the ice of temperate glaciers, the ice is remarkably clean and free of debris. Because the ice is well below freezing point, meltwater is restricted to the glacier surface and flows off the glacier edge. These characteristics contrast with alpine glaciers in the temperate latitudes, which are at their melting point throughout and are not frozen to their bed. Instead, they slide over the substrate. Meltwater enters the body of temperate glaciers, flows internally, and finally emerges from subglacial tunnels at the glacier edge (Fountain and Walder 1998).

The mass exchange of these polar glaciers is relatively small, compared to temperate glaciers. Our observations indicate that about 10 to 30 cm of snow accumulates in the upper zones and about 6 to 15 cm is lost from the ablation zone. These values are consistent with results from previous studies in the adjacent Wright Valley (Bull and Carnein 1970; Chinn 1980). The snow in the upper reaches of the glaciers is cold and dry, and no snowmelt has been observed directly. Mass loss in the upper snow zone is by sublimation because no melting occurs. In the lower elevation (ice-exposed) zone of the glacier, the mass loss is dominated by sublimation and melting. Results from the Canada Glacier indicate that during the summer, sublimation accounts for 40–80% of the ice mass loss, the remainder being lost to melting (Lewis et al. 1998). Since 1993, meltwater production has been limited to the lower fringe of the glaciers within a few hundred meters of the ice edge. Certainly, no meltwater has been observed on the glaciers at elevations above 250–500 m.

Hydrology

Meltwater flows off the glacier to form streams at the base of the ice cliffs. These streams are ephemeral and typically flow for 4–10 weeks a year (McKnight et al. 1999). They transport water, sediment, and nutrients to terminal lakes (figure 16.4). As previously mentioned, snowfall on the valley floor does not contribute significantly to the streams because it usually sublimates before melting (Chinn 1980). However, winter drifts of snow piled against the glacier termini or in stream channels contribute to streamflow in early spring before disappearing by early summer (Fountain et al. 1998). The streams are channelized and flow over continuous permafrost, which occurs at shallow depths of a few tens of centimeters (McKnight et al. 1999). Therefore, groundwater flow is probably limited to the near-surface hyporheic zones, the saturated zone adjacent to and under the stream channel. Because of the shallow depth and lack of lateral groundwater inflow, the hyporheic zone can extend laterally for several meters on either side of the channel.

The glacial meltwater that feeds the streams typically has a very low solute content, on the order of 1–10 micro Molar (μM) per chemical species (Lyons et al. 1998). The solute content of the glaciers is controlled, in part by the chemistry of the snow accumulation but is otherwise dominated by recycled solutes blown onto the glacier surfaces from the valley floor (Lyons et al. in press). Preliminary analysis of the spatial pattern of glacier meltwater chemistry suggests that it is controlled by the pattern of sediment on the glacier (M. Tranter, unpubl. data, 2001). The mass flux of the stream water increases by one to two orders of magnitude before it reaches the lake as a result of salt dissolution and chemical weathering processes in the channel and within the hyporheic zone (Lyons et al. 1998; Gooseff 2002). Evaporative losses in the stream channel also help to increase the concentration of solutes (B. Vaughn, unpubl. data, 1993).

Three main lakes occupy the bottom of Taylor Valley (lakes Fryxell, Hoare, and

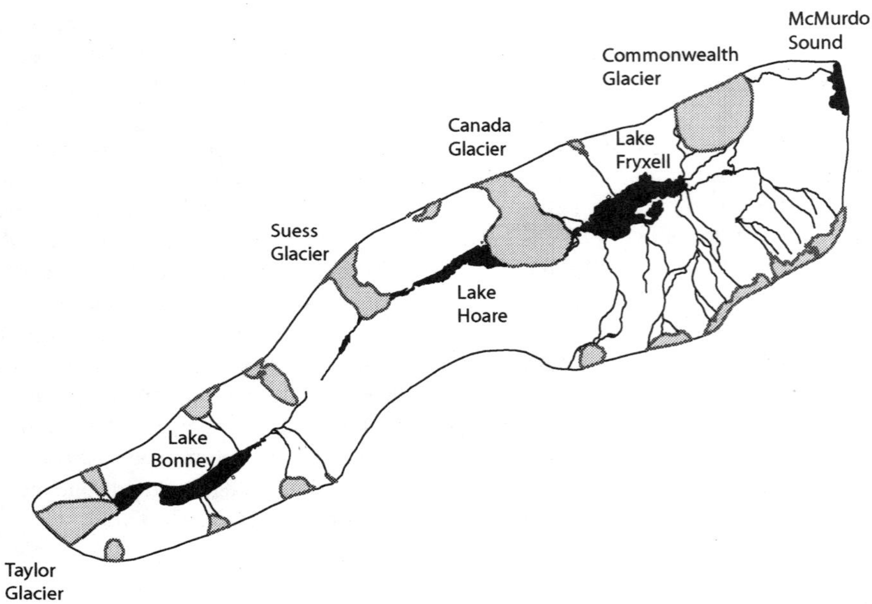

Figure 16.4 Map of the streams in Taylor Valley, draining from the glacier (gray) into the lakes (black). McMurdo Sound is the ocean area.

Bonney) in addition to numerous smaller lakes and ponds that dot the landscape. All lakes and ponds have permanent ice cover (3–6 m thick), and the smaller water bodies are probably completely frozen during winter. As previously mentioned, the lakes are terminal, and lose water only through sublimation of the ice surface and by evaporation from the narrow fringe of open water around the perimeter of each lake in late summer. As with terminal lakes elsewhere (e.g., Great Salt Lake), these lakes are sensitive to small changes in water inflow. Since the beginning of measurement recording in the 1970s by Chinn (1993), generally all lakes in the dry valleys have been rising. Only recently have the lakes slowed or stopped rising (figure 16.5). For Lake Hoare, the level has dropped over the past few years in response to cooler summer weather and increased summer snowfall (Fountain et al. 1999; Doran et al. 2002a). Because the perennial ice effectively covers the lake water, there are no wind-generated currents (Hawes 1983). The exchange of gases between the lake water and the atmosphere is therefore restricted (Wharton et al. 1986), and light penetration into the lake water is reduced (Howard-Williams et al. 1998). The major lakes in Taylor Valley are density stratified because of salinity gradients (Spigel and Priscu 1998). Some mixing occurs in summer at the lake edge, where the ice melts and localized turbulence can propagate into the lake (Miller and Aiken 1996). The motion of the lake water is limited to slow horizontal movement. The melted lake fringe also allows a limited exchange of gases with the atmosphere. In addition, streamflow enters the lake through the melted fringe, supplying the lake with nutrients and water (Tyler et al. 1998).

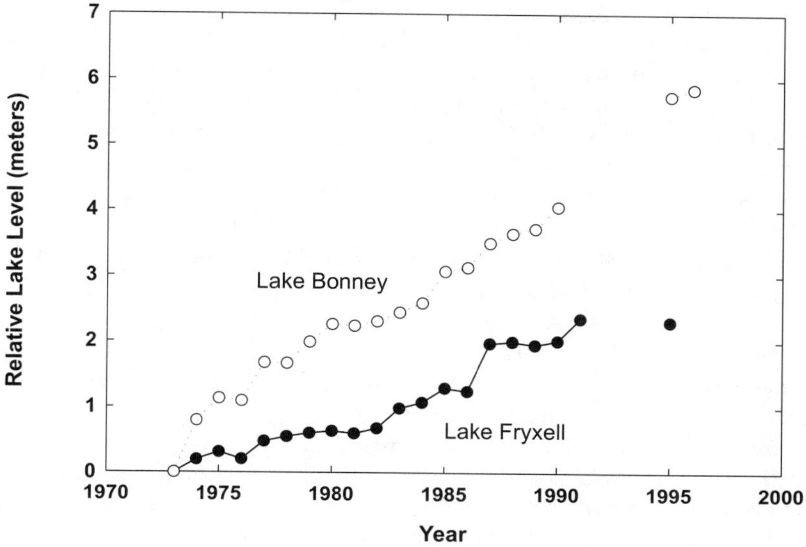

Figure 16.5 Lake level change in Taylor Valley.

Soils

A pebble to boulder (1 to 20 cm diameter) surface pavement covers a coarse, but variable, soil texture with little cohesion and little organic content (Bockheim 1997; Campbell et al. 1998). Permafrost is ubiquitous in the valleys, and in soils it is typically found under a 10- to 30-cm-thick active layer. In Taylor Valley, the soil texture is dominated 95–99% by coarse-grained sand (Burkins et al. 2000). The high variability of the soil texture results from glacial action in the valley, as described subsequently. The organic content of the soils does not exceed 0.03% by weight of organic carbon content (Burkins et al. 2000). The moisture content of the soil is low and varies with respect to the distance from the coast. In the coastal region, soil moisture averages about 1% by weight; farther inland on the valley floor soil moisture drops to 0.5% (Campbell et al. 1998). Smaller scale patterns of soil moisture depend on proximity to water sources such as lakes and ephemeral streams. In addition, winter snow accumulations, which are typically thin and quite patchy, provide a transient water source. However, the effect of these sources does not extend beyond 10 m (Campbell et al. 1998; McKnight et al. 1999), and 95% of the soil does not receive liquid water (Campbell et al. 1998). Moisture for these soils depends on either brief accumulations of snow, which typically sublimate before melting, or sublimation of ice-cemented permafrost at depth (Campbell et al. 1998; McKay et al. 1998).

Ecosystems

Ecosystems in each of the dry valleys is relatively isolated because high ice-capped mountains separate the valleys and water drains internally within each valley. The lack of transport between valleys is suggested by mitochondrial DNA analysis of the soil nematode, *Scottnema lindsayae*. It was thought that the genetics of this nematode would be fairly uniform because it is easily dispersed by winds and streams, but soil samples from five different valleys exhibited significant mitochondrial differences, suggesting that the gene flow among the populations is restricted (Courtright et al. 2000). Within valleys, linkages among soil, stream, and lake ecosystems are relatively weak compared to valleys in temperate regions. In Taylor Valley, overland flow and groundwater are nonexistent, therefore the soils are connected to the streams or lakes only through the hyporheic zone interactions during the few weeks of streamflow each year.

One integrating process, although weak, is the wind. Eolian processes transport sediment and organic material across the valley. The wind entrains organisms (Virginia and Wall 1999) from the soil and erodes algal mats and mosses from the stream channels (McKnight et al. 1999). Benthic algal mats in the lakes rise from the lake bottom as a result of the buoyancy caused by accumulated gases in the mats (Parker et al. 1982) and become frozen in the lake ice at the surface. Through continual freezing on the bottom and sublimation at the top, these mats are transported through the ice and become exposed at the surface where they are wind eroded. All these materials are transported across the valley and are redeposited in soils, stream channels, and lakes. It is common to find bits of algal mat on Commonwealth Glacier at 300 m in altitude. We hypothesize a net transport of sediment and organic material down valley toward the coast due to the high velocity wind events caused by katabatic winds flowing off the ice sheet. Although a down-valley gradient of increasing organic carbon is observed (Virginia and Wall 1999), local modifications, as explained by climatic events (which we discuss later in this chapter), also can contribute to this trend.

The polar climate of the dry valleys includes a long winter of darkness, very cold temperatures, and no surface water. During this period the linkages between the ecosystems are severed. Certainly winds continue to disperse organic matter, but the ecosystems are no longer integrated and the only functioning ecosystem, the lakes, is completely isolated. Each system adopts a strategy to survive the winter in isolation. Therefore, in such an environment the biological linkages between systems must necessarily be weak because they are severed seasonally. The soil nematodes enter a state of anhydrobiosis, which allows them to survive a winter of no water and subfreezing temperatures (Crowe 1971). They have been known to survive in this state for over 60 years (Freckman 1986). Stream mosses and algal mats also become freeze-dried after the meltwater supply ceases and the streams stop flowing. After rewetting, some mats began to photosynthesize in 10–20 minutes (Vincent and Howard-Williams 1986; Hawes et al. 1992). Recent experiments show that stream algal mats that have not experienced water for 25 years reactivated a full microbial ecosystem over a 1.5-km stream reach within 2 weeks of rewetting (McKnight et al. 1999). In winter, the lakes become isolated from the external environment as the fringe of open water around the lakes and within the perennial lake ice freezes. For the rest of the winter, the lake environments are entirely isolated from any exchange with the outside. Moreover, little light exists in the polar autumn and spring, whereas complete darkness occurs in the middle of winter. Phytoplankton in the lakes, which depend on light for photosynthesis, turn to alternative sources of energy. Studies on Antarctic lakes, particularly those in the dry valleys, show that some phytoplankton species will ingest bacteria, a source of organic carbon, during low light conditions (Roberts and Laybourn-Parry 1999; Laybourn-Parry et al. 2000). This mixotrophic strategy allows the phytoplankton to endure the winter darkness. Our understanding of midwinter biologic processes in the lakes is incomplete because of the logistical limitations that to date have precluded over-winter studies. In spring phytoplankton growth is triggered solely by increased solar radiation because vertical mixing is largely absent. Biomass and phytoplankton production generally increase throughout the summer through late January.

Century- to Millennial-Scale Climate Changes

Climate change in the dry valleys has been largely inferred from geomorphic evidence of past glacier positions (e.g., Stuiver et al. 1981; Denton et al. 1989) and lake level heights (e.g., Stuiver et al. 1981; Hall and Denton 1995, 2000). In addition, profiles of chemical concentrations in the lakes have been used to infer past lake drawdowns (Wilson 1964; Lyons et al. 1998). More recently, isotopic results from ice cores (Steig et al. 2000) and temperature measurements in boreholes in the bedrock (G. Clow, unpubl. data, 1993) and in glaciers (Clow and Waddington 1996) have been used to more directly infer climate changes in the region. We summarize the millennial-scale climate changes based on all this evidence, using the ice core data as our baseline chronology.

An ice core was obtained from Taylor Dome, about 140 km from Taylor Valley. Although a number of cores have been obtained from the continent (e.g., Jouzel et al. 1987; Morgan et al. 1997; Mayewski et al. 1996; Legrand and Mayewski 1997), this particular core is closest to the dry valleys and provides better information on regional climatic variations than cores from much farther away. In fact, the results from the Taylor Dome core contrast with results from cores at Vostok and Byrd (Blunier et al. 1997, 1998), which reside on opposite sides of the Antarctic continent away from Taylor Dome. Climatic data inferred from the other cores are out of phase with that from Greenland, whereas data from the Taylor Dome core are in phase. These differences have been attributed to the spatially variable deep ocean currents that transport heat between the southern ocean and the lower latitudes (Steig et al. 1998).

A variety of geochemical analyses have been completed on the Taylor Dome core, but we concentrate here on those results of importance to the climate of the dry valleys as summarized by Steig et al. (2000). Figure 16.6 shows the inferred air temperatures from oxygen isotopes and the snow accumulation record from the ice core. Starting about 60 thousand (kyr) ago (calendar years) a slow cooling began, which culminated in the last glacial maximum. During the same time period, snow accumulation decreased, reaching a minimum during the last glacial maximum. Rapid warming, known as the Bølling-Allerød event, occurred about 15 kyrs ago at the termination of the glacial period. Subsequently, the Younger-Dryas cooling

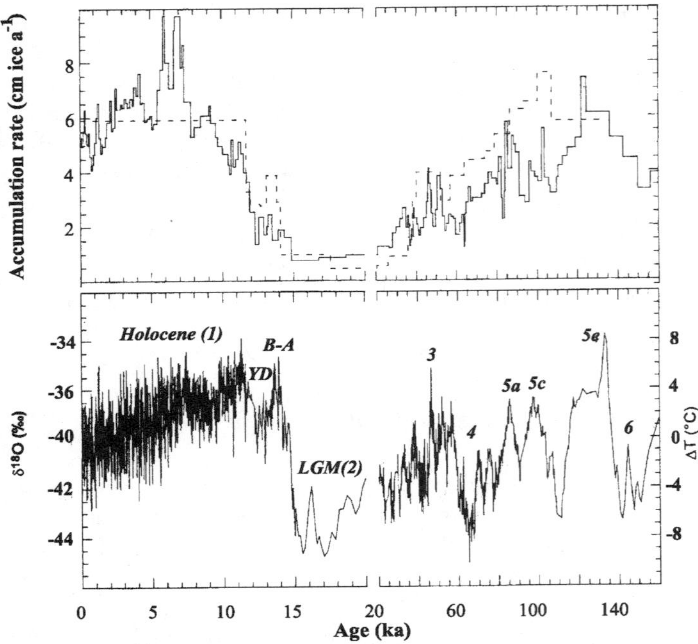

Figure 16.6 Top: Accumulation rate in cm of ice per year (cm ice a⁻¹) at Taylor Dome based on ¹⁰Be data. Bottom: Oxygen isotope values (δ¹⁸0) from Taylor Dome (left axis) and inferred temperature change (ΔT) (right axis). The timescale is in thousands of years before present. Note the shift in timescale at 20,000 years. These data are from Steig et al. 2000, figure 7 and figure 6, respectively.

event dropped air temperatures by about 4°C before returning to the warmer Bølling-Allerød values by about 11 kyr ago. Substantial cooling events occurred at 9.5 kyrs and 6.5 kyrs ago. From the Bølling-Allerød event through the cooling event of 6.5 kyrs ago, accumulation generally increased. A general cooling trend and decreasing accumulation has dominated the late Holocene. Details of the temperature trend for the late Holocene (past few kyrs) are not available from the Taylor Dome ice core at this time because of potential nontemperature influences in the isotope data (Waddington, pers comm., 2000). Fortunately, subsurface earth temperatures were measured in a borehole in Taylor Valley, and based on temperature variations with depth, a record of surface temperatures was obtained (Clow 1998). A general cooling persisted in the valley from about 4 kyrs ago to about 1 kyr. Since that time the air temperatures have warmed.

These climatic trends have caused changes in the regional ice sheet extent and in lake levels. With the cooling of the last glacial maximum, the Antarctic Ice Sheet began to enlarge and advance. By about 23.8 (C¹⁴) kyrs ago (~ 27 kyrs), the Ross Ice Shelf entered Taylor Valley from the ocean (Hall and Denton 2000; Stuiver et al. 1981). The ice shelf blocked the seaward opening of the valley and dammed an inland lake, known as Lake Washburn, to a level of about 200 m deep. The lake existed until about 8.3 (C¹⁴) kyrs ago (~9.5 kyrs) when the ice shelf receded from the valley and the lake waters drained from the valley. The valley has cooled 1.2°C since about 4 kyrs ago; this probably led to the major drawdown of the lakes in the dry valleys, which reached their maximum drawdown 0.9–1.2 kyrs ago (Wilson 1964; Lyons et al. 1998). Lake Hoare either completely desiccated or did not exist prior to ~2 kyrs ago (Lyons et al. 1998). Since that time, the lakes have refilled as air temperatures increased about 2°C in the past 1000 years. These recent changes in the valleys are well correlated with other inferred climate changes in the region. For example, the "penguin optimum," associated with rookery abundance and a warmer climate, occurred between 3 and 4 kyrs ago and terminated about 3 kyrs ago (Baroni and Orombelli 1994). The rookeries were reoccupied starting about 1200 years ago.

Humans first visited the dry valleys in 1903 when Robert F. Scott's western journey included a side trip down the Taylor Glacier to Lake Bonney. Members of the party measured the width of the lake at its narrowest point, which was later used by Chinn (1993) to estimate a lake level. Since 1903 the lakes have risen, indicating a generally warmer climate. The trend has continued from the 1970s, when Chinn (1993) started taking measurements, through the 1980s. Thinning of the lake ice suggest a 2°C warming during the 1980s, although other factors may have been important (McKay et al. 1985). This warming may have caused increased melt and runoff, which in turn caused increasing lake levels. Between 1986 and 2000, air temperatures have decreased by 0.7°C per decade, runoff from glaciers has slowed, and the lake levels have generally stopped rising and in some cases are falling (Doran et al. 2002a; Welch et al., chapter 10 this volume).

Ecosystem Response

The formation of Lake Washburn about 27 kyrs ago created a lake that extended from the valley mouth to Taylor Glacier at the western end (figure 16.7). Phytoplankton in the water column and benthic algal mats apparently inhabited this enlarged lake because Burkins et al. (2000) found that the pattern of lacustrine and marine sources of organic carbon in soil transects correlated with the former lake extent. The total lacustrine biomass during the time of Lake Washburn was much larger than it is at present. The organic carbon from this biomass is still present in the soils as indicated by Burkins's work. Therefore, the current carbon production in the valley is very slow. Heretofore, it was thought that wind-transported organic carbon (lacustrine microbial mats, endolithic communities, and stream-based mats and mosses) in the valley was a major source of carbon transfer. Although real-time observations suggest that this is true, the significance of this process is disputed by the correlation of the lacustrine-derived organic carbon and the extent of Lake Washburn (Burkins et al. 2000). Given the lack of higher plants and animals, soil organisms are apparently sustained by the organic carbon in the soils left by Lake Washburn. Therefore, the dry valley lakes are connected to the soils across time rather than across space.

Once the Ross Ice Shelf retreated from Taylor Valley about 9.5 kyrs ago, the ice no longer blocked the valley, and Lake Washburn drained, leaving smaller lakes

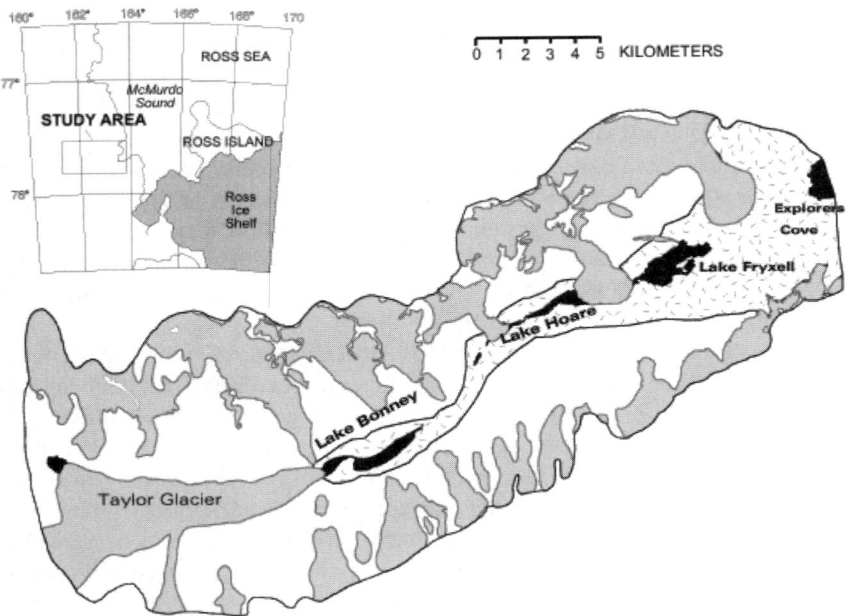

Figure 16.7 Extent of Lake Washburn and the distribution of organic carbon in Taylor Valley.

ponded in the low spots along the valley floor. The climate began to cool about 4 kyrs ago and the streamflow from the glaciers slowed below that required to maintain the lakes. Because of this, the lakes began to decrease in size. They reached a minimum by ~1000 years ago. If Lake Hoare existed prior to that time, it had completely evaporated and the residual salts were blown away (Lyons et al. 1998). We presume that the stream ecosystems were largely absent and the soil ecosystem had been significantly reduced by this time. Whether all terrestrial biota entered a long-term anhydrobiosis cannot be ascertained. Lakes Fryxell and Bonney evaporated down to a brine pool, concentrating the salts and nutrients such as organic and inorganic carbon. Since that time, the climate warmed and the greater meltwater fluxes increased in the lake. The fresh meltwater did not mix with the concentrated brine pools, due to density differences, and floated over the top, trapping the brine at depth. Diffusion fluxes from the concentrated brines at depth have been used to obtain the refilling age of ~1 kyr bp (Wilson 1964; Lyons et al. 1998).

Studies of the phytoplankton in Lake Bonney indicate that its productivity and biomass increase when solar radiation penetrates the ice cover early in the season (Priscu et al. 1999). This increase first occurs in the shallow waters just underneath the ice surface. Overall, less than 0.1% of the dissolved carbon pool in the trophogenic zone is contributed by streamflow (Priscu et al. 1999). As light intensity increases with the approach of the austral summer solstice, the main region of productivity shifts from the surface 15 m down the water column to the bottom of the trophogenic zone (figure 16.8). This shift results from sufficient light reaching

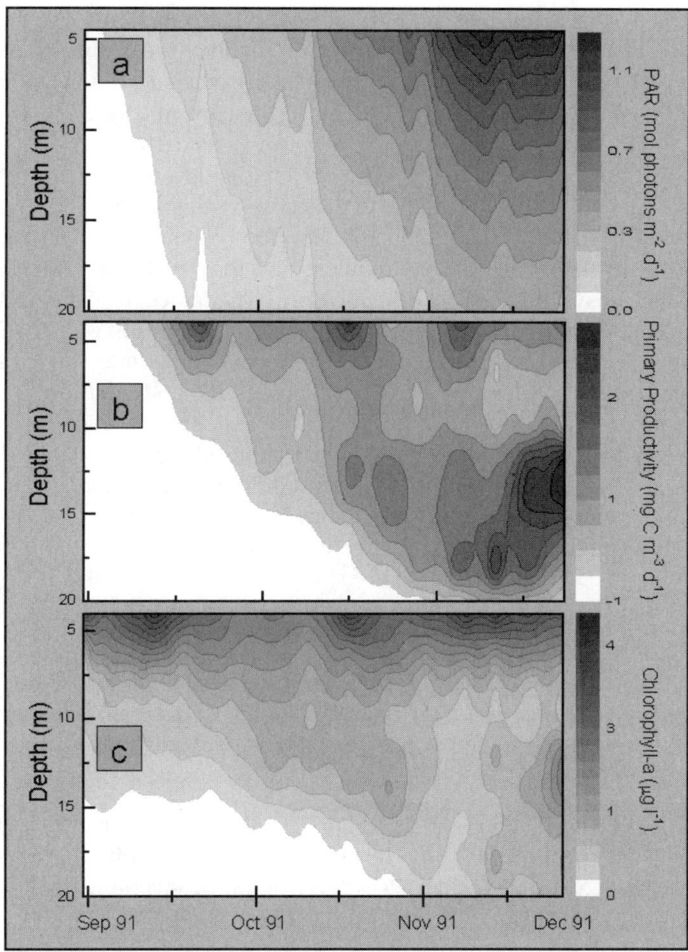

Figure 16.8 Water column properties for the east lobe of Lake Bonney. (a) Photosynthetically available radiation. (b) Primary productivity, (c) Chlorophyll a. (From Priscu et al. 1999)

the depth of relatively high flux of inorganic and organic carbon, through molecular diffusion, from the pools of concentrated nutrients below (Priscu 1995; Priscu et al. 1999). We presume that the phytoplankton continue to be productive at this depth into the Antarctic autumn after the streams stop flowing. The depth of productivity may change to more shallow levels again in response to low light levels of the Antarctic winter. Unfortunately, we lack measurements during this period and during the winter because of logistical constraints. In any case, it is clear that the current lake ecosystem is strongly conditioned by past climatic events. The drawdown of the lakes about 1000 years ago concentrated the lake nutrients into pools on which modern phytoplankton now depend.

The effect of past imprints on current ecosystem functioning is known as a legacy (Vogt et al. 1997). In our case, the legacy involves the change in climate and its impact on lake size. Formerly large lakes supply nutrients to lowland soils, and the remnants of paleolakes supply nutrients to the current lakes. The vital importance of climatic legacy in the dry valleys is due to its extreme environment, low biodiversity, and short food chains. The extreme polar environment reduces the amount of water and energy available to the ecosystem, greatly slowing the rate of nutrient cycling. The low biodiversity and short food chains make the ecosystem directly dependent on the physical environment such that few buffers exist and the response of the ecosystem to slight climatic change is immediate. This is illustrated by the current decrease in lake primary productivity as a result of a cooling in climate (Doran et al. 2002a). The cooling trend decreased the magnitude of meltwater produced on the glaciers, which in turn decreased the streamflow and nutrient flux to the lakes. Lake levels have stopped rising and, in some cases, are falling. It should be emphasized that the ice core and borehole temperature data indicate that these massive changes in lake size have been driven by annual changes of 2°C or less. Again, we emphasize that in more temperate settings these are small changes, but in the polar environment these relatively small variations are amplified to bring about very large environmental responses. In the Taylor Valley this is due, in part, to the fact that this ecosystem is so closely dependent on the change of state of water and that, during the austral summer, very small temperature changes can greatly influence the hydrologic system.

General Ecosystem Questions and Summary

At the Workshop on Century- to Millennial-Scale Climate Change and Ecosystem Response, as part of the Long-Term Ecological Research program's All Scientist Meeting in Snowbird, Utah (2–4 August 2000), several general ecosystem questions were asked. These are summarized here for the McMurdo Dry Valleys LTER.

What preexisting conditions affect the significance of the climatic event? Our food web is populated by microbiota responding to microenvironmental variations that change significantly over short distances and over time. The lack of higher plant communities tie our terrestrial ecosystems more closely to geologic structures than in more temperate regions. Also, a nutrient-poor environment with weak spatial linkages and slow geomorphic change enhances the linkages over time (legacy) (Morehead et al. 1999).

Are the effects direct or cascading? Our current understanding is that the effects are direct because of the short food chains. Poor linkages that would otherwise buffer the direct effects are absent.

Is the ecosystem response completed before the next event? Events of different magnitudes occur over different timescales, each with their own response time. However, the current legacy of the Lake Washburn event (27,000 years ago) is contemporaneous with the current legacy of the lake drawdown 1200 years ago. Given the extremely slow cycling of nutrients and the pace of geomorphic change, we suspect that ecosystem responses are overprinted on each other and are not completed before the next event occurs.

Does the system return to its original state after the event? We don't actually know, but the answer depends on whether we are currently in a state of perturbed static equilibrium, whereby a constant state exists to which the system can return, or whether the system is in a state of dynamic equilibrium, where there is no constant state but the system varies about some constant value. Given the relatively short record of direct observations, the patchy record of past events, and the slow rate of change in the valleys, it is impossible to tell at this time. However, we can say that the current system is perturbed, given the important legacies recently revealed.

Conclusions

The ecosystem of the McMurdo Dry Valleys exhibits a strong and clear dependence on past climatic conditions. Relatively large variations in climate have concentrated nutrients that cannot be attained under "normal" or static conditions. Because of the nutrient-poor, energy-limited environment of the dry valleys, past concentrations of nutrients play a major role in current ecosystem structure and function. This finding draws attention to the need for understanding past climatic conditions, at least in extreme polar environments, to interpret current ecosystems. The ecosystems within the dry valleys are certainly poorly linked, and the soils and lakes appear to behave completely independently. Only through our understanding of legacy do we realize that the systems are linked across time in a very real and vital manner.

In addition, we observe a polar amplification whereby small changes in climate produce large variations in ecosystem response. The energy limitation in the dry valleys keeps much of the potential water supply locked in the frozen reservoirs of the surrounding glaciers. Small changes in the summer surface energy balance of the glaciers, resulting in the surface temperature rising above the melting point, create a large water flux. Conversely, a small change that results in lowering the temperature below the melting point significantly reduces water production. The summer temperatures of the dry valleys hover around the melting point, and this highly nonlinear response to changes in temperature is a normal condition of the valleys. Although our examination of past climatic effects on ecosystem response does not explicitly address this process, we know from our field experiences that this nonlinear response is at the heart of all observed changes. The creation of Lake Washburn must have occurred when local energy levels allowed the ice temperature to warm and sustain melting. Conversely, the energy levels during lake drawdown must have dropped sufficiently to reduce the ice temperatures below the freezing point for much of the summer periods.

What do these research findings have to offer toward understanding ecosystems in more energy- and water-rich environments? Legacies, including climatic, anthropogenic, and volcanic events, are probably important at all spatial and temporal scales. The significance of each legacy is dependent on the magnitude of the change and response time of the landscape and ecosystem to accommodate that change and return the entire system back to preexisting conditions. Perhaps by understanding the impact and residence time of natural changes, we can anticipate the effect of anthropogenic change. In addition, our research may help to frame how life might survive extreme events such as a bollide impact or a "snowball" earth condition. When ecological linkages are severed, some resilience is provided by legacy effects that store nutrients that would otherwise not be available.

Acknowledgments This chapter would not have been possible without the insight and labors of our LTER colleagues and graduate students. This chapter is dedicated to them. We also appreciate the superb effort of Antarctic Support Associates, who provided much of the logistical support that made this project possible. Funding for this project was provided by the National Science Foundation, Office of Polar Programs, grant McMurdo Dry Valleys: A Cold Desert Ecosystem, OPP 9211773, OPP 9813061; and The Role of Natural Legacy on Ecosystem Structure and Function in a Polar Desert: The McMurdo Dry Valleys LTER Program OPP 9810219, OPP 0096250.

References

Arz, H. W., S. Gerhardt, J. Patzold, and U. Rohl. 2001. Millennial-scale changes of surface and deep water flow in the western tropical Atlantic linked to Northern Hemisphere high-latitude climate during the Holocene. *Geology* 29:239–242.

Baroni, C., and G. Orombelli. 1994. Abandoned penguin rookeries as Holocene paleoclimatic indicators in Antarctica. *Geology* 22:23–26.

Behrendt, J. C., and A. Cooper. 1991. Evidence of rapid Cenozoic uplift of the shoulder escarpment of the Cenozoic West Antarctic rift system and a speculation on possible climatic forcing. *Geology* 19:315–319.

BioScience. 1999. (Special section on the McMurdo Dry Valleys). *BiosScience* 49:959–1019.

Blunier, T., and 9 others, 1997. Timing of the Antarctic cold reversal and atmospheric CO_2 increase with respect to the Younger Dryas event. *Geophysical Research Letters* 24:2683–2686.

Blunier, T., and 10 others, 1998. Asynchrony of Antarctic and Greenland climate change during the last glacial period. *Nature* 394:739–743.

Bond, G., and 13 others, 1993. Evidence for massive discharges of icebergs into the North Atlantic Ocean during the last glacial period. *Nature* 360:245–249.

Bockheim, J. G. 1997. Properties and classification of cold desert soils from Antarctica. *Journal of the Soil Science Society of America* 61:224–231.

Bond, G., W. Showers, M. Cheseby, R. Lotti, P. Almasi, P. de Menocal, P. Priore, H. Cullen, I. Hajdas, and G. Bonani. 1997. A pervasive millennial-scale cycle in North Atlantic Holocene and glacial climates. *Science* 278:1257–1266.

Bromley, A. A. 1985. Weather observations: Wright valley, Antarctica. Informational Publication 11. New Zealand Meteorological Service. Wellington, New Zealand.

Bull, C., and C. R. Carnein. 1970. The mass balance of a cold glacier: Meserve Glacier, south Victoria Land, Antarctica. *International Association of Hydrological Sciences* 86:429–446.

Campbell, I. B., G. G. C. Claridge, D. I. Campbell, and M. R. Balks. 1998. The soil environment of the McMurdo Dry Valleys, Antarctica. In Priscu, J. C. (editor) *Ecosystem Dynamics in a Polar Desert.* American Geophysical Union. Antarctic Research Series 2:297–322.

Burkins, M. B., R. A. Virginia, C. P. Chamberlain, D. H. Wall. 2000. The origin of soil organic matter in Taylor Valley, Antarctica. *Ecology* 81:2377–2391.

Chinn, T. J. 1980. Glacier balances in the dry valleys area, Victoria Land, Antarctica. *International Association of Hydrological Sciences* 126: 237–247.

Chinn, T. J. 1988. The "Dry Valleys" of Victoria Land, in R. S. Williams, and J. G. Ferrigno. *Satellite image atlas of glaciers of the world, Antarctica.* U.S. Geological Survey Professional Paper 1386-B:B39–B41.

Chinn T. J. 1993. Physical hydrology of the Dry Valley lakes. Pages 1–52 in W. J. Green and E. I. Friedmann, editors. *Physical and Biological Processes in Antarctic lakes.* Antarctic Research Series 59. American Geophysical Union, Washington, DC.

Clow, G. D. 1998. Reconstructing past climatic changes in Greenland and Antarctica from borehole temperature measurements [abs]. *EOS—Transactions of the American Geophysical Union* 79:F833.

Clow, G. D., C. P. McKay, G. M. Simmons, Jr., and R. A. Wharton, Jr. 1988. Climatological observations and predicted sublimation rates at Lake Hoare, Antarctica. *Journal of Climatology* 1:715–728.

Clow, G. D., and E. D. Waddington. 1996. Acquisition of borehole temperature measurements from Taylor Dome and dry valleys for paleoclimate reconstruction. *Antarctic Journal of the United States* 31:71–72.

Courtright, E. M., D. H. Wall, R. A. Virginia, L. M. Frisse, J. T. Vida, and W. K. Thomas. 2000. Nuclear and Mitochondrial DNA Sequence Diversity in the Antarctic Nematode *Scottnema lindsayae. Journal of Nematology* 32:143–153.

Crowe, J. H., 1971. Anhydrobiosis: An unsolved problem. *American Naturalist* 105:563–571.

Denton, G. H., J. G. Bockheim, S. C. Wilson, and M. Stuiver. 1989. Late Wisconsin and early Holocene glacial history, inner Ross Embayment, Antarctica. *Quaternary Research* 31:151–182.

Doran, P. T., J. C. Priscu, W. B. Lyons, J. E. Walsh, A. G. Fountain, D. M. McKnight, D. K. Moorhead, R. A. Virginia, D. H. Wall, G. D. Clow, H. G. Fritsen, C. P. McKay, A. P. Parsons. 2002a. Antarctic climate cooling and terrestrial ecosystem response. *Nature* 415:517–520.

Doran, P. T., C. P. McKay, G. D. Clow, G. L. Dana, A. G. Fountain, T. Nylen, W. B. Lyons. 2002b. Climate observations from the McMurdo Dry Valleys, Antarctica, 1986–2000. *Journal of Geophysical Research-Atmospheres.* 107:1–12.

Drewery, D. J., S. R. Jordan, and E. Jankowski. 1982. Measured properties of the Antarctic ice sheet: Surface configuration, ice thickness, volume and bedrock characteristics. *Annals of Glaciology* 2:834–891.

Fitzgerald, P. G., M. Sandiford, P. J. Barrett, and A. J. W. Gleadow. 1986. Asymmetric extension associated with uplift in the Transantarctic Mountains and Ross Embayment. *Earth and Planetary Science Letters* 81:67–78.

Fountain, A. G, G. L. Dana, K. J. Lewis, B. L. Vaughn, and D. M. McKnight. 1998. Glaciers of the McMurdo Dry Valleys, Southern Victoria Land, Antarctica. Pages 65–76 in J. C. Priscu, editor. *Ecosystem Dynamics in a Polar Desert: The McMurdo Dry Valleys, Antarctica.* Antarctic Research Series 72, American Geophysical Union, Washington, D.C.

Fountain, A. G., and 12 others. 1999. Physical controls on the Taylor Valley ecosystem, Antarctica. *BioScience* 49:961–971.

Fountain, A. G., and J. S. Walder. 1998. Water flow through temperate glaciers. *Reviews of Geophysics* 36:299–328.

Freckman, D. 1986. The ecology of dehydration in soil organisms. Pages 157–168 in A. C. Deopold, editor. *Membranes, metabolism, and dry organisms.* Cornell University Press, Ithaca, New York.

Freckman, D., and R. Virginia. 1998. Soil biodiversity and community structure in the McMurdo Dry Valleys, Antarctica. Pages 323–336 in J. C. Priscu, editor. *Ecosystem Dynamics in a Polar Desert: The McMurdo Dry Valleys, Antarctica.* Antarctic Research Series 72, American Geophysical Union, Washington, D.C.

Gooseff, M. N., W. B. Lyons, A. E. Blum, and D. M. McKnight. 2002. Weathering reactions and hyporheic exchange controlling stream water chemistry in a glacial meltwater stream in the McMurdo Dry Valleys. *Water Resources Research,* 38:15-1–15-17.

Grimm, E.C., G. L. Jacobson, Jr., W. A. Watts, B. C. S. Hansen, and K. A. Maasch. 1993. A 50,000 record of climate oscillations from Florida and its temporal correlation with the Heinrich events. *Science* 261:198–200.

Hall, B., and G. H. Denton. 1995. Late Quaternary lake levels in the Dry Valleys, Antarctica. *Antarctic Journal of the United States* 30:52–53.

Hall, B., and G. H. Denton. 2000. Radiocarbon chronology of Ross Sea Drift, eastern Taylor Valley, Antarctica: Evidence for a grounded ice sheet in the Ross Sea at the last glacial maximum. *Geografiska Annaler* 82A:305–336.

Hawes, I. 1983. Turbulence and its consequences for phytoplankton development in two ice covered Antarctic lakes. *British Antarctic Survey Bulletin* 60:69–81.

Hawes, I., C. Howard-Williams, and W. F. Vincent. 1992. Desiccation and recovery of Antarctic cyanobacterial mats. *Polar Biology* 12:587–594.

Howard-Williams C., A. Schwarz, I. Hawes, and J. C. Priscu. 1998. Optical properties of lakes of the McMurdo Dry Valleys. Pages 189–204 in J. C. Priscu, editor. *Ecosystem Dynamics in a Polar Desert: The McMurdo Dry Valleys, Antarctica.* Antarctic Research Series 72, American Geophysical Union, Washington, D.C.

Jouzel, J., C. Lorius, J. R. Petit, C. Genthon, N. I. Barkov, V. M. Kotlyakov, and V. M. Petrov. 1987. Vostok ice core: A continuous isotope temperature record over the last climatic cycle (160,000 years). *Nature* 329:403–407.

Keys, J. R. 1980. *Air temperature, wind, precipitation and atmospheric humidity in the McMurdo region.* Department of Geology Publication No. 17 (Antarctic Data Series No. 9), Victoria University of Wellington, New Zealand.

Laybourn-Parry, J., E. C. Roberts, and E. M. Bell. 2000. Mixotrophy as a survival strategy among planktonic protozoa in Antarctic lakes. Pages 33–40 in W. Davidson, C. Howard-Williams, and P. Broady, editors. *Antarctic ecosystems: Models for wider ecological understanding.* Caxton Press, Christchurch, New Zealand.

Legrand, M., and P. A. Mayewski. 1997. Glaciochemistry of polar ice cores: A review. *Reviews of Geophysics* 35:219–243.

Lewis, K. J., A. G. Fountain, and G. L. Dana. 1998. Energy balance studies of Canada Glacier, Taylor Valley, McMurdo Dry Valleys, Antarctica. *Annals of Glaciology* 27:603–609.

Lyons, W. B., P. A. Mayewski, L. R. Bartek, and P. T. Doran. 1997. Climate history of the McMurdo Dry Valleys since the Last Glacial Maximum. Pages 15–22 in C. Howard-Williams, W. B. Lyons, and I. Hawes, editors. *Ecosystem Processes in Antarctic Ice-Free Landscapes.* A.A. Balkema, Rotterdam, Netherlands.

Lyons, W. B., S. W. Tyler, R. A. Wharton, Jr., D. M. McKnight, and B. H. Vaughn. 1998. A late Holocene desiccation of Lake Hoare and Lake Fryxell, McMurdo Dry Valleys, Antarctica. *Antarctic Science* 10: 247–256.

Lyons, W. B., K. A. Welch, A. G. Fountain, G. D. Dana, B. V. Vaughn, and D. M. McKnight, in press, Surface glaciochemistry of Taylor Valley, Southern Victoria Land, Antarctica, and its relationship to stream chemistry. *Hydrological Processes.*

Lyons, W. B., K. A. Welch, C. A. Nezat, D. M. McKnight, K. Crick, J. K. Toxey, J. A. Mastrine. 1998. Chemical weathering rates and reactions in the Lake Fryxell Basin, Taylor Valley: Comparison to temperate river basins. Pages 147–154 in C. Howard-Williams, W. B. Lyons, and I. Hawes, editors. *Ecosystem Processes in Antarctic Ice-Free Landscapes.* A.A. Balkema, Rotterdam, Netherlands.

Marchant, D. R., and G. H. Denton. 1996. Miocene and Pliocene paleoclimate of the Dry Valley region, southern Victoria Land: A geomorphological approach. *Marine Micropaleontology* 27:253–271.

Mayewski, P. A., and 13 others. 1996. Climate change during the last deglaciation in Antarctica. *Science* 272:1636–1638.

Mayewski, P.A., L. D. Meeker, M. S. Twickler, S. I. Whitlow, Q. Yang, W. B. Lyons, and M. Prentice. 1998. Major features and forcing of high-latitude northern hemisphere atmospheric circulation using a 110,000 year long glaciochemical series. *Journal of Geophysical Research* 102:26345–26366.

McKay, C. P., G. Clow, R. A. Wharton, Jr., and S. W. Squyres. 1985. Thickness of ice on perennially frozen lakes. *Nature* 313:561–562.

McKay, C. P., M. T. Mellon, and E. I. Friedmann. 1998. Soil temperatures and stability of ice-cemented ground in the McMurdo Dry Valleys, Antarctica. *Antarctic Science,* 10:31–38.

McKnight, D. M., D. K. Niyogi, A. S. Alger, A. Bomblies, P. A. Conovitz, and C. M. Tate. 1999. Dry valley streams in Antarctica: Ecosystems waiting for water. *BioScience* 49:985–995.

Miller, L. G., and G. R. Aiken. 1996. Effects of glacial meltwater inflows and moat freezing on mixing in an ice-covered Antarctic lake as interpreted from stable isotope and tritium distributions. *Limnology and Oceanography* 41:966–976.

Moorehead, D. L., P. T. Doran, A. G. Fountain, W. B. Lyons, D. M. McKnight, J. C. Priscu, R. A. Virginia, and D. H. Wall. 1999. Ecological legacies: Impacts on ecosystems of the McMurdo Dry Valleys. *BioScience* 49:1009–1019.

Morgan, V. I., C. W. Wolley, J. Li, T. D. van Ommen, W. Skinner, and M. F. Fitzpatrick. 1997. Site information and initial results from deep drilling on Law Dome, Antarctica. *Journal of Glaciology* 48:3–10.

O'Brien, S. R., P. A. Mayewski, L. D. Meeker, D. A. Meese, M. S. Twickler, and S. I. Whitlow. 1995. Complexity of Holocene climate as reconstructed from a Greenland ice core. *Science* 270:1962–1964.

Parker, B. C., G. M. Simmons, Jr., G. F. Love, R. A. Wharton, Jr., and K. G. Seaburg. 1982. Removal of organic and inorganic matter from Antarctic lakes by aerial escape of blue-green algal mats. *Journal of Phycology* 18:72–78.

Petit, J. R., and 18 others. 1999. Climate and atmospheric history of the past 420,000 years from the Vostok ice record, Antarctica. *Science* 286:2141–2144.

Priscu, J. C. 1995. Phytoplankton nutrient deficiency in lakes of the McMurdo dry valleys, Antarctica. *Freshwater Biology* 34:215–227.

Priscu, J. C. (Editor). 1998. *Ecosystem dynamics in a polar desert, the McMurdo Dry Valleys, Antarctica.* Antarctic Research Series 72. American Geophysical Union, Washington, D.C.

Priscu, J. C., C. H. Fritsen, E. E. Adams, S. I. Giovannoni, H. W. Paerl, C. P. McKay, P. T. Doran, D. A. Gordon, B. I. Lanoil, and J. L. Pinckney. 1998. Perennial Antarctic lake ice: An oasis for life in a polar desert. *Science* 280:2095–2098.

Priscu, J. C., C. F. Wolf, C. D. Takacs, C. H. Fritsen, J. Laybourn-Parry, E. C. Roberts, B. Sattler, and W. B. Lyons. 1999. Carbon transformations in a perennially ice-covered Antarctic lake. *BioScience* 49:997–1008.

Roberts, E. C., and J. Laybourn-Parry. 1999. Mixotrophic cryptophytes and their predators in the Dry Valley lakes of Antarctica. *Freshwater Biology* 41:737–746.

Spigel, R. H., and J. C. Priscu. 1998. Physical limnology of the McMurdo Dry Valleys lakes. Pages 153–187 in J. C. Priscu, editor, *Ecosystem dynamics in a polar desert: The McMurdo Dry Valleys, Antarctica.* Antarctic Research Series 72, American Geophysical Union, Washington, D.C.

Steig, E. J., E. J. Brook, J. W. C. White, C. M. Sucher, M. L. Bender, S. J. Lehman, E. D. Waddington, D. L. Morse, and G. D. Clow, 1998. Synchronous climate changes in Antarctica and the North Atlantic. *Science* 282:92–95.

Steig, E. J., D. L. Morse, E. D. Waddington, M. Stuiver, P. M. Grootes, P. A. Mayewski, M. S. Twickler, and S. I. Whitlow. 2000. Wisconsinan and Holocene climate history from an ice core at Taylor Dome, Western Ross Embayment, Antarctica. *Geografiska Annaler* 82A: 213–235.

Stuiver M., G. H. Denton, T. J. Hughes, and J. L. Fastook. 1981. History of the marine ice sheet in West Antarctica during the last glaciation: A working hypothesis. Pages 319–436 in G. H. Denton and T. H. Hughes, editors. *The last great ice sheets.* Wiley-Interscience, New York.

Swanson, F. J., T. K. Kratz, N. Caine, R. G. Woodmansee. 1988. Landform effects on ecosystem patterns and processes. *BioScience* 38:92–98.

Tyler, S. W., P. G. Cook, A. Z. Butt, J. M. Thomas, P. T. Doran, and W. B. Lyons. 1998. Evidence of deep circulation in two perennially ice-covered Antarctic lakes. *Limnology and Oceanography* 43:625–635.

Vincent, W. F., and C. Howard-Williams. 1986. Antarctic stream ecosystems: Physiological ecology of a blue-green algal epilithon. *Freshwater Biology* 16:219–233.

Virginia, R. A., and D. H. Wall. 1999. How soils structure communities in the Antarctic dry valleys. *BioScience* 49:973–983.

Vogt, E. A. 1997. *Ecosystems: Balancing science with management.* Springer-Verlag, New York.

Wharton, R. A., Jr., C. P. McKay, G. D. Clow, and D. T. Andersen. 1993. Perennial ice covers and their influence on Antarctic lake ecosystems. Pages 53–70 in W. J. Green, and E. I. Friedmann, editors. *Physical and biological processes in Antarctic lakes.* Antarctic Research Series 59, American Geophysical Union, Washington, D.C.

Wharton, R. A., Jr., C. P. McKay, B. C. Parker, and G. M. Simmons, Jr. 1986. Oxygen budget of a perennially ice-covered Antarctic dry valley lake. *Limnology and Oceanography* 31:437–443.

Wharton, R. A., Jr., C. P. McKay, G. M. Simmons, Jr., and B. C. Parker. 1985. Cryoconite holes on glaciers. *BioScience* 35:499–503.

Wilch, T. R., G. H. Denton, D. R. Lux, and W. C. McIntosh. 1993. Limited Pliocent glacier extent and surface uplift in Middle Taylor Valley, Antarctica. *Geografiska Annaler* 75A:331–351.

Wilson, A. T. 1964. Evidence from chemical diffusions of climatic changes in the McMurdo Dry Valley 1200 years ago. *Nature* 201:176–177.

17

Millennial-Scale Climate Variability and Ecosystem Response at the Jornada LTER Site

H. Curtis Monger

The Jornada Long-Term Ecological Research (JRN LTER) program consists of studies superimposed on three research entities, the Jornada Experimental Range, the Chihuahuan Desert Rangeland Research Center, and the Desert Soil-Geomorphology project (figure 17.1). The JRN site is in the northern part of the Chihuahuan Desert and represents, for the LTER network, the desert shrubland and desert grassland ecosystems of the southwestern United States. Climate data at the Jornada site and surrounding area span the last 110 years. Ecological data span the last 144 years.

Despite having over 100 years of data, researchers at the Jornada LTER have struggled to answer the focal question of this book: How have ecosystems responded to climatic variability? This is because, simultaneous with climate, another important factor has had a major impact on ecosystems—human land use. Cattle grazing, brush control, and habitat fractionation have merged with climate to produce external pressures on Jornada ecosystems (Schlesinger et al. 1990; Havstad et al. 2000). Even more uncertain is the cause-and-effect relationship between climate and ecosystems in prehistoric times. Here evidence is limited to indicators, such as former lake shorelines, plant fossils in packrat middens, fossil pollen, $^{13}C/^{12}C$ ratios in paleosols, and erosion rates. When some indicators are used by themselves, circularity arises if a conclusion about ecosystem response to climate change is based on an inference about climate change, which is based, in turn, on ecosystem change. For example, grasslands increased at the end of the middle Holocene as the result of increased rainfall, where the interpretation of increased rainfall is based on increased grass pollen in the middle Holocene sediments (Freeman 1972).

Although focusing on millennial-scale climate and ecosystem variability, this

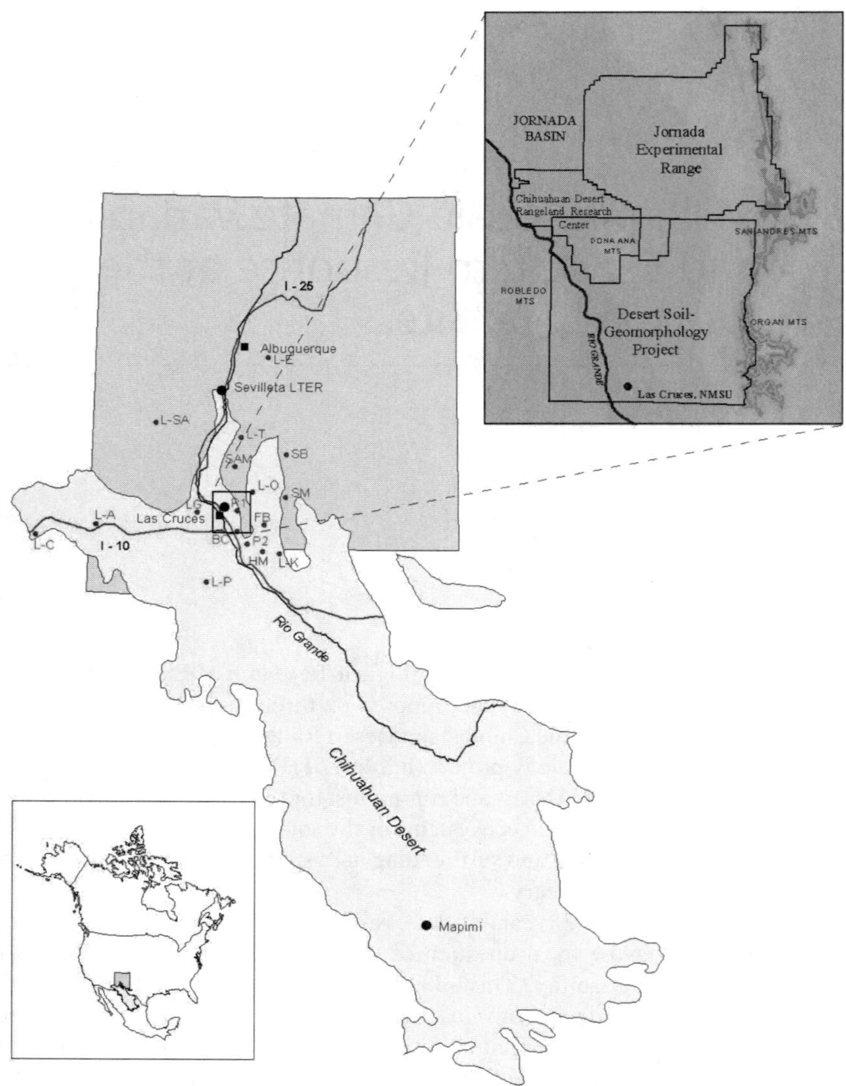

Figure 17.1 Location of the Jornada LTER site in relation the Chihuahuan Desert as defined by Schmidt (1979). Jornada LTER is superimposed on the Jornada Experimental Range, the Chihuahuan Desert Rangeland Research Center, and the Desert Soil-Geomorphology Project area. Pluvial lakes are Lake Otero (L-O), Lake Estancia (L-E), Lake King (L-K), Lake Palomas (L-P), Lake Animas (L-A), Lake Cochise (L-C), Lake Goodsite (L-G), Lake San Agustin (L-SA), and Lake Trinity (L-T). The location of the nearest alpine glacier was Sierra Blanca Mountain (SB). Packrat data are from San Andres Mountains (SAM), Sacramento Mountain (SM), Hueco Mountain (HM), and Bishsop's Cap (BC). Pollen data are from piedmonts of the San Andres Mountains (P1) and Organ Mountains (P2). Carbon isotope samples are from Fort Bliss eolian basin floor (FB) and from the Organ Mountains piedmont (P2).

Table 17.1 Assigned ages of Quaternary temporal terms used in this article

Time period[a]	Years[b]
Historical	A.D. 1850 to present
Holocene	
Late	Present to 4,000
Middle	4,000 to 8,000
Early	8,000 to 10,000
Pleistocene	
Latest	10,000 to 20,000
Late	20,000 to 125,000
Middle	125,000 to 790,000
Early	790,000 to 1,650,000

[a] These periods are based on time intervals described by Bull (1991) and Gile et al. (1981).

[b] Radiocarbon ages in this paper are not calibrated to calendar years.

chapter briefly discusses historic variability for comparison and as a means for describing the setting. The historic-prehistoric boundary for the Jornada area has been set at A.D. 1850 (table 17.1).

Setting of the Jornada LTER Site

Located at 32.5° N and 106.8° W, in New Mexico, USA, the Jornada LTER site is in the Basin and Range province (Peterson 1981), which is characterized by parallel mountain ranges separated by structural basins filled with Cenozoic sediments (Hawley 1986). Elevations at the Jornada range from 1,180 m (3,870 ft) in the Rio Grande floodplain to 2,749 m (9,012 ft) in the Organ Mountains.

The modern terrain is mainly the product of the Rio Grande Rift tectonic system that has been active since the Oligocene (Mack et al. 1998). Bedrock units of Mesozoic volcanic and Paleozoic sedimentary rock have been tilted, creating block-faulted mountain ranges whose eroded sediments fill neighboring basins. In the Jornada basin, the depth of these sediments exceeds 2134 m (7,000 ft) (Seager et al. 1987). Beginning about 5 million years ago, sediments from adjacent mountains were supplemented with sediments from the ancestral Rio Grande, which spilled into several basins of the Jornada region, filled the basins with river deposits, and alternately spilled into adjoining basins as those basins became topograpically lower (Mack et al. 1997). The Rio Grande floodplain as it exists today was formed after the river downcut through its previously deposited sediments. This downcutting probably began sometime between 780,000 years ago (the age of the Matuyama-Brunhes paleomagnetic boundary) and 760,000 years ago (the age of the Bishop volcanic ash) (Mack et al. 1993, 1996, 1998).

Vegetation at Jornada consists of shrublands, desert grasslands, and, to a minor extent in higher elevations, juniper savanna (Allred 1988, 1993). Shrublands and grasslands occur in elevations below about 1,678 m (5,500 ft), which roughly corresponds to 330 mm (13 in.) annual rainfall (USDA-NRCS 1999; Dick-Peddie

1993). Juniper savannas, depending on aspect, occur above this elevation-rainfall boundary. Shrublands are commonly dominated by mesquite (*Prosopis glandulosa*) on sandy soils of the basin floor, by creosotebush (*Larrea tridentata*) on many soil types of piedmont slopes, and by tarbush (*Flourensia cernua*) on clayey soils of lower piedmont slopes. Fourwing saltbush (*Atriplex canescens*) and broom snakeweed (*Gutierrezia sarothrae*) are components of shrublands on a variety of soils and landforms (Gardner 1951). Grasslands are commonly dominated by black grama (*Bouteloua eriopoda*), with lesser amounts of mesa dropseed (*Sporobolus flexuosus*), and red threeawn (*Aristida* spp.) on sandy and loamy soils of basin floors and piedmont slopes, and by tobosa (*Pleuraphis mutica*) and burrograss (*Scleropogon brevifolius*) on clayey soils of lower piedmont slopes and basin-floor depressions. Savanna vegetation is composed of red-berry juniper (*Juniperus erythrocarpa*) and various grasses, shrubs, and Mexican pinyon pine (*Pinus cembroides*) (Allred 1993) on mountain slopes and valleys.

Historical Climate Variability at the Jornada LTER Site

Measured Climate Variability (1892–1993)

The climatic record at the Jornada Experimental Range began July 1915 (Ares 1974) and is summarized in figure 17.2. A statistical treatise of historical climate at the Jornada LTER site by Wainwright is forthcoming in a synthesis volume on the Jornada LTER site to be published by the Oxford University Press. The climatic record at New Mexico State University (NMSU) in Las Cruces (figure 17.1), 45 km south of the Jornada weather station, began January 1892 (Malm 1994). Based on the record at NMSU, Malm (1994) compiled the following statistics. The mean annual temperature is 16°C (60.3°F). The mean annual rainfall is 222 mm (8.73 in.), with greater than 50% falling in the summer monsoonal season of July, August, and September. The annual pan evaporation is 2393 mm (94.2 in.), which is more than ten times the rainfall amount. Highest pan evaporation occurs in June, with an average of 341 mm.

Differences between rainfall at New Mexico State University and rainfall at the Jornada Experimental Range are statistically significant (Conley et al. 1992). Though little differences occurred for winter rainfall (53 mm at NMSU vs. 54 mm at Jornada), larger differences occurred for summer rainfall (135 mm at NMSU vs. 151 mm at Jornada). Temperature differences between New Mexico State University and the Jornada Experimental Range are also statistically significant, with NMSU showing a slight and linear warming trend since 1892 in contrast to the Jornada which shows a warming trend until about 1950 after which it shows a slight cooling trend (Conley et al. 1992).

The wettest year at NMSU received 498 mm (19.60 in.) in 1941, whereas the driest year received 87 mm (3.44 in.) in 1970. The period around 1900 was very wet, but was followed by a very dry period around 1910. The longest drought began in 1945 and continued through 1956. Temperature at NMSU ranged from a high of 43°C (109°F) on 8 July 1951 to a low of −23°C (−10°F) on 11 January 1962. On av-

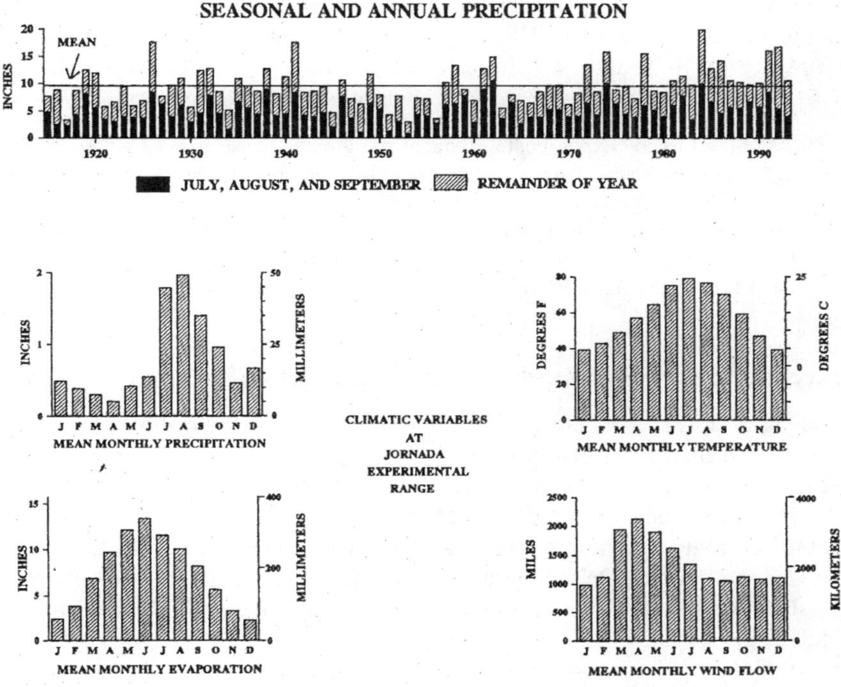

Figure 17.2 Climate variables at the Jornada Experimental Range from 1915 to 1993 (USDA-ARS, 2000).

erage, 9 days per year reach 38°C (100°F), mostly in June and July. There have been 4 years between 1892 and 1991 when the temperature did not reach 38°C (100°F) (1927, 1938, 1941, 1988), and 3 years when more than 30 days were at or above 38°C (100°F) (1951, 32 days; 1978, 33 days; 1980, 32 days). Temperatures of −18°C (0°F) or lower are rare, with only eight occurrences between 1892 and 1991 (Malm 1994).

Other important climatic factors impacting the ecosystem are light intensity and wind speed. Light intensity has an annual average of 21.6 megajoules per square meter per day (MJ/m²/day) (516 langleys/day). Solar radiation is lowest in December with a monthly mean of 11.8 MJ/m²/day (282 langleys/day) and is highest in June with a monthly mean of 30.5 MJ/m²/day (729 langleys/day) (Malm 1994). Wind speeds are greatest in March, April, and May (figure 17.2), when occasional gusts can reach 145 km/hr (90 mi/hr).

Vegetation Change (1858–1963)

In 1857, the Jornada region was first surveyed by the United States General Land Office. The surveyors included in their notes observations about vegetation, soils, and topography. From these notes, the 1858 vegetation map of the Jornada Experimental Range was constructed (Buffington and Herbel 1965). After establishment of the Jornada Experimental Range in 1912, subsequent vegetation maps were made in 1915, 1928, and 1963. Based on these maps, Buffington and Herbel (1965, p. 61) made the following conclusions about vegetation change:

> In 1858 the Jornada Experimental Range was a great expanse of grass with only isolated spots of mesquite. On the higher areas along the mountains, [shrubs] brush was present; however, grass was also good in most places. A few tarbush plants were present in some of the lower lying areas. Since 1858 the grass cover has decreased tremendously, and the brush has increased to the point that it was present on the entire study area in 1963. Less than 25% of the study had a fair stand of grass in 1963.

Anecdotal Evidence of Vegetation Change (1598–1885)

Grazing of domestic livestock at the Jornada LTER site probably spans a longer time period than most places in the United States because the Jornada Basin is crossed by one of the oldest roads in the United States—El Camino Real or "Royal Road," which connected Mexico City with Santa Fe (Hallenbeck 1950). Don Juan de Oñate's route from Zacatecas north to the land of the Pueblos in 1598 became the Camino Real. Oñate's entourage in 1598 was approximately 4 miles long and consisted of 170 families, 129 soldiers, 80 wagons, and 7,000 head of livestock (Hallenbeck 1950).

After Oñate's initial trip, caravans traveled the road about once every 3 years from Mexico bringing supplies to Santa Fe. In the 1700s and early 1800s, the road was increasingly used for trade between Chihuahua and Santa Fe, including livestock trade. When the Santa Fe Trail reached the northern terminus of the Camino Real in 1821, trading between the United States and Mexico increased. By the 1850s, however, traffic on the Camino Real diminished as shorter trade routes through Texas connected markets in the United States with Mexico. By 1882 the road ceased to be a major travel route when the railroad linked Chicago, El Paso, and Mexico City.

Oñate and other Spanish explorers mentioned plants cultivated or used by Native Americans, but made few comments on vegetation in general (Gardner 1951). Yet by the mid-1800s, descriptions of vegetation were more common, as compiled by Fredrickson et al. (1998, 196–197):

> Writing about his passage through the Jornada del Muerto in 1846, Dr. Frederick Adolph Wislizenus . . . stated, 'The wide country through which we have to travel, in the elevation of from four to five thousand feet above the sea, [has] dry, hard soil, tolerable grass, and an abundance of mezuite and palmillas.' Two years later, journalist John Cremony . . . described the same area, '. . . is a large desert, well supplied with grama grass in some portion, but absolutely destitute of water or shade for ninety-six miles.' Beale . . . described the Jornada Plain as, '. . . a level plain, covered thickly with the most luxurious grass, and filled with wildflowers . . . Hundreds and hundreds of thousands of acres containing the greatest abundance of the finest grass in the world, and the richest soil are here lying vacant, and looked upon by the traveler with dread because of its want for water.' . . . Albert Fountain [1885] wrote that the Jornada Plain was treeless and waterless but covered with rich, nutritious grass.

Anecdotal descriptions of vegetation by travelers and military expeditions in the 1800s are also contained in Gardner (1951). This evidence, combined with interviews of long-term residents in southern New Mexico, led Gardner to conclude that there is little room for doubt that grass cover has markedly decreased and shrubs greatly increased during the past hundred years.

Causal Factors

Two major hypotheses have been put forth as causal factors in the shift from dominantly grassland vegetation to dominantly shrubland vegetation—(1) overgrazing by domestic livestock and (2) climate variability. The climate-variability hypothesis was tested by analyzing climatic data using several statistical methods (Conley et al. 1992). Based on these analyses, Conley et al. rejected the climate-variability hypothesis and accepted the overgrazing hypothesis as the cause of the vegetation shift.

To Buffington and Herbel (1965), however, the rapid increase of shrubs was caused by both climate variability and overgrazing. This combination involved the effects of seed dispersal, periodic droughts, and selective grazing of grass by livestock. The equilibrium between creosote bush and black grama, for example, was shifted in favor of the shrub with selective grazing and because black grama is sensitive to soil erosion and burial by sediments. Even if heavy grazing is discontinued, the return of the grass may require a very long time as compared to more humid grasslands because of the low rainfall, the loss of soil, the scarcity of propagules, and the presence of shrubs (Gardner 1951).

The invasion of grasslands is not limited to shrub expansion; there has been concurrent expansion of woodlands into grasslands (Dick-Peddie 1993). Both invasions cause a concentration of soil resources beneath the woody plants (Wright 1982; Schlesinger et al. 1990; Davenport et al. 1996). This redistribution of soil resources becomes an important factor for continued vegetation change, as do changes in fire regime, changes in vertebrate and invertebrate distribution, changes in atmospheric CO_2, changes in water redistribution, and changes in competition by nonnative species (Gibbens et al. 1983; Ludwig 1987; Pieper and Beck 1990; McAuliffe 1994; Herbel et al. 1994; Huenneke and Noble 1996; Whitford 1996; Herrick et al. 1997; McPherson and Weltzin 2000).

Millennial-Scale Climate Variability and Ecosystem Response

Conclusions about millennial-scale climate and ecosystem variability at the Jornada LTER site are based on physical, biotic, and soil-geomorphic evidence. Physical evidence includes features made by pluvial lakes, alpine glaciers, rock glaciers, and groundwater. Biotic evidence includes packrat middens, fossil pollen, and carbon isotopes. Soil-geomorphic evidence includes geomorphic surfaces and soil profile characteristics.

The most direct evidence concerning prehistoric climate are marks produced by physical entities, such as lake shorelines produced by Pleistocene lakes (figure

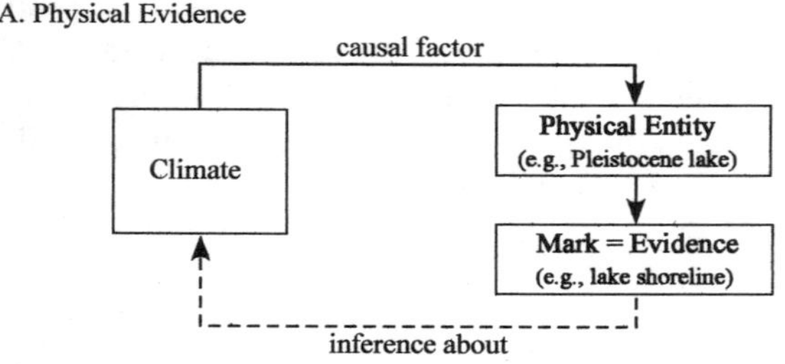

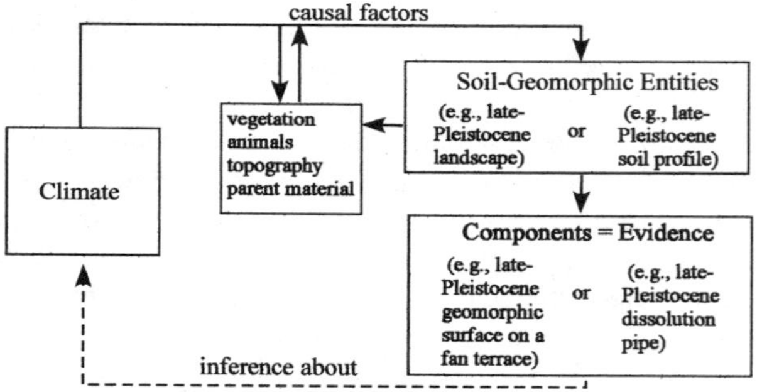

Figure 17.3 Sources of proxy evidence for climate inferences based on physical, biotic, and soil-geomorphic entities and the marks and components produced by those entities. Solid lines show causal factor links. Dashed lines show uniformatarian inferences made about climate based on marks and components.

17.3A). Biotic fossils are also evidence of prehistoric climate because climate is a causal factor for the development of biological entities such as grassland. A climate-vegetation linkage, however, is confounded because fire, erosion, atmospheric CO_2, seed dispersal by animals, successional stages, and nutrient limitations are also factors that affect vegetation (figure 17.3B). The least direct evidence of prehistoric climate is based on soil-geomorphic components because both climate and vegetation (as well as animals, topography, and parent material) are causal factors for soil and geomorphic entities (figure 17.3C). Nevertheless, a late-Pleistocene geomorphic surface on a terrace, for example, is a vestige of a landscape formed in the late Pleistocene with a higher base level (e.g., Hawley 1975). Moreover, dissolution pipes through a petrocalcic horizon are vestiges of soil formation in a wetter climate (Gile et al. 1981). As the complexity of these systems increases, uncertainty about cause-and-effect relationships also increases.

Physical Evidence

Lakes. Though dry now, many large depressions in the Chihuahuan Desert were filled with lakes at various times in the Quaternary. Lake Jornada (Gile 2002), for example, probably existed during much of the early and middle Pleistocene. More recent lakes, however, existed in the late Pleistocene and early Holocene, and they provide higher resolution evidence about climate change. Nine such pluvial lakes surrounding the Jornada LTER site (figure 17.1) have been reported in the literature: Lake Otero 40 km to the northeast (Hawley 1993), Lake Estancia 230 km to the northeast (Allen and Anderson 1993), Lake King 160 km to the southeast (Wilkins and Currey 1997), Lake Palomas 100 km to the south (Reeves 1969), Lake Animas 190 km to the southwest (Fleischhauer and Stone 1982), Lake Cochise 290 km to the southwest (Waters 1989), Lake Goodsite 35 km to the west (Hawley 1965), Lake San Agustin 200 km to the northwest (Markgraf et al. 1984), and Lake Trinity 110 km to the north (Neal et al. 1983).

Of these lakes, Lake Estancia, Lake King, Lake San Agustin, and Lake Cochise have the best chronologic information about water levels. High stands occurred at all four lakes during the latest Pleistocene. At Lake Estancia two major high stands occurred: one beginning at ca 19,700 ^{14}C years B.P. and another at ca 13,700 ^{14}C years B.P. (Allen and Anderson 1993). At Lake King, high stands occurred at ca 22,600, 19,100, and 17,200 ^{14}C years B.P. (Wilkins and Currey 1997). At Lake San Agustin, lacustrine ostracodes suggest a high stand from ca 22,000 to 19,000 years ago (Forester 1987). At Lake Cochise, a high stand occurred between 13,750 and 13,400 ^{14}C years B.P., after which the lake level dropped until about 9,000 ^{14}C years B.P. when it experienced a renewed rise (Waters 1989).

After the early-Holocene lake level rise at Lake Cochise (ca 9,000 years B.P.), the lake progressively dried. There is no evidence of lakes during the middle Holocene, although two small lakes probably formed shortly after the middle Holocene, but dates are uncertain (Waters 1989). At Lake San Agustin, ostracodes indicate that, at about 5,000 years ago, rainfall became so low the lake changed into its present dry playa (Forester 1987). Lake Estancia was dry in the middle Holocene based on deflation of lake sediments and dune formation during that period (Hawley 1993).

Alpine Glaciers. During the last full glacial period (marine-O-isotope 2, ca 25,000 to 10,000 B.P.) alpine glaciers formed on high mountain peaks of New Mexico (Richmond 1986). The nearest glaciated peak to the Jornada LTER site is 130 kilometers to the northeast at Sierra Blanca (elevation 3,660 m, 12,003 ft) (figure 17.1). This mountain is the southernmost late Pleistocene-age glaciated peak in the United States (Smith and Ray 1941). Glacial features found there include a cirque on the northeastern side of the mountain, a steeply sloping glaciated valley, well-defined moraines, and a protalus rampart within the cirque.

Rock Glaciers. Rock glaciers consist of poorly sorted angular boulders, fine material, and interstitial ice in permafrost areas of high mountains (Bates and Jackson 1987). The formation of interstitial ice is responsible for their downslope creep. In New Mexico, mountains located both northeast and northwest of the Jornada LTER site contain some 116 rock glaciers (Blagbrough 1994). They have been used as paleoclimatic indicators because, based on modern rock glaciers, they are most active a few hundred meters below the orographic snow line. Elevations of the New Mexico rock glaciers indicate that the mean annual temperature during the last full glacial period was approximately 7 to 8°C cooler than today (Blagbrough 1994).

This range is fairly close to the 5 to 7°C cooler-than-present range estimated by Phillips et al. (1986) and the 5.5°C cooler-than-present temperature estimated by Stute et al. (1995). Both were based on groundwater studies in northwestern New Mexico. This technique is based on the principle that solubility of noble gases in the atmosphere is a function of temperature. Thus, noble gases dissolved in the groundwater of a confined aquifer record the temperature at the water table (Stute et al. 1995).

Biotic Evidence

Packrat Middens. Packrat middens are clumps of plant materials gathered by *Neotoma* woodrats (Betancourt et al. 1990). The plant material is gathered by rats from a surrounding area of about 30 meters to construct dens and is cemented by dried urine. These nests, when in a protected shelter, such as limestone caves or ledges, can exist for tens of thousands of years and thereby provide a fossil record of changing plant communities (e.g., Spaulding 1991). Because the age of plant fossils can be determined with radiocarbon dating, a chronology can be assigned to biotic changes.

In the vicinity of the Jornada LTER site, packrat middens have been studied in the San Andres Mountains 60 km to the north, in the Sacramento Mountains 90 km to the east, in the Hueco Mountains 120 km to the southeast, and in the Franklin Mountains (Bishop's Cap) 60 km to the south (Van Devender 1990) (figure 17.1). One of the more complete records occurs at the Hueco Mountain site. Here, the fossil record suggests a four-step climatic-vegetative shift: (1) A pinyon-juniper-oak woodland grew in the area from 42,000 to 10,800 ^{14}C years B.P. (2) After this period, pinyon disappeared, but oaks and junipers persisted through the early Holocene until about 8000 ^{14}C years B.P. when oaks disappeared. (3) At this time, honey mesquite and prickly pear appeared in a gradual transition to desert grassland. (4) The shift to modern desert scrub conditions was completed before about 3650 ^{14}C years B.P., the time creosotebush appeared (Van Devender, 1990).

Although packrat middens provide firm evidence that certain plants were present at certain times, the limitations of using this method for broad paleoclimatic reconstruction have been reviewed by Hall (1997). First, woodrats gather vegetation only within a range of about 30 meters. Second, woodrats do not randomly collect plants from their 30-m home range, but select specific plants for food storage, nesting, and den construction. Third, rocky-escarpment vegetation may differ significantly from vegetation occupying broad piedmont slopes and basin floors. Fourth, field workers have been known to target middens containing exotic plants, therefore, plant species found in the modern flora could be bypassed.

Still, packrat middens provide important, well-dated information about biotic changes from which climatic inferences can be made. For the Chihuahuan Desert at large, four major climatic inferences have been made (Van Devender 1990). (1) The last full glacial period was a time of equable climates with cooler summers than today, greater rainfall, and mild winters with few freezes. (2) By early Holocene, winter rainfall continued to be greater than today, but summer temperatures increased beyond what they were previously. (3) By the middle Holocene, winter rainfall had been replaced by biseasonal rainfall that was dominated by the summer monsoon. Also, during the middle Holocene, summer temperatures dramatically increased over what they had been in the early Holocene and latest Pleistocene. (4) The late Holocene was a period when the modern climatic regime was established. This included fewer winter freezes, monsoonal summer rainfall, and increased droughts (Van Devender 1990, p. 126).

Fossil Pollen. The dominant shrubs at the Jornada LTER site—mesquite, creosotebush, and tarbush—are insect pollinated, and, therefore, do not release copious amounts of pollen into the wind. In contrast, plants that do release large amounts of pollen include the Gramineae and members of the genus *Atriplex* (Horowitz 1992). At the Jornada site, several grass types and *Atriplex canescens* (Fourwing saltbush) are common components of the modern flora. After pollen has fallen on the land surface, some of these silt-size particles illuviate into the upper soil horizons and some are transported laterally by runoff water into playas.

Although fossil pollen provides a picture of regional vegetation, it also has several limitations (Hall 1997), especially when the pollen record is from arid paleosols. These limitations include imprecise chronologic control in which pollen ages can only be bracketed between strata having radiocarbon-dated charcoal. Sparseness of fossil pollen is also a limitation, especially for recently deposited eolian and alluvial sediments. Some plants, such as pines, produce proportionally more pollen than their abundance in the landscape, whereas insect pollinated plants produce proportionally less pollen than their abundance in the landscape (Horowitz 1992). Also, differential destruction of pollen grains in soils is another limitation (Moore et al. 1991). Poor preservation of pollen occurs in all arid-land depositional environments in which the sediments have been subject to weathering processes (Hall 1997). Pollen grains in arid soils are usually corroded, due either to oxidation or to bacterial or fungal activity. In addition to chemical decomposition, mechanical degradation occurs in cemented soil horizons (Horowitz 1992).

Pollen analysis is further handicapped at detecting change within grasslands by taxonomic limitations, in which case phytoliths have been useful (e.g., Fredlund and Tieszen 1997). Phytolith studies at the Jornada LTER are rare, but they offer opportunities for future research because phytoliths have been identified for the major grasses, yucca leaves, and conifer needles in the Jornada region (Pease 1967).

Given these limitations, comparative changes in pollen have, nevertheless, provided insight into biotic changes. Most of the fossil pollen studies in the vicinity of Jornada have been collected from buried paleosols on piedmont slopes. At the Jornada LTER site, fossil pollen was studied on the piedmont slope of the San Andres Mountains 5 km east of the Jornada Experimental Range (Freeman 1972) and on the piedmont slope of the Organ Mountains 65 km to the southeast (Monger et al. 1998) (figure 17.1). For making inferences about bioclimatic changes, the main pollen categories of these studies were the Cheno-Am and the Gramineae taxonomic units. The Cheno-Am taxon includes all members of the Chenopodiaceae (except *Sarcobatus*, greasewood) and *Amaranthus* (in the Amaranthaceae plant family). A prominent member of the Cheno-Am taxon is fourwing saltbush (*Atriplex canescens*). The Gramineae family includes all grasses.

Two major inferences about bioclimatic change were made from the piedmont slope site of San Andres Mountains: (1) Desert scrub vegetation and aridity was greatest in the middle Holocene. (2) The middle Holocene desert scrub and aridity gave way to increased grasslands and a more mesic climate by the late Holocene (Freeman 1972). Three bioclimatic inferences were made from the piedmont slope site of the Organ Mountains: (1) Grasslands were prominent in the early Holocene and latest Pleistocene; (2) by middle Holocene time, there was increased desert scrub vegetation; and (3) as with the San Andres site, desert scrub and aridity gave way to increased grasslands and the more mesic climate of the late Holocene.

Bioclimatic inferences based on a synthesis of fossil pollen data in the Southwest as a whole were made by Hall (1997, p. 36): "By 14,000 years ago, as the climate warmed, pinyon dropped out of the low-elevation terrain and were replaced by a drier Chenopodiaceae-Asteraceae shrub grassland. Throughout the Holocene, low-elevation desert regions were dominated by desert-shrub grassland vegetation. The mid-Holocene was characterized by a decrease in grasses and an increase in shrubs due to hot, dry climate during that period."

Ratios of $^{13}C/^{12}C$ in Soil Organic Matter and Pedogenic Carbonate. At the Jornada LTER site, shrub species use the C_3 photosynthetic pathway—with the exception of fourwing saltbush, which uses the C_4 pathway (Syvertsen et al. 1976). Grass species use the C_4 photosynthetic pathway. Cacti, which are a minor component of both shrublands and grasslands at the Jornada site, use the CAM photosynthetic pathway.

Atmospheric CO_2 has ^{13}C values that range from −7.8 to −12 ‰ (Boutton 1991). Because C_4 plants incorporate more atmospheric $^{13}CO_2$ into their biomass than C_3 plants, C_4 plants have higher $\delta^{13}C$ values. The range for C_4 plants is from −7 to −15 ‰, in contrast to C_3 plants that range from −20 to −35 ‰ (Cerling and Quade 1993). Of the plants measured at the Jornada LTER site, mesquite, creosotebush, tarbush, and soaptree yucca fall into the C_3 range, whereas tobosa grass, black grama grass, prickly pear cactus, and fourwing saltbush fall into the C_4 range (figure 17.4).

Because soil organic matter maintains the same $\delta^{13}C$ signature as the vegetation from which it was derived, $\delta^{13}C$ values of soil organic matter have been used to make inferences about vegetation growing on landscapes of the past (e.g., Kelly et al. 1993). At the Jornada LTER site, $\delta^{13}C$ values of soil organic matter reflect the invasion of desert shrubs into grasslands during the middle Holocene (Cole and Monger 1994; Monger et al. 1998) and the last 150 years (Connin et al. 1997a,b).

In addition to soil organic matter, soil carbonate ($CaCO_3$) carries a $\delta^{13}C$ signa-

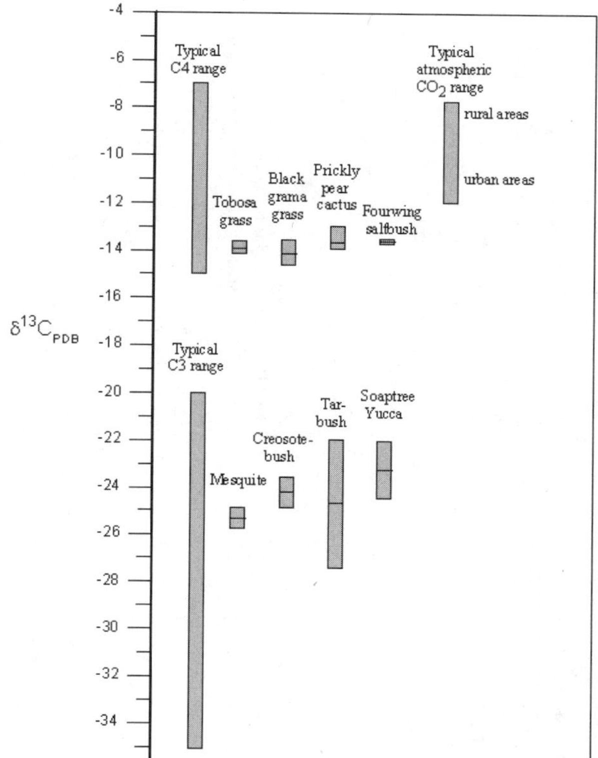

Figure 17.4 Ranges of carbon isotope ($\delta^{13}C$) values at the Jornada LTER site. Typical $\delta^{13}C$ ranges are from Boutton (1991) and Cerling and Quade (1993). Ranges for plant types are for species gathered at the Jornada LTER site (Gallegos 1999).

ture of vegetation (Cerling 1984; Wang et al. 1996). Inferences made about vegetation from soil carbonate involve the presumption that $\delta^{13}C$ values of carbonate are enriched 14 to 16 ‰ with respect to the vegetation controlling soil CO_2 (Cerling et al. 1989; Amundson et al. 1998). In theory, if carbonate precipitates in isotopic equilibrium with CO_2 respired by a pure C_4 grassland, carbonate should have a $\delta^{13}C$ value of +2 ‰. In contrast, if carbonate precipitates in isotopic equilibrium with CO_2 respired by a pure C_3 shrubland, carbonate should have a $\delta^{13}C$ value of −12 ‰ (Boutton 1991).

At the Jornada LTER site, a record of $\delta^{13}C$ values in pedogenic carbonate is contained in buried paleosols and overlying modern soils. On piedmont slopes, buried paleosols are common in alluvial deposits. On basin floors, buried paleosols are common in eolian deposits. The Organ allostratigraphic unit, a major deposit of middle Holocene age, occurs on both the piedmont slope and basin floor landforms (table 17.2). This unit began to be laid down around 7,500 ^{14}C years B.P. based on dated charcoal (Gile 1975). Because it disconformably buries older soils (paleosols), carbon isotopes in paleosols are considered to be a component of the vegetation growing on those paleosols until they were buried by Organ sediments.

Table 17.2 Geomorphic surfaces at the Jornada LTER site and surrounding area that occur on areas astride the Rio Grande floodplain (i.e., valley border), the piedmont slope, and the basin floor landforms[a]

| | Geomorphic surface | | Carbonate stage | | |
Valley border	Piedmont slope materials	Basin floor materials	Nongravelly	Gravelly	Estimated soil age (years B.P. or epoch)
Coppice dunes	Coppice dunes	Whitebottom			Historical (since A.D. 1850)
		Lake Tank			Present to 150,000
Fillmore	Organ	Organ	0, I	I	Middle and late Holocene (100 to 7,000)
	III		I	I	(100(?) to 1,000)
	II		I	I	1,100 to 2,100
	I		I	I	2,200 to 7,000
Leasburg	Isaacks' Ranch		II	II, III	Latest Pleistocene (10,000–15,000)
Late Picacho	Late Jornada II		III	III	Latest to late Pleistocene (15,000–75,000)
Picacho	Jornada II	Petts Tank	III	III, IV	Late to middle Pleistocene (75,000–150,000)
Tortugas			III	IV	Late to middle Pleistocene (150,000–250,000)
Jornada I	Jornada I	Jornada I	III	IV	Middle Pleistocene (250,000–400,000)
	Dona Ana			IV	Middle Pleistocene (>400,000)
Lower La Mesa			III, IV		Middle to early Pleistocene (780,000)
JER La Mesa			IV, V		Early Pleistocene (780,000–1,600,000)
Upper La Mes			V		Late Pliocene (2,000,000–2,500,000)

[a] After Gile 2002.

On the piedmont slope, a major isotopic shift was observed across the contact between paleosols and the overlying Organ unit (figure 17.5). This shift ranged from -2 ‰ in the paleosols to -8 ‰ in the younger Organ unit. The $\delta^{13}C$ values accorded well with an increase of Cheno-Am pollen, and erosion that indicated a change from a C_4 grassland to a C_3 shrubland in the middle Holocene (Cole and Monger 1994; Monger et al. 1998).

On the basin floor, a similar isotopic shift was observed in Organ eolian sediments (figure 17.5). This shift ranged from -1 ‰ in the paleosols to -7 ‰ in the Organ unit and was also interpreted as indicating a change from a C_4 grassland to a C_3 shrubland (Buck and Monger 1999). Unlike sites on the piedmont slope, however, the upper strata of two of the eolian sites (figure 17.5D and 17.5G) suggest a gradual return of grasses in that landform. In four of the profiles (figures 17.5A, B, D, E), the lower and older strata may indicate increased C_3, possibly a juniper savanna at about the last full glacial period.

Oxygen isotopes ($^{18}O/^{16}O$) in soil carbonates have been used to make inferences about paleotemperatures (Cerling 1984; Cerling and Quade 1993) and rainwater sources (Amundson et al. 1996; Liu et al. 1996). However, analysis of $\delta^{18}O$ values in the vicinity of the Jornada LTER site shows no consistent trend. In some cases, the $\delta^{18}O$ values change little despite major shifts in carbon isotopes (figures 17.5A, B, C). In other cases, the $\delta^{18}O$ values have trends similar to those of carbon isotopes (figures 17.5D, E, H).

Soil-Geomorphic Evidence

In 1957, the U.S. Department of Agriculture began a study of soil-geomorphic relationships in a 400-square-mile (1024 km²) area surrounding Las Cruces (Gile et al. 1981). This project, termed the Desert Project, includes an area that overlaps the southern portion of the Jornada Experimental Range and Chihuahuan Desert Rangeland Research Center (figure 17.1). A major objective of the Desert Project was to map soils and geomorphic surfaces. In addition to Aridisol, Entisol, and Mollisol soil types, geomorphic surfaces were identified for three physiographic units: the Rio Grande valley border, the piedmont slope, and the basin floor (table 17.2). Ages of the geomorphic surfaces are based on a combination of radiocarbon dates of charcoal and pedogenic carbonate (Gile et al. 1981), K-Ar dates of lava flows (Seager et al. 1984), Ar-Ar dates of volcanic ash and pumice (Mack et al. 1996), paleomagnetism dates (Mack et al. 1993), and megafauna fossils (Hawley et al. 1969; Tedford 1981; Morgan et al. 1998).

Geomorphic surfaces in the valley border astride the Rio Grande floodplain occur as stepped fan terraces. Each surface rises from the Rio Grande floodplain, which acts as the local base level. Progressively higher steps are progressively older and contain progressively more pedogenic carbonate (Gile et al. 1966; Machette 1985). The piedmont slope surfaces also occur as progressively older steps in some areas. In other areas, younger sediments bury older surfaces, resulting in stacked sequences of buried paleosols.

Erosion as an Indicator of Bioclimatic Variability. Based on modern measurements, erosion is greater in shrublands than in grasslands (Abrahams et al. 1995)

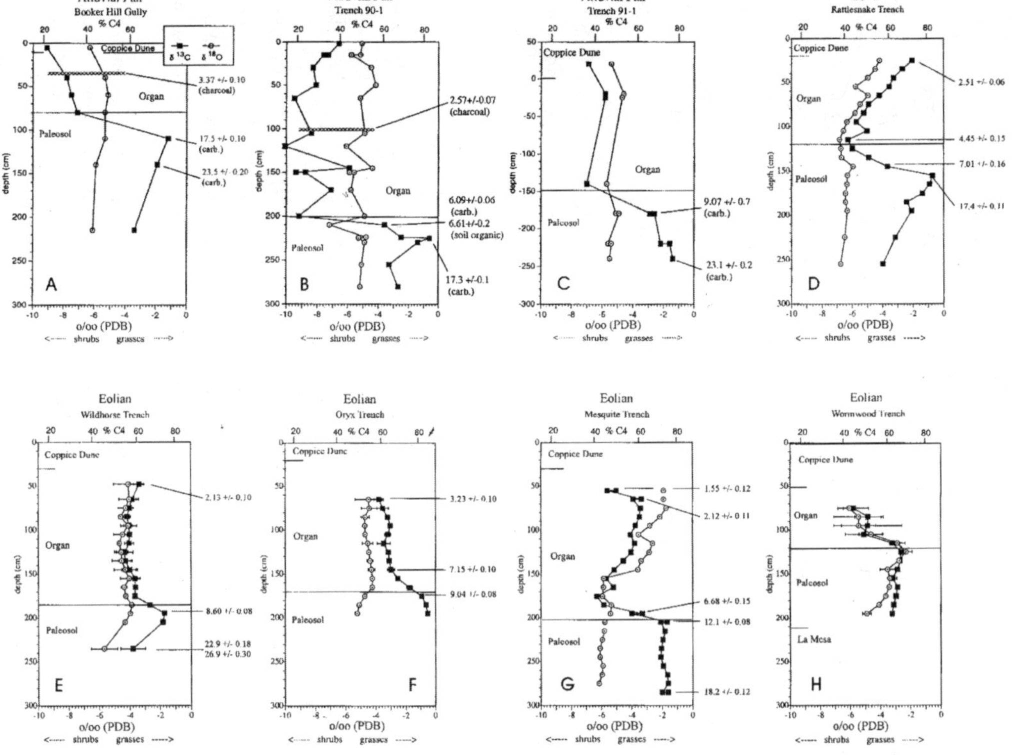

Figure 17.5 Comparison of $\delta^{13}C$ and $\delta^{18}O$ values of pedogenic carbonates across the Organ-Paleosol boundary on piedmont slope (i.e. Alluvial Fan) and basin floor (i.e. Eolian) sites. Isotopic compositions are expressed as per mil relative to the PDB standard. Percent C_4 vegetation is based on a model by Cerling (1984). The ^{14}C ages, expressed as years B.P. x 1000, are of carbonate unless stated otherwise. The ages of carbonates are less reliable than dates of charcoal (Gile et al. 1981) (after Buck and Monger 1999).

because more bare soil is exposed in shrublands. Though arid, there is enough rain at the Jornada LTER site to cause significant amounts of erosion, especially in the summer monsoon season when thunderstorms are common. Consequently, a working hypothesis to explain episodic base-level changes and buried paleosols was given by Ruhe (1962) and Gile and Hawley (1966). The hypothesis states that periods of erosion and sedimentation correspond to periods of aridity when shrubland dominate, in contrast to periods of landscape stability and soil formation when grasslands or woodlands dominate.

This hypothesis follows the Langbein-Schumm curve, which emphasizes the importance of both vegetation and rainfall on erosion (Langbein and Schumm 1958). In their curve, sediment yield is low in extremely arid lands (<100 mm rainfall), but rises steeply and reaches a maximum near the arid-semiarid boundary (at about 300 mm rainfall). At progressively greater amounts of rainfall, erosion and sediment yield decline. At the Jornada Experimental Range, the average annual rainfall is 247 mm (9.72 in), which is near the Langbein-Schumm maximum of 300 mm, though some years are less (e.g., 90 mm) and some years are more (e.g., 475 mm) (figure 17.2). Thus, applying the Langbein-Schumm model to the paleosol record at the Jornada piedmont slope site, periods of high erosion-sedimentation and burial of soils downslope may represent climatic conditions similar to today. Periods of landscape stability and soil formation probably represent climatic conditions wetter than today.

In addition to buried paleosols, arroyo cutting can provide insight about past climates. In southeast Arizona, for example, a major period of arroyo cutting occurred in the middle Holocene Altithermal period when temperatures were at their postglacial high and effective moisture at its postglacial low (Waters and Haynes 2001). This period of major arroyo cutting in Arizona corresponds to deposition of the Fillmore and Organ unit at the Jornada site (table 17.2). After about 4000 years ago, less pronounced, but more frequent, arroyo cutting occurred in Arizona, which Waters and Hayes (2001) attributed to the development of increased El Niño frequency and strength.

At a longer timescale, the geomorphic surfaces astride the Rio Grande have been linked to climate variability by the following cyclic-entrenchment model (Hawley 1975). (1) The Rio Grande downcut during glacial periods, when greater rainfall gave rise to more water in the river and denser vegetative cover on the land. During this interval, the river channel carried a lower sediment load and had a greater capacity to entrench. (2) The Rio Grande backfilled during waning glacial and early interglacial times, when reduced rainfall gave rise to less water in the river and less vegetative cover on land. During this interval, the river carried a greater sediment load, lost its ability to entrench, and aggraded. (3) Aggradation ceased and the base level stabilized during the latter parts of interglacial times, until (4) the cycle began again with renewed downcutting in response to the renewed waxing phase of the next glacial cycle. Although this model provides a general explanation of links between rainfall and erosion, base level changes are also a function of stream slope, sediment size, hydraulic roughness, stream discharge, and bed load (Bull 1991). In some areas water table depths are also important.

A model, similar to the one for the Rio Grande, has been used to explain geomorphic surfaces on piedmont slopes (e.g., Hawley et al. 1976; Gile et al. 1981). (1) During periods of aridity that correspond to interglacial periods, vegetative density declined, bare ground increased, and erosion-sedimentation increased. (2) During periods of more effective precipitation during glacial periods, vegetative cover increased, bare ground decreased, and erosion-sedimentation decreased. (3) With the discontinuance of erosion-sedimentation, landscape stability ensued, and pedogenesis produced soil horizons.

Coppice Dune Formation as an Indicator of Vegetation Change. In historical times, the relationship between vegetation change and wind erosion caused the formation of coppice dunes. Coppice dunes are mounds of sand held together by shrubs, mainly mesquite. These dunes form when laterally blowing sand encounters and accumulates around mesquite plants. As the sand aggrades, the mesquite plant grows through the rising dune. During this process, mesquite stems send out roots and exploit the resources of the accreting sand.

Ecosystems on sandy soils are some of the most fragile ecosystems because sandy soils are more vulnerable to wind erosion than rocky or clayey soils. Land surveyor notes for 1857 relating to sandy sites in southern New Mexico lacked any mention of coppice dunes at section corners west of Las Cruces. When the same section corners were revisited in the 1960s, the level, smooth black grama land recorded for 1857 had changed to steep-sided coppice dunes (Gile 1966). The earliest aerial photography of the Jornada region was taken in 1936. Coppice dunes are present in these photographs, so the dunes must have formed between the late-1850s and 1936 (Gile 1966). Land survey notes indicate similar relations in the Jornada Experimental Range (Buffington and Herbel 1965).

Pedogenic Carbonate as an Indicator of Climate Variability. Climate, as one of the five soil-forming factors, exerts major control on the mineralogical nature of the soil profile (Dokuchaev 1883; Jenny 1941). This is particularly the case with horizons of pedogenic carbonate because (1) its presence is generally restricted to soils of dry climates and (2) its depth is a function of rainfall (Arkely 1963; McFadden and Tinsely 1985; Mayer et al. 1988). With respect to carbonate's presence in the soil, the boundary between carbonate-accumulating soils and non–carbonate-accumulating soils in the Western United States corresponds to an annual rainfall of about 500 mm (20 in) (Birkeland 1999). In the Midwest, however, carbonate can occur in limestone soils that receive up to 800–900 mm of mean annual rainfall (Jenny and Leonard 1934; Nordt et al. 1998).

Carbonate is present in modern soils and buried paleosols at the Jornada LTER site. Trenches and natural exposures reveal multiple buried paleosols with calcic horizons that date to at least 400,000 years B.P. based on correlation to the Jornada I surface (Gile et al. 1981). The presence of carbonate in these stratigraphic sections suggests, by deduction, that wet periods received less than 500 mm of annual rainfall during this 400,000-year period.

With respect to the depth of carbonate in soil, the zone of carbonate accumulation not only depends on the amount of rainfall, but also on texture, parent material, and erosion. Nevertheless, some trends have been observed for soils in general (Jenny and Leonard 1934) and for soils at the Jornada LTER site (Gile 1977). By tracing carbonate depth in soils of the Organ geomorphic surface from arid lower elevations to semiarid higher elevations, Gile (1977) documented a progressive deepening of carbonate at the Jornada site. At the lower elevations (220 mm of rainfall), carbonate depth was 25 cm below the surface. At the higher elevations (350 mm of rainfall), carbonate depth was 104 cm below the surface.

With respect to carbonate depth as a paleoclimatic indicator, carbonate in ancient soil profiles at the Jornada LTER probably reflect the wet-dry cycles of much of the late Quaternary. Such soils exist on stable geomorphic surfaces of middle Pleistocene age and older. These ancient polygenetic soils would have experienced multiple glacial-interglacial climates based on a 100,000-year glacial-interglacial periodicity (Morrison 1991). Vertical pipes crosscut many petrocalcic horizons in these ancient soils (figure 17.6). These pipes, according to the current working hypothesis, originated as a result of dissolution during wet glacial periods (Gile et al. 1981). In this sense, the pipes are small-scale versions of karst topography—that is, they are pedokarst. According to this hypothesis, greater rainfall would lead to deeper water penetration through the profile, movement across the laminar cap of the petrocalcic horizon, and down the pipes. As a test of this hypothesis, radiocarbon ages of both carbonate and organic matter in layers of the laminar cap have been measured (Gile et al. 1981, p. 122). Radiocarbon ages of these materials range from about 32,000 to 21,000 ^{14}C years B.P., near the time of glaciopluvial conditions based on lake studies.

Moreover, a shift from a wetter to a drier climate would cause an upward shift in the depth of wetting. This would lift the zone of carbonate formation from the laminar zone of the petrocalcic horizon to a higher zone in the soil. This shallower zone would extend horizontally across pipes (figure 17.6) and would be at a similar depth to carbonate accumulation in neighboring soils known to be middle Holocene based on radiocarbon dating. As a test of this hypothesis, radiocarbon ages of this shallow-zone carbonate have been measured and are middle Holocene in age, from about 4,940 to 1,600 ^{14}C years B.P. (Gile et al 1981, p. 149; Monger et al. 1998, p. 152).

Summary of Climatic Variability and Ecosystem Response at the Jornada LTER Site

Inferred climate variability at the Jornada LTER site, based on several types of evidence, is presented in figure 17.7. This graph is a working hypothesis of the bioclimatic changes during the last 20,000 years. The graph plots aridity, C_4 grasslands, C_3 woodlands, and C_3 shrublands across 9 bioclimatic time intervals. Because vegetation is laterally diverse today, it was probably laterally diverse in the prehistoric past. Basin floor vegetation is commonly different than piedmont slope vegetation which, in turn, is different from mountain vegetation (Dick-Peddie 1993). Therefore, to simplify the inferences, the curve in figure 17.7 is limited to the piedmont slopes (bajadas) at the LTER site.

The first and earliest interval represented by figure 17.7 is the last full glacial pe-

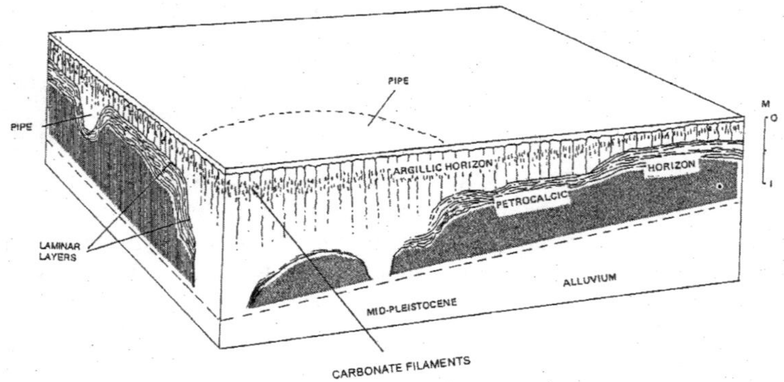

Figure 17.6 Dissolution pipes in a petrocalcic horizon and carbonate filaments of Holocene age in the upper part of the soil profile. (Modified from Gile and Grossman 1979)

riod (about 20,000–18,000 years B.P.). Before this period, less paleoclimatic information is available than after this period for the Jornada region. Notable exceptions include faunal remains in caves in southern New Mexico (Harris 1987). Based on these studies, late Pleistocene climates before the last glacial maximum were relatively mild, cool, and moist compared with today's climate, yet not as cool and moist as the climate during the last full glacial. Still farther back in time, the previous interglacial period (O-isotope stage 5e, about 120,000–130,000 years B.P.) was probably a warm, dry period like the Holocene (Hawley 1993). During the Quaternary as a whole, all of the multiple interglacial periods, each about 100,000 years apart, probably had warm, dry climates during which erosion and sedimentation were active. Glacial periods, in contrast, were each probably periods of landscape stability and soil development.

Interval 1 (20,000 to 18,000 years B.P.) in figure 17.7 represents the last glacial maximum. During this time, the climate was wetter and cooler than today at the Jornada LTER site. The evidence that climate was wetter is based on highstands that occurred at Lake Estancia about 19,700 ^{14}C years B.P. (Allen and Anderson 1993), at Lake King about 19,100 ^{14}C years B.P. (Wilkins and Currey 1997), and at Lake San Agustin about 22,000 to 19,000 years ago (Forester 1987). Radiocarbon ages of carbonate and organic matter in pipes through petrocalcic horizons suggest deeper wetting in soil profiles (Gile et al. 1981). The climate was cooler based on alpine glaciers that occurred at Sierra Blanca during the last full glacial interval (marine-O-isotope stage 2; Hawley 1993). Elevations of rock glaciers in the mountains suggest temperatures 7 to 8°C cooler than today (Blagbrough 1994). These temperatures roughly correspond to temperature estimates of 5 to 7°C cooler than today based on groundwater in northwest New Mexico (Phillips et al. 1986; Stute et al. 1995). Packrat middens record pinyon-juniper-oak vegetation along rock escarpments that are now desert grassland and shrubland (Van Devender 1990). Packrat middens also suggest greater rainfall than today, rather than cold, dry conditions suggested by Galloway (1970, 1983) and Brackenridge (1978). Fossil pollen indi-

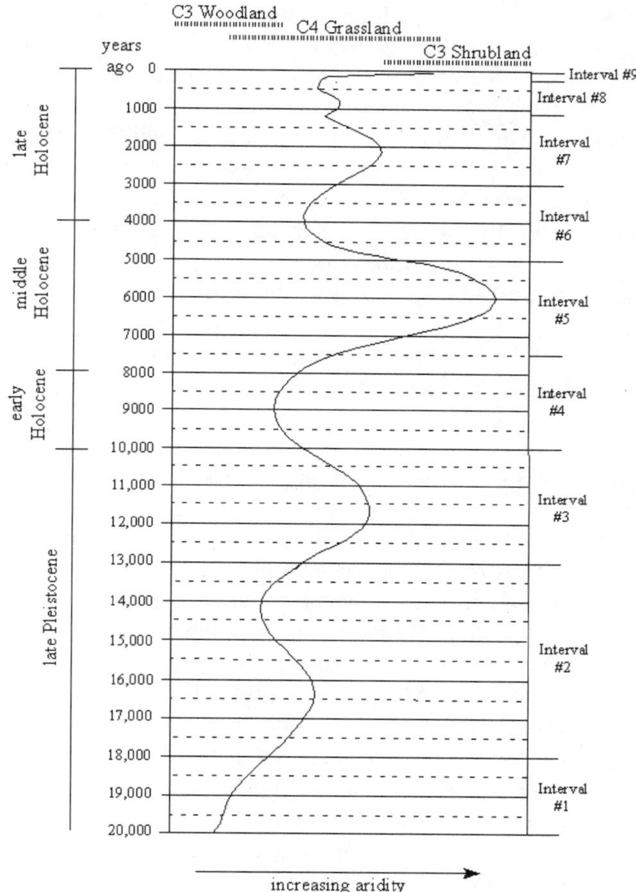

Figure 17.7 Inferred bioclimatic changes on piedmont slopes (bajadas) at the Jornada LTER site (see text for explanation).

cates pinyon at lower elevations (Hall 1997). Carbon isotopes of pedogenic carbonate in both piedmont slope and basin floor soils suggest some C_3 woodland at this time (figure 17.5).

Interval 2 (18,000 to 13,000 years B.P.) in figure 17.7 represents a period having a slight increase in aridity followed by a return to more humid conditions based on the observation that lakes desiccated but returned in the latter part of this interval. At Lake Estancia a highstand is recorded at about 13,700 ^{14}C years B.P. (Allen and Anderson 1993). At Lake Cochise, a highstand is recorded at about 13,700 and 13,400 ^{14}C years B.P. (Waters 1989). Packrat middens indicate that pinyon-juniper-oak vegetation was still present on rock escarpments (Van Devender 1990). Values of $\delta^{13}C$ record peak C_4 grassland at four sites during this period (figure 17.5).

Interval 3 (13,000 to 10,000 years B.P.) represents a period of increased aridity based on lake desiccation. In addition, both packrat and pollen records indicate the loss of pinyon (Van Devender 1990, Hall 1997). Pollen records also suggest the establishment of drier Chenopodiaceae-Asteraceae shrub grassland. Formation of the Isaacks' Ranch geomorphic surface on piedmont slopes during most of this interval suggests landscape instability, erosion, and sedimentation (Hawley 1975, Gile 1987). Compared to the more extensive Organ surface, however, the Isaacks' Ranch surface suggests that this interval had less aridity than the middle Holocene.

Interval 4 (10,000 to 7,500 years B.P.) represents a return to more effective moisture based on a highstand at Lake Cochise at about 9,000 ^{14}C years B.P. (Waters 1989). Packrat middens suggest that winter rainfall was greater than today (Van Devender 1990).

Interval 5 (7,500 to 5,000 years B.P.) represents a major period of aridity. Deflation and dune formation occurred at lake sites (Hawley 1993). Packrat middens record the disappearance of oaks along rock escarpments. Packrat middens also suggest that this was a time when winter rainfall was replaced by biseasonal rainfall dominated by the summer monsoon (Van Devender 1990). Fossil pollen indicates aridity based on the high amount of Cheno-Am pollen (Freeman 1972; Hall 1985; Monger et al. 1998). A major loss of C_4 grassland and an increase in C_3 shrubs is suggested by δ^{13}C values on both piedmont slopes and basin floors (figure 17.5). A major period of aridity is also based on the deposition of Organ I sediments on piedmont slopes (Gile 1975) and Fillmore sediments along the Rio Grande (Gile et al. 1981). Radiocarbon dates of carbonate in the upper zones of soil profiles suggest shallow wetting fronts (Monger et al. 1998). This period corresponds, roughly, to Antevs' hot, dry Altithermal (Antevs 1955) and was a period of major arroyo cutting in southeast Arizona (Waters and Haynes 2001).

Interval 6 (5,000 to 3,000 years B.P.) represents a decrease in aridity based on the return of two small lakes at Lake Cochise after the middle Holocene (Waters 1989). Fossil pollen indicates an increase of grass and a more mesic climate than the middle Holocene (Freeman 1972). Packrat middens, however, suggest that this was a time when the modern climate was established, with fewer winter freezes, monsoonal summer rainfall, and increased droughts, as well as the appearance of creosotebush (Van Devender 1990). On the other hand, landscape stability is suggested by soil development in Organ I sediments during this period (Gile et al. 1981). El Niño events became stronger and more frequent based on an increased frequency of arroyo cutting in southeast Arizona (Waters and Haynes 2001).

Interval 7 (3,000 to 1,100 years B.P.) represents another interval of aridity, albeit less profound than middle Holocene aridity. Evidence for aridity in this interval is based on the deposition of gravelly Organ II sediments that had occurred by 2,200 ^{14}C years B.P. (Gile et al. 1981). Pollen also indicates increased aridity, but less than the aridity of middle Holocene (Freeman 1972). Toward the end of this period, aridity waned as indicated by soil formation in Organ II sediments (Gile et al. 1981).

Interval 8 (1,100 years B.P. to A.D. 1850) represents a still later period of aridity based on erosion and deposition of Organ III sediments (Gile et al. 1981). But this period of erosion and sedimentation was of lower magnitude than even Organ II sediments, which were minor compared to Organ I sediments of middle Holocene age. Toward the end of this period, grasses were much more abundant than today, based on anecdotal evidence (Gardner 1951).

Interval 9 (since A.D. 1850) represents the historical record of bioclimatic change, characterized by the progressive increase of shrublands and the progressive decrease of grasslands based on land survey records of the late-1850s and vegetation surveys after 1915 (Buffington and Herbel 1965). Soil-geomorphic studies also record a major period of wind erosion and formation of mesquite coppice dunes in response to the loss of grasslands (Gile 1966). Unlike bioclimatic changes, this change has been affected by humans in ways ranging from direct land use to increased atmospheric CO_2 from fossil fuel emissions.

Acknowledgments   Funding for various aspects of fossil pollen, carbon isotopes, and radiocarbon analyses was provided by the Jornada LTER program supported by the National Science Foundation, DEB-0080412 and DEB-94111971. Other support was provided by the USDA Jornada Experimental Range, the New Mexico State University Agricultural Experiment Station, and Fort Bliss Military Reservation. Discussions with David R. Cole on terrestrial isotopic records and with Leland H. Gile and John W. Hawley on landscape evolution are greatly appreciated. Grateful acknowledgment is made to Sue Ann Monger, Rebecca Kraimer, Marco Inzunza, and Barbara Nolen for their help with the manuscript.

References

Abrahams, A.D., A.J. Parsons, and J. Wainwright. 1995. Controls and determination of resistance to overland flow on semiarid hillslopes, Walnut Gulch. *Journal of Soil and Water Conservation* 50:457–460.

Allen, B.D., and R.Y. Anderson. 1993. Evidence from western North America for rapid shifts in climate during the last glacial maximum. *Science* 260:1920–1923,

Allred, K.W. 1988. *A field guide to the flora of the Jornada Plain.* Agricultural Experiment Station, Bulletin 739, Las Cruces, New Mexico.

Allred, K.W. 1993. *A field guide to the grasses of New Mexico.* Agricultural Experiment Station, Department of Agricultural Communications, Las Cruces, New Mexico.

Amundson, R., O. Chadwick, C. Kendall, Y. Wang, and M. DeNiro. 1996. Isotopic evidence for shifts in atmospheric circulation patterns during the late Quaternary in mid-North America. *Geology* 24:23–26.

Amundson, R., L. Stern, T. Baisden, and Y. Wang. 1998. The isotopic composition of soil and soil-respired CO_2. *Geoderma* 82:83–114.

Antevs, E. 1955. Geologic-climatic dating in the West. *American Antiquity* 20:317–335.

Ares, F.N. 1974. *The Jornada Experimental Range.* Range Monograph No. 1. Society for Range Management, Denver, Colorado.

Arkley, R.J. 1963. Calculation of carbonate and water movement in soil from climatic data. *Soil Science* 96:239–248.

Bates, R.L., and J.A. Jackson. 1987. *Glossary of geology.* American Geological Institute, Alexandria, Virginia.

Betancourt, J.L., T.R. Van Devender, and P. S. Martin, 1990. *Packrat Middens: The last 40,000 years of biotic change.* The University of Arizona Press, Tucson.

Birkeland, P.W. 1999. *Soils and geomorphology.* Oxford University Press, New York.

Blagbrough, J.W. 1994. Late Wisconsin climatic inferences from rock glaciers in south-central and west-central New Mexico and east-central Arizona. *New Mexico Geology* 16:65–71.

Boutton, T.W. 1991. Stable carbon isotope ratios of natural materials, II. Atmospheric, terrestrial, marine, and freshwater environments. Pages 173–186 in C. Coleman and B. Fry (eds.), *Carbon isotope techniques.* Academic Press, New York.

Brakenridge, G.R. 1978. Evidence for a cold, dry full-glacial climate in the American Southwest. *Quaternary Research* 9:22–40.

Buck, B.J., and H.C. Monger. 1999. Stable isotopes and soil-geomorphology as indicators of Holocene climate change, northern Chihuahuan Desert. *Journal of Arid Environments* 43:357–373.

Buffington, L.C., and C. H. Herbel. 1965. Vegetation changes on a semidesert grassland range. *Ecological Monographs* 35:139–164.

Bull, W.B. 1991. *Geomorphic response to climatic change.* Oxford University Press, New York.

Cerling, T.E. 1984. The stable isotopic composition of soil carbonate and its relationship to climate. *Earth and Planetary Science Letters* 71:229–240.

Cerling, T.E., and J. Quade. 1993. Stable carbon of oxygen isotopes in soil carbonates. Pages 217–231 in P.K. Swart, K.C. Lohmann, J. McKenzie, and S. Savin (eds.), Climate Change in Continental Isotopic Records, American Geophysical Union, Geophysical Monograph 78.

Cerling, T.E., J. Quade, Y. Wang, and J.R. Bowman. 1989. Carbon isotopes in soils and palaesols as paleoecologic indicators. *Nature* 341:138–139.

Cole, D. R., and H. C. Monger. 1994. Influence of atmospheric CO_2 on the decline of C_4 plants during the last deglaciation. *Nature* 368:533–536.

Conley, W., M.R. Conley, and T.R. Karl. 1992. A computational study of episodic events and historical context in long-term ecological processes: Climate and grazing in the northern Chihuahuan desert. *Coenoses* 7:55–60.

Connin, S.L., R.A. Virginia, and C.P. Chamberlain. 1997a. Carbon isotopes reveal soil organic matter dynamics following arid land shrub expansion. *Oecologia* 110:374–386.

Connin, S.L., R.A. Virginia, and C.P. Chamberlain. 1997b. Isotopic study of environmental change from disseminated carbonate in polygenetic soils. *Soil Science Society of America Journal* 61:1710–1722.

Davenport, D.W., B.P. Wilcox, and D.D. Breshears. 1996. Soil morphology of canopy and intercanopy sites in a Pinon-Juniper Woodland. *Soil Science Society of America Journal* 60:1881–1887.

Dick-Peddie, W.A. 1993. *New Mexico vegetation, past, present, and future.* University of New Mexico Press, Albuquerque.

Dokuchaev, V.V. 1883. *Russian Chernozem (Russkii chernozem).* Translated from Russian by Israel Program for Science Translations, 1967, Jerusalem.

Fleischhauer, H.L., Jr., and W.J. Stone. 1982. Quaternary geology of Lake Animas, Hidalgo County, New Mexico. New Mexico Bureau of Mines and Mineral Resources, Socorro, New Mexico. Circular 174.

Forester, R.M. 1987. Late Quaternary paleoclimate records from lacustrine ostracodes. Pages 261–276 in W. F. Ruddiman, and H.E. Wright, Jr. (eds.). *North America and adjacent oceans during the last deglaciation.* Geological Society of America, Boulder, Colorado.

Fredlund, G.G., and L.L. Tieszen. 1997. Phytolith and carbon isotope evidence for late Quaternary vegetation and climate change in the southern Black Hills, South Dakota. *Quaternary Research* 47:206–217.

Fredrickson, E., K.M. Havstad, R. Estell, and P. Hyder. 1998. Perspectives on desertification: Southwestern United States. *Journal of Arid Environments* 39:191–207.

Freeman, C.E. 1972. Pollen study of some Holocene alluvial deposits in Dona Ana County, southern New Mexico: *Texas Journal of Science* 24:203–220.

Gallegos, R.A. 1999. Biogenic carbonate, desert vegetation, and stable carbon isotopes. M.S. thesis, New Mexico State University, Las Cruces.

Galloway, R.W. 1970. The full-glacial climate in the southwestern United States. *Annals of the Association of American Geographers* 60:245–256.

Galloway, R.W. 1983. Full-glacial southwestern United States: Mild and wet or cold and dry. *Quaternary Research* 19:236–248.

Gardner, J.L. 1951. Vegetation of the creosotebush area of the Rio Grande Valley in New Mexico. Ecological Monographs 21:379–403.

Gibbens, R.P., J.M. Tromble, J.T. Hennessy, and M. Cardenas. 1983. Soil movement in mesquite dunelands and former grasslands of southern New Mexico from 1933 to 1980. *Journal of Range Management.* 36:145–148.

Gile, L.H. 1966. Coppice dunes and the Rotura soil. *Soil Science Society of America Proceedings* 30:657–660.

Gile, L.H. 1975. Holocene soils and soil-geomorphic relations in an arid region of southern New Mexico. *Quaternary Research* 5:321–360.

Gile, L.H. 1977. Holocene soils and soil-geomorphic relations in a semi-arid region of southern New Mexico. *Quaternary Research* 7:112–132.

Gile, L.H. 1987. A pedogenic chronology for Kilbourne Hole, southern New Mexico: II. Time of the explosions and soil events before the explosions. *Soil Science Society of America Journal* 51:752–760.

Gile, L.H. 2002. Lake Jornada, an early-middle Pleistocene lake in the Jornada del Muerto Basin, southern New Mexico. *New Mexico Geology* 24:3–14.

Gile, L.H., and R.B. Grossman.1979. *The Desert Project soil monograph.* Soil Conservation, U.S. Department of Agriculture, National Soil Survey Center, Lincoln, Nebraska.

Gile, L.H., and J. W. Hawley. 1966. Periodic sedimentation and soil formation on an alluvial-piedmont in southern New Mexico. *Soil Science Society of America Proceedings* 30:261–268.

Gile, L.H., J.W. Hawley, and R.B. Grossman. 1981. *Soils and geomorphology in the Basin and Range area of southern New Mexico—Guidebook to the Desert Project.* New Mexico Bureau of Mines and Mineral Resources, Memoir 39.

Gile, L.H., F.F. Peterson, and R.B. Grossman. 1966. Morphological and genetic sequences of carbonate accumulation in desert soils. *Soil Science* 101:347–360.

Hall, S.A. 1985. Quaternary pollen analysis and vegetational history of the Southwest. Pages 95–123 in V.M. Bryant, Jr., and R.G. Holloway (eds.). *Pollen Records of Late-Quaternary North American Sediments.* American Association of Stratigraphic Palynologists Foundation, Dallas, Texas.

Hall, S.A. 1997. Pollen analysis and woodrat middens: Re-evalutation of Quaternary vegetation. *Southwestern Geographer* 1:25–43.

Hallenbeck, C. 1950. *Land of the conquistadors.* Caxton Printers, Ltd., Caldwell, Idaho.

Harris, A.H. 1987. Reconstruction of mid-Wisconsin environments in southern New Mexico. *National Geographic Research* 3:142–151.

Havstad, K.M., W.P. Kustas, A. Rango, J.C. Ritchie, and T.J. Schmugge. 2000. Jornada Experimental Range: A unique arid land location for experiments to validate satellite systems. *Remote Sensing of Environment* 73:13–25.

Hawley, J.W. 1965. Geomorphic surfaces along the Rio Grande valley from El Paso, Texas, to Caballo Reservoir, New Mexico. Pages 188–198 in New Mexico Geological Society, Guidebook 16th field conference, Socorro, New Mexico.

Hawley, J.W. 1975. Quaternary history of Dona Ana County region, south-central New Mexico. Pages 139–150 in W.R. Seager, R.E. Clemons, and J.F. Callender (eds.), New Mexico Geological Society, Guidebook 26th field conference, Socorro, New Mexico.

Hawley, J.W. 1986. Physiographic provinces (and) land forms of New Mexico. Pages 28–31 in Williams, J. (ed.), *New Mexico in Maps*. The University of New Mexico Press, Albuquerque, New Mexico.

Hawley, J.W. 1993. Geomorphic setting and late Quaternary history of pluvial-lake basins in the southern New Mexico region. New Mexico Bureau of Mines and Mineral Resources, Open-File Report 387, Socorro, New Mexico.

Hawley, J.W., G.O. Bachman, and K. Manley. 1976. Quaternary stratigraphy in the Basin and Range and Great Plains provinces, New Mexico and western Texas. Pages 235–274 in W.C. Mahaney (ed.), *Quaternary stratigraphy of North America*, Dowden Hutchinson, and Ross, Inc. Stroudsburg, Pennsylvania.

Hawley, J.W., F.E. Kottlowski, W.S. Strain, W.R. Seager, W.E. King, and D.V. LeMone. 1969. The Santa Fe Group in the south-central New Mexico border region. Pages 52–76 in *Border stratigraphy symposium*. F.E. Kottlowski and D.E. Lamone (eds.). New Mexico Bureau of Mines and Mineral Resources, Circular 104.

Herbel, C.H., L.H. Gile, E.L. Fredrickson, and R.P. Gibbens. 1994. Soil water and soils at soil water sites, Jornada Experimental Range. Soil Survey Investigations Report No. 44. Soil Conservation Service, Lincoln, Nebraska.

Herrick, J.E., K.M. Havstad, and D.P. Coffin. 1997. Rethinking remediation technologies for desertified landscapes. *Journal of Soil and Water Conservation* 52:220–224.

Horowitz, A. 1992. *Palynology of arid lands*. Elsevier, Amsterdam.

Huenneke, L.F. and I. Noble. 1996. Ecosystem function of biodiversity in arid ecosystems. Pages 99–128 in H.A. Mooney, J.H. Cushman, E. Medina, O.E. Sala, and E.D. Schulze (eds.) *Functional Roles of Biodiversity: A Global Perspective*. John Wiley and Sons Ltd., New York.

Jenny, H. 1941. *Factors of soil formation*. McGraw-Hill, New York.

Jenny, H., and C.D. Leonard. 1934. Functional relationships between soil properties and rainfall. *Soil Science* 38:363–381.

Kelly, E.F., C. Yonkers, and B. Marino. 1993. Stable carbon isotope composition of paleosols: An application to Holocene. Pages 233–239 in P.K. Swart, K.C. Lohmann, J. McKenzie, and S. Savin (eds.), *Climate Change in Continental Isotopic Records*. American Geophysical Union, Geophysical Monograph 78.Washington, D.C.

Langbein, W.B., and S.A. Schumm. 1958. Yield of sediment in relation to mean annual precipitation. *American Geophysical Union Trans.* 39:1076–1084.

Liu, B., F.M. Phillips, and A.R. Campbell. 1996. Stable carbon and oxygen isotopes of pedogenic carbonates, Ajo Mountains, southern Arizona: Implications for paleoenvironmental change. *Palaeogeography, Palaeoclimatology, Palaeoecology* 124:233–246.

Ludwig, J.A. 1987. Primary productivity in arid lands: Myths and realities. *Journal of Arid Environments* 13:1–7.

Machette, M.N. 1985. Calcic soils of the southwestern United States. Pages 1–21 in D.L. Weide (ed.), *Soils and Quaternary geology of the southwestern United States*. Geological Society of America Special Paper 203. Boulder, Colorado.

Mack, G.H., D.W. Love, and W.R. Seager. 1997. Spillover models for axial rivers in regions of continental extension: The Rio Mimbres and Rio Grande in the southern Rio Grande rift, USA. *Sedimentology* 44:637–652.

Mack, G.H., W.C. McIntosh, M.R. Leeder, and H.C. Monger. 1996. Plio-Pleistocene pumice floods in the ancestral Rio Grande, southern Rio Grande rift, USA: *Sedimentary Geology* 103:1–8.

Mack, G.H., S.L. Salyards, and W.C. James. 1993. Magnetostratigraphy of the Plio-Pleistocene Camp Rice and Palomas Formations in the Rio Grande rift of southern New Mexico. *American Journal of Science* 293:49–77.

Mack, G.H., S.L. Salyards, W.C. McIntosh, and M.R. Leeder. 1998. Reversal magnetostratigraphy and radioisotopic geochronology of the Plio-Pleistocene Camp Rice and Palomas Formation, southern Rio Grande Rift. Pages 229–236 in G.H. Mack, G.S. Austin, and J.M. Baker (eds). New Mexico Geological Society, Guidebook of Las Cruces Country II, Socorro.

Malm, N.R. 1994. Climatic guide, Las Cruces, 1892–1991, 100 years of weather records. Agricultural Experiment Station, Research Report 682. New Mexico State University, Las Cruces.

Markgraf, V., J.P. Bradbury, R.M. Forester, G. Singh, and R.S. Sternberg. 1984. San Agustin Plains, New Mexico: Age and paleoenvironmental potential reassessed. *Quaternary Research.* 22:336–343.

Mayer, L., L.D. McFadden, and J.W. Harden. 1988. Distribution of calcium carbonate in desert soils: A model. *Geology* 16:303–306.

McAuliffe, J.R. 1994. Landscape evolution, soil formation, and ecological patterns and processes in Sonoran Desert Bajadas. *Ecological Monographs* 64:111–148.

McFadden, L.D., and J.C. Tinsley. 1985. Rate and depth of pedogenic-carbonate accumulation in soils: Formation and testing of a compartment model. Pages 23–42 in D.L. Weide (ed.), *Soils and Quaternary Geology of the Southwestern United States.* Geological Society of America Special Paper 203, Boulder, Colorado.

McPherson, G.R., and J.F. Weltzin. 2000. Disturbance and climate change in United States/Mexico borderland plant communities: A state-of-the-knowledge review. General Technical Report RMRS-GTR-50, U.S. Forest Service. U.S. Government Printing Office. Washington, D.C.

Monger, H.C., D.R. Cole, J.W. Gish, and T.H. Giordano. 1998. Stable carbon and oxygen isotopes in Quaternary soil carbonates as indicators of ecogeomorphic changes in the northern Chihuahuan Desert, USA. *Geoderma* 82:137–172.

Moore, P.D., J.A. Webb, and M.E. Collinson. 1991. Pollen analysis. Blackwell Scientific Publication, London.

Morgan, G.S., S.G. Lucas, and J.W. Estep.1998. Pliocene (Blancan) vertebrate fossils from the Camp Rice formation near Tonuco Mountain, Dona Ana County, southern New Mexico. Pages 237–249 in G.H. Mack, G.S. Austin, and J.M. Baker (eds.), New Mexico Geological Society, Guidebook of Las Cruces Country II, Socorro.

Morrison, R.B. 1991. Introduction. Pages 1–12 in R.B. Morrison (ed.), *Quaternary Nonglacial Geology: Conterminous U.S.* Geological Society of America, Volume K-2. Boulder, Colorado.

Neal, J.T., R.E. Smith, and B.F. Jones. 1983. Pleistocene Lake Trinity, an evaporite basin in the northern Jornada del Muerto, New Mexico. Pages 285–290. New Mexico Geological Society, 34th Annual Field Conference, Guidebook, Socorro.

Nordt, L.C., C.T. Hallmark, L.P. Wilding, and T.W. Boutton. 1998. Quantifying pedogenic carbonate accumulations using stable carbon isotopes. *Geoderma* 82:115–136.

Pease, D.S. 1967. Opal phytoliths as indicators of paleosols. M.S. thesis, New Mexico State University, Las Cruces. 81 pp.

Peterson, F.F. 1981. Landforms of the Basin & Range province defined for soil survey. Technical Bulletin 28, Nevada Agricultural Experiment Station, University of Nevada, Reno.

Phillips, F.M., L.A. Peeters, M.K. Tansey, and S.N. Davis. 1986. Paleoclimatic inferences from an isotopic investigation of groundwater in the central San Juan Basin, New Mexico. *Quaternary Research* 26:179–193.

Pieper, R.D., and R.F. Beck. 1990. Range condition from an ecological perspective: Modifications to recognize multiple use objectives. *Journal of Range Management* 43:550–552.

Reeves, C.C., Jr. 1969. Pluvial Lake Palomas, northwestern Chihuahua, Mexico. Pages 143–154. New Mexico Geological Society, 20th Annual Field Conference, Guidebook, Socorro.

Richmond, G.M. 1986. Stratigraphy and correlation of glacial deposits of the Rocky Mountains, the Colorado Plateau, and ranges of the Great Basin. Pages 99–127 in V. Sibrava, D.Q. Bowen, and G.M. Richmond (eds.), *Quaternary Glaciation in the Northern Hemisphere*. Pergamon Press, New York.

Ruhe, R.V. 1962. Age of the Rio Grande valley in southern New Mexico. *Journal of Geology* 70:151–167.

Schlesinger, W.H., J.F. Reynolds, G.L. Cunningham, L.F. Huenneke, W. M. Jarrell, R.A. Virginia, and W. G. Whitford. 1990. Biological feedbacks in global desertification. *Science* 247:1043–1048.

Schmidt, R.H., Jr. 1979. A climatic delineation of the "real" Chihuahuan Desert. *Journal of Arid Environments* 2:243–250.

Seager, W.R., J.W. Hawley, F.E. Kottlowski, and S.A. Kelley. 1987. Geology of east half of Las Cruces and northeast El Paso 1E x 2E sheets: New Mexico Bureau of Mines and Mineral Resources, Geologic Map 57, 1:125,000.

Seager, W.R., M. Shafiqullah, J.W. Hawley, and R. Marvin. 1984. New K-Ar dates from basalts and the evolution of the southern Rio Grande rift. *Geological Society of America* 95:87–99.

Smith, H.T.U., and L.L. Ray. 1941. Southernmost glaciated peak in the United States. *Science* 93:209.

Spaulding, W.G. 1991. A middle Holocene vegetation record from the Mojave Desert of North America and its paleoclimatic significance. *Quaternary Research* 35:427–437.

Stute, M., J.F. Clark, P. Schlosser, and W.S. Broecker. 1995. A 30,000 yr continental paleotemperature record derived from noble gases dissolved in groundwater from the San Juan Basin, New Mexico. *Quaternary Research* 43:209–220.

Syvertsen, J.P., G.L. Nickell, R.W. Spellenberg, and G.L. Cunningham. 1976. Carbon reduction pathways and standing crop in three Chihuahuan Desert plant communities. *Southwest Naturalist* 21:311–320.

Tedford, R.H. 1981. Mammalian biochronology of the late Cenozoic basins of New Mexico. *Geological Society of America Bulletin* 92:1008–1022.

USDA-ARS. 2000. *http://www.usda-ars.nmsu.edu.*

USDA-NRCS, 1999. New Mexico annual precipitation 1961–1990. USDA-NRCS National Cartography and Geospatial Center, Fort Worth, Texas.

Van Devender, T.R. 1990. Late Quaternary vegetation and climate of the Chihuahuan Desert, United States and Mexico. Pages 104–133 in J.L. Betancourt, T.R. Van Devender, and P.S. Martin, (eds.), *Packrat Middens: The last 40,000 years of biotic change.* The University of Arizona Press, Tucson.

Wang, Y., E. McDonald, R. Amundson, L. McFadden, and O. Chadwick. 1996. An isotopic study of soils in chronological sequences of alluvial deposits, Providence Mountains, California. *Geological Society of America Bulletin* 108:379–391.

Waters, M.R. 1989. Late Quaternary lacustrine history and paleoclimatic significance of pluvial Lake Cochise, southeastern Arizona. *Quaternary Research* 32:1–11.

Waters, M.R., and C.V. Haynes. 2001. Late Quaternary arroyo formation and climate change in the American Southwest. *Geology* 29:399–402.

Whitford, W.G. 1996. The importance of biodiversity of soil biota in arid ecosystems. *Biodiversity and Conservation.* 5:185–195.

Wilkins, D.E., and D.R. Currey. 1997. Timing and extent of late Quaternary paleolakes in the Trans-Pecos closed basin, west Texas and south-central New Mexico. *Quaternary Research* 47:306–315.

Wright, R.A. 1982. Aspects of desertification in Prosopis dunelands of southern New Mexico, U.S.A. *Journal of Arid Environments* 5:277–284.

18

Millennial and Century Climate Changes in the Colorado Alpine

Scott Elias

Introduction

Ecosystems are the products of regional biotic history, shaped by environmental changes that have occurred over thousands of years. Accordingly, ecological changes take place at many timescales, but perhaps none is more significant than the truly long-term scale of centuries and millennia, for it is at these timescales that ecosystems form, break apart, and reform in new configurations. This is certainly true in the alpine regions, where glaciations have dominated the landscape for perhaps 90% of the last 2.5 million years (Elias 1996a). In the alpine tundra zone, the periods between ice ages have been relatively brief (10,000–15,000 years), whereas glaciations have been long (90,000–100,000 years). Glacial ice has been the dominant force in shaping alpine landscapes. Glacial climate has been the filter through which the alpine biota has had to pass repeatedly in the Pleistocene.

This chapter discusses climatic events during the last 25,000 years (figure 18.1). At the beginning of this interval, temperatures cooled throughout most of the Northern Hemisphere, culminating in the last glacial maximum (LGM), about 20,000–18,000 yr B.P. (radiocarbon years before present). The Laurentide and Cordilleran ice sheets advanced southward, covering most of Canada and the northern tier of the United States. Glaciers also crept down from mountaintops to fill high valleys in the Rocky Mountains. In the Southern Rockies, the alpine tundra zone crept downslope into what is now the subalpine, beyond the reach of the relatively small montane glaciers. By about 14,000 yr B.P., the glacier margins began to recede, leading eventually to the postglacial environments of the Holocene. It is now becoming apparent that the climate changes that drove these events were sur-

Geologic Epoch	Physical environment	MCR temperature reconstructions	Vegetation history	Chronology (^{14}C yr BP)
Holocene	Glaciers expand during Little Ice Age	TMAX oscillating with warm intervals at 3500 and 900 yr BP; cold intervals at 3000 and 200 yr BP	Treeline descends after 4000 yr BP, then remains ± stable after 2000 yr BP	
	Glaciers retreat to high mountain cirques; postglacial alpine soils begin to develop	TMAX > modern from 10,000 to 5300 yr BP	Treeline rises to modern levels and beyond from 9000–4000 yr BP	5000 / 10,000
Late Glacial Interstadial	Glacial margins retreating	Beetles indicate rapid warming by 13,200 yr BP	No upslope movement of forests	15,000
Late Pleistocene	Mountain glaciers cover regions above 2450 m; widespread permafrost in high elevations	Very cold, dry climate on Eastern Slope; TMAX depressed by 10–11°C	Treeline depressed by 300–700m; TMAX depressed by 2–5°C	20,000

Figure 18.1 Summary of geologic epochs in the Late Quaternary, associated paleoenvironmental changes at high elevations in Colorado, and radiocarbon chronology. TMAX stands for the mean temperature of the warmest month of the year. Glaciological data are from Madole and Shroba (1979). MCR temperature reconstructions are from this chapter. Vegetation history data is from Fall (1997).

prisingly rapid and intense. This chapter examines the evidence for these climatic changes and the biotic response to them in the alpine zone of Colorado.

To reconstruct the environmental changes of this period, we must rely on proxy data, that is, the fossil record of plants and animals, combined with geologic evidence, such as the age and location of glacial moraines in mountain valleys. As of this writing, the principal biological proxy data that have been studied in the Rocky Mountains are fossil pollen and insects. This chapter focuses mainly on the fossil insect record because it has supplied quantitative estimates of past climates. For the most part, pollen analysis in this region has provided only qualitative climate reconstructions, although there are exceptions (i.e., Fall 1997).

Methods

The remains of beetles are very valuable as proxy data, that is, as indirect evidence for past environmental conditions. Beetles are the largest order of insects. They have been the main insect group studied from Quaternary sediments, and, in fact, they are the most diverse group of organisms on Earth, with more than one million species known to science (Crowson 1981). In addition, their exoskeletons, reinforced with chitin, are extremely robust and are commonly preserved in large numbers in lake sediments, peats, and other types of deposits. In most cases, beetles have quite specialized habitats that apparently have not changed appreciably during the Quaternary (Elias 1994). This characteristic makes them excellent environmental indicators. The exoskeletons of beetles and some other insects are covered with exquisite microsculpture, enabling paleontologists to identify fossil exoskeletons to the species level in at least half of all preserved specimens, even though insect exoskeletons are most often broken up into the individual plates in fossil specimens.

Beetles are very quick to colonize a region when suitable habitats become available. They often respond more quickly than plants, which, until recently, were relied on almost exclusively as indicators of environmental change on land. Like plant macrofossils, insect fossils are generally deposited in the catchment basin in which the specimens lived. Thus they provide a record of local conditions, in contrast to pollen, which can be carried many miles on winds and often gives a more regional "signal."

Studies of insect fossils in two-million-year-old deposits from the high arctic have failed to show any significant evidence of either species evolution or extinction. Beetle species have apparently remained constant for as many as several million generations (Elias 1994)

Insect fossils are generally extracted from organic-rich lake or pond sediments or peats. Ancient stream flotsam, deposited in fluvial sediments and later exposed along stream banks, is often a rich source of insect fossils.

Insect fossil data are usually presented as minimum numbers of individuals for each species identified. Paleoclimatic reconstructions are generally made on the basis of the climatic conditions in the region where the species in a given assemblage can be found living together today, that is, the climate of the region where their modern distributions overlap. This method has recently been refined, by focusing on the climatic conditions associated with beetle species' modern ranges (the "climate envelope" of the species), rather than on the geographic overlap of their modern distributions. This is called the Mutual Climatic Range (MCR) technique.

The MCR Technique

The MCR technique is based on the assumption that the present climatic tolerance range of a species can be applied to its Quaternary fossil record, so that fossil occurrences of a given species imply a paleoclimate that was within the same tolerance range. MCR studies focus on predators and scavengers, because these groups are assumed to show the most rapid response to climate change. The predators are nearly all generalists that prey on a wide variety of small arthropods. Plant-feeding groups are not considered, because these species cannot become established in new regions until their host plants arrive. In contrast to this, predators and scavengers have been shown to be able to shift distributions at a continental scale in a few tens or hundreds of years. This has been demonstrated in the fossil record of western Europe (Coope 1977) and of North America (Elias 1994), and has been clearly demonstrated for fossil faunas from the Rocky Mountain region (Elias 1991). The fossil assemblages considered in the Rocky Mountain regional study include 74 species in the families Carabidae (ground beetles), Dytiscidae (predaceous diving beetles), Hydrophilidae (water scavenger beetles), Staphylinidae (rove beetles), Scarabaeidae (dung beetles), and Coccinellidae (ladybird beetles).

To determine the climatic tolerances of the beetles in the fossil assemblages, I developed a climate envelope for each species, based on the mean July and mean January temperatures of all the North American locations where the species presently occur. The temperature regimes of these localities were plotted on a diagram of the mean July temperature versus the difference between the mean July and mean January temperature, based on a 25-km-grid North American climate database (Bartlein et al. 1994). This database was used to pair temperature regimes with the modern beetle collection sites, using the geographically nearest grid location to each collecting site. To test MCR accuracy for North American fossil beetle assemblages, I developed a linear regression model (Elias et al. 1996) that tested predicted versus observed modern temperatures at sites with meteorological stations. The regression equations were used to calibrate paleotemperature estimates made by the MCR method. I used the MCR method to predict the modern mean July and mean January temperatures of 35 sites in North America, based on the overlap of climate envelopes of beetle species that live at these sites. The climate envelopes were developed for species found in the Wisconsin-age fossil assemblages (Elias et al. 1996), all of which are extant. A linear regression of observed versus predicted mean July temperatures yielded an r^2 value of 0.94. A regression of observed versus predicted mean January temperatures yielded an r^2 value of 0.82. The slopes of predicted versus actual temperatures were $0.78\pm.03$ and $0.72\pm.06$, respectively. The linear regressions of predicted on observed TMAX and TMIN values yielded the following equations:

$$\text{TMAX [calibrated]} = [\text{median predicted TMAX} \times 0.787] + 3.4$$
$$\text{TMIN [calibrated]} = [\text{median predicted TMIN} \times 0.716] - 4.9$$

The standard errors of the regressions were $\pm0.7°C$ for TMAX and $\pm10°C$ for TMIN. This indicates that MCR estimates of mean July temperature are probably far more reliable than MCR estimates of mean January temperature. This makes sense from an ecological standpoint, because mountain-dwelling beetles are only active during the summer months.

In this study, I applied MCR analysis to 21 fossil beetle assemblages from eight sites spanning the interval 14,500–400 yr B.P. (table 18.1) and used the linear regression equations (given previously) to calibrate the data. The sites are in the Colorado Front Range region of the Rocky Mountains (figure 18.2). Although there

Table 18.1 Site data and summary of modern and paleoclimatic data

Site	Elevation (m asl)	Sample Age (^{14}C yr BP)	Late Quaternary				Modern		Change in Temperature	
			TMAX (°C)	TMAX calibrated (°C)	TMIN (°C)	TMIN calibrated (°C)	TMAX (°C)	TMIN (°C)	July ΔT	January ΔT
Lamb Spring	1731	14,500 ±500	10–11	11.7	−31 to −27	−25.8	21.4	−1.3	−9.7	−24.5
Mary Jane	2882	13200	9.8–10.2	11.3	−29.3 to −27.6	−25.4	13.4	−8.6	−2.1	−16.8
Mary Jane	2882	12800	10–10.2	11.4	−29.1 to −27.6	−25.3	13.4	−8.6	−2	−16.7
Sky Pond	3320	10000	8.3–10.0	10.6	−33.2 to −27.2	−26.6	10.0	−10.7	+0.6	−15.9
La Poudre Pass	3100	9850±300	15–18	16.4	−17.5 to −7	−13.7	11.3	−7.7	+5.1	−6.0
Lake Isabelle Delta	3323	9000±285	11.75–14.5	13.8	−31.25 to −15	−21.6	10.8	−8.2	+3.0	−13.3
Sky Pond	3320	8950	9–15.5	13.1	−33 to −17.5	−23.1	10.0	−10.7	+3.1	−12.4
La Poudre Pass	3100	8800±90	13.5–16.5	15.2	−19.5 to −9	−15.2	11.3	−7.7	+3.9	−7.5
Lake Isabelle Delta	3323	8500	10.5–13	12.7	−23.5 to −16	−19.1	10.8	−8.2	+1.9	−10.9
Lake Isabelle Delta	3323	7800±255	11–13	12.9	−14 to −9	−13.2	10.8	−8.2	+2.1	−5
Lake Isabelle Fen	3325	7080±90	10.25–13	12.6	−14.75 to −7.5	−12.9	10.8	−8.2	+1.8	−4.7
Sky Pond	3320	6500	11–15.5	13.9	−27.5 to −17	−20.9	10.0	−10.7	+3.9	−10.2
La Poudre Pass	3100	5360±90	12.5–13.5	13.7	−12.5 to −11.5	−13.5	10.8	−8.2	+2.9	−5.3
Sky Pond	3320	5250	7.5–10.3	10.4	−29.5 to −17	−21.6	10.0	−10.7	+0.4	−10.9
La Poudre Pass	3100	3485±180	11.75–15	14.0	−21.25 to −14.5	−17.8	11.3	−7.7	+2.7	−10.1
Lake Isabelle Fen	3325	3000	10.25–13	12.6	−14.5 to −7.5	−12.8	10.8	−8.2	+1.8	−4.6
Longs Peak Inn	2732	2965±75	12–15.5	14.2	−26.5 to −14	−19.5	14.5	−5	−0.3	−14.5
Longs Peak Inn	2732	2680±80	13.5–15.5	14.9	−24 to −15	−18.9	14.5	−5	+0.4	−13.9
Roaring River	2800	2400±130	14.25–14.75	14.9	−18.25 to −16.5	−17.4	14.3	−5.5	+0.6	−11.9
Mount Ida Bog	3520	900±150	10.25–12	12.2	−14.75 to −9	−13.4	9.3	−11.5	+2.9	−1.9
Longs Peak Inn	2732	395±100	13.5–15.5	14.9	−24.5 to −15		14.5	−5	+0.4	−14.1

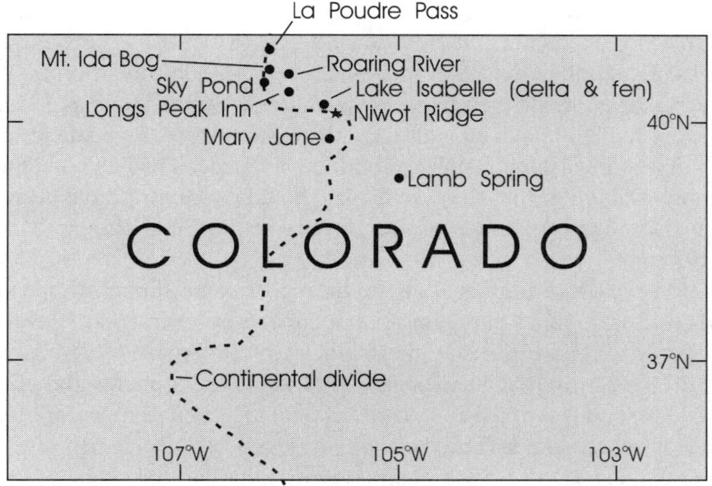

Figure 18.2 Map of Colorado, showing location of fossil sites discussed in the text and the Niwot Ridge LTER site.

are many gaps in the transect (both spatially and temporally), the available sites provide sufficient information to allow an initial paleoclimatic reconstruction.

Results and Discussion

The MCR reconstructions provided estimates of mean July and January temperatures since the last glaciation. The TMAX estimates are generally well constrained, and these results are discussed in detail here. The TMIN estimates are generally poorly constrained, and so they are only referred to occasionally in the following discussion.

Late Pleistocene History of the Front Range Region

The Wisconsin Glaciation is called the Pinedale Glaciation in the Rocky Mountain region, after terminal moraines near the town of Pinedale, Wyoming. The Pinedale Glaciation began about 110,000 yr B.P. and included at least two major ice advances and retreats in most regions of the Rocky Mountains. The history of glaciation is not as well worked out for the Colorado Front Range region as it is for regions in the Central and Northern Rockies. For example, geologists have documented three separate ice advances in the Teton Range during Pinedale times (Pierce and Good 1992). In northern Colorado we know that there were earlier and later Pinedale ice advances, but we do not know when the earlier advance (or multiple advances) took place. However, based on geologic evidence (Madole and Shroba 1979), the early Pinedale glaciation was more extensive than that of the late Pinedale. Early Pinedale moraines can be seen near the western edge of the town of Estes Park, whereas late Pinedale ice formed moraines several kilometers up-valley.

Geologic evidence indicates that during the last glaciation, Pinedale glaciers flowed out of high mountain cirques, down-valley to elevations between 2440 and 2470 m, the elevation of the lower montane forests (Madole and Shroba 1979). On the western slope, Pinedale glaciers extended downslope as much as 33 km. These glaciers were fed by Pacific moisture. On the eastern slope, Pinedale ice advanced only 14–15 km downslope from the continental divide. Then as now, this region was in a rain shadow for westerly moisture. Pinedale ice may have been as much as 450 m thick near the heads of the glaciers in the Front Range (Madole and Shroba 1979).

We now have good estimates of the timing of some Pinedale glacial events, based on radiocarbon ages of organic-rich sediments at several high-elevation sites. At the Mary Jane ski area, near Winter Park, Colorado (figure 18.2), an excavation for a ski lift tower exposed a series of alternating lake sediments and glacial tills. The oldest lake bed was dated at about 30,000 radiocarbon years before present (yr B.P.) (Nelson et al. 1979). This bed was overlain by glacial till, and the next youngest lake bed yielded a radiocarbon age at the base of about 13,750 yr B.P. Based on the Mary Jane sequence, it appears that the last major ice advance of the Pinedale Glaciation took place between the time of deposition of the older and younger lake beds, so between 30,000 and 13,750 yr B.P.

At Devlins Park, near Lake Isabelle in the Indian Peaks Wilderness area (figure 18.2), Legg and Baker (1980) studied sediments from a lake that was dammed by late Pinedale ice. During the time that Glacial Lake Devlin existed (22,400–12,200 yr B.P.), ice covered the Devlins Park region. The lake drained as the ice retreated, and the youngest sediments from this lake provide a limiting age for this event. Presumably the late Pinedale glacier that advanced downslope from the continental divide west of Devlins Park area took some centuries to reach that elevation (2953 m), so the glacial advance began before 22,400 yr B.P. The terminal moraine of this glacier is located 2.3 km downslope from the study site.

The commencement of alpine peat bog growth has been used to date the retreat of late Pinedale ice from montane valleys back to the alpine zone where they originated. At La Poudre Pass (figure 18.2), Madole (reported in Elias 1983) obtained a radiocarbon date of 10,000 yr B.P. from peat that formed in a bog after Pinedale ice retreated. The pass, which is located at modern treeline, was free of ice prior to that. In the Indian Peaks Wilderness, sediments began accumulating in Long Lake, in the upper subalpine zone north of Niwot Ridge, by 12,000 yr B.P. (Short 1985). Studies in the San Juan Mountains of southern Colorado suggest that the melting of mountain glaciers began after 14,000 yr B.P. (Carrara et al. 1984), so the process of deglaciation was relatively rapid, probably because the glaciers were not very thick compared to glaciers farther north, and the relatively low latitude of the Southern Rockies (the Rocky Mountain ranges from southern Wyoming to central New Mexico) is associated with greater insolation than that of more northerly regions.

Late Pinedale Environments of the Front Range

Paleoclimatic reconstructions for the Rocky Mountain region indicate that the Colorado Front Range received less moisture than ranges to the north (in the Yellowstone region) and the south (the San Juan Mountains). Consequently, the mountain glaciers of the Front Range region were small, and glaciers from most drainages did not coalesce to form larger glaciers or ice sheets. Paleoclimatic reconstructions based on fossil insect assemblages from the Front Range region (Elias 1986, 1996b) indicate that mean July temperatures were as much as 10–11°C colder than modern temperatures as late as 14,500 yr B.P., and mean January temperatures were depressed by as much as 26°C, compared with modern climate. The fossil insect data, therefore, suggests that the temperature regime during the last glaciation was cold enough to foster the growth of glacial ice. The lack of substantive glaciers in the Front Range appears to have been caused by a lack of sufficient winter precipitation to develop the necessary alpine snow pack.

The oldest Pinedale site in the region that has yielded paleoenvironmental data is the Mary Jane site, near Winter Park. Pollen in lake sediments laid down during an interstadial interval before the last major Pinedale ice advance (circa 30,000 yr B.P.) records a sequence of vegetation beginning with open spruce-fir forest with herbs and shrubs, adjacent to the lake. This was followed by a colder phase, in which alpine tundra replaced the subalpine forest (Short and Elias 1987). The youngest (uppermost) sediments in this lake bed reflect climatic amelioration, as indicated by the return of spruce forest to the vicinity before the advance of late Pinedale ice. The Mary Jane site is at an elevation of 2882 m, in the lower part of the modern subalpine forest. The existence of alpine tundra at this site in mid-Pinedale times translates into a depression of tree line by more than 500 m. This, in turn, corresponds to a climatic cooling of at least 3°C from average modern summer temperatures.

Paleotemperature Estimates from the MCR Study

The earliest indications of climatic amelioration were found at the Mary Jane site, where peat layers were deposited after the retreat of late Pinedale ice. Short and Elias (1987) reported on pollen and insect remains from peat layers ranging in age from 13,740 to 12,350 yr B.P. Fossil evidence from layers dated 13,740–12,700 yr B.P. suggest open ground environments with flora and insect fauna associated with alpine tundra habitats. Elias (1996b) performed an MCR reconstruction of mean July and January temperatures from a fossil beetle assemblage dated 13,200 yr B.P. and 12,800 yr B.P., respectively. These assemblages showed that TMAX values had risen quite dramatically from previous full glacial conditions. Mean July temperatures reconstructed for these Mary Jane assemblages were only 3.2–3.6°C cooler than present, although mean January temperatures remained 19–20°C cooler than present.

Unfortunately, there is a temporal gap in the fossil insect data between about 12,500 and 10,000 yr B.P. From a paleotemperature perspective, this is one of the most interesting and potentially oscillating intervals of the late glacial period. Elsewhere in the Northern Hemisphere, for instance, in Alaska (Elias 2000) and in Northwest Europe (Coope and Lemdahl 1995), a major climatic oscillation (the Younger Dryas interval in Europe) occurred between 11,000 and 10,000 yr B.P. It remains to be seen whether this oscillation took place in the Rocky Mountains, although there is some evidence for glacial readvances in the Canadian Rockies at this time (Reasoner et al. 1994).

Early Holocene Environments

During the Holocene, the Colorado Front Range experienced a series of climatic fluctuations. Insect assemblages from several sites are indicative of warmer-than-present summer temperatures and colder-than-present winter temperatures. The earliest Holocene records in the insect fossil study transect come from Sky Pond and La Poudre Pass. Sky Pond is an alpine pond in Rocky Mountain National Park. A fossil beetle assemblage from 10,000 yr B.P. yielded a calibrated MCR estimate of mean July temperature that is approximately 3°C warmer than modern. The La Poudre Pass site is a peat bog situated near tree line near Cameron Pass, just north of Rocky Mountain National Park. Here, an assemblage dated 9850 yr B.P. yielded a calibrated MCR estimate of mean July temperature that is about 5°C warmer than modern. This assemblage represents the greatest degree of summer warming of the entire 14,000-year record in the Rocky Mountain region. Winter temperatures were as much as 10°C colder than modern temperatures, however, so the degree of continentality also reached a peak at this time. These predictions based on fossil beetle data agree well with Berger's (1978) reconstruction of incoming solar radiation (insolation), based on the Milankovitch insolation model, which predicts a summer insolation maximum and winter insolation minimum in the midlatitudes of the Northern Hemisphere from about 9000-12,200 ^{14}C yr B.P. (10,000–14,200 calendar yr B.P.) (figure 18.3). This peak in summer insolation coincides precisely with the MCR estimates of the postglacial warming in the Colorado Rockies. The fossil insect record is the only fossil data source from the Rocky Mountain region to register this degree of warming in early postglacial times, consistent with glaciological data that suggest rapid melting of regional glaciers before 12,000 yr B.P. (Madole and Shroba 1979). Evidence from the San Juan Mountains indicates that the major glaciers in that region had melted as early as 15,000 yr B.P. (Carrara et al. 1984).

By 9000 yr B.P., the fossil insect data indicate that summer temperatures were already declining from an early Holocene peak, though they were still above modern values (figure 18.3). A fossil insect assemblage from Lake Isabelle, a subalpine lake in the Indian Peaks Wilderness area, yielded calibrated MCR reconstructions indicating mean July temperatures 3°C warmer than modern temperatures and mean January temperatures well below modern levels.

Mid-Holocene Environments

From 7800 to 3000 yr B.P., insect fossil assemblages from La Poudre Pass and Lake Isabelle show a gradual summer cooling trend. The 7800 yr B.P. assemblage from Lake Isabelle yielded a calibrated MCR estimate of mean July temperature 2.1°C warmer than modern temperatures. The 5250 yr B.P. assemblage from Sky Pond yielded a mean July temperature estimate 0.4°C warmer than modern levels. This is the oldest Holocene assemblage that yielded a TMAX range that dipped near the

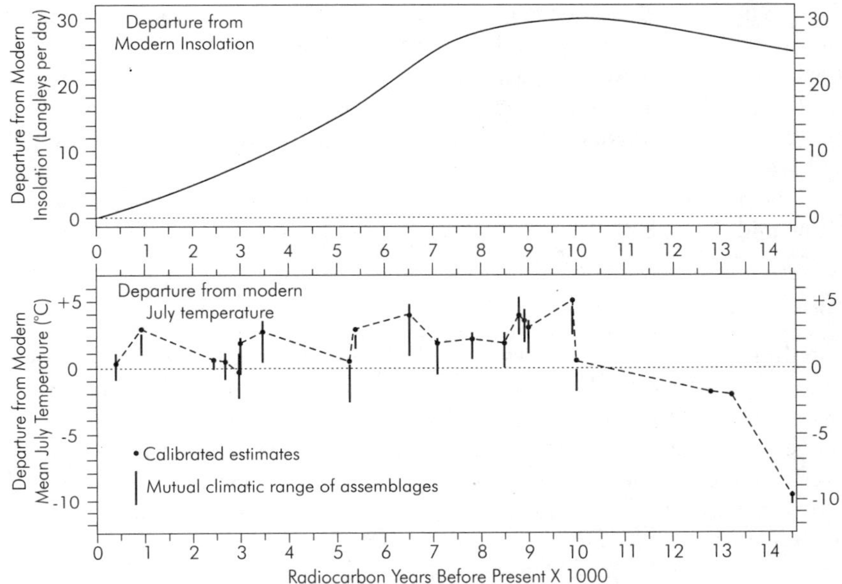

Figure 18.3 (Top) Insolation curve for 45° N latitude during the last 15,000 years. Data from Berger (1978). (Bottom) MCR reconstruction of late Pinedale and Holocene mean July temperatures, as indicated by fossil beetle assemblages from the Colorado Front Range region, shown as departures from modern mean temperatures at the study sites.

modern baseline level. However, most assemblages that date between 7000 and 3000 yr B.P. yielded TMAX estimates that were warmer than modern values by 2–3°C. According to the MCR reconstructions of TMIN values, winter temperatures remained below modern levels throughout the mid-Holocene. Winter temperatures below modern mean values persisted in the study region until the last 1000 years. Again, this is in agreement with the insolation curve for midlatitudes in the Northern Hemisphere.

The fossil insect MCR reconstructions shed new light on the question of a mid-Holocene thermal event. The concept of a hot, dry, "altithermal" climatic regime from 7500 to 4000 yr B.P. was first invoked by Antevs (1948), based on archaeological evidence from the northern Great Basin. Benedict (1979) called for a mid-Holocene altithermal period from 7500–5000 yr B.P., based on shifting land-use patterns in the Archaic cultural period. Human adaptations may have been driven more by drought cycles than by changes in temperature. The fossil insect record of the Rockies does not provide estimates of past moisture regimes, but it provides paleotemperature estimates that contradict the theory of a mid-Holocene thermal maximum. In terms of summer temperatures, the thermal maximum for the postglacial period in the Rocky Mountains of Colorado took place between 11,000 and 9000 yr B.P.. There was a second peak in summer temperatures between about 7000 and 5000 yr B.P., but on a slightly lesser scale than the earlier peak.

Pollen evidence from the eastern slope of the Colorado Rockies suggests that the tree line shifted upslope to elevations beyond its modern limit during the interval from 7000 to 3500 yr B.P. (Short 1985; Fall 1985). Evidence from the western slope suggests that trees migrated above modern tree line there from 9000–4000 yr B.P. (Fall 1997). These upward shifts in tree line have generally been interpreted as being driven by climatic warming. The conflicting interpretations of insect, pollen, and archaeological data during the mid-Holocene interval are puzzling, but highlight the need for additional regional studies to clarify the climatic reconstruction of the mid-Holocene.

Late Holocene Environments

Late Holocene insect records from the Colorado Front Range show a progression from warmer-than-modern to cooler-than-modern summers, and back to warm again. At 3000 yr B.P., the calibrated TMAX estimate from Lake Isabelle was 1.8°C above the modern value. An assemblage just a few decades younger (and in fact, overlapping in radiocarbon age) from Longs Peak Inn yielded a calibrated TMAX estimate 0.3°C cooler than modern levels. Mean summer temperatures apparently remained near modern levels until sometime after 2400 yr B.P. A brief warming pulse was inferred from a 900-yr-B.P. assemblage from Mount Ida Bog, then temperatures returned to near-modern levels by 400 yr B.P. Winter temperatures finally warmed to near modern levels at 900 yr B.P., then cooled again by 400 yr B.P. The 900-yr-B.P. warming may correspond to what historians refer to as the "Medieval warm period." The subsequent cooling, or "Little Ice Age," is suggested by the cooling in summer temperatures, but it is more strongly indicated by a cooling of mean January temperatures by perhaps 14°C below modern levels. Additional well-dated late Holocene insect assemblages are needed to clarify the timing and intensity of climatic change during the last few thousand years.

It would be appropriate to compare the paleotemperature estimates derived from the fossil insect data to estimates derived from other proxies, such as tree rings, pollen, and glacier mass balance reconstructions. However, this is not possible at the present time for the following reasons. Tree-ring research in the Colorado Front Range has largely been limited to studies of the past few centuries (Mast et al. 1998; Veblen et al. 2000) and has focused more on reconstruction of drought episodes than on paleotemperatures (Woodhouse 2001). Pollen studies from this region have only provided general outlines of changes in temperature regime; few quantitative temperature estimates have been attempted. One exception to this was Fall's (1997) study of tree line fluctuations on the western slope of the Colorado Rockies. Fall estimated that temperatures near the upper tree line in her study region were 2–5°C cooler than modern temperatures before 11,000 yr B.P. She also estimated that mean July temperatures were 1–2°C warmer than modern levels between 9000 and 4000 yr B.P. She attributes a downslope shift in tree line from 6000–4000 yr B.P. to a decrease in effective moisture, rather than to changing temperatures. She interpreted pollen records from the last 2000 years as being indicative of essentially modern climatic conditions.

The reconstruction of paleotemperatures in the Colorado Rockies, based on past glacial limits, is hindered by several factors. First, this sector of the Rocky Mountains has apparently been relatively dry throughout at least the Late Pleistocene, so montane glaciers here have never been as extensive as they have been in the Central and Northern Rockies (Elias 1996a). Second, even in locations where past glacial limits have been mapped and dated (Leonard 1989), it is quite difficult to derive paleotemperature estimates from these limits. The reason for this difficulty is that the size of glaciers is controlled by multiple factors, including temperature, precipitation, slope, and aspect. Without sufficient moisture, large glaciers cannot become established, even during intervals of prolonged low temperatures, such as glacial stadials.

There are some important biotic lessons to be gleaned from the fossil record of the Colorado Front Range. First, it is evident that the Pinedale glaciation exerted long-term effects on the shaping of biotic communities. For instance, postglacial communities were limited to the species able to survive the Pinedale glaciation and become reestablished in the alpine zone following deglaciation. This means that the current group of species in the alpine ecosystem may not be the best fit for the environment—they are simply the best fit among those species able to persist regionally through the last glacial cycle. We have no measure of past versus present species diversity in alpine tundra plant communities, because the pollen of the alpine plant species that is preserved in the fossil record can only be identified to the generic or family level, in most cases. However, because alpine tundra now exists in "habitat islands" rather than in continuous belts along the Rockies, postglacial colonization by alpine tundra species would have been made more difficult. This is in contrast to the montane and subalpine vegetation, which exists in more-or-less continuous belts along elevational zones in the Rockies.

Second, there is some fossil and modern evidence that the ecotone between the alpine and subalpine ecosystems is not in equilibrium with the modern climate, but rather is a relict of a prior warming event in which the tree line migrated upslope to its current elevation. Burned patches of forest that occur near tree line have been very slow to recover. New seedling establishment in these areas appears to be much poorer than it would be if these upper forest stands were truly in equilibrium with modern climate.

There is also good evidence that postglacial warming took place 500–1000 years in advance of the ultimate upslope migration of the tree line in the early Holocene. At sites such as La Poudre Pass and Long Lake, the evidence for the establishment of subalpine forest stands near the elevation of the modern tree line begins at about 9000 yr B.P. On the western slope of the Rockies, however, Fall (1997) found pollen evidence for subalpine trees becoming established at modern tree line elevations as early as 10,000 yr B.P. Perhaps differences in precipitation account for the differences in the timing of establishment of trees near their elevational limit between the western and eastern slopes of the Colorado Rockies. In some regions, however, it appears that at the century to millennial timescale, the response of major components of the vegetation in high altitude ecosystems of the Colorado Front Range lags behind major temperature changes.

Finally, the fossil insect record indicates that during the last 14,000 years, regional climates have often changed abruptly, almost in a stepwise fashion between major thermal regimes. The more gradual temperature changes previously interpreted from regional palynological studies now appear to be an artifact of vegetation response lag, specifically the lag in response of trees growing near tree line to changing temperature regimes.

Acknowledgments   I thank Elyse Ackerman-Salazar, who prepared the Sky Pond samples for fossil insect identification, and Dr. Mel Reasoner, Brunel University, London, who coordinated the collection of sediment cores from Sky Pond. Kathy Anderson prepared the climate envelopes for beetle species found in the fossil assemblages. Financial support for Front Range paleoecological research has come from Long-Term Ecological Research grants from the National Science Foundation, DEB-9211776 and DEB-9810218.

References

Antevs, E. 1948. Climatic changes and pre-white man. *University of Utah Bulletin* 38: 168–191.

Bartlein, P. J., B. Lipsitz, and R. S. Thompson. 1994. Modern climate data for paleoenvironmental interpretations. *American Quaternary Association Thirteenth Biennial Meeting, Program and Abstracts*, 197.

Benedict, J. B. 1979. Getting away from it all: A study of man, mountains and the two-drought altithermal. *Southwestern Lore* 45: 1–12.

Berger, A. L. 1978. Long-term variations in caloric insolation resulting from the earth's orbital elements. *Quaternary Research* 9: 139–167.

Carrara, P. E., W. N. Mode, M. Rubin, and S. W. Robinson. 1984. Deglaciation and postglacial timberline in the San Juan Mountains, Colorado. *Quaternary Research* 21: 42–55.

Coope, G. R. 1977. Fossil Coleopteran assemblages as sensitive indicators of climatic changes during the Devensian (Last) cold stage. *Philosophical Transactions of the Royal Society of London, Series B* 280: 313–340.

Coope, G. R., and G. Lemdahl. 1995. Regional differences in the Lateglacial climate of northern Europe based on coleopteran analysis. *Journal of Quaternary Science* 10: 391–395.

Crowson, R. A. 1981. *The Biology of the Coleoptera*. Academic Press, New York.

Elias, S. A. 1983. Paleoenvironmental interpretations of Holocene insect fossil assemblages from the La Poudre Pass site, northern Colorado Front Range. *Palaeogeography, Palaeoclimatology, Palaeoecology* 41: 87–102.

Elias, S. A. 1986. Fossil insect evidence for Late Pleistocene paleoenvironments of the Lamb Spring site, Colorado. *Geoarchaeology* 1: 381–386.

Elias, S. A. 1991. Insects and climate change: Fossil evidence from the Rocky Mountains. *BioScience* 41: 552–559.

Elias, S. A. 1994. *Quaternary Insects and Their Environments*. Smithsonian Institution Press, Washington D.C.

Elias, S. A. 1996a. *Ice-Age Environments of National Parks in the Rocky Mountains*. Smithsonian Institution Press, Washington, D.C.

Elias, S. A. 1996b. Late Pleistocene and Holocene seasonal temperatures reconstructed from fossil beetle assemblages in the Rocky Mountains. *Quaternary Research* 46: 311–318.

Elias, S. A. 2000. Late Pleistocene climates of Beringia, based on fossil beetle analysis. *Quaternary Research* 53: 229–235.

Elias, S. A., K. H. Anderson, and J. T. Andrews. 1996. Late Wisconsin climate in northeastern USA and southeastern Canada, reconstructed from fossil beetle assemblages. *Journal of Quaternary Science* 11: 417–421.

Fall, P. L. 1985. Holocene dynamics of the subalpine forest in central Colorado. *American Association of Stratigraphic Palynologists Contribution Series* 16: 31–46.

Fall, P. L. 1997. Timberline fluctuations and late Quaternary paleoclimates in the Southern Rocky Mountains, Colorado. *Geological Society of America Bulletin* 109: 1306–1320.

Legg, T. E., and R. G. Baker. 1980. Palynology of Pinedale sediments, Devlins Park, Boulder County, Colorado. *Arctic and Alpine Research* 12: 319–333.

Leonard, E. M. 1989. Climatic change in the Colorado Rocky Mountains—Estimates based on modern climate at Late Pleistocene equilibrium lines. *Arctic and Alpine Research* 21: 245–255.

Madole, R. F., and R. R. Shroba. 1979. Till sequence and soil development in the North St. Vrain drainage basin, east slope, Front Range, Colorado. Pages 124–178 in F. G. Ethridge, editor. *Guidebook for Postmeeting Field Trips Held in Conjunction with the 32nd Annual Meeting of the Rocky Mountain Section of the Geological Society of America*, May 26–27, 1979, Colorado State University. Geological Society of America, Boulder, Colorado.

Mast, J. N., T. T. Veblen, and Y. B. Linhart. 1998. Disturbance and climatic influences on age structure of ponderosa pine at the pine/grassland ecotone, Colorado Front Range. *Journal of Biogeography* 25: 743–755.

Nelson, A. R., A. C. Millington, J. T. Andrews, and H. Nichols. 1979. Radiocarbon-dated upper Pleistocene glacial sequence, Fraser Valley, Colorado Front Range. *Geology* 7: 410–414.

Pierce, K. L., and J. D. Good. 1992. *Field guide to the Quaternary geology of Jackson Hole, Wyoming*. U.S. Geological Survey Open File Report 92-504, 54 pp.

Reasoner, M. A., G. Osborn, and N. W. Ruter. 1994. Age of the Crowfoot advance in the Canadian Rocky Mountains: A glacial event coeval with the Younger Dryas oscillation. *Geology* 22: 439–442.

Short, S. K. 1985. Palynology of Holocene sediments, Colorado Front Range: Vegetation and treeline changes in the subalpine forest. *American Association of Stratigraphic Palynologists Contribution Series* 16: 7–30.

Short, S. K., and S. A. Elias. 1987. New pollen and beetle analysis at the Mary Jane site, Colorado: Evidence for Late-Glacial tundra conditions. *Geological Society of America Bulletin* 98: 540–548.

Veblen, T. T., T. Kitzberger, and J. Donnegan. 2000. Climatic and human influences on fire regimes in ponderosa pine forests in the Colorado Front Range. *Ecological Applications* 10: 1178–1195.

Woodhouse, C. A. 2001. A tree-ring reconstruction of streamflow for the Colorado Front Range. *Journal of the American Water Resources Association* 37: 561–569.

Century to Millennial Timescale—Synthesis

Douglas G. Goodin
Raymond C. Smith

At longer timescales, the interaction among climate, ecosystems, and the abiotic components of the environment become increasingly important. These relationships are apparent in the three chapters in part IV. Fountain and Lyons (chapter 16), examining the McMurdo Dry Valleys (MCM) ecosystem in Antarctic, provide an excellent example of a case where past climatic variations truly dictate current ecosystem status. The relatively large climate variations at MCM have concentrated nutrients that could not have been attained without this climate variability. Fountain and Lyons infer climate change from geomorphic evidence of past glacier positions and lake level heights as well as more recent isotopic results from ice cores and temperature measurements from boreholes. They focus on evidence from the most recent 60,000 years. Monger (chapter 17) provides an analysis of millennial-scale climate and ecosystem variability at the Jornada LTER site in southern New Mexico. Monger notes the difficulty of untangling prehistoric climate/ecosystem interactions, where researchers must rely on indirect proxy indicators in lieu of measured data. Monger analyzes a number of proxy data sources, including paleolake levels, plant remnants preserved in packrat middens, fossil pollens, carbon isotope ratios in paleosols, and erosion rates. Although noting the danger of circular reasoning in using proxy data (i.e., ecosystem response used to infer information about climatic change, which is in turn inferred from ecosystem response) Monger uses these data to construct a cogent picture of climate change at the Jornada site (JRN) since the Last Glacial Maximum (LGM) about 18,000–20,000 years B.P. Using remains of beetles, Elias (chapter 18) constructs a temperature history of the Colorado Alpine since the LGM. These late Holocene insect records show a progression from warmer-than-modern to cooler-than-modern summers, and back to warm again. All the authors in this section provide examples to show that it is at century to millennial timescales that ecosystems form, are broken apart and imprinted by the past, and reformed in new configurations.

The McMurdo Dry Valleys is the most poleward-terrestrial ecosystem where streams, lakes, and soil are interconnected. In this polar desert, the biotic system must adopt a strategy to survive the winter in isolation, and the disturbance and formation of the landscape has been primarily dictated by climate and associated abiotic processes. During the last glacial period, the Ross Ice shelf entered Taylor Valley, damming the valley and forming a 200-m-deep lake (23.8 kyrs). Rapid warming occurred about 15 kyrs ago at the termination of the glacial period, and the lake remained until about 8.3 kyrs ago when the ice shelf retreated and the lake drained, leaving smaller lakes in low spots along the valley floor. The former large lake provided nutrients to the soils, and the drawdown of the smaller lakes 1000 years ago concentrated nutrients into pools on which the current ecosystem depends. It is because of the nutrient-poor and energy-limited environment of this polar desert that the past concentrations of nutrients play such a dominant role in the current structure and function of the ecosystem. In addition, the sensitivity of this system to the presence or absence of liquid water and the nonlinear response to changes in temperature near the melting point of water create a system where small changes in climate produce large variations in ecosystem response.

Based on paleoclimatic reconstruction, Monger identifies 9 intervals of climate variability at the Jornada LTER site over the past 20,000 years. His inferences are couched in terms of relative abundance of vegetation by life-form type (i.e., C_4 grassland, C_3 woodland, C_3 shrubland). In general, his reconstruction shows a trend toward increased abundance of C_3 shrubs, displacing C_4 grasses and C_3 woodlands. This trend represents a general increase in aridity consistent with regional changes in climate from the close of the Pleistocene. Although the trend is toward increased aridity, the changes are not monotonic. Particularly evident is the hot, dry Altithermal period (see figure 17.7). This climatic period, culminating about 6000 years B.P., was characterized by an expansion of grasslands in North America well east of their current ranges. Causes for the Altithermal warming are not entirely clear, but probably represent the additive effects of return to preglacial atmospheric CO_2 levels and solar radiation fluxes higher than their modern values (Kutzbach et al. 1996). Forcing by Milankovitch mechanisms has also been suggested as a cause for Altithermal warming (Kutzbach and Street-Perrott 1985; Gillespie et al. 1983). These changes probably weakened the North American monsoon circulation, causing reduction in precipitation and a shift toward predominant zonal westerly flows. Decreased precipitation and increased temperature signals show clearly in the Monger's reconstruction as a decline in lake level (or lake disappearance) and a marked increase in shrubland. Climate intervals 8 and 9 show a recovery of conditions from their mid-Holocene state, but they are still periods of aridity. Historical records for period 9 (since 1850) show a progressive increase of shrubland and a loss of grassland, consistent with continued postglacial aridity. This interval is strongly influenced by human activities, thus complicating the determination of cause and effect.

Using fossil insect records, Elias (table 18.1) provides a temperature reconstruction and vegetation history since the LGM. During the Holocene the Colorado

Front Range (NWT) has experienced a series of climatic fluctuations that have shifted glacial margins and biotic communities. Elias's predictions based on fossil beetle data agree well with a reconstruction of solar radiation based on Milankovitch insolation models (Berger 1978) at millennial scales. However, conflicting interpretations of insect, pollen, and archaeological data during the mid-Holocene suggest the need for additional regional studies. Elias notes that glacial ice has been the dominant force in shaping alpine landscapes, with postglacial communities limited to those able to survive and become reestablished after deglaciation. Elias (chapter 18, p. 466) suggests that "the current group of species in the alpine ecosystem may not be the best fit for the environment, they are simply the best fit among those species able to persist regionally through the last glacial cycle." Elias also notes that the response of major components of vegetation in high-altitude ecosystems may lag behind major climatic changes.

Relationship to Framework Questions

The results discussed in this section clearly show the presence of climate variability at millennial timescales, although (as pointed out previously) they must be interpreted cautiously to avoid circular reasoning. Monger's results coincide with those of other paleoclimate analyses both in the U.S. Southwest (e.g., Hall and Scurlock 1991) and elsewhere (Gillespie et al. 1983). Elias's temperature reconstruction is consistent with Milankovitch forcing, but it differs in details from some other reconstructions. This may be, in part, because of the regional specificity of the Colorado Front Range. Evaluation of some of the other framework questions is complicated by both the nature and timescale of the changes considered here. Use of proxy data always involves inferences about the relationship between the proxies and the climate data they represent; the certainty of these relationships decreases as the inferences extend further into prehistoric time. Nevertheless, results in this section do fit into some of the framework questions. At the millennial timescale, the LGM is an important defining preexisting condition.

Fountain and Lyons show the dominant influence of preexisting conditions, in this case a paleolake and its subsequent contribution of nutrients and organic carbon to the structure and function of the current ecosystem. There are cascades at shorter timescales through the aquatic part of this polar desert ecosystem (Welsh et al., chapter 10) that are driven by factors influencing the presence of liquid water, but the legacy effects in this environment are on the order of thousands of years. Superimposed on this legacy is the nonlinear response at the melting point of ice, which is "at the heart of all observed changes." This melting transition point is critical to discussion of the flow of material and energy through, and the direction of evolution of, this system. A consideration of cycles within this context must take note of this critical transition point.

Monger's results showing changes in vegetation life-form accompanying climate change represents a cascade effect. Monger notes that climate changes can result in vertical reallocation of water by runoff, resulting in the increased availability of moisture downslope. Another cascade can be inferred in the activities of rock glaciers during the late Pleistocene that form new geomorphic surfaces on which new ecosystems develop. These results may be cyclical, but if so the cycle occurs over very long time periods, extending even beyond the millennial timescales considered in this section. The climate events in this case show little evidence of reversal, at least at the timescales considered here. Monger's analysis also hints at the importance of preexisting conditions in the dynamics of the arid ecosystem. Much of his analysis is presented in terms of effects relative to the local topography (i.e., piedmont vs. basin floor, see figure 17.5), suggesting that the "lie of the land" is a crucial influence in this climate/ecosystem. Similar climate change might result in a different outcome given some other geomorphic surface.

Elias draws particular attention to the possible lags between climate variability and the response of trees growing near tree line to changing temperature regimes. This is an important observation, particularly in light of the current rate of climate variability and efforts to understand and predict the response of forests to this relatively rapid change. He notes that the current group of species in the alpine zone consist of those able to survive glaciation and become reestablished in the alpine zone. These species are not necessarily the best "fit" among all possibilities; instead, they are the best fit among those species persisting through the last glacial cycle. Elias further states that present-day ecotones in alpine and subalpine ecosystems are not in equilibrium with the current climate, but are instead a relict of an earlier warm period. Both of these facts point to an important role for legacy effects in alpine climate/ecosystem interaction. If glaciation is viewed as a climate "disturbance," then Elias's findings also suggest that the climate/vegetation interaction does not return to its previous state (i.e., a hysteresis effect) when a climatic disturbance event is completed. The lag effect between climate variation, which often occurs abruptly, and ecosystem response, which lags in response, results in a system where feedback mechanisms associated with previous climate cycles might often overlap. Thus, simple correspondence between climate "event" and ecosystem response is not a suitable framework for analysis of this ecosystem at millennial timescales.

Martinson and coauthors (1998), in presenting a science plan for decade to century-scale climate variability and change, note that the paradigm used for the study of climate variability at seasonal to decadal timescales may not be applicable to decadal and longer timescales. Paleoclimate and historical records are often too short to apply the process of generating hypotheses and quickly evaluating them. Martinson et al. (1988) argue that making progress at these longer timescales will require improved and faster climate models, and expanded paleoclimate data bases. Understanding processes at these longer timescales is essential because it is at these timescales that, as Elias (p. 387) notes, "ecosystems form, break apart, and reform in new configurations." Also, Martinson et al. (1998) note that it is over these time periods that the life prospects of future generations are defined by climatic variability. They argue that informed stewardship of Earth's resources requires a sustained effort to understand processes on these longer timescales.

References

Berger, A. L. 1978. Long-term variations in caloric insolation resulting from the earth's orbital elements. *Quaternary Research* 9: 139–167.

Gillespie, R., F. A. Street-Perrott, and R. Switsur. 1983. Post-glacial arid episodes in Ethiopia have implications for climate prediction. *Nature* 306: 680–683.

Hall, D. O., and J. M. O. Scurlock. 1991. Climate change and productivity of natural grasslands. *Annals of Botany* 67: 49–55.

Kutzbach, J. E., G. Bonan, J. Foley, and S. P. Harrison. 1996. Vegetation and soil feedbacks on the response of the African monsoon to orbital forcing in the early to middle Holocene. *Nature* 384: 623–626.

Kutzbach, J. E., and F. A. Street-Perrott. 1985. Milankovitch forcing of fluctuations in the level of tropical lakes from 18 to 0 kyr BP. *Nature* 317: 130–134.

Martinson, D. G., K. Bryan, M. Ghil, M. M. Hall, T. R. Karl, E. S. Sarchik, S. Sorooshian, and L. D. Talley. 1998. Decade-to-century-scale climate variability and change: A science strategy. National Research Council, National Academy Press. Washington, D.C.

Part V

Climate Variability and Ecosystem Response at Selected LTER Sites at Multiple Timescales

Introductory Overview

David Greenland
Douglas G. Goodin

The timescale structure of this book has served well to keep the attention of investigators focused on specific aspects of climate variability and ecosystem response. Indeed, judging by the responses received by the editors of this volume, when given a choice between focusing on one timescale or several timescales, the LTER community was far more comfortable dealing with just one scale. There are obvious reasons for this, not the least of which is that focusing on a single scale greatly simplifies things. The real world, however, does not focus on one timescale. Climatic events and ecosystem responses occur simultaneously at a variety of scales. We wished to explore the climatic variability and ecosystem responses at LTER sites across several different timescales, and the two chapters in this part attempt such an exploration. The chapters consider the temperate rainforest of the H. J. Andrews LTER site in Oregon and the tallgrass ecosystem of the Konza Prairie LTER in Kansas.

For the Andrews rainforest, and to some extent the Pacific Northwest (PNW) in general, Greenland et al. (chapter 19) discuss climate variability and ecosystem response at the daily, multidecadal, and century to millennial scales. This discussion for the PNW is supplemented in chapters 6 and 13 of this volume by a consideration of the quasi-quintennial scale and an additional ecosystem response at the decadal scale. The forest ecosystem is more complex than the grassland ecosystem. Greenland et al. cover a wide variety of potential ecosystem responses for the PNW Forest, ranging from severe weather events, to pine cone production, to century- and millennial-scale forest fire frequency regimes and their variation. The focus of chapter 19 is on some of the framework questions of this volume. The questions specifically addressed include the following: What preexisting conditions affect the impact of the climatic event or episode? Is the climatic effect on the ecosystems direct or cascading? Does the system return to its original state? The authors also consider potential future climate change and its possible ecosystem effects. They found that timescale becomes important in addressing some of these questions. For example, at century to millennial timescales, it is suggested that there are likely to be no identical past analogs to the ecosystem at any point in time. It is unlikely that an ecosystem will return to its "original" state at this longer timescale, and the concept of "original" state itself has little meaning.

In chapter 20, Goodin et al. examine how interannual, quasi-quintennial, and interdecadal variation in annual precipitation and mean annual temperature at a tallgrass prairie site (Konza Prairie Biological Station) may be related to various climatic indexes and phenomena. They examine solar activity, the El Niño–Southern Oscillation (ENSO), the North Atlantic Oscillation (NAO), and the North Pacific Index (NP), as well as how these indexes may be related to aboveground net primary productivity (ANPP). The authors present (1) period-spectrum analyses to characterize the predominant timescales of temperature and precipitation variability at Konza Prairie; (2) correlation analyses between quantitative indices of the major atmospheric processes and Konza temperature and precipitation values; and (3) the implications of variation in major atmospheric processes for seasonal and interannual patterns of ANPP. The key finding of this analysis is that the historic temperature and precipitation record at Konza Prairie displays periodicities similar to those for ENSO, NAO, and NP. Periods of stronger NAO (i.e., larger positive index values) are associated with warmer winters, periods of stronger ENSO with wetter winters, and periods of stronger NP with warmer summers. The course of the growing season as represented by aboveground biomass accumulation appears to be limited initially by temperature, then later by soil moisture.

Goodin et al. find that the effects of variation in some climatic indexes are indirect, whereas others are direct, and this relates to the type of cascade of ecosystem responses that come into play. Their analysis shows that different periodicities of varying importance make up the total temporal variation in the values of the climate indexes. The different periodicities form part of a hierarchy of climate variation. Focusing on the degree to which these hierarchical periods of temperature and precipitation variability reinforce (or oppose) each other may shed more light on the regulation of variability of ANPP or other ecosystem characteristics than considering a single periodicity or time scale separately. In addition, although, grasslands are the simpler system (compared to forests), Goodin et al. suggest the grasslands are poised in a dynamic equilibrium that makes them especially sensitive to both biotic and abiotic disturbances, including climate variability.

Both of these chapters hint, in their different ways, at the multidisciplinary, multitimescale, and multidimensional considerations that will have to be confronted in future stages of LTER studies.

19

Climate Variability and Ecosystem Response at the H. J. Andrews Long-Term Ecological Research Site

David Greenland

Frederick Bierlmaier

Mark Harmon

Julia Jones

Arthur McKee

Joseph Means

Frederick J. Swanson

Cathy Whitlock

Introduction

The H. J. Andrews (AND) Long-Term Ecological Research (LTER) site represents the temperate coniferous forest of the Pacific Northwest (PNW) of the United States. The general climate of the area is highly dynamic, displaying variability at a variety of timescales ranging from daily to millennial. AND, and its surrounding region, is therefore an ideal site for examining some of the guiding questions of climate variability and ecosystem response addressed by this volume (see chapter 1). A legacy of more than 50 years of research at the site and its surrounding area ensures that several of the questions can be investigated in some depth. Here we organize our discussion within a timescale framework that is consistent with the structure of this volume. Thus, following a brief description of the general climate of the site, we discuss climate variability and ecosystem response at the daily, multidecadal, and century to millennial scale. This discussion for the PNW is supplemented in chapters 6 and 13 by a consideration of the quasi-quintennial scale and an additional ecosystem response at the decadal scale.

Having described some of the climate variability and ecosystem response at the selected timescales, we will consider what this information can tell us regarding some of the guiding questions of this book. The questions that we specifically address include the following: What preexisting conditions affect the impact of the

"

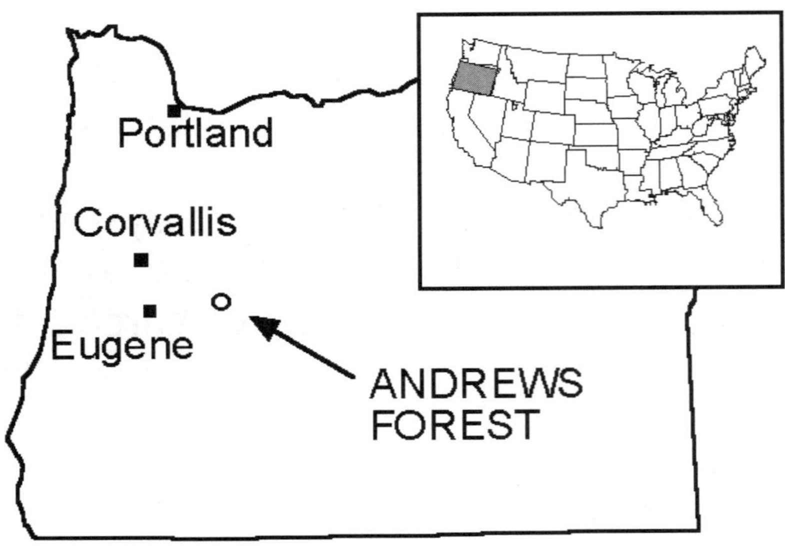

Figure 19.1 Location map of the H. J. Andrews Experimental Forest LTER site.

climatic event or episode? Is the climatic effect on the ecosystems direct or cascading? Does the system return to its original state? We also consider potential future climate change and its possible ecosystem effects.

The Climate of the H. J. Andrews Experimental Forest

Located at latitude 44.2° N and longitude 122.2° W, the Andrews Forest is situated in the western Cascade Range of Oregon in the 6400-ha (15,800-acre) drainage basin of Lookout Creek, a tributary of the Blue River and the McKenzie River (figure 19.1). Elevation ranges from 410 m (1350 feet) to 1630 m (5340 feet). Broadly representative of the rugged mountainous landscape of the Pacific Northwest (PNW), the Andrews Forest contains excellent examples of the region's conifer forests and associated wildlife and stream ecosystems. Lower elevation forests are dominated by Douglas-fir (*Pseudotsuga menziesii*), western hemlock (*Tsuga heterophylla*), and western red cedar (*Thuja plicata*). Upper elevation forests contain noble fir (*Abies procera*), Pacific silver fir (*Abies amabilis*), Douglas-fir, and western hemlock. Low- and midelevation forests in this area are among the tallest and most productive in the world. As elevation increases, Douglas-fir and western red cedar decline in importance and western hemlock is gradually replaced by Pacific silver fir.

The climate is controlled by its close midlatitude proximity to the Pacific Ocean and by the perpendicular orientation of the Coast and Cascade mountain ranges to the prevailing westerly flow. The Andrews Forest is located near the border between temperate maritime and temperate continental climates as a result of these mountain barriers to passage of air masses from the west. Temperatures are moderated at almost all times of the year by maritime air, particularly in winter.

Winter precipitation is high, averaging 287 mm (11.3 in.) per month between January and March. Low-pressure areas and associated storms are steered into the area by the polar jet stream. Long-duration but generally low-intensity storms result from the passage of strongly occluded fronts that are slowed by the mountains. Daily precipitation is significantly autocorrelated up to 14 days (Post and Jones 2001). Temperatures associated with these storms are often mild enough that rain falls at lower elevations of the Andrews Forest while snow falls at higher elevations. This usually results in deep (2 to 4 m), long-lasting, snowpacks above approximately 1000 m. Occasional strong storms can have severe ecological consequences such as windthrow—the toppling of trees by the force of the wind. Late summer and early fall wind from the central Oregon desert may also drive large forest fires. Summertime precipitation is usually low to nonexistent, averaging 38 mm (1.5 in.) per month between June and August. The North Pacific anticyclone intensifies and expands to the northeast along the coast. This blocks the passage of cyclonic storms and stabilizes the air.

Summer drought, mild wet winters, a heavy snowpack above 1000 m, and light to nonexistent snowpack below 800 m are factors affecting the flora and fauna. Late summer moisture stress of the forest has an important part in determining the composition and structure of various forest communities. This moisture stress also helps to give rise to the coniferous nature of the Pacific Northwest forest (Waring and Franklin 1979). Snow and lower temperatures at upper elevations play an important role in the formation of a distinctly different forest zone—the Pacific silver fir (*Abies amabilis* Dougl. ex Forbes)—through mechanical force and modification of temperature and moisture regimes. Large animals, such as elk and deer, are forced to lower elevations by the heavy upper elevation snowpack, whereas smaller animals use it for shelter and cover. At lower elevations, the mildness and wetness of the winters, combined with little snow, produces a nearly stress-free environment for plants and animals. The mild climate also results in a long growing season. Water use by evapotranspiration in the old growth forests is greatest during the spring and fall and is limited by the low precipitation of the summer months.

Superimposed on this general picture is considerable temporal variability in the climate. At a daily scale there can be severe storms. The El Niño–Southern Oscillation (ENSO) operates at a 2–7 year (quasi-quintennial) scale and provides a context for warmer and drier (El Niño) or cooler and wetter (La Niña) conditions. The Pacific Decadal Oscillation (PDO) functions at a multidecadal timescale that is also characterized by warmer and drier or cooler and wetter periods. Evidence exists for similar climate variability at century, subcentury, and millennial timescales, and these signals have varied in strength over time. There is also the possibility of climatic trends at century and longer timescales. Change is one of the few certainties in the dynamic environment of the Pacific Northwest.

Daily to Annual Scale: Severe Storms and Floods, Insect Outbreaks, and NPP

The most dramatic climatic events at the daily timescale are those concerning severe storms that are accompanied by floods and in some cases windthrow events. Daily precipitation and streamflow values vary by more than two orders of magnitude within each year. Fifteen-minute precipitation and streamflow values can vary by the same amount within a few days. Climatic/meteorologic events related to the ignition and spread of forest fire might also be considered in this category, although these events also incorporate effects of preceding droughts and associated drying of fuels. At the Andrews Forest, a small number of daily timescale events can have a large impact. Snyder (2000), for example, found that in the 50 years of records, most flood-related action and landslides occurred during only three major storm events. Similarly, most windthrow events in the northern Cascade Range, Oregon, since 1890 are associated with just three major individual storms (Sinton et al. 2000).

A very large 50-year-return period flood in 1996 led to a large direct response from the ecosystem. This February flood resulted from 290 mm of precipitation over 4 days that melted a large amount of already accumulated snow. Swanson et al. (1998) and Nakamura et al. (2000) list and document landslides and channel erosion and related disturbance of aquatic and riparian organisms and their habitats as responses to this flood. The hydrographic response varied with altitude because of the varying snowpack dynamics. At least 35 debris flows severely disturbed stream and riparian environments. There was a large amount of fluvial erosion. In some areas riparian vegetation was entirely removed in larger channels, and boulder and coarse woody debris movement was common (Johnson et al. 2000). Scouring in places uncovered objects that had long been buried. Wood samples exposed along the northeast side of Watershed 3 turned out to be over 46,000 years old (http:// www.fsl.orst.edu/lter/pubs/spclrpfr.htm). Many stream restoration project structures were washed away. Some biotic responses were very fast. Benthic algae recovered from the event within weeks. Again, this web page on the 1996 flood provides details on these effects: http://www.fsl.orst.edu/lter/pubs/spclrpfr.htm.

At the Andrews LTER site, windthrow events result mainly from southeasterly winds associated with storms arriving from the Pacific (Gratkowski 1956). Windthrow events in winter in the northern Cascade Range of Oregon were found, in some cases, to highlight the importance of preexisting conditions. Sinton et al. (2000) found such events occurred particularly when winds were from the north or east with a preceding period of dry weather. High-pressure conditions in winter gave rise to icing on the branches of trees prior to some windthrow events (D. S. Sinton, pers. comm., 1996). As is well known, windthrow events cause forest gaps that subsequently undergo a cascade of successional events leading to the reestablishment of the forest. However, canopy gaps, especially those with fresh, clear-cut edges, are particularly prone to additional windthrow (Gratkowski 1956; Sinton et al. 2000).

In some cases ecosystems respond to the coincidence of two climatic events. One such example is the occurrence of Douglas-fir bark beetle outbreaks (Powers et al. 1999). In western Oregon and Washington, these insects are usually sapro-phytic, reproducing in freshly downed Douglas-fir trees. In rare instances, how-ever, this species can kill live trees, and for a 2- to 3-year period can increase over-all mortality in forests by a factor of 3 to 10. The coincidence of two climatic events is necessary for this to happen. First, a major windstorm or incidence of ice damage is necessary to create a large amount of breeding habitat. This allows the popula-tion to expand to sufficient numbers to attack living trees and kill them. Curiously, at least for this species of beetle (and spruce bark beetle as well), fire-killed timber is not a suitable enough habitat to increase the population. Second, the trees must be under stress during the growing season. This stress is usually caused by drought and reduces the trees' ability to respond to the beetle attack. Even in large numbers, the Douglas-fir bark beetle has little ability to overwhelm trees. Although the bee-tles can reproduce in live trees, they are unable to increase numbers in this habi-tat; therefore, outbreaks in live trees rarely last for more than 3 years despite the length of the drought. The rare coincidence of these two sets of climatic conditions means that Douglas-fir bark beetle outbreaks are rare events for western Oregon and Washington forests. Although the historical record of outbreaks is not long, outbreaks appear to occur at an average frequency of 50 years. These outbreaks do have important impacts: They alter forest composition (ironically by removing a more drought-resistant species), speed the rate shade-tolerant species dominate stands, temporarily increase the amount of detritus, and reduce the Net Primary Productivity (NPP) of the forest, with the end result of creating a temporary source of CO_2 to the atmosphere.

Year-to-year oscillations in precipitation are responsible for variations in NPP, decomposition, and Net Ecosystem Productivity (NEP). By examining tree cores and litter fall records, Fraser (2001) found that tree growth and litter fall varied ±30% from year to year. Given lag of 4 to 5 years between leaf production and lit-ter fall, the amount of combined variation is not clear; however, it is likely to be in a similar range. Year-to-year variation in decomposition rates has not been studied extensively, but fortuitous studies carried out during extremely dry and wet years indicate a range of ±30% (Valachovic 1998; Harmon, unpubl. data, 1992). This re-sponse is not likely to be mirrored in other forms of detritus, however, because their rates of drying, and response to moisture, differ substantially. Fine litter, for exam-ple, dries quickly and, because of its high ratio of surface area to volume, is rarely limited by excessive moisture. In contrast, large wood dries slowly (Harmon and Sexton 1995) and has a low enough surface-area-to-volume ratio that diffusion of oxygen can become limiting for decomposition when moisture content is high (Harmon et al. 1986). This means that summers with high precipitation can lead to fast decomposition of fine litter, but slow decomposition of large wood. Conversely, in summers that are dry, fine litter decomposition can be slow and that of large wood fast. As a result, the year-to-year variation in overall decomposition is likely to be dampened as the detritus pools are "decoupled" temporally from each other. By combining these sources of variation in NPP and decomposition, preliminary estimates are that NEP (the net exchange of carbon with the atmosphere) could vary as much as 2 Mg ha^{-1} year^{-1} in Douglas-fir/western hemlock old-growth forests (Harmon et al., in press). This is a substantial level of variation. Although these forests are thought to have an NEP close to zero over the long term, this variation means that in some years they are uptaking as much carbon as a forest in the peak carbon accumulation phase (Janisch and Harmon in press). Clearly, a more thorough examination of the cause of this year-to-year variation is necessary, but the key lessons that ecosystem processes are not responding to the same climatic variable in the same way and that the sign of the response could differ, even within a process, are likely to hold.

Quasi-Quintennial Scale: ENSO

The atmospheric manifestation of El Niños and La Niñas in the PNW is well documented. El Niños are correlated with warmer winter temperatures, reduced precipitation (Redmond and Koch 1991), and reduced snowpack and streamflow (Cayan and Webb 1992) in the region. The reverse tends to be true for La Niña years. Heavy-rain-bearing storms tend to be a feature of La Niña years. The large flood of February 1996 is a case in point. January, February, and March 1996 were marked by La Niña conditions. An earlier major flood in 1964 also occurred when El Niño conditions were changing over to La Niña conditions.

Recent forest fire history at the Andrews LTER site is probably, at least in part, a response to climate variability. El Niño years tend to lead to drier and warmer winter conditions in the PNW. The strong El Niño year of 1987 was accompanied by numerous large forest fires in the PNW in 1987 and 1988. The El Niño years of the early 1990s were also accompanied by large fires in 1992 and 1994.

Multidecadal Timescale

Multidecadal changes for the Pacific Northwest (PNW) are related to the PDO. Taylor and Southards (1997) (http://www.ocs.orst.edu/reports/climate_fish.html) noted a cool, wet period from 1896 to 1914, a warm and dry period from 1915 to 1946, a cool and wet period from 1947 to 1975, and a warm and dry period from 1976 to 1994. Mantua et al. (1997) have shown these periods to be related to changes in the synoptic-scale climate indices that have reversal times during the period 1900–1996 in 1925, 1947, and 1977. The climate regime shifts related to the PDO were first noticed after the 1976 shift because of the correspondence in numerous ecosystem and environmental responses in the PNW (Ebbesmeyer et al. 1991). These responses include variables such as the numbers of goose nests, crab production, mollusk abundance, and the path of returning salmon and salmon catch (see chapter 13). There are suggestions that another climate regime shift may have occurred in the mid-1990s (JISAO CIG 1999).

Given these regional changes, one might expect clear evidence of such climate variability and ecosystem responses in the Andrews ecosystems. However, an unequivocal variability before and after 1976 is not immediately apparent in the values of some variables where it might be expected, such as winter water year precipitation, stream discharge, or in the percentage change of water yield relative to that in the 12 years before 1976. There are, however, suggestions of some interesting multidecadal changes in "cyclic" behavior of parts of the system. As described subsequently, such changes in cone production, the number of peak streamflows per year, and possibly in debris flow frequency need about another 100 years of records to establish the reality or absence of cyclic behavior, but these changes do raise some interesting research questions.

Cone production records from high elevations above about 1000 m for noble fir, silver fir, and mountain hemlock at the AND and in other parts of the Cascade Range commence in 1962. There is some evidence to suggest that temporal patterns of cone production by upper slope noble fir, silver fir, and mountain hemlock in the Oregon and Washington Cascades may be associated with variability in the PDO (figure 19.2). Cones counted on canopy trees on 14 plots in the Cascades show a marked 3-year periodicity from 1962 to 1974, as exemplified by Pacific silver fir (*Abies amabilis*) in figure 19.2. This was during a period of lower than average PDO (cool phase). For at least the next two decades, this periodicity ends with the 1971–1974 cone production period for Pacific silver fir, noble fir, and mountain hemlock when the north Pacific sea surface temperature rises above normal (PDO positive, warm phase). This loss of this 3-year periodicity may be caused by loss of a trigger, common to these species, needed for cone production. Preliminary analyses indicate that, until the mid-1970s, warmer than average summer temperatures in the Cascades preceded by one year the large cone crops seen in figure 19.2. No warming or cooling can be seen, however, in average summer (June, July, August) temperatures after 1976. After 1976, another change in cone production pattern occurred: There was less synchronicity both among and within species (figure 19.2). If a PDO regime shift in the late 1990s does prove true, a return to the 3-year cyclic cone production will be one test of the PDO/cone productivity relationship. More research is needed to explore potential links between PDO change and cone production response.

Cone production also displays the importance of preexisting conditions at a monthly timescale. In the case of Douglas-fir, a warm sunny dry June 15 months before cone maturation, cool moist March and April 17 and 18 months before cone maturation, and cool moist summer months 25 to 27 months before cone maturation are all associated with increased cone production.

Although no clear evidence of a 1976 climate regime shift is seen in precipitation and stream discharge records at the Andrews Forest, peak streamflows show an interesting pattern. Five-year running means of the number of peak streamflows per year in unharvested, high-elevation basins at the Andrews site between 1952 and 1996, as counted by storm matching techniques, show two complete and similar "cycles" with a period of about 10 years (Jones and Grant 1996). The "cyclic" nature of these data stops at approximately 1976 and is not seen in the later part of the time series.

Another geomorphologic and ecosystem response to PDO climate fluctuations may involve the occurrence of debris flows, rapid mass movements of 100 to greater than 1000 m^3 of soil and organic debris down steep headwater stream channels. Snyder (2000) examined the inventory of 91 debris flows occurring between 1946 and 2000 in a 125-km^2 study area including the Andrews Forest. Debris flows

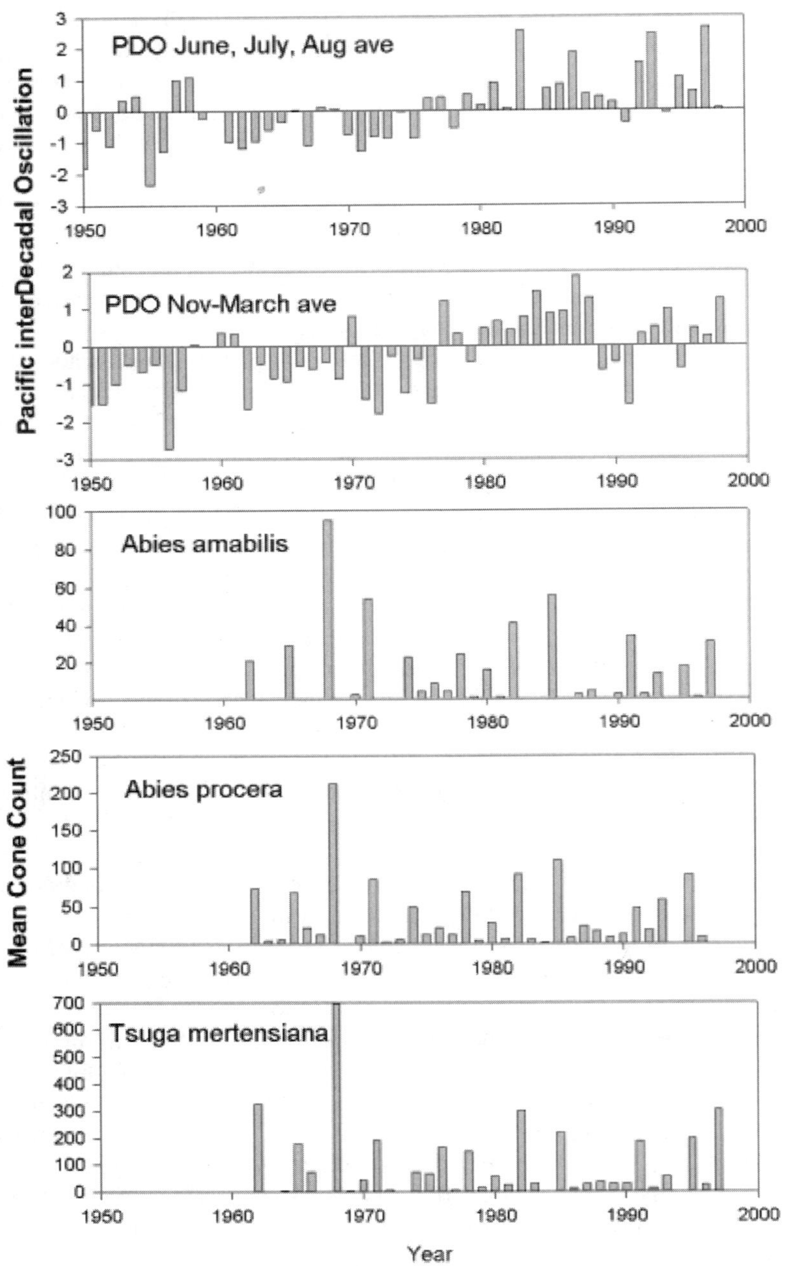

Figure 19.2 Relationship between Pacific Decadal Oscillation and cone production of three upper slope conifer species in the Cascades of Oregon and Washington. (Courtesy of Joseph Means)

initiated in unharvested forested areas occurred at a rate of 0.38 events per year in the wet phases of the PDO (before 1976 and after 1994) and only 0.05 events per year in the intervening dry phase. The majority (81%) of events in forested areas occurred in just three storms during this greater-than-50-year period, which raises the issue of whether these records represent the vagaries of storm history or a true PDO signal.

Once again, many future years of data and fairly precise methods of identifying and dating debris flows of the past 150–200 years are necessary to see whether the PDO/debris flow relationship may be established. Preexisting conditions may be important in this possible relationship. No matter what meteorological and climatological circumstances occur, debris flows can take place only if the potential debris material has already been rendered into a potentially movable condition.

Some interesting research questions are raised by these data. First, what kind of changes in system subvariables manifest themselves as a result of a multidecadal climate-driving cycle in a system? Can there be changes in system subvariables that show themselves as cyclic but with higher frequencies? Can changes in system subvariables be represented by the absence of a "response cycle" altogether in one phase of the driver cycle? More interesting, what are the complex steps in the system cascade that could give rise to this state of affairs, assuming the answers to these questions are positive. At least three other possibilities exist besides climatic cause and effect. First, there may be some other nonclimatic drivers at work such as land management and road construction. Second, nonclimatic drivers interact with climatic drivers. Third, the ecosystem events are random and there is no cause and effect.

Century to Millennial Scale

The absence of direct meteorological observations for most time periods and geographic areas at the century to millennial scale forces investigators to use proxy evidence from which to infer information concerning the variability of climate. At this scale, therefore, the ecological response is being used to provide information concerning the climate. We thus admit, in this section, to engaging to a certain degree in circular argument while discussing the "inferred" climate variability (as a "cause") and ecosystem response (as a "result"). The fields for which we have the most information at this timescale are those related to tree-ring thickness variability (Graumlich and Brubaker 1986; Graumlich 1987; Holmes et al. 1986; Buckley et al. 1992; Wiles et al. 1995; Garfin and Hughes 1996) and tree-ring–based (Weisberg and Swanson 2003) and lake-charcoal–based forest fire histories (Long et al. 1998), as well as vegetation change noted from pollen analysis (Warona and Whitlock 1995; Sea et al.1995; Grigg and Whitlock 1998).

Warm periods from 1400 to about 1575 and from 1800 to about 1925 were associated with widespread forest fires at the Andrews site and in the western Cascades (Weisberg and Swanson 2003). During the cool period from 1700 to about 1775, there was a marked decrease in the extent of forest fires. Forest fire histories based on tree rings at the Andrews site and in other study areas in western Oregon indicate widespread forest fires during the periods of 1475–1550 and 1850–1900. The most recent period of widespread fire is associated, among other factors, with a warmer, drier climate beginning about 1840, as noted in the tree-ring record (Weisberg and Swanson 2003). They point to anthropogenic factors acting synchronously with climate variability to produce the overall fire history.

The Weisberg and Swanson study suggests that as for crown fire-driven landscapes in general, the PNW may have exhibited high, spatiotemporal variability at any spatial scale. Climate variability at the century to millennial scale operating through the provision of periods for variable forest fire frequency leads to a highly dynamic ecosystem. Swanson has noted that forest establishment after fire may take place in periods of unusually stressful climate. He speculates that this may have affected succession and ultimately the development of present-day old growth forests in ways unlike the potential consequences of forests established by natural processes or management actions in areas with other climate conditions.

Forest fire histories based on lake charcoal for a site about 140 km west of the Andrews Forest complied by Long et al. (1998) extend our information on the interaction of climate and forest fire back even further. Climate models and known changes in the timing of the perihelion and the tilt of Earth's axis (Kutzbach et al. 1993) indicate that, between about 9000 and 6850 years before present (B.P.), the climate was warmer and drier than it is today. During this time fire intervals in the Oregon Coast Range averaged 110 ± 20 years. From about 6850 to 2750 B.P., there was an onset of cool, humid conditions, and, although there was an increase in the abundance of fire-sensitive species, the fire interval lengthened to 160 ± 20 years. From 2750 B.P. to the present, cool, humid conditions resulted in mesophytic taxa, and the mean fire interval increased to 230 ± 30 years. Although the actual fire intervals may be different in the Cascades and near the Andrews site, the overall pattern of changing climate and the ecological response in terms of relative fire intervals might have been similar.

The same overall climate changes that affected fire regime led to pronounced vegetation changes in the PNW. The long-term record shows that the composition of the forests has not been static, but instead has changed continuously with climate changes. Records from Little Lake (central Oregon Coast Range), Indian Prairie (Oregon Western Cascades), and Gold Lake Bog (central Oregon Cascades), for example, show changes in forest composition in the past that were most likely a response to shifts in summer drought and winter precipitation (Worona and Whitlock 1995; Sea et al. 1995). These, in turn, were driven by changes in the seasonal amplitude of insolation, the position of winter storm tracks, and the strength of the northeast Pacific subtropical high-pressure area.

The paleoecological record also suggests that forest communities in this region can change fairly rapidly with climate change. One episode of rapid vegetation change occurred at Little Lake around 14,850 years ago (Grigg and Whitlock 1998) when the pollen record shows that spruce forest was replaced by forest dominated by Douglas-fir in less than a century. A douglas-fir forest then persisted for about 350 years, when it reverted back to spruce forest. The increase in Douglas-fir at Little Lake was preceded by a prominent charcoal peak, which suggests that one fire or several closely spaced fires helped trigger the vegetation change by killing spruce and creating soil conditions suitable for Douglas-fir establishment. Warmer conditions than before allowed Douglas-fir to remain competitive for several decades or even centuries before spruce returned. Additional records of comparable resolution are necessary to determine whether this event is of regional significance. Nonetheless, the Little Lake data suggest that vegetation changes can occur rapidly when climate alters disturbance and tree regeneration conditions.

The PNW is a topographically complex area. Whitlock (1992) has described the Holocene vegetation history for this area as a response by plants to a hierarchical set of environmental controls of which climate is but one. Vegetation changes at the millennial timescale appear to respond to warming and, in this part of the world, drying associated with the retreat of the main Laurentide ice sheet. From 20,000 to 16,000 B.P., there was an influx of xerothermic subalpine vegetation (*Picea engelmanni* and *Artemisia*). Mesophytic subalpine vegetation appeared (*Tsuga mertensiana, Picea sitchensis*, and *Alnus sinuata*) after 16,000 B.P. when the main storm tracks are believed to have shifted northward. The later establishment of warm-loving and drought-adapted species from 12,000 to 6000 B.P. is associated with greater solar radiation and an expansion of the subtropical high-pressure zone. *Pseudotsuga* and *Alnus* then dominated the forests. Prairies and grasslands also appeared. At shorter timescales, Whitlock notes that fires were probably more frequent in the early Holocene warm dry period, so early successional and forest-opening species would have been more abundant. Also, at a smaller geographic scale, substrate conditions became important in influencing vegetation type. Prairie and oak woodland in the Puget Sound area favored summer drought conditions on the coarse-textured soils found there today and presumably throughout the Holocene. The modern forests of the Pacific Northwest are believed to have formed only in the last few millennia when the climate became wetter and solar radiation was reduced. Whitlock (1992, p. 22) concludes, "modern communities are loose associations composed of species independently adjusting their ranges to environmental changes on various time scales."

Future Climate Variability and Ecosystem Response

There continues to be concern about the possible effects on global climate change related to increased greenhouse gases in the atmosphere. If such climate change does occur, it is difficult to conceive of a potentially more important example of climate variability and ecosystem response. Future climate change will have complex, cascading, and, in some cases, detrimental effects on the ecosystems of the PNW.

Global General Circulation Models (GCMs) of the atmosphere and ocean are being used to investigate the question of what possible climate change might occur. Many caveats accompany the use of these models to estimate the potential changes of climate that might occur in a particular region. Apart from model deficiency, one of the most important caveats relates to the uncertainty in the rate of greenhouse gas emissions over the next 100 years. Also the range of temperature and precipitation projected by different models is quite large. For example, projected monthly temperatures vary by about 4°C (7°F) for different models. However, the models have been able to reproduce the increase of global temperatures in the twentieth century and the possibly anthropogenic part of that increase since 1970 (JISAO CIG 1999).

Output from the Canadian Centre for Climate Modeling and Analysis (CCC) GCM was employed by the JISAO group to suggest possible changes of climate in the PNW in the twenty-first century. This model suggests an increase in the cool season (Oct.–Mar.) temperature of about 5.5°C (10°F) and about 4°C (7°F) in the warm season (Apr.–Sept.) by the year 2100. A decrease of cool season temperature of about 5.5°C might eliminate subfreezing temperatures in some parts of the PNW. This might greatly decrease the snowpack and lead to a nonlinear response. The CCC model also suggested an increase in precipitation of about 330 mm (13 in.) in the cool season and 25 mm (1 in.) in the warm season. Despite the projected increase in precipitation, the rise may not be beneficial to forest environments because the large addition in the cool season will increasingly fall as rain, as opposed to snow, because of the higher temperatures. Much of this precipitation will run off and not add to the winter snowpack, if it still exists, for later release as snowmelt. The consensus of opinion of the University of Washington Climate Impacts Group, based on the output of seven GCMs, was that "the models are generally in agreement that winters will be warmer and wetter, but are divided about whether summers will be wetter or drier" (JISAO CIG 1999, p. 20). If some of these scenarios come to pass, the effect on PNW forests might not be favorable. The higher temperatures and possibly decreased amount of warm season soil moisture might increase the possibility of forest fires. Directly, these changes will lead to changes in the rates of growth, seed production, and seedling mortality. Indirectly, they will influence the disturbance regimes of fire, insect infestation, landslides, and disease (Franklin et al. 1992). The fossil record suggests that climate change coupled with disturbance will lead to disequilibrium between vegetation and climate as species adjust to new conditions and competitive interactions change.

Modeling studies (Urban et al. 1993), using an increase in temperature of 2.0–5.0°C, showed some altitudinal zonal and plant composition changes in Cascade ecosystems, but these studies used models that were set to run for 1000 years. Other model studies of biome and hydrologic response currently take an equilibrium approach, so they do not provide information on how, when, or even whether the vegetation/hydrosphere can respond to climate changes of a 2.0–5.0°C magnitude in the next century. Nonetheless, the equilibrium changes in hydrology and vegetation in the West are dramatic. An assessment by Thompson et al. (1998) suggests that it is unlikely that biotic adjustments can be accomplished in the next century for several reasons. First, vegetation responds more slowly than the projected climate change, especially long-lived species such as those in the PNW. The best paleoecological estimates for plant migration rates in the past are 40 times slower than those needed to keep pace with a doubled CO_2-related climate change in the twenty-first century. The plant species that predate humans did not have to contend with human land-use alteration that set up impediments to migration and dispersal. Second, species may not be able to migrate without assistance across a landscape fragmented by past land use. Third, the models only describe what potential, as opposed to actual, vegetation could occupy the climate space. Nonclimatic factors, including competition, will slow the migration process. Disturbance will probably be the catalyst of vegetation change, and the increase in severe fires during the last decade may already be the harbinger of effects of climate conditions to come.

Discussion

Present knowledge of climate variability and ecosystem response at a range of timescales provides a variety of answers to some of the guiding questions of this volume. New questions also emerge. Sometimes, it is difficult to specify the most important timescale at which causes and consequences are operating. For example, is forest fire occurrence at the Andrews LTER site more related to a seasonal scale or to a decadal to century scale? What is the interaction, if any, or relative importance between scales? A schematic representation of the characteristic timescales of some of the ecosystem responses to climatic disturbances helps to conceptualize the temporal variability (figure 19.3).

Preexisting conditions appear to be particularly important at the shorter timescales considered in this chapter. Important preexisting conditions can occur because of natural and/or anthropogenic-derived variability. Natural factors, such as fuel buildup, emphasize preexisting conditions with respect to fire frequency at the century scale. The need for suitable antecedent soil moisture conditions and potentially movable debris as a precursor for debris flow is another example. Swanson et al. (1998) also note that the preexisting condition of the geography of controls on debris flow occurrence causes some headwater streams to experience repeated, severe disturbance, whereas others may never have debris flows. Anthropogenic factors, such as forest management practices, may also be regarded as establishing preexisting conditions. By far the largest area of windthrow in the Bull Run basin in the northern Cascades of Oregon over a 100-year period, for example, was found to have occurred only after forest harvesting began in 1958 (Sinton et al. 2000).

The situation regarding preexisting conditions may be different at the longer timescales. The speed with which plant communities can be altered at the millennial scale in the PNW region, as represented by vegetation changes at Little Lake, implies that the exact nature of the preexisting communities is less important at this scale. The vegetation history in the PNW suggests that the nature of the vegetation existing previous to a climate change plays a minor role in determining the type of ecosystem response in terms of the new vegetation community that takes over a given location. For example, the Little Lake pollen record near 14,500 years B.P., which shows relatively rapid changes from spruce to Douglas-fir and back again (Grigg and Whitlock 1998), gives little evidence, except possibly that related to seed availability, that the later vegetation affected the type of the newer vegetation.

A consideration of longer timescales leads investigators to examine the timing of ecosystem response. Neglecting, for the moment, disturbance- and succession-related vegetation change, as far as the forests of the PNW are concerned, evidence suggests that climate-induced vegetation change can show response to climatic episodes at timescales of as little as 500 to 1000 years (Whitlock 1992). Paleoeco-

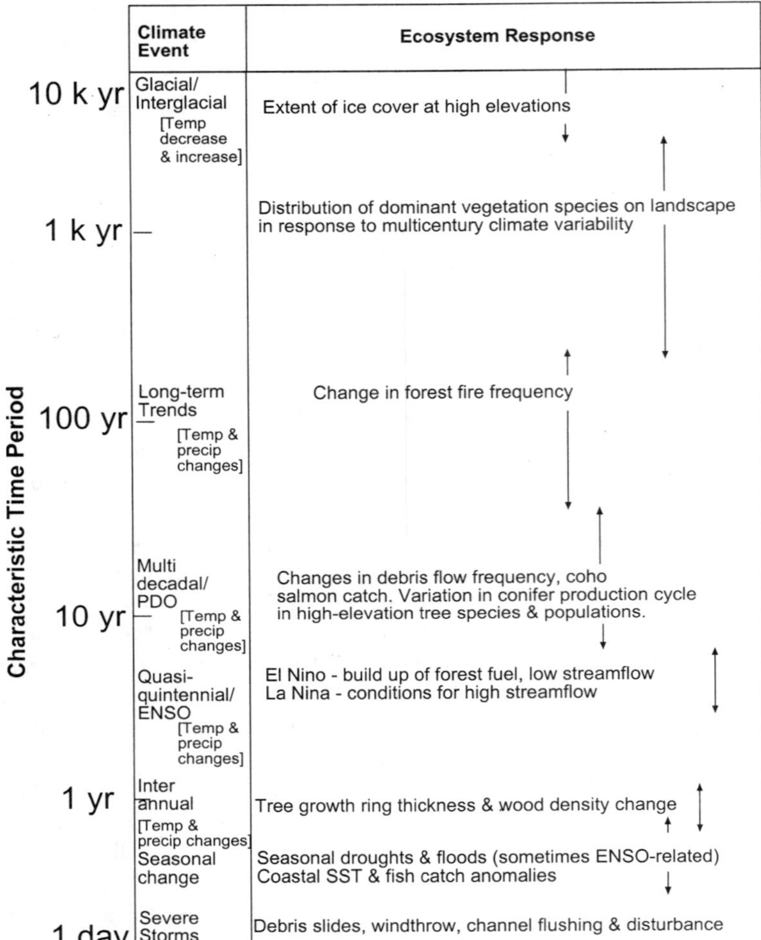

Figure 19.3 Characteristic timescales of some of the ecosystem responses to climatic disturbances.

logic records reveal the relatively ephemeral nature of modern communities. Modern forests represent an association that has existed for less than 3–6 millennia, and in the Cascade Range only a few generations of the forest dominants have been present in some sites at this timescale (Sea and Whitlock 1995). Species apparently have responded individualistically to Holocene environmental changes rather than as whole communities, and in the process, plant associations have been dismantled and reformed at a millennial pace.

There is no doubt that climate change and variability in the ecosystems of this region go far beyond an individual cause and result. There is almost always a cascade of resulting effects. Nakamura et al. (2000) explicitly employed the cascade concept for short-term events. They have established common sequences of events related to a cascade of hydrological and geomorphological events associated with the floods at the Andrews Forest. A biotic example of a cascade of events acting through time is evident in the effect of a windstorm, which may topple small trees or patches of trees. In some cases, at the edge of clear-cuts, the toppling event may be followed by drought that favors local eruptions of bark beetles, who emerge from the fallen trees to attack nearby live trees (Powers et al. 1999). At a longer timescale, Long et al. (1998) suggest that throughout the Holocene, changes in both vegetation and fire frequency were controlled by climate in finally determining the species composition and distribution of Coast Range forests. Studies from this region also show that besides cascading effects, extra factors can act as additional forcing functions alongside climate forcing. Factors such as soil texture and humans both causing, and suppressing, fires are examples. This is not a surprising conclusion, but it does emphasize the continued need to establish the importance of climate relative to other kinds of ecosystem forcing.

In this region we find examples at the quasi-quintennial and the multidecadal scales where the event, such as climate stage of ENSO, and the response, such as stream discharge, return to their "original" state by the time of the next event. This might not be true if vegetation or other environmental conditions have changed in the meantime. For example, since stream discharge is affected by water use by the vegetation, a lagged response in vegetation to specific climate variability may produce a lagged response in stream discharge. Furthermore, the concept of vegetation communities being "loose associations composed of species independently adjusting their ranges to environmental changes on various timescales" (Whitlock 1992, p. 22) suggests that at the century and millennial timescales there are likely to be no identical past analogs to the ecosystem at any point in time. It is unlikely that an ecosystem will return to its "original" state at this longer timescale, and the concept of "original" state itself has little meaning.

Acknowledgment   The material in this chapter is based on studies related to the H. J. Andrews Experimental Forest Long-Term Ecological Research program, which is supported by the National Science Foundation and the USDA Forest Service Pacific Northwest Research Station.

References

Buckley, B. M., R. D. D'Arrigo, and G. C. Jacoby. 1992. Pages 35–45 in K. T. Redmond and V.L. Tharp, editors. *Proceedings of the Eighth Annual Pacific Climate (PACLIM) Workshop, March 10–13, 1991*. California Department of Water Resources, Interagency Ecological Studies Program, Technical Report 31.

Cayan, D. R., and R. H. Webb. 1992. El Niño Southern Oscillation and streamflow in the western United States. Pages 29–68 in H. F. Diaz and V. Markgraf, editors. *El Niño: Historical and Paleoclimatic Aspects of the Southern Oscillation*. Cambridge: Cambridge University Press.

Ebbesmeyer, C. C., D. R. Cayan, D. R. McLain, F. H. Nichols, D. H. Peterson, and K. T. Redmond. 1991. 1976 Step in the Pacific Climate: Forty environmental changes between

1968–1975 and 1977–1984. Pages 115–126 in J. L. Betancourt and V. L. Tharp, editors. *Proceedings of the Seventh Annual Pacific Climate (PACLIM) Workshop, April 1990.* California Department of Water Resources. Interagency Ecological Studies Program. Technical Report 26. 235 pp.

Franklin, J. F., F. J. Swanson, M. E. Harmon, D. A. Perry, T. A. Spies, A. Thomas, V. H. Dale, A. McKee, W. K. Ferrell, J. E. Means, E. Joseph, S. V. Gregory, J. D. Lattin, T. D. Schowalter, and D. Larsen. 1992. Effects of global climatic change on forests in Northwestern North America. Pages 244–257 in R. L. Peters and T. E. Lovejoy, editors. *Global Warming and Biological Diversity.* New Haven, Connecticut: Yale University Press.

Fraser, V. D. 2001. *Biomass and productivity in an old-growth Douglas-fir/western hemlock stand in the western Cascades of Oregon.* New Haven, Connecticut: Yale University; project report for a non-thesis M.S. degree. 15 pp.

Garfin, G. M., and M. K. Hughes. 1996. *Eastern Oregon Divisional Precipitation and Palmer Drought Severity Index from Tree-Rings.* Final Report to U.S.D.A. Forest Service Cooperative Agreement PNW 90-174.

Gratkowski, H. J. 1956. Windthrow around staggered settings in old-growth Douglas-fir. *Forest Science* 2:60–74.

Graumlich, L. J. 1987. Precipitation variation in the Pacific Northwest (1675–1975) as reconstructed from tree rings. *Annals of the Association of American Geographers* 77:19–29.

Graumlich, L. J., and L. B. Brubaker. 1986. Reconstruction of annual temperature (1590–1979) for Longmire Washington, derived from tree-rings. *Quaternary Research* 25:223–234.

Grigg, L. D., and C. Whitlock. 1998. Late-glacial vegetation and climate change in western Oregon. *Quarternary Research* 49:287–298.

Harmon, M. E., K. Bible, M. J. Ryan, D.Shaw, H. Chen, J. Klopatek, and Xia Li. In press. Production, respiration, and overall carbon balance in an old-growth Pseudotsuga/Tsuga forest ecosystem. *Ecosystems.*

Harmon, M. E., and J. Sexton. 1995. Water balance of conifer logs in early stages of decomposition. *Plant and Soil* 172:141–152.

Harmon, M. E., J. F. Franklin, F. J. Swanson, P. Sollins, J. D. Lattin, N H. Anderson, S. V. Gregory, S. P. Cline, N. G. Aumen, J. R. Sedell, G. W. Lienkaemper, K. Cromack, Jr., and K. W. Cummins. 1986. The ecology of coarse woody debris in temperate ecosystems. *Recent Advances in Ecological Research* 15:133–302.

Holmes, R. L., R. I. C. Adams, and H. C. Fritts. 1986. *Tree-ring chronologies of western North America: California, eastern Oregon, and northern Great Basin.* Chronology Series VI, Laboratory of Tree-ring Research, University of Arizona, Tucson, Arizona. 182 pp.

Janisch, J. E., and M. E. Harmon. Successional changes in live and dead wood stores: Implications for Net Ecosystem Productivity. *Tree Physiology* 22:77–89.

JISAO CIG. 1999. *Impacts of climate variability and change in the Pacific Northwest. Pacific Northwest Contribution to the National Assessment in the Potential Consequences of Climate Change for the United States.* The Joint Institute for the Study of Atmosphere and Ocean Climate Impacts Group, Box 354235. University of Washington, Seattle, Washington. JISAO contribution #715. 109 pp.

Johnson, S. L., F. J. Swanson, G. E. Grant, and S. M. Wondzell. 2000. Riparian forest disturbances by a mountain flood—The influence of floated wood. *Hydrological Processes* 14:3031–3050.

Jones, J. A., and G. E. Grant. 1996. Peak flow responses to clear-cutting and roads in small and large basins, western Cascades, Oregon. *Water Resources Research* 32:959–974.

Kutzbach, J. E., P. J. Guetter, P. J. Behling, and R. Selin. 1993. Simulated climatic changes: Results of the COHMAP climate-model experiments. Pages 24–93 in H. E. Wright, J. E. Kutzbach, T. Webb III, W. F. Ruddiman, F. A. Street-Perrott, and P. J. Bartlein, editors. *Global Climates of the Last Glaciation.* University of Minnesota Press, Minneapolis.

Long, C. J., C. Whitlock, P. J. Bartlein, and S. H. Millspaugh. 1998. A 9000-year fire history from the Oregon Coast Range, based on a high-resolution charcoal study. *Canadian Journal of Forest Research* 28:774–787.

Mantua, N. J., S. R. Hare, Yuan Zhang, J. M. Wallace, and R. C. Francis. 1997. A Pacific interdecadal climate oscillation with impacts on salmon production. *Bulletin of the American Meteorological Society* 78:1069–1079.

Nakamura, F., F. J. Swanson, and S. M. Wondzell. 2000. Disturbance regimes of streams and riparian systems—A disturbance-cascade regime. *Hydrological Processes* 14:2849–2860.

Post, D. A., and J. A. Jones. 2001. Hydrologic regimes of forested, mountainous, headwater basins in New Hampshire, North Carolina, Oregon, and Puerto Rico. *Advances in Water Resources* 24:1195–1210.

Powers, J. S., P. Sollins, M. E. Harmon, and J. A. Jones. 1999. Plant-pest interactions in time and space: A Douglas-fir bark beetle outbreak as an example. *Landscape Ecology* 14:105–120.

Redmond, K. T., and R. Koch. 1991. ENSO vs. surface climate variability in the western United States. *Water Resources Research* 27:2381–2399.

Sea, D. S., and C. Whitlock. 1995. Postglacial vegetation and climate of the Cascade Range, Central Oregon. *Quaternary Research* 43:370–381.

Sinton, D. S., J. A. Jones, J. L. Ohmann, and F. J. Swanson. 2000. Windthrow disturbance, forest composition, and structure in the Bull Run basin, Oregon. *Ecology* 81:2539–2556.

Snyder, K. U. 2000. *Debris flows and flood disturbance in small, mountain watersheds.* M.S. thesis, Oregon State University. Corvallis.

Swanson, F. J., S. L. Johnson, S. V. Gregory, and S. A. Acker. 1998. Flood disturbance in a forested mountain landscape. *BioScience* 48:681–689.

Taylor, G. H., and C. Southards. 1997. Long-term Climate Trends and Salmon Population. http://www.ocs.orst.edu/reports/climate_fish.html.

Thompson, R. S., S. W. Hostetler, P. J. Bartlein, and K. H. Anderson. 1998. *A Strategy for Assessing Potential Future Changes in Climate, Hydrology, and Vegetation in the Western United States.* U.S. Geological Survey Circular 1153. 20 pp.

Urban, D. L., M. E. Harmon, and C. B. Halpern. 1993. Potential response of Pacific Northwestern forests to climatic change, effects of stand age and initial composition. *Climatic Change* 23:247–266.

Valachovic, Y. S. 1998. *Leaf litter chemistry and decomposition in a Pacific Northwest coniferous forest ecosystem.* M.S. thesis, Oregon State University, Corvallis.

Waring, R. H., and J. F. Franklin. 1979. Evergreen coniferous forests of the Pacific Northwest. *Science* 204:1380–1386.

Worona, M. A., and C. Whitlock. 1995. Late Quaternary vegetation and climate history near Little Lake, central Coast Range, Oregon. *Geological Society of America Bulletin* 107:867–876.

Weisberg, P. J., and F. J. Swanson. 2003. Regional synchroneity in fire regimes of western Oregon and Washington, USA. Forest Ecology and Management 172:17–28.

Whitlock, C. 1992. Vegetational and climatic history of the Pacific Northwest during the last 20,000 years: Implications for understanding present-day biodiversity. *The Northwest Environmental Journal* 8:5–28.

Wiles, G. C., R. D'Arrigo, and G. C. Jacoby. 1995. Modeling North Pacific temperatures and pressure changes from coastal tree-ring chronologies. Pages 67–78 in K. T. Redmond and V. L. Tharp, editors. *Proceedings of the Eleventh Annual Pacific Climate (PACLIM) Workshop*. Interagency Ecological Studies Program for the Sacramento-San Joaquim Estuary. Tech. Report. 40. California Department of Water Resources.

20

Climate Variability in Tallgrass Prairie at Multiple Timescales: Konza Prairie Biological Station

Douglas G. Goodin
Philip A. Fay
Maurice J. McHugh

Climate is a fundamental driver of ecosystem structure and function (Prentice et al. 1992). Historically, North American grassland and forest biomes have fluctuated across the landscape in step with century- to millennial-scale climate variability (Axelrod 1985; Ritchie 1986). Climate variability of at decadal scale, such as the severe drought of the 1930s in the Central Plains of North America, caused major shifts in grassland plant community composition (Weaver 1954, 1968). However, on a year-to-year basis, climate variability is more likely to affect net primary productivity (NPP; Briggs and Knapp 1995; Knapp et al. 1998; Briggs and Knapp 2001). This is especially true for grasslands, which have recently been shown to display greater variability in net primary production in response to climate variability than forest, desert, or arctic/alpine systems (Knapp and Smith 2001).

Although the basic relationships among interannual variability in rainfall, temperature, and grassland NPP have been well studied (Sala et al. 1988; Knapp et al. 1998; Alward et al. 1999), the linkages to major causes of climate variability at quasi-quintennial (~5 years) or interdecadal (~10 year) timescales in the North American continental interior, such as solar activity cycles, the El Niño–Southern Oscillation (ENSO), the North Atlantic Oscillation (NAO), and the North Pacific Index (NP), are less well understood.

In this chapter, we will examine how interannual, quasi-quintennial, and interdecadal variation in annual precipitation and mean annual temperature at a tallgrass prairie site (Konza Prairie Biological Station) may be related to indexes of solar activity, ENSO, NAO, and NP, and in turn how these indexes may be related to aboveground net primary productivity (ANPP). Specifically, we present (1) period-spectrum analyses to characterize the predominant timescales of temperature and precipitation variability at Konza Prairie, (2) correlation analyses of quantitative indexes of the major atmospheric processes with Konza temperature and precipitation records, and (3) the implications of variation in major atmospheric processes for seasonal and interannual patterns of ANPP.

Konza Prairie Biological Station

The Konza Prairie Biological Station (KNZ), which lies in the Flint Hills (39°05' N, 96°35' W), is a 1.6-million-ha region spanning eastern Kansas from the Nebraska border to northeastern Oklahoma (figure 20.1). This region is the largest remaining tract of unbroken tallgrass prairie in North America (Samson and Knopf 1994) and falls in the more mesic eastern portion of the Central Plains grasslands. Konza's climate falls within well-recognized temperature and rainfall parameters for grassland biomes. The mean temperature for Konza is 12°C. Total rainfall averages 835 mm y^{-1}, with 75% falling during the growing-season months of April through October. Growing-season rainfall is bimodal, with high monthly rainfall totals during May and June, low rainfall and high temperatures in July and August, and a second rainy period in September. High variability is common in yearly rainfall totals and seasonal distribution (Hayden 1998). Because Konza Prairie is located in the transition zone from mesic tallgrass to more xeric midgrass prairie and has inherently variable climate patterns and productivity responses (Knapp and Smith 2001), it is well suited for examination of possible linkages among ENSO, NAO, NPI, or other large-scale climate mechanisms and ecosystem responses.

Regulation of ANPP in Tallgrass Prairie

Climate variability is one of several important biotic and abiotic factors regulating ANPP in tallgrass prairie. Multiple factors, including fire, nutrients, grazing by large ungulates, and topography, are involved in the regulation of ANPP in tallgrass prairie. For example, a synthesis of a 20-year record of ANPP at Konza Prairie showed that, in general, early growing-season fire and moderate-intensity grazing increased ANPP (Knapp et al. 1998). Herbivores and fire in some ways have similar effects, with both removing the plant canopy and detritus layers, allowing increased penetration of light to the soil surface, which warms the soil and enhances plant growth. Interactions among these factors are pervasive. For example, topographic position influenced ANPP most strongly in annually burned sites, where deep-soil lowlands were more productive than more shallow-soil uplands. In contrast, at long-term unburned sites, there was little topographic effect on ANPP. The simultaneous presence of these multiple interacting controls on ANPP means that there is considerable temporal variation in limitations on ANPP in tallgrass prairie and that ANPP depends strongly on the degree to which the multiple controlling factors reinforce each other or cancel each other out (Knapp et al. 1998). Climate variability can be viewed as the backdrop against which these other productivity-limiting factors operate.

The climate variability influencing ANPP in tallgrass prairie operates in a larger

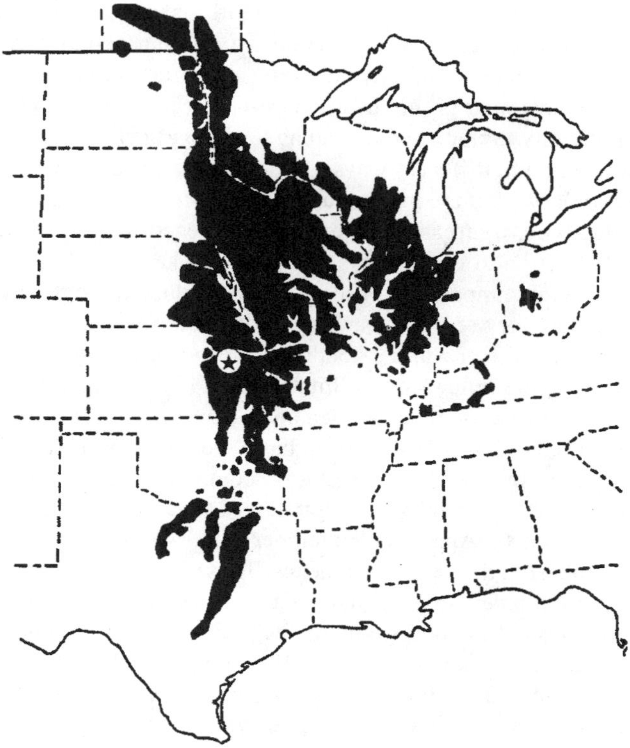

Figure 20.1 Location of Konza Prairie Biological Station (KPBS); site of the KNZ LTER indicated by the star. The shaded area indicates the estimated extent of tallgrass prairie prior to European settlement.

spatial context. The Central Plains is a vast west-to-east gradient in grassland species composition and ANPP, with a strong increase in ANPP ($r^2= 0.90$) following the eastward increase in annual rainfall (Küchler 1974; Sala et al. 1988). For individual locations, relationships between grassland ANPP and rainfall quantity are strongest in the drier western portions of the Central Plains (Epstein et al. 1997) and weaker in the more mesic (and more variable) eastern portions of the Great Plains (Knapp et al. 1998).

The basic patterns of plant community structure in modern grasslands provide the foundation for a mechanistic understanding of productivity responses to climate variability in tallgrass prairie. Native plant communities in the Central Plains grasslands are composed of species from several functional groups (Körner 1994). These include warm-season C_4 grasses, cool-season C_3 graminoids (grasses and sedges), and a diverse array of other C_3 herbaceous dicots (hereafter referred to as "forbs"), nitrogen-fixing leguminous species, and woody species. The C_4 grasses consist of relatively few species but are abundant, widely distributed, temporally stable, and they account for roughly 80% of the biomass and canopy cover (Briggs and Knapp 1995; Knapp and Medina 1999). Conversely, forbs constitute a small fraction of the biomass but a large fraction of the species, and forb species are temporally dynamic in presence and abundance (Collins and Glenn 1991; Hartnett and Fay 1998).

The impact of climate variability on ecosystem structure and function depends on species and functional group differences in morphology and physiology, and their resultant ability to track and acclimate to changing climate conditions. Forbs and grasses respond in different ways to interannual variation in rainfall (Briggs and Knapp 2001). Most of the interannual variability in ANPP is due to fluctuations in grass productivity, whereas forbs tend to be unresponsive to these climate elements. This stability of forbs in terms of annual productivity stands in contrast to the dynamic species composition of the forb assemblage through time, potentially reflecting compensatory responses among various members of the forb functional group in response to the prevailing conditions of each growing season.

When studying the influence of climate on ANPP, it is important to carefully consider how the term *climate* is defined and quantified. Most studies of tallgrass response have used indexes derived from basic meteorological data (i.e., temperature, precipitation), but they also take into account climate/vegetation feedbacks. Briggs and Knapp (1995) used total precipitation (1 January–31 December), growing-season precipitation (1 April–30 September), and summer pan evaporation (1 July–30 September) as their climate indexes. These indexes integrate both temperature and precipitation effects, and in the case of the growing-season precipitation and pan evaporation, they concentrate on the period of the year when ANPP is determined. Using these indexes, Briggs and Knapp (1995) found correlations with ANPP ranging from 0.53 to 0.65 for productivity at all site at Konza. When sites were differentiated by treatment type (burned vs. unburned) and topographic position (upland vs. lowland), correlations as high as 0.87 were found between growing-season precipitation and total ANPP in annually burned uplands. Correlation values for other combinations of climate variable and treatment/position ranged from 0.40 to 0.85.

Drivers of Climate Variability

The energy that drives the earth's climate originates with the sun. Variability in the output of solar energy occurs at a variety of timescales from interannual to millennial. At the timescales investigated here, sunspots are the major mechanism of variation in solar irradiance (Landscheit 1983). During active sun periods (i.e., periods of increased sunspots), solar irradiance increases. Systematic human observations of sunspot cycles have been made for over 300 years, and they indicate an 11-year cycle (Lean et al. 1995). The sunspot cycle causes a variation in solar irradiance of about 0.15% (~2 Wm^{-2}). This modest variability in solar irradiance may have links to climate variability and drought (Eddy 1983), although direct solar influences on interdecadal climate response are controversial (Ruddiman 2001).

Climate variability also arises from atmospheric processes internal to the earth's system. These processes are defined by coherent and correlated patterns of atmospheric temperature, pressure, winds, and circulation at characteristic locations across the globe. Although associated with specific locations, these processes exhibit teleconnections capable of influencing weather and climate patterns across the globe, including in the continental interior of North America.

One of the best known and most studied of these atmospheric teleconnection processes is the El Niño–Southern Oscillation (ENSO). ENSO is a pattern of pressure change across the Pacific (the Southern Oscillation), accompanied by a flow of warm water from the equatorial western Pacific Ocean toward South America (El Niño). Presence of the warm equatorial countercurrent off western South America, in conjunction with lower atmospheric pressures, is referred to as the ENSO warm phase. The cold phase occurs when cool currents and increased atmospheric pressures predominate. ENSO events are quantified by several indexes, mostly derived from atmospheric pressure and sea surface temperature (SST) within various regions in the Pacific. ENSO-related climate variability occurs mainly at quasi-quintennial timescales. The NINO1+2 and NINO3 indexes show significant periods at about 3.5 and 7.0 years (Kaplan et al. 1998; Reynolds and Smith 1994; Latif et al. 1998).

El Niño events are correlated with temperature and precipitation in the central United States, although the strength of association may depend somewhat on the index used (Philander 1990; Greenland 1999). Sittel (1994) found that warm ENSO events were associated with warmer average temperatures; however, fall and winter were dry during warm events. Cold events did not alter temperature patterns as strongly as did warm events, but they did result in lower spring and fall precipitation.

The North Atlantic Oscillation (NAO) also affects climate in the interior of North America. NAO is a low-frequency oscillation in atmospheric mass and, consequently, pressure across the North Atlantic Ocean. In its positive phase, the NAO is characterized by higher subtropical pressure and lower subpolar pressure, strengthening midlatitude westerly winds. In the negative phase, both these pressure features are weakened, and westerly wind speeds decrease. The amplitude and phase of the NAO are highly variable, ranging from interseasonal to interdecadal (van Loon and Rogers 1978; Wallace et al. 1996; Hurrell 1995). The NAO exhibits variability at both interannual and interdecadal timescales. Spectral analysis of the NAO index values show prominent oscillations at about 5 and 9 years (Hurrell 1995). NAO effects are observed during all seasons, accounting for 33% of the variance in winter sea level pressure (Cayan 1992). The NAO is positively correlated with spring and summer temperature and precipitation in the Central Plains (van Loon and Rogers 1978); however, the strength of association is weaker than for other regions of North America.

The North Pacific (NP) index is a third teleconnection pattern affecting the North American continental interior. The NP is the area-weighted mean sea level pressure over the region 30° N to 65° N and 160° E to 140° W, and it exhibits periodicities at 12–14 years and at 50 years (Trenberth and Hurrell 1994). The NP indexes the intensity of the Aleutian low-pressure cell and is related to climate events in North America through its influence on the downstream predominance of zonal or meridional circulation. Temperature and precipitation effects related to NP have been observed in the North American interior (Trenberth and Hurrell 1994). Negative NP values are associated with below-average precipitation and above-average

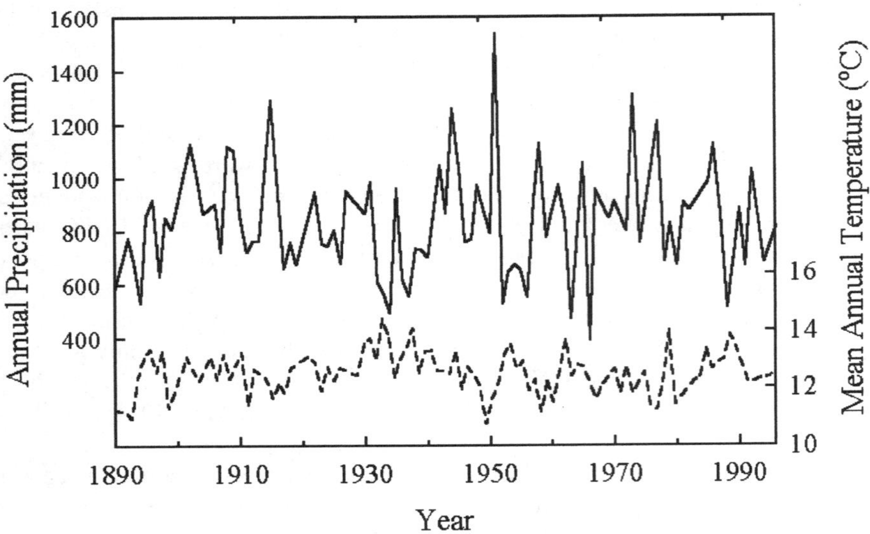

Figure 20.2 Time series of climate variables used in this chapter for the period 1891–1998. Annual precipitation is shown by the solid line and mean annual temperature by the dashed line.

temperature in the North American interior. Patterns reverse when NP is positive. The index is also associated with changes in synoptic storm tracks (Rogers and Rohli 1991). The NP is highly negatively correlated with another well-known teleconnection pattern, the Pacific North American (PNA) index (Trenberth 1990).

Observed Temperature and Precipitation Variability at Konza Prairie

To examine how sunspot cycles, ENSO, NAO, and NPI may influence temperature and precipitation variability in a tallgrass prairie, we conducted period-spectrum analyses of a 108-year weather record (1891–1999) from Manhattan, Kansas, approximately 12 km north of Konza Prairie. Annual mean temperature and annual precipitation totals were tabulated and expressed as anomalies. Two spectral analyses were conducted, one filtered to emphasize interdecadal-scale periodicities, the second filtered to emphasize quasi-quintennial periodicities.

Annual mean temperature and precipitation values between 1891 and 1999 (figure 20.2) show interannual variability characteristic of the continental interior. Prominent peaks in precipitation occurred between 1900 and 1920, and in the 1940s and 1950s, whereas periods of drought were apparent in the 1930s and 1950s. The interdecadal-filtered spectra indicate significant 14.1- and 22.8-year cycles in precipitation and 11.1- and 18.5-year periodicities in temperature (all $p<0.05$; figure 20.3A,B). The quasi-quintennial–filtered spectra suggested 3.3- and 6.9-year precipitation periods and an 8.4-year temperature period (all $p < 0.05$; figure 20.3C,D).

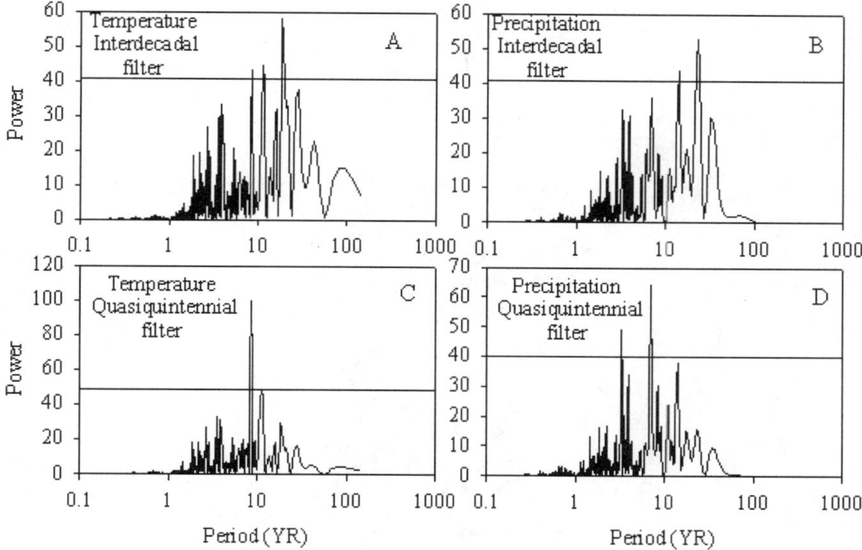

Figure 20.3 Power spectra of 108-year climate data record at Manhattan, Kansas. (A) Temperature data, filtered to emphasize interdecadal-scale cycles. (B) Precipitation data, filtered to emphasize interdecadal-scale cycles. (C) Temperature data, filtered to emphasize cycles at the quasi-quintennial scale. (D) Precipitation data, filtered to emphasize cycles at the quasi-quintennial cycles.

The periodicities in the Manhattan temperature and precipitation spectra correspond to periods reported for sunspots (11 and 22 years), ENSO (3.5 and 7 years), NAO (9 years), and NP (12 years). To test for an association between teleconnections and climate pattern, we performed correlation analyses of seasonal Manhattan temperature and precipitation means over the 108-year record with sunspot counts, and we reconstructed solar irradiance (Lean et al. 1995), ENSO (NINO1+2 and NINO3; Kaplan et al. 1998; Reynolds et al. 1994; Latif et al. 1998), NAO (van Loon and Rogers 1978), and NP (Trenberth and Hurrell 1994) indexes. The strongest correlations were found between NAO and winter (January–March) temperatures (r = 0.47, p< 0.0001), ENSO activity and winter precipitation (NINO1+2 r = 0.31, p = 0.0004; NINO3 r = 0.33, p = 0.0001), and the NP index and summer (July–August) temperature (r = 0.28, p = 0.0057). Other significant precipitation or temperature correlations with atmospheric processes were found, but correlation coefficients were 0.25 or less. No significant correlations were found between solar activity cycles and climate variables. This observation is consistent with the results of Baliunas and Jastrow (1993), who note that the variation in energy output in active versus quiet sun phases is small and unlikely to have detectable interdecadal effects on ecosystems.

Taken together, this information suggests possible links between the ENSO, NAO, NP, indexes and interdecadal and quasi-quintennial temperature and precipitation variability at Konza Prairie. Similar NAO/temperature relationships and El

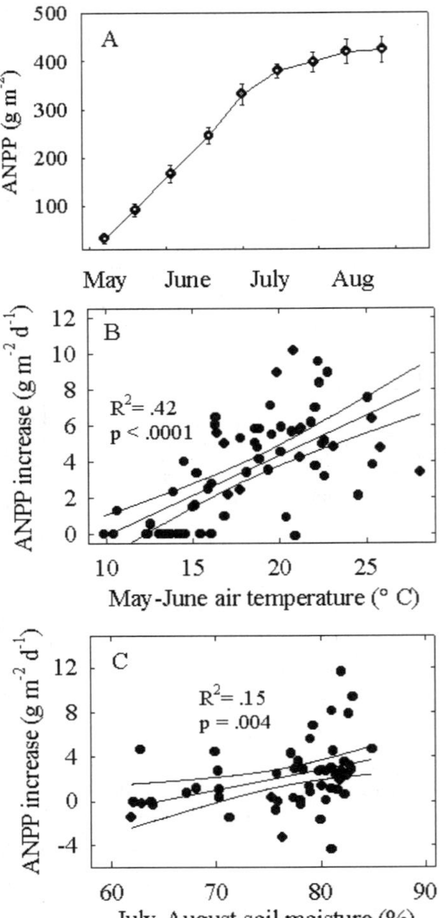

Figure 20.4 Growing-season biomass accumulation and climatic correlates in tallgrass prairie. (A) Seasonal course of aboveground biomass accumulation based on 14 years of bi-weekly ANPP sampling (1984–1997) at Konza Prairie (means ± 1 SE). (B) Linear regression analysis of the May–June NPP increase (in g m^{-2} d^{-1}) versus air temperature (± 95% confidence intervals). (C) Linear regression of the July–August NPP increase versus soil moisture.

Niño/precipitation relationships have been reported (e.g., Ropelewski and Halpert 1986; Sittel 1994; Hurrell and van Loon 1997). Our results indicate a stronger influence by ENSO indexes at Konza than that noted by Greenland (1999). Greenland's analysis used a 33-year monthly data series correlated to the Southern Oscillation Index (SOI), whereas our analysis used a longer time series of coarser resolution (seasonal) data correlated to the two El Niño indices.

How might climate processes impact ANPP at KNZ? To address this question, we looked at patterns of aboveground biomass accumulation derived from 14 years of biweekly harvests from two Konza Prairie watersheds (1984–1997; figure 20.4). We calculated the rate (in g m^{-2} d^{-1}) of accumulation for each harvest interval and conducted regressions of accumulation rate with air temperature and soil moisture. Aboveground biomass accumulation occurred in two distinct phases, one of rapid accumulation from the beginning of the growing season (late April) through early summer (late June) and a second slower phase through the hotter, drier remainder of the growing season through September. The rate of biomass accumulation during the faster first phase was correlated with air temperature (figure 20.4B), whereas the slower second phase was better correlated with soil moisture (figure 20.4C). This suggests that ANPP in tallgrass prairie is limited by a temporally shifting set of climate factors. This result also implies that the seasonal timing of precipitation and temperature are critical factors in determining ANPP.

Synthesis

A key finding of this analysis is that the historic temperature and precipitation record at Konza Prairie displays periodicities similar to those for ENSO, NAO, and NP. Periods of stronger NAO (i.e. larger positive index values) were associated with warmer winters, periods of stronger ENSO with wetter winters, and periods of stronger NP with warmer summers. The growing-season course of aboveground biomass accumulation appears to be limited initially by temperature, then later by soil moisture.

Warmer or wetter winters caused by ENSO or NAO activity may thus influence aboveground biomass accumulation by promoting an earlier start to the growing season than in colder, drier winters, thereby lengthening the rapid first phase of biomass accumulation. Warmer summers caused by a strengthened NP index may reduce ANPP by lowering accumulation rates during summer, when rainfall is relatively sparse, evaporative demands are at their maximum, and soil moisture has been depleted from its springtime maximum. This scenario implies that the juxtaposition of ENSO, NAO, and NP events may be as important as the strength of any single process because of their potential to either augment or offset each other. For example, a strong NAO/ENSO combined with a weak NP influence may promote peak ANPP years; the reverse (weak NAO/ENSO with strong NP influence) could cause low-productivity years.

The key synthesis question is then, How might these climate teleconnections, acting in concert, affect ecological response in tallgrass prairie? A comprehensive treatment of this question is beyond the scope of this chapter, but the results presented here provide some insight. Our results imply that the primary effects of atmospheric teleconnections on tallgrass prairie productivity are indirect. For example, the ENSO indexes were correlated with precipitation during fall and winter (table 20.1), which recharges the soil profile with moisture but which is otherwise weakly linked to ANPP compared to growing-season precipitation (Briggs and Knapp 1995). Similarly, since NAO correlates with non–growing-season temperatures, its effect on ANPP is also indirect (table 20.1).

In contrast, the NP may have more direct impacts on ANPP because it was cor-

Table 20.1 Seasonal correlations between climate variables and teleconnection indices

| | Season | | | |
Index	JFM	AMJ	JAS	OND
Temperature				
Niño 1,2	0.00	−0.08	−0.03	−0.05
Niño 3	−0.01	−0.07	−0.01	0.00
NAO	0.47*	0.25*	0.15	0.26*
Sunspots	−0.09	−0.07	0.03	−0.01
Insolation	0.12	0.12	−0.02	0.08
NP	0.10	0.22*	0.28*	0.02
Precipitation				
Niño 1,2	0.31*	0.06	0.01	0.16
Niño 3	0.33*	−0.01	−0.02	0.19*
NAO	−0.09	−0.19*	−0.07	0.13
Sunspots	0.10	−0.04	−0.04	0.00
Insolation	0.17	0.09	0.01	0.11
NP	−0.04	0.05	0.01	0.23*

*Indicates significant correlation ($p < 0.05$).

related with temperature during the active growing season (table 20.1). Temperature is significantly related to NAO throughout most of the year, including the early growing season (April–June; $r = 0.25$, $p < 0.05$). Temperature is also positively correlated with NP during these months ($r = 0.23$, $p < 0.05$). The NAO operates mainly at quasi-quintennial timescales (periods ~5 and 9 years), whereas NP is an interdecadal-scale oscillation (period ~12 years). The similar correlations values between NAO and temperature and NP and temperature are surprising, given the differing timescales of these indexes. However, NP and NAO are significantly correlated during June ($r = 0.21$, $p < 0.005$). This correlation suggests that early growing-season temperature is responding to an atmospheric circulation process that affects both indices.

For the precipitation data, significant negative correlations are found with NAO during the growing season ($r = −0.19$, $p < 0.05$). Early growing-season (May, June) precipitation should have little effect on ANPP accumulation (see previous discussion), however, precipitation dynamics in July might affect total productivity. Correlation between July NAO and precipitation is significant ($r = −0.20$, $p < 0.05$). The negative correlation indicates that higher values of NAO associate with low precipitation totals and vice versa. Although the negative correlation values for NAO and precipitation are opposite in sign from those between NAO and temperature, they may actually represent a coherent influence on productivity. Early season temperatures (positively related to NAO phase) would affect the timing of peak values, whereas moisture conditions later in the season (negatively correlated to precipitation) would limit upper ANPP values (see figure 20.4A).

Conclusions

The relationships between climate variability and ecological response proposed here are speculative. The brevity of the available ANPP data precludes any direct comparison between climate variability and ecosystem response at any except the briefest time spans. Nevertheless, the results we present are important for two reasons. First, they provide insight into interpreting current patterns of ANPP. The tallgrass prairies (and grasslands in general), more so than other ecosystems (forest, desert, arctic/alpine), are poised in a dynamic equilibrium that makes them especially sensitive to both biotic and abiotic disturbances, including climate variability. Increased understanding of climate effects on the system provides a framework for proposing specific hypotheses about climate-ecosystem interactions at all temporal scales.

Second, our results might form the basis for predicting the response of tallgrass prairies to climate change. Climate change scenarios that predict changes in periodic circulation phenomena can be linked to probable changes in productivity patterns. Although we detected periodicities of 3.5 to 12 years in temperature and precipitation at Konza Prairie and categorized them as quasi-quintennial and interdecadal, in fact these different periods occur as part of a hierarchical system of climate variation. Focusing on the degree to which these hierarchical regimes of temperature and precipitation variability reinforce (or oppose) each other may shed more light on the regulation of variability in ANPP or other ecosystem characteristics than considering them independently. Clearly, much more research will be needed before these complex links can be fully understood, but the results presented here provide direction for such research.

References

Alward, R. D., J. K. Detling, and D. G. Milchunas. 1999. Grassland vegetation changes and nocturnal global warming. *Science* 283:229–231.

Axelrod, D. I. 1985. Rise of the grassland biome, central North America. *The Botanical Review* 51:163–201.

Baliunas, S., and R. Jastrow. 1993. Evidence on the climate impact of solar variations. *Energy* 18:1285–1295.

Briggs, J. M., and A. K. Knapp. 1995. Interannual variability in primary production in tallgrass prairie: Climate, soil moisture, topographical position, and fire as determinants of aboveground biomass. *American Journal of Botany* 82:1024–1030.

Briggs, J. M., and A. K. Knapp. 2001. Determinants of C_3 forb growth and production in a C_4 dominated grassland. *Plant Ecology* 152:93–100.

Cayan, D. R. 1992. Latent and sensible heat flux anomalies over the northern oceans: The connection to monthly atmospheric circulation. *Journal of Climate* 5: 354–369.

Collins, S. L., and S. M. Glenn. 1991. Importance of spatial and temporal dynamics in species regional abundance and distribution. *Ecology* 72:654–664.

Eddy, J. A. 1983. Keynote address: An historical review of solar variability, weather, and climate. Pages 1–27 in B. M. McCormac, editor. *Weather and Climate Response to Solar Variations*. Colorado Associated University Press, Boulder, Colorado.

Epstein, H. E., W. K. Lauenroth, I. C. Burke, and D. P. Coffin. 1997. Productivity patterns of C_3 and C_4 functional types in the U.S. Great Plains. *Ecology* 78:722–731.

Greenland, D. 1999. ENSO-related phenomena at long-term ecological research sites. *Physical Geography* 20:491–507.

Hartnett, D. C., and P. A. Fay. 1998. Plant populations: Patterns and processes. Pages 81–100 in A. K. Knapp, J. M. Briggs, D. C. Hartnett, and S. L. Collins, editors. *Grassland Dynamics: Long-Term Ecological Research in Tallgrass Prairie.* Oxford University Press, New York.

Hayden, B. P. 1998. Regional climate and the distribution of tallgrass prairie. Pages 19–34 in A. K. Knapp, J. M. Briggs, D. C. Hartnett, and S. L. Collins, editors. *Grassland Dynamics: Long-Term Ecological Research in Tallgrass Prairie.* Oxford University Press, New York.

Hurrell, J. W. 1995. Decadal trends in the North Atlantic Oscillation: Regional temperatures and precipitation. *Science* 269:676–679.

Hurrell, J. W., and H. van Loon. 1997. Decadal variations in climate associated with the North Atlantic Oscillation. *Climatic Change* 36:301–326.

Kaplan A., M. Cane, Y. Kushnir, A, Clement, M. Blumenthal, and B. Rajagopalan. 1998. Analysis of global sea surface temperature. *Journal of Geophysical Research* 103: 18,567–18,589.

Knapp, A. K., J. M. Briggs, J. M. Blair, and C. L. Turner. 1998. Patterns and controls of aboveground net primary productivity in tallgrass prairie. Pages 193–221 in A. K. Knapp, J. M. Briggs, D. C. Hartnett, and S. L. Collins, editors. *Grassland Dynamics: Long-Term Ecological Research in Tallgrass Prairie.* Oxford University Press, New York.

Knapp, A. K., and E. Medina. 1999. Success of C_4 photosynthesis in the field: lessons from communities dominated by C_4 plants. Pages 251–283 in R. F. Sage and R. K. Monson (eds.), C_4 *Plant Biology.* Academic Press, New York.

Knapp, A. K., and M. D. Smith. 2001. Variation among biomes in temporal dynamics of aboveground primary production. *Science* 291:481–484.

Körner, C. 1994. Scaling from species to vegetation: The usefulness of functional groups. Pages 117–140 in E.-D. Schulze and H. A. Mooney, editors. *Biodiversity and Ecosystem Function.* Springer-Verlag, Berlin.

Küchler, A. W. 1974. A new vegetation map of Kansas. *Ecology* 55:586–604.

Landscheidt, T. 1983. Solar oscillations, sunspot cycles, and climatic change. Pages 348–378 in *Weather and Climate Response to Solar Variations.* Colorado Associated University Press, Boulder, Colorado.

Latif, M., D. Anderson, T. Barnett, R. Cane, R. Kleeman, A, Leetmaa, J. J. O'Brien, A. Rosati, A., and E. Schneider. 1998. A review of predictability and prediction of ENSO. *Journal of Geophysical Research* 103:14,365–14,393.

Lean, J., J. Beer, and R. Bradley, R. 1995. Reconstruction of solar irradiance since 1610: Implications for climate change. *Geophysical Research Letters* 22:3195–3198.

Philander, S. G. H. 1990. *El Niño, La Niña, and the Southern Oscillation.* Acdemic Press, San Diego, California.

Prentice, C. I., W. Cramer, S. P. Harrison, R. Leemans, R. A. Monserud, and A. M. Soloman. 1992. A global biome model based on plant physiology and dominance, soil properties and climate. *Journal of Biogeography* 19:117–134.

Reynolds, R. W., and T. M. Smith, 1994. Improved global sea surface temperature analysis using optimum interpolation. *Journal of Climate* 7:929–948.

Ritchie, J. C. 1986. Climate change and ecological response. *Vegetatio* 67:65–74.

Rogers, J. C., and R. V. Rohli. 1991. Florida citrus freezes and polar anticyclones in the Great Plains. *Journal of Climate* 5:127–139.

Ropelewski, C. F., and M. S. Halpert. 1996. North American precipitation and temperature trends associated with the El Niño Southern Oscillation (ENSO). *Monthly Weather Review* 114:2352–2362.

Ruddiman, W. F. 2001. Earth's Climate: Past and Future. W.H. Freeman & Co., New York.

Sala, O. E., W. J. Parton, L. A. Joyce, and W. K. Lauenroth. 1988. Primary production of the central grassland region of the United States. *Ecology* 69:40–45.

Samson, F. B., and F. L. Knopf. 1994. Prairie conservation in North America. *Bioscience* 44:418–421.

Sittel, M. 1994. Differences in the Means of ENSO Extremes for Temperature and Precipitation in the United States. COAPS Technical Report 94-2. Center for Ocean-Atmosphere Prediction Studies. Florida State University, Tallahassee, Florida.

Trenberth, K. E. 1990. Recent observed interdecadal climate changes in the Northern Hemisphere. *Bulletin of the American Meteorological Society* 71:988–993.

Trenberth, K. E., and J. W. Hurrell. 1994. Decedal atmosphere-ocean variations in the Pacific. *Climate Dynamics* 9:303–319.

van Loon, H., and J. C. Rogers. 1978. The seesaw in winter temperatures between Greenland and northern Europe. Part I: General description. *Monthly Weather Review* 106:296–310.

Wallace, J. M., Y. Zhang, and L. Bajuk. 1996. Interpretation of interdecadal trends in Northern Hemisphere surface air temperature. *Journal of Climate* 9:249–259.

Weaver, J. E. 1954. *North American Prairie*. Johnsen Publishing Co, Lincoln, Nebraska.

Weaver, J. E. 1968. *Prairie Plants and Their Environment: A Fifty-Year Study in the Midwest.* University of Nebraska Press, Lincoln, Nebraska.

21

Climate Variability and Ecosystem Response—Synthesis

David Greenland

Douglas G. Goodin

Raymond C. Smith

Frederick J. Swanson

At the outset we identified the theme of this book as how ecosystems respond to climate variability. We have examined this theme at a variety of LTER sites and at a variety of timescales. The subject matter of the book was also to be focused on a series of framework questions. We noted that the theme of climate variability and ecosystem response is inherently deterministic and implicitly carries with it the notion of climate cause and ecosystem result. The analyses in this volume demonstrated that this is a valid and fruitful working assumption. However, the idea of a simple single climate cause and effect might be true in some cases, but it is obviously simplistic. More realistically, the effects of climate variability cascade through ecosystems. In almost all cases there is the probability of many secondary and associated effects accompanying the primary effects. As an example, the possible results of potential warming in the Pacific Northwest forests include changes in global carbon dioxide input, nutrient cycling between the plants and the soil, and feedback links between the plant and soil organisms (Perry and Borchers 1990).

In general there seem to be at least three broad classes of interaction between climate and ecosystems. First, the ecosystem simply responds to individual climate events or episodes that exceed some threshold for response. Second, ecosystems may buffer climate variability. In this sense they are filtering the effect of the climate event or episode. The same component in an ecosystem can sometimes act as a buffer and sometimes not, according to the nature of the climate event. Thus a riparian environment might provide soil moisture that acts as a buffer to a drought, but the whole environment might be destroyed by a large flood event. Third, we hypothesize that the ecosystem may move into resonance with the climate variability with positive and negative feedbacks that produce a strong ecosystem response. The relationship between fire and the Southern Oscillation indicates that the Southwest United States (Swetnam and Betancourt 1990) may provide an example of such resonance. Other examples of resonance, discussed subsequently, may exist in the forests of Interior Alaska and Puerto Rico. If there is indeed an ecosystem response to climate variability, the response tends to occur in cascades. The cascades and intermediate cascade elements may act as gateways, filters, and/or catalysts in response to the climate signal. The senses in which we use these terms will be explained here.

In this concluding chapter, we will develop these themes within the context of the framework questions with which we embarked. Emerging from this discussion are some general propositions that seem to hold for our examples from LTER sites and could be tested in other contexts and in later LTER studies.

The Framework Questions Revisited

The framework questions (figures 1.3, 1.4) have proved useful in making comparisons between the climate variability and ecosystem responses of all the LTER sites considered in this volume. The framework first called for an identification of the type of climate variability involved. The framework then poses the following questions, stated here in abbreviated form. What preexisting conditions will affect the impact of the climate event or episode? Is the effect direct or cascading? Is the effect completed by the time of the next event or episode? Does the event or episode return to an original state? Does the event or episode have an upper or lower limit? Does the climate and/or ecosystem exhibit chaos?

Identification of the Climate Variability

On first consideration, the identity of climate variability is clear in most of the studies in this book. Sometimes not so clear is the interaction of climate events at some timescales in relation to events and episodes at other timescales. Also, it is helpful to attempt to distinguish between process, which occurs at a particular timescale, and pattern, which represents how a process manifests itself in space.

A wide variety of climate variability has been addressed. Specific hurricane and drought periods and processes are both considered in part I on short-term climate events. A study of the frequency of storms in the twentieth century (chapter 14) identifies the pattern of some areas such as those of the western U.S. LTER sites (CAP, JRN, SEV, NWT, SGS), where large increases in storm frequency have occurred. Our investigations at the quasi-quintennial timescale (part II) focus almost exclusively on the process of the ENSO events. ENSO gives rise to important geographical patterns of climate response across the world. ENSO events in the Western Antarctic Peninsula occur within the context of an almost 60-year warming trend. In this example, two timescales and processes must be considered simultaneously. Periodicities at the quasi-triennial timescale appear in LTER site growing-season mean and maximum temperature data (McHugh and Goodin, chapter 11). The Pacific Decadal Oscillation (PDO) is the principal process of climate variability considered in part III on the interdecadal timescale. Juday et al. (chapter 12) have established climate regime shifts for Interior Alaska that extend back over 200 years and are most likely related to the PDO. However, the interdecadal-scale studies display linkages to both the quasi-quintennial scale and the century scale. Indeed, El Niño events in Interior Alaska are seen as amplifications of PDO-related episodes as far as reproduction of white spruce is concerned (Juday et al., chapter 12). Investigators at the Sevilleta LTER site identify a 55- to 62-year periodicity in precipitation values (chapter 15). McHugh and Goodin (chapter 11) also confirm a 50-year periodicity in maximum temperature records from some LTER sites. The processes behind these periodicities are not yet clear. The studies at the century to millennial scale in part IV focus on the last 25,000 years in the Colorado Rockies and the Dry Valleys of the Antarctic and further into the Pleistocene for southern New Mexico. This time period includes the episodes of the Late Glacial Maximum and the return to warmer periods in the Holocene. It also includes the colder Younger Dryas period, as well as the Holocene Altithermal and Medieval Warm and Little Ice Age periods of the last thousand years. It is becoming clear that the pattern of the effects of some of these events may not be as globally homogeneous and intense as once was thought (Mann 2001). In all of these cases, we are more concerned with the climate signal that arises from these phenomena, such as ENSO and PDO, than the mechanisms of the phenomena themselves.

Many of our studies show that it is rare that an ecosystem is dominated by climate variability at one specific timescale. More likely, ecosystems are responding to a suite of climate variability occurring at a variety of timescales. For example, Goodin et al. (chapter 20) have identified several different timescales that affect the prairie ecosystem at the Konza Prairie, Kansas. Occasionally, the events at one timescale are clearly dominant, but even in these cases climate variability at longer and shorter timescales is still important. In addition, Goodin et al. (chapter 20) point out that the juxtaposition of climate events at different timescales may be as important as the strength of any single process because of their potential to either augment or offset each other. Ideally, we would like to have a climate record and a record of responses so we can distinguish variation in driver signal from the responder signal. This variation might involve lags and dampening or amplification of the driver signal. Rarely do we find such a clear-cut situation in the real world.

In some instances the identification of climate variability and the time and space scales at which it is operating poses some difficulty (McHugh and Goodin, chapter 11). The first area of difficulty is in distinguishing trends and discontinuities. At issue are how fast the trend is and when a shift in trend becomes a jump. The answer is scale-dependent. Also, of course, a cycle with a periodicity of two decades or more could be said to be composed of a series of alternating trends. The second difficulty arises because the interpretation of climate at one point (e.g., an LTER site) is dependent partially on the operation of climate at distant locations and the multiple interactions existing in the climate system as a whole. The third difficulty is that two, or more, timescales may be important to the same type of climate event. Storms, for example, primarily act at a daily timescale, but both the extratropical storms (chapter 14) and hurricanes (chapter 2) display interdecadal variation of frequency at a timescale of a century.

Bryson (1997) pointed out that climate is multidimensional, a vector property as opposed to a scalar datum. Thus we should recognize that climate variability seldom occurs in one climate variable alone. Brazel and Ellis (chapter 7) illustrate this point when they establish that, at the Central Arizona and Phoenix LTER site, an El Niño event is likely to be accompanied by change in the values of at least eight climate-related variables. Similar situations exist with climate variability at other LTER sites.

We should also point out that negative results may often be as interesting as positive findings. It is important, for example, to note that McHugh and Goodin (chapter 11) find no important periodicities in growing-season precipitation at LTER sites. Growing season precipitation is a critical variable for ecosystems, as shown by Gage (chapter 4). However, the absence of any periodicity in precipitation totals does not necessarily mean an absence of periodicity in drought occurrence—particularly in months that are very important to crop growth, as shown by the work of Gage (chapter 4). Another example is storm frequency. Hayden and Hayden (chapter 14) find no change in storm frequency nationally during the last century. Certain subregions of the country do, however, manifest some very important climate changes in terms of an increase of storm frequency over the century—sometimes by as much as 300%. Similarly, those authors also note no substantive difference between storminess during El Niño or La Niña years and an average year for the entire period of record.

Several factors require investigators to use a degree of caution when studying climate variability in the context of ecosystem response. These include the interactive hierarchy of time and space scales at which climate variability occurs, the multiplicity of climate variables, and the potential for some types of climate variability to be defined without sufficient rigor. Climatologists working in the LTER program also sometimes note that ecologists occasionally award climate data more accuracy than is warranted. Among other issues calibration drift in instrumentation can sometimes be a problem. Occasionally this is related to the dynamics of the ecosystems themselves. Changing tree cover around meteorological observing sites is a common problem in the forested LTER sites. Still, in most of the studies in this book, the climate variability under discussion is fairly clear.

Preexisting Conditions

Whether a climate event or episode exceeds the threshold to produce an ecosystem response, as well as the degree of buffering to the climate signal the ecosystem may provide, both depend on preexisting conditions in the ecosystem and, sometimes, the climate system. Certainly, we see frequent examples of preexisting conditions themselves acting as gateways, filters, and even catalysts of ecosystem response to a climate driver. Many of the studies presented in this book contain examples of the importance of preexisting conditions to later ecosystem response.

A straightforward, simple example concerned the windthrow events in winter in the northern Cascade Range of Oregon. Such events were found to occur particularly when winds were from the east after a period of dry weather. High-pressure conditions in winter gave rise to icing on the branches of trees, setting the stage for some windthrow events (chapter 19; Sinton and Jones 2002). Another straightforward example involves the white spruce in Interior Alaska, which must have a sufficient level of growth reserves as a precondition to a successful seed production event that itself would be triggered by a climate episode and the consequences thereof (chapter 12). A second example from Oregon shows that one climate event can act as a preexisting condition for a second climate event that, in turn, gives rise to an important ecosystem response. This is the case of the bark beetle outbreak following a drought that was preceded by a windthrow event. In these examples the preexisting conditions act as gateways to permit additional ecosystem response. The existence of ice on the trees might be regarded as a third element that acts as a catalyst and makes the windthrow event worse than it might otherwise have been. The other two elements present in this case are dry conditions and wind.

Other examples of the importance of preexisting conditions are less straightforward. For example, the previous disturbance and land-use history are very important in determining the exact effects of a new hurricane storm event (chapter 2). In another case, some preexisting conditions can increase the certainty that plants will suffer adverse effects, but there are also situations when the adverse effects can occur anyway if a detrimental climate event occurs. This is shown in the example of drought in the North Central Region (chapter 4). Low corn yields occurred in 1988 because of a high heat/precipitation value in July. But in this case the stage had virtually been set for low yields because of the physiological stress that had occurred in the previous May and June. On the other hand, in Michigan in 2001, May and June precipitation values were well above normal, but an exceptionally dry July and August period led to low yields for this year as well. Shallow roots for corn and soybeans were unable to make use of preexisting soil moisture at the lower level of the soil. In these agricultural examples the preexisting conditions could be said to be acting as filters to further ecosystem response.

In general, the effect of preexisting conditions is more marked at the shorter timescales, as in the previous examples. When dealing with decadal and longer timescales, the climate and related biophysical conditions become part of the preexisting conditions for the next climate episode. In the Palmer LTER example, the 60-year warming trend itself becomes part of a preexisting condition on which quasi-quintennial variation is superimposed (chapter 9). A similar pattern is seen at the Arctic LTER in Alaska, except that in this case an interannual variation is superimposed on an 11-year (so far) warming trend (chapter 5). An extreme example of how preexisting conditions do play an important role at the longer scale is at the McMurdo Dry Valleys LTER site. Here, in Taylor Valley (77.5° S), Fountain and Lyons (chapter 16) observe a strong climatic legacy whereby past climate conditions strongly imprint current ecosystem structure, function, and biodiversity. Specifically, shifting precipitation and temperature patterns caused the Ross Ice Shelf to enter Taylor Valley and impound a valley-wide lake beginning about 27,000 years ago. Ice sheet retreat, again due to changes in precipitation pattern, about 9,500 years ago caused the lake to drain. Relic benthic algal mats from the ancient lake locally increase the organic carbon content of the Taylor Valley soils. Fountain and Lyons believe the current soil communities depend on this organic carbon matter as a primary carbon source. At the longest timescale considered in this book, Monger (chapter 17) describes how climate variability between glacial and interglacial periods for the area around the Jornada LTER site can actually give rise to new geomorphic surfaces on which ecosystems develop. During times of higher precipitation in the southern New Mexico area, erosion and sedimentation markedly altered the surfaces of both river valley and piedmont areas. In addition, at this timescale, climate and vegetation actually work together to control erosion rates on piedmont slopes.

Some conceptual, as well as real biophysical, elements can be considered as preexisting conditions. We consider three examples. First, as was found in some of the ENSO cases (chapter 6), there must be a specific plant physiological linkage available for plants to respond to unusual extreme climate conditions. The rooting depth of species is a recurrent example of a physiological linkage. Such a linkage could, in some ways, be considered a preexisting condition. Second, preexisting conditions may also be thought of as a set of nonclimate-related processes that form a backdrop on which climate variability operates, or the opposite is equally true. For example, in the Konza Prairie such factors or processes include fire, nutrients, grazing by large ungulates, soil characteristics, and topography (chapter 20). Another example relates to the timing of preexisting conditions. For some tree species at Coweeta, the degree of effect of a drought is in part determined by the stage of the cycle of the Southern Pine Beetle population (chapter 3). Third, we should also consider the issue of preexisting conditions in terms of the fact that most of our ecosystems are chaotic systems. It is an inherent characteristic of chaotic systems that a small change in initial conditions may lead to large changes in subsequent conditions. With respect to climate warming, Shaver et al. (2000, p. 880) point out that "the same temperature change applied to different ecosystems will illicit different responses depending on initial position on the temperature response surface . . . and on initial biogeochemical conditions and composition. . . ." Analogous situations are found throughout the natural world and in the examples given in this volume.

System Cascades

Of all the guiding questions for this book, research on cascades in systems has been the most fruitful. This is because it strikes to the heart of explaining how the systems operate. Indeed, the cascades are the ecosystem responses of our title. The more we know about system operation, the more we will understand the true nature of the system. The complexity and extent of cascades in ecosystems caused by climate impact is due especially to process connections between the living components of ecosystems. In some cases abiotic components also affect the cascades. Initial and intermediate cascade elements may act as gateways, filters, and/or catalysts to the climate signal. Gateways can be open or closed. They can either permit, or not permit, the passage of material, energy, or information. Filters may pass a variable amount of material, energy, or information along through the cascade. The amount varies from all to none and includes all the possibilities in between. Thus, the filters in the system provide a buffering function to a climate disturbance. Catalysts occur where the presence of one component greatly enhances the interaction of two or more other system components. Another consideration is that the climate event or episode having a potential effect on an ecosystem is often itself a part of a cascade in the climate system. For example, the variability in the values of many surface climate variables at the CAP LTER during ENSO events (chapter 7) occurs at the end of a cascade that started with anomalous sea surface temperatures (SSTs) in the equatorial Pacific Ocean. These SST anomalies entered a series of atmospheric processes that altered the upper part of the atmospheric flow and eventually affected the surface in the Southwest United States.

The complexity and extent of the cascades of climate effects within ecosystems is best illustrated through a detailed case study. The example selected here deals with the cascade of events in the hydrologic system of the Andrews LTER site following a large rain, or rain on snow, event such as that which occurred in February 1996. This is an interesting example for many reasons. First, it is well documented. Second, not only does the example deal with the abstract concept of a cascade, but it also is a very real cascade of water working in a real time sequence over a series of spatially linked channel and topographic components arranged on the landscape from higher to lower elevations. Third, the example focuses on how different processes may display some sequential linkage. Identification of the sequencing of different processes in a cascade is very important. Within the context of a global-scale temperature increase, Shaver et al. (2000) note that the dominant controls over ecosystem response will change over time as different processes change at different rates. Furthermore, the changing sequence will not necessarily be the same in all ecosystems. The following description of what the authors call a "disturbance cascade" is derived from the analysis by Nakamura et al. (2000).

The Andrews precipitation-forced disturbance cascade is a parallel cascade consisting of two initial drivers. The first is small, rapid debris slides from hillslopes. The second is large, slow-moving earthflows. In the first path, debris slides move into steep, headwater stream channels and move through the channels as debris flows. The movement delivers sediments and logs to larger streams. On entering fourth- and fifth-order channels, the debris can be entrained and float along on rafts of coarse woody debris (CWD) or can cause jams at the confluence with larger channels. The jams may break during floods, causing a surge that pushes the debris further downstream. The CWD transport may terminate in areas of accumulations of wood or may be dissipated gradually, as wood levees, along stream banks. In the second path, slow earthflows gradually constrict stream channels. This increases the potential for stream bank erosion and stream side slides during high-flow events. The slides can deliver sediment and trees that form temporary dams. The breakup of the dams triggers flood surges downstream. Associated CWD may move in a congested manner sometimes disturbing riparian vegetation.

The sequence of processes may be interrupted at any point along the flowpath. Sometimes preexisting conditions, such as a change in the channel slope, may halt or alter the nature of the sequence. Roads may intersect the cascade flow path and act as filters or have other effects (Wemple et al. 2001). Occasionally, the cascade sequence will not occur at all. Streamside slides may merely alter a channel location with few downstream consequences. Similar outcomes, such as flood surges, may result from different processes. These disturbance cascades were found to produce a gradient of decreased overall severity of impact on the ecosystem and increased variability of severity of stream and riparian disturbance in the downstream direction. For the February 1996 event, instances of changing processes from one to another and halting in particular parts of the cascade were quantified. Especially noteworthy is that only a very small fraction of all the initial mass movement events resulted in the full cascade sequence. Thus, filtering in this cascade was strongly marked.

This example suggests a case in which the climate event had to pass through a geomorphological cascade before beginning an ecosystem cascade affecting the flora and fauna. The immediate ecosystem effect varied from complete removal of alluvium, soil, and vegetation on steep, narrow, low-order channels to localized patches of toppled trees.

We learn much from the comparison of system cascades identified at the various LTER sites discussed in this book. To start with, cascades can be short or long, intuitively obvious or less obvious, linear or nonlinear. Some short cascades are found in the Antarctic. The remarkable responses of fauna in the glacial lakes of the McMurdo Dry Valleys LTER site to the freezing of the lake surface and the extreme low temperatures represents some extraordinary cascades through the aquatic ecosystem, as described by Fountain and Lyons (chapter 16). Although the responses are extraordinary, the cascades are short because the food chains are short. Fountain and Lyons (p. 334) point out that "the low biodiversity and short food chains make the ecosystem directly dependent on the physical environment such that few buffers exist and the response of the ecosystem to slight climate change is immediate." Short cascades are also seen at a daily timescale at the Arctic LTER, where there is approximately a direct response in Net Ecosystem Production (NEP) to an increase of photosynthetically active radiation levels, among other things (chapter 5). Cascades on agricultural crops also tend to be short (chapter 4). Many of the other cascades described in the preceding chapters are long ones. Often the more we learn about the way the ecosystem operates, the longer the cascades become. So, for example, LTER investigations have shown that a simple relation between high water flow events and increase productivity in the lakes of the arctic tundra actually involves changes in the degree of mixing of lake water and variations in available nutrient content (chapter 5).

Some cascades are simple and intuitively obvious such as the increase of glacial meltwater and streamflow in response to higher radiation input values at the McMurdo Dry Valleys LTER site (chapter 10). Others, such as the increase of traffic accidents in and near Phoenix initiated by a La Niña event, gene switching and the production of new phenotypes in species at the Coweeta LTER site in response to drought (chapter 3), or the hurricane-initiated increase of forest fire danger and possible extensive logging that itself can create huge ecosystem effects (chapter 2), are certainly not intuitively obvious. Neither is the fact that some ecosystem processes may respond in different ways to a given climate episode, in the case of NEP levels in Oregon forests in relation to summer precipitation values (chapter 19).

Many cascades are linear, but we have also recognized nonlinear responses. For example, because summer temperatures are close to freezing point at MCM, the change between liquid and solid water in the hydrologic systems is delicately balanced. Welch et al. (chapter 10) report that small changes in temperature and radiant energy are amplified by large, nonlinear changes in the hydrologic budgets that can cascade through the system. Another example is the response at Luquillo, Puerto Rico, where 75% of the sediment export occurs during the 1% of the days that have the greatest rainfall (chapter 8). Such a nonlinear response can be exacerbated even more when the heavy rain events give rise to debris flows that course down hillslopes and through streams, as sometimes happens both at Luquillo and at the Andrews LTER site in Oregon. In an entirely different environment, the Sevilleta LTER site of New Mexico, another surprising, nonlinear response to decadal-length drought is hypothesized to be economic collapse and a large number of changes of landownership toward the end of a drought (chapter 15). We learn something new about climate variability and ecosystem response from almost every different cascade. Drought in the corn crop ecosystem may lead to a cascade in which the plant system suffers mortality or becomes weakened and susceptible to insect herbivory or disease (chapter 4). A lesser recognized cascading effect, pointed out by Gage, for agricultural systems subjected to drought is the establishment of more irrigation systems with their subsequent effect on local and regional water tables. In these two last examples, the cascades existing in the ecosystem represent catalysts for later major changes in the human dimensions of local and regional change.

Scientists at the Palmer LTER site believe they have identified an important cascade in their ecosystem. Pygoscelid Penguins are representatives of the higher trophic level in this ecosystem. Population variations at quasi-quintennial and decadal timescales in Pygoscelid penguins have to be understood via a cascade that starts with an entrainment of phytoplankton in newly forming sea ice of the previous autumn. In some senses this could be regarded as a preexisting condition. The cascade continues with the growth of sea ice communities during the winter and the spring release of a potential bloom inoculum of particulate organic matter in the water column. These events are related, in turn, to the survival of larval krill that depend on the algal food source in the sea ice. In summary, there are strong linkages among sea ice, phytoplankton, and krill. The foraging ecology of the penguins is dependent on krill recruitment and abundance, indirectly through habitat changes that mediate the availability of krill (Smith et al. chapter 9). This type of cascade best illustrates the concepts of gateways and filters. Because the sea ice is necessary for phytoplankton development, the sea ice extent represents a gateway. Whether this gateway is open depends on the delicate balance of the sea water temperatures near freezing point. If the sea ice exists, it acts as an open gate. If sea ice is not present, its absence acts as a closed gate and does not permit the development of the phytoplankton that are the first level of the cascade. If the cascade is established then each step in the food chain filters the passage of chemical energy to subsequent steps in the cascade.

The concept of gateways existing in cascades has been developed very explicitly by Juday et al. (chapter 12) in the case of white spruce reproduction in Interior Alaska. These workers identify five climate-mediated gateways in the overall white spruce reproduction cascade. The first gateway is the preexisting condition of the need for a sufficient level of growth reserves. The second gateway is the need for a drought stress signal at the time of the formation of the bud primordia, which occurs at the end of vegetative shoot elongation. A third gateway is the requirement of a lack of severe pruning of reproductive shoots by wind and canopy snow loading in the fall and winter of the first year of seedling growth. A fourth gateway is the requirement of high growing degree-day heat sums to promote the maturation of the pollen and cone buds in time for the remainder of the steps to be completed before frost ends the growing season. Finally, a double fifth gateway requires both the survival of pollen and cone buds in early stages (e.g., lack of killing frost) and a heavy pollen flight (e.g., lack of persistent rains) to ensure high levels of cross pollination. In this example a suite of different aspects of climate variability plays a role in the final successful, or otherwise, species reproduction.

There are at least two temporal elements implicit in the concept of gateways. The first is sequencing. Second, the timing of the open gate must match the timing of other possible constraints. The gate will be open or closed at certain discrete times, and the timing of the opening is important if the ecosystem cascade is to be followed. So, for example, as mentioned previously in the context of preexisting conditions, in the Coweeta forest whether a drought has an effect related to the population dimensions and impacts of the Southern Pine Beetle (SPB) depends on the stage of beetle population. The drought has to coincide with the open gateway of a high SPB population for the cascade that ends in tree mortality to be completed.

The diversity of LTER sites presents a huge variety of potential cascades and the events occurring in these cascades. Nowhere is this more true that in the urban Central Arizona–Phoenix (CAP) site. Brazel and Ellis (chapter 7) list multiple resulting cascades and effects that are strongly driven by ENSO-related climate episodes. Some of the effects are very surprising. A case in point is the increase of traffic accidents associated with the frequent dust storms of La Niña years. ENSO events also partially control the intensity of the Phoenix urban heat island. Even more importantly, Brazel and Ellis point out that many of these cascading effects feed back into the urban ecosystem. The multiple effects of both El Niño and La Niña events on the CAP urban ecosystem suggest an extension of the cascade concept. Much of our attention in this book has been directed to a single cascade in the ecosystem following a climate event or episode. The ENSO climate driver establishes parallel cascades through its precipitation and temperature signals. The analysis in chapter 6 shows that sometimes the temperature effect results in an ecosystem response and sometimes the precipitation effect does. The CAP (chapter 7) case makes us recognize that there can be multiple, separate, parallel cascades to a single climate driver. Indeed, the reality is most likely that multiple climate drivers produce multiple parallel cascades, some of which interact and some of which do not. The MCM case (chapter 10) also identifies several parallel cascades. These relate first to algal mats and stream nitrogen uptake, second to salinity and stability of lake water columns as well as the type of phytoplankton species that are dominant, and third to soil invertebrates. The salmon catch case (chapter 13) refers to parallel cascades that can occur in both coastal and deep-sea ocean waters. The hurricane case (chapter 2) also can have parallel cascades associated not only with wind damage but also possibly to related primary effects such as river floods or salt water inundation, or secondary impacts such as landslides or fires. It is the task of the LTER, and other, researchers to identify the strands of the cascades and their interactions. This is particularly important because Shaver et al. (2000) suggest (in other words) that the longer the time of operation of the ecosystem the greater will be the chance of interaction of parallel cascades. In the case of ecosystem warming, these authors note many changes that will only take place during very long time periods. Such changes include soil profile development, organic matter accumulation, changes in fire regimes, or long-distance movement of herbivores or timberlines. Many such changes will be due to the interaction of parallel cascades.

An inherent characteristic of cascades is the temporal dimension. The importance of timing on the degree of effectiveness of the Southern Pine Beetle in killing trees at the Coweeta LTER site was mentioned in the context of cascade gateways. At the same site shoestring root rot fungus also has a greater impact during drought stress than at times of other climate conditions. The cross-site ENSO study (chapter 6) made it clear that the timing of a climate episode or event is critical if a subsequent cascade of events is to follow. Also the occurrence of hurricanes in New England in October and November when a minimum number of leaves is on the trees makes the forest less prone to hurricane damage (chapter 2). Thus, the timing of an event or episode may also act as a gateway. Studies at the North Temperate Lakes LTER (Robertson et al. 1994) clearly show that a difference of one month in the timing of an El Niño signal can be critical in determining whether that signal will have an effect on the ecosystem. In another example Gage (chapter 4) shows how important early growing-season precipitation is to the eventual corn yield of the North Central Region. The critical importance of the timing of snowmelt at the Arctic site and development and decay of sea ice at Palmer are additional examples (chapters 5 and 9). Whether the ecosystem response gateway is open often depends in these cases on the timing of the climate driver. An extension to the cascade principle and its temporal element is that an ecosystem response may be driven in sequence by two, or presumably more, climate drivers. For example, KNZ NPP is most highly correlated to air temperature in the early part of the growing season, whereas later in the growing season it is better correlated with precipitation values acting through soil moisture conditions (chapter 20).

The geography of the LTER network is important. Geographical considerations demonstrate the importance of cross-LTER site studies. The same initial climate driving function may have totally different effects in different areas. This is especially true when large-scale diving factors such as ENSO or PDO variability are considered. In the case of ENSO, the climate precipitation signal in the Pacific Northwest (PNW) is opposite that of the Southwest. This appears to be the case throughout the twentieth century at the decadal timescale (Schmidt and Webb 2001). In another example, McHugh and Goodin (chapter 11) give an analysis of the way in which the climate, particularly growing-season mean, maximum, and minimum temperature, is inversely associated between the Andrews and the Bonanza Creek LTER sites, respectively located in the Pacific Northwest and Alaska. Times of higher than average values of growing season maximum and minimum temperature in interior Alaska tend to correspond with times of lower than average growing season maximum and minimum temperatures in the Pacific Northwest. This is quite consistent with the salmon catch data referred to in chapter 13 and relates to the distinct manner of operation of the PDO. Intersite comparisons raise exciting new questions. A possible linkage between cone production in the PNW (chapter 19) and Interior Alaska (chapter 12) and the relation to the state of the PDO demands more investigation. In another geographical contrast, the same climate variable ENSO has a larger effect in the SW than in the PNW. In the PNW the difference in precipitation values between El Niño and La Niña years is not so significant as in the SW because the variation is around a higher mean value. As noted previously, the dry, forest-grassland types of vegetation in the SW seem to be "tuned" to ENSO variations, which stimulate grass and fuel production during wet phases and burning during dry phases (Swetnam and Betancourt 1998; chapter 15). Fuel variation is not so dynamic in the PNW conifer forests.

Yet another dimension to the consideration of cascades is exhibited at the Sevilleta LTER site (chapter 15). Here it is hypothesized that an increase in creosote bush shrub is somewhat self-enforcing because the shrubs emit nonmethane hydrocarbons that act as local greenhouse gasses, keeping minimum temperatures as much as 4°C higher than they would be without the shrubs. Our discussion of cascades so far has been in unidirectional terms. This example shows that we must also consider the possibility of the cascade turning back on itself with a positive feedback. Cases of negative feedback are also conceivable. Yet, we tend to see the cases of cascade elements acting as catalysts in the situations of positive feedback.

Completion of Ecological Response

Our framework question asks, Is the ecosystem effect or response completed by the time of the start of the next climate event or episode? This question can be asked in different ways such as in the three questions posed by Boose (chapter 2) in his hurricane study. The question can also be posed implicitly in different contexts. For example, Parmesan et al. (2000, p. 446) have stated "the initial resistance, trajectory of response, and extent to which a system returns to original conditions (resilience) after a disturbance depend on the frequency, intensity, duration, and extent of disturbance, as well as the inherent properties of the biological system, including evolutionary history. . . . " LTER studies confirm this in many cases of climatological or meteorological disturbance.

It is also likely that different parts of an ecosystem will have different recovery times. The example of the February 1996 flood at the Andrews LTER site is interesting because the range of recovery times for different parts of the ecosystem have been documented for this event (Swanson et al. 1998). The recovery times ranged from less than 3 months for aquatic algae, through 1–3 years for cutthroat trout, to more than 30 years for coniferous trees. Boose (chapter 2) recognizes an important caveat to our thinking on recovery times when he points out that some adaptive responses, such as the creation of new foliage and branches following a hurricane, cannot be repeated indefinitely at short intervals. In contrast to an individual flood or hurricane event, suggested ecosystem process and component response times to decade- to century-long temperature increases have a wide range. The range extends from one day for leaf photosynthesis and respiration, through a decade for litter mass change, to over 1000 years for soil organic matter development and plant migration and invasion processes (Shaver et al. 2000).

LTER studies reveal some cases where the ecological response to a drought is completed and other cases where the response is not completed by the time the next drought occurs. Gage (chapter 4) points out that the effects of a one-year drought, such as that of the 1988, on annual rotational agronomic systems are minimal. However, the same 1988 drought had long-lasting, documented effects on some of the natural vegetation species at the Cedar Creek LTER site in southern Minnesota. Citing Tilman and Downing's (1994) work, Gage notes that the effects of the 1988 drought were evident in the oak savanna complex 5 years later. About 30% of the pin oaks and 19% of bur oaks died. As a result of such episodes, some parts of the ecosystem may return to their original state, whereas other parts are affected for many years to come or even permanently. It is certainly true that the landscapes of most LTER sites exhibit long-term legacies after a severe, relatively short-term, climate episode. This raises the question, What is it that determines which parts of the ecosystem will be most negatively affected? Based on our small number of examples, it seems that vegetation with longer life spans such as trees, as opposed to grasses, is most vulnerable.

The Cedar Creek finding is reminiscent of the fact that dead junipers still reside on the landscape at the Sevilleta LTER site, the result of La Niña-related droughts of the 1950s (chapter 15). However, studies based on tree rings, which provide information for almost 400 years at the Sevilleta site, place this result in an even more surprising context. Sevilleta researchers hypothesize that the 1950s drought was one of a series of droughts that recur at an interval of 55–62 years. Partly due to the fact that it was accompanied by the introduction of cattle ranching, the ecosystem has not yet, and possibly never will, recover from the previous cyclic drought of the 1890s. At both Sevilleta and the Jornada LTER site in southern New Mexico, shrubland took over from grassland, and there is no sign of a return before the beginning of the next drought period. Thus, although in some senses, these semiarid ecosystems may return to "normal" in terms of biomass productivity levels, for example, after an El Niño–related season or two of above average winter and spring precipitation, they still may not be returning to "normal" at the multicentury timescale.

Cross-timescale considerations are also important at the Palmer Antarctic site, where ENSO-scale events govern key biophysical interactions and many of the interactions are complete by the time of the next ENSO event. However, the circa 60-year warming trend at this location complicates matters such that Palmer researchers find it difficult to envisage an "end scenario" (chapter 9). Nevertheless, it is very interesting that the 600-year fossil record at Palmer shows the current presence of chinstrap and gentoo penguins to be unprecedented and that the site was dominated throughout most of the 600 years by Adélies. In some cases, long-term trends, or their operation in association with oscillatory climate phenomena, can set new "preexisting" conditions for each cycle of "cyclic" climate variability as in the case of variability in PNW salmon abundance at the decadal timescale.

Two of the studies in this book extend to the timescale of 25,000 years. Beetle assemblage evidence from Elias (chapter 18) suggests that certainly at this timescale the ecosystem has made its necessary adjustments by the time of the next ice advance or return to warmer climates. However, these "adjustments" in the insect assemblage are to a "lowest common denominator." As Elias (p. 381) expresses it, "the current group of species in the alpine ecosystem may not be the best fit for the environment—they are simply the best fit among those species able to persist regionally through the last glacial cycle." In this case we have a climate filter acting at the millennial timescale. Furthermore, Elias (p. 381) believes ". . . that at the century to millennial timescale, the response of major components of the vegetation in high altitude ecosystems of the Colorado Front Range lags behind major temperature changes." The lag is in the order of 500–1000 years. Analogous, lagged responses are described by Monger (chapter 17) for the arid environments of the Southwest as they change between glacial and interglacial times. In the Antarctic Dry Valleys, however, the ecological response to warming conditions is still dependent on events that started at least 24,000 years ago, as described previously and in chapter 16. In this case, even at these large timescales, the ecological response to Holocene warming cannot be said to have been completed. This is because it remains to be seen what would happen to the current soil communities if the carbon source derived from relic benthic algal mats were ever to be completely depleted. Indeed, Fountain and Lyons (p. 334) suggest "given the extremely slow cycling of nutrients and the pace of geomorphic change, we suspect that ecosystem responses are overprinted on each other and are not completed before the next event occurs."

Return to Original State of Climate Variable and Ecosystem Response

We have asked, Does the climate event or episode and the ecosystem response return to an original state? Many of the issues that fit appropriately in this section have been discussed in the previous section on "Completion of Ecological Response" concerning whether the ecosystem will return to its original state before the onset of the next climate event. We have not, however, addressed the manner of return. The nature of the return to the original state is important. In some cases, whether the cascade reverses depends largely on whether some component of the system has been destroyed, such as in the complete removal of topsoil, or simply made temporarily unavailable, such as the temporary absence of phytoplankton under certain conditions in the PNW coastal ocean. We also acknowledge that some would argue that the concept of "return to original climate and ecosystem conditions" is misleading because the systems do not operate in those ways. For example, a "return to an original condition" might occur only in the most superficial of senses. Some of this debate is a matter of discipline. For example, palynologists have little sense of returning to an original position. On the other hand, the language and metric of dendrochronological studies of fire history sometimes seem to assume cyclic system behavior and a return toward predisturbance conditions. We certainly agree with the available evidence that suggests that the longer the timescale being considered, the less likely it is that one will find a perfect analog to the current or any single past or suggested future condition. Despite these considerations, we believe that it is worthwhile to examine the concept of "return to original climate and ecosystem conditions" using the information provided by the case studies of this volume.

If we assume there is a return by the atmosphere and ecosystem to an original state, what is the manner of return? We have examples where the return seems to be rather smooth. The case of the oscillating salmon catch in the Pacific Northwest (chapter 13) or grassland NPP levels (chapter 20) are good examples. A relatively long life span sometimes helps a somewhat smooth return to original conditions. For example, the grayling in the Arctic aquatic ecosystem benefit from their 20-year life span. This longevity helps to filter out and dampen the effects of interannual climate variability on their population numbers (chapter 5). In other cases, the return to some original state by no means follows the path of the original change and sometimes is dependent on the timescale involved. The cases of dead trees resulting from drought periods on the landscapes of the Cedar Creek, Sevilleta, and Coweeta LTER sites suggests that the ecosystem did not immediately return to its original state following the climate event. The relatively slow growth rates of juniper, for example, render it impossible to replace the dead trees with mature new trees in a decade or two. A longer term perspective on the issue is that intermittent drought and standing necromass, are, to a certain extent, part of the long-term original and natural state of these ecosystems.

We have also seen examples where the ecosystem does not return to its original state after a climate disturbance. The Palmer Antarctic ecosystem appears to be under a strong directional change driven by warming, and at the decadal and century timescales shows no sign of returning to the state found by LTER researchers when they began their studies in the early 1990s. Interior Alaska also seems to be experiencing a marked warming at the century timescale. Of interest here is the suggestion that, although the white spruce may be adapted to decadal-scale climate variability, a continued warming trend might lead to this species losing its sensitivity to the relationship between summer temperature values and growth rate (chapter 12). Increasingly, a smaller amount of growth is exhibited for a given increase in temperature. One possible reason for this increasing lack of sensitivity might be that the environment of the white spruce is moving closer to the upper cardinal temperature limit for the species.

The concept of returning to an original state tends to lose its meaning at the millennial timescale. Changes between glacial maxima and minima in the Quarternary defy the definition of an original state. In addition, the time spans start to be so long that evolutionary processes begin to make their mark on flora and fauna. However, knowledge of climate variability and ecosystem response at this timescale can have important implications. For example, the fact that the paleorecord shows the existence of arid desert shrub in the middle Holocene (presumably unimpacted by intense human activity) before grasses developed in the later Holocene is vital (chapter 17). This fact greatly informs the debate on the possible causes of the change over the last 150 years from grassland to shrubland at the Jornada LTER site and in other places in the Southwest.

The return period of climate events is an important factor in determining whether the ecological response has been completed by the time of the next event. In addition, the fact that some climate features have characteristic return periods of ecologically important events is related to the concept of the ecosystem entering into resonance with the climate variability. Resonance seems to exist at a variety of timescales. For example, at shorter timescales, it is fairly easy to see that agriculture resonates closely with seasonal and interannual climate variations (chapter 4). Similarly, in natural ecosystems resonance is clear. See, for example, the close relationship between net ecosystem production (NEP) for acidic tundra at Toolik Lake and photosynthetically active radiation (PAR) (figure 5.7). However, we have only just begun to hypothesize that resonance also exists at the longer timescales.

Let us first reexamine the case of the tropical rainforest LTER site at Luquillo, Puerto Rico. In this forest, stream water export of potassium and nitrate ions increased following the disturbance caused by Hurricane Hugo and remained elevated until the canopy leaf cover returned (chapter 8). We might speculate that the response of this part of the ecosystem to the hurricane event will depend on how long it is before the next hurricane passes over this site. Interestingly, in other Luquillo studies, it has been noted that the time for maturity of a tabonuco forest stand is approximately the same as the 60-year average recurrence interval of category 4–5 hurricanes in Puerto Rico (Scatena 1995). Boose (chapter 2) quotes a similar time period (50 years for category 3 storms). Boose also identifies one of the negative feedback processes at work during the forest recovery, noting that because of their reduced stature, heavily damaged stands are naturally protected from subsequent wind damage for a period of years or decades. This represents another example of the ecosystem moving into resonance with the period of climate variability. The example of Pacific Northwest salmon catch is clearer than the case of Puerto Rico forests. Because the life span of the Coho salmon—about 3 years—is much shorter than the decadal-scale climate regime shift of about 20–30 years, the immediate ecosystem response, in terms of population numbers, should be complete before the next climate episode occurs. However, we should be aware that some demographic models suggest that population systems have a memory. Whether the memory, in this case, would extend beyond 20–30 years requires further investigation. Also requiring more research is the question of whether the decadal- scale variation of climate and Coho salmon population represents the latter moving into resonance with the former.

Another possible example of resonance is in the U.S. Southwest. There is a strong fire response to the El Niño–Southern Oscillation (ENSO), with wet times leading to fuel buildup and dry times associated with a higher frequency of fire. The fires cause the release of nutrients that encourage growth/fuel buildup in the next wet episode. Thus, the fuel-fire cycle can operate well within the pace of the ENSO (climate) dynamic and without long lags. Swetnam and Betancourt (1990) noted a close relationship between ENSO and the fire regime in the U.S. Southwest. We can speculate that the resonance in the ecosystems of this region may operate at both the quasi-quintennial timescale and at the circa 52-year periodicity identified in chapter 15.

Yet another possible example of an ecosystem moving into resonance with climate variability is that of white spruce reproduction in Interior Alaska. Here, the episodic cone production of the species is suggested as an evolutionary adaptive strategy to a climate regime with a decadal variability. Juday et al. (chapter 12, p. 245) write "it appears that the described reproductive timing of white spruce maximizes the odds that seeds will be released into a landscape in which fires have occurred recently." The fire has usually removed a thick organic mat that otherwise prevents seed success, and the reproductive timing of white spruce maximizes the odds that seeds will be released into a landscape in which fires have occurred recently. Other biotic elements may play roles in ecosystems moving into resonance with the climate, but more investigation is needed to identify the subtleties. For example, irregular seed crops may protect a plant from a high level of seed predation. It would be interesting to see whether there is a linkage among climate variability, seed crop production, and seed predator population numbers.

Four more concepts and issues emerge. First, if ecosystems come into resonance with climate variability, then what are the processes they use to do so? Second, if it is possible to identify the timescales and processes at which the resonance takes place, then we can identify other potential climate and other disturbances that might have an even larger impact on an ecosystem because the system is not in resonance with them. Human insults to ecosystems are the best example of this. Management, or other human activities, may well interrupt natural resonance. However, natural examples exist as well, such as the stochastic nature of earthquakes, volcanic eruptions, tsunamis, and meteorite impacts. Third, the pace and magnitude of ecosystem response in a "resonating" system will be important. If the frequency of climate variability and that of ecosystem response is well matched, the magnitude of the response will be at its most efficient. Conversely, if the frequency of the climate variability becomes higher or lower over time, then the magnitude of the ecosystem response may be muted because feedback mechanisms may not act effectively. Fourth, we must recognize the difference between the ecosystem response to a single disturbance on one hand and the response to a disturbance regime on the other. Cascades of effects may be fairly easily identifiable in the first case, whereas in the second case we must consider the timing, severity, and spatial patterning of one or more disturbance processes over time. Consideration of disturbance regimes raises questions of how frequency and severity affects species composition at a site. The concept of resonance might be more difficult to apply in the case of disturbance regimes.

Greenland (chapter 6) introduces the concept of a "characteristic timescale" for an ecosystem. Identifying such a timescale would help address the questions related to completion of ecosystem response and return to original state. It also raises the interesting issue of whether there is such a thing as a characteristic timescale for an ecosystem. At least for "simple" biological responses to a climate event or episode, both Clark (1985) and Woodward (1987) have provided quantitative analyses. Some of our studies indicate what the characteristic timescale may be. Boose (chapter 2, p. 28), for example, proposes "long-term impacts of hurricanes on forests can be understood only at a scale of centuries." On the other hand, the reality is that ecosystems are made up of many different components, and each component will have its own characteristic timescale of response. Using the concept of recovery times,

Swanson et al. (1998) and Scatena (1995) have documented this for temperate and tropical rainforests, respectively. Yet perhaps some characteristic timescales, such as those related to cases where the ecosystem moves into resonance with particular types of ecosystem variability, dominate over other potential characteristic timescales for a particular ecosystem and are a fundamentally important part of the nature of that ecosystem. If this is the case, then the characteristic timescale is another way of describing the resonance between the climate and the ecosystem.

Limits of Climate Variability and Ecosystem Response

Few authors in this volume attempted to explicitly address the question of whether the climate event or episode and the ecosystem response had identifiable upper and lower limits. This is because it is important to study climate events that cause severe ecosystem change so the limits of the ecosystem response are clearly bounded. Not all of the subject matter of our chapters meets this criterion. In retrospect, the reason for the lack of consideration of limits is that the subject is partially timescale-related. If an individual investigator is not familiar with all the timescales at which their ecosystem operates, he or she will not have all the information needed to answer this question. The interdisciplinary approach of LTER research is often helpful in extending an individual's knowledge of an ecosystem, so we can expect more answers to this question to appear as the LTER program further matures.

Although the answer to this question of limits is partially related to the timescale for which we have information, it is also related to the physics or biophysics of the climate event and ecological response in question. This is well demonstrated for white spruce in Interior Alaska (chapter 12), where the gateways in the suggested model are specifically related to certain limiting values of climate variability (e.g., summer temperature and its relation to drought) and ecosystem response or preexisting condition (e.g., growth reserves). In another case, Schaefer (chapter 8, p. 154) states "although there is no fixed upper limit to the amount of rain that can fall within a 24-hour period, there are no records that . . . it has exceeded 600 mm in Puerto Rico." A longer record might produce a higher 24-hour record. Particular wind velocities are used to define the strength category of a hurricane, but the highest category, 5, is open ended at a wind velocity exceeding 69 m/sec (155 mi/hr). It should be possible to use physical principles and information on sea surface and air temperature extremes, as well as maximum and minimum storm wind velocities, to make a fairly good estimate of the maximum possible precipitable water for the location. Alternatively, the theory of extreme statistics could be applied. Neither of these approaches has yet been used much at LTER sites. One sense of a limit to ecosystem response to a hurricane is implicit in chapter 2, where Boose describes a range of responses from partial defoliation to complete blowdown of a mature forest. The latter might be taken as the upper limit of ecosystem response in the hurricane context.

The Palmer LTER site (chapter 9) shows some situations where climate variability and ecosystem response display limits and other cases where it does not.

The limits are clearly shown in the ENSO-dominated timescale, but they are completely unknown in relation to the multidecadal warming trend because it is not possible to say where and when this warming trend will end. In an analogous fashion, the limits of climate (precipitation) variability seem to be well established at the quasi-quintennial scale and, to some extent, at the century scale at the Sevilleta site. But we know less about them at the millennial scale. Additionally, where the ecosystem is subject to alteration by human activity, the response to a particular climate event may be quite different from one event to another. This was shown by the emerging dominance of shrubland over grassland in the southwestern LTER sites partly related to cattle grazing in the 1890s.

The millennial timescale, which here is taken to be the Holocene but which also can involve the Pleistocene and some of its preceding geological epochs, does not play such an important role for present-day ecosystem managers. But it does help to be aware of changes at this timescale for two reasons. First, changes at the millennial scale can give information concerning the extremes to which the system can move and/or give some feel for its degree of homeostasis. The millennial-scale changes set the "limiting values" on the natural system changes in a practical and hierarchical sense. It is conceivable that human influence can help exceed these limits, but it is useful to have some idea of where the limits are or have been in the past. Second, it is also important to recognize that many of the floral and faunal species presently found in the ecosystem, or close relations of current species, have survived throughout all these extremes. This helps us to understand the degree of resilience of the ecosystem to natural changes. Third, it helps to recognize that many atmospheric phenomena that are important today have been present for a long time. For example, radiolarian records from the Santa Barbara basin indicate that El Niños have been occurring for at least 5.5 million years (Casey et al. 1989).

There have been warmer and cooler periods throughout the Pleistocene. Yet the extreme climates from the Last Glacial Maximum (LGM) 20,000 years ago to the warmer climates of the Holocene treated in chapter 18 may, in many ways, be regarded as representing, or approaching, limits of values of climate variables that modern Rocky Mountain and semiarid southwestern ecosystems may have to withstand in the absence of human influences. One suite of ecosystem responses to these changes is the varying assemblages of beetles. Elias (p. 370) argues that "ecological changes take place at many timescales, but perhaps none is more significant than the truly long-term scale of centuries and millennia, for it is at these timescales that ecosystems form, break apart, and reform in new configurations." Vegetation response in the Colorado Front Range took the form of a change from alpine tundra to subalpine forest and a decrease of the tree-line elevation of 500 m during the colder times of the mid-Pinedale glaciation. Elias' statement is also applicable to the ecosystem changes described by Monger for southern New Mexico.

Chaos

Only one author in this volume elected to address the question of whether chaos is exhibited in the climate or ecosystem. We believe this is due to a number of reasons. First, most investigators in climate and ecosystem sciences are unprepared to address the topic rigorously. Second, in many cases, the quantitative understanding of our systems is not yet advanced enough to apply large parts of chaos theory to LTER study sites.

We believe, however, that we should endeavor in future years to position ourselves to be able to apply chaos theory to our systems to discover new insights. Calls by the National Science Foundation for investigations into ecosystem complexity are consistent with this. Phillips (1999) has pointed to ways in which a qualitative analysis of partially specified dynamical Earth surface systems (ESS) can be made. In addition, virtually all that he says about ESS applies to ecosystems. We have mentioned previously, for example, that it is the very nature and definition of a chaotic system that small changes in initial conditions will often give rise to large changes in subsequent effects.

McHugh and Goodin are the sole authors who address the topic of chaos and complexity (chapter 11). Among other things they emphasize the large number of nonlinearities in the climate system. Other parts of LTER literature, such as the development of desertification theory at the Jornada LTER site (Scheslinger et al. 1990), suggest that nonlinearities are plentiful in our ecosystems. Sooner or later we will have to address the presence of nonlinearities, complexity, and chaos directly because these aspects are part of the real nature of our systems. We speculate that one or two decades from now a future LTER meeting on climate variability and ecosystem response will be couched in a framework of chaos theory.

We also note that ecosystem science is not alone in its failure to address chaos theory. Although Lorenz (1963) established a major part of the theory of chaos in atmospheric science, meteorologists have not pursued the theory with much vigor. On the other hand, Lorenz' discovery led to a paradigm shift in atmospheric science, which stemmed from the realization that because the atmospheric system was chaotic we could never hope to realize the dream of making a perfect weather forecast. We wonder what comparable paradigm shift, or shifts, await the field of ecology when we examine our systems with a focus on their nonlinear nature.

Emerging Concepts and Principles of Climate Variability and Ecosystem Response

In this synthesis, the following recurrent principles begin to emerge:

Issues of time and space scale are pervasive throughout the field of climate variability and ecosystem response. It is not always possible to separate the effects on ecosystems of climate events and episodes of different timescales.

At each LTER site climate events and episodes operate at different timescales. Consequently, these scales cannot be viewed in complete isolation.

Some timescales, like that on which the ENSO operates, show patterns with a broad spatiotemporal coherence that therefore encompass responses across a wide range of ecosystems.

Most LTER sites show evidence on their landscape of some past climate event or episode.

Timescales of climate variability and ecosystem response determine, in large part, whether the response is complete by the time of the next climate episode or event.

Some ecosystems return to an original state following climate disturbance, whereas others do not. When ecosystems return to an original state, sometimes the return is linear and sometimes it is nonlinear.

For a climate event or episode to be effective, there must be some identifiable, usually physiologically related, link to the flora and/or fauna of the ecosystem. Some proportion of climate variability will not have an effect on the ecosystem.

In some cases, for a climate event or episode to be effective it may involve a nonlinear amplification in forcing that later has an impact on the ecosystem.

Most ecosystem responses to climate events and episodes are not simple, single-cause, single-effect responses. Rather, the response takes the form of a cascade of effects.

The response cascades may be short or long, intuitively obvious or not, and linear or nonlinear or both. The nature of the cascade often depends on the complexity of the ecosystem. The LTER network includes some relatively simple ecosystems such as MCM and some very complex ecosystems such as LUQ.

Response cascades may take place both in time and space. Shaver et al. (2000) point out the need for improved models of the temporal sequence of ecosystem response because long-term responses may be very different from initial responses and responses will not be uniform in space.

Cascades that result from climatic impact in ecosystems often take time to manifest themselves and can result in legacies within ecosystems that condition subsequent climate impacts. Because cascading climate-driven impacts within ecosystems are often lagged in time, efforts to identify fixed time correlations are sometimes ineffective.

An initial climate driver may cause parallel cascades acting through several different climate variables.

There may be many parallel cascades, sometimes interacting with each other and sometimes not interacting.

Many of our studies focus on a single process. A focus on cascades leads us to concentrate more on the sequential linkage of one process to the next.

Whether upper and lower limits of the values of climate events and episodes and resulting ecosystem responses can be identified depends on both the degree of our knowledge of the relevant biophysical processes and the amount of empirical data available.

Cascades or parts of cascades in the atmosphere and ecosystem may act as gateways, filters, and catalysts to additional ecosystem response.

There seem to be at least three broad classes of interaction between ecosystems and climate:

1. The ecosystem buffers climate variability.
2. The ecosystem system simply responds to individual climate events and episodes that exceed some threshold for response. This threshold is often crossed or triggered by a nonlinear process.

3. We hypothesize that the ecosystem can move into resonance with the climate variability with positive and negative feedbacks that produce strong ecosystem response.

Future Research

From this discussion it is clear that the LTER program provides a platform from which a huge amount of information emerges on the topic of climate variability and ecosystem response. Based on the information in this book, many avenues of research on this topic will be important in the future.

1. We must continue to obtain more information at each LTER site on climate as a disturbance factor of ecosystems. Each new piece of information on this topic alters our perspective of the principles that emerge from this field. We need to develop tools that are sensitive to both atmospheric and ecosystem variability. In addition, we must attempt to anticipate the correct *combination* of system properties to be observed to be able to demonstrate in detail system resonance or another kind of behavior.

2. We must continue to be aware of the need for cross-site comparison. One corollary to this is that we must strive to design our experiments, and to collect our data, in such a manner to facilitate intersite comparison. The LTER network has a unique infrastructure for being able to make such comparisons as long as a certain amount of preplanning is accomplished. We have the potential to formulate hypotheses related to climate variability and ecosystem response for groups of LTER sites with common properties. The network has an ever-increasing number of site-years of sampling for different disturbance processes. The hurricane event is a case in point. Currently, nine LTER sites in the Caribbean and the East Coast of the United States are well positioned to observe the effects of tropical storms that make landfall. Four LTER, or former LTER, sites have been directly impacted by tropical storms. In 1938 an unnamed hurricane passed over the Harvard Forest LTER site. Hurricane Hugo passed over the Luquillo, Puerto Rico, LTER site and the North Inlet, South Carolina, former LTER site in September 1989 and Hurricane Opal passed over the Coweeta, North Carolina, site in October 1995. At the time of this writing, the subnetwork of hurricane-vulnerable LTER sites collectively represented about 160 site-years of direct, LTER-supported observation of both hurricanes and ecosystem responses and perhaps more than 2000 site-years of archival records of hurricane occurrence. This mininetwork is well configured geographically and temporally to obtain a good sample of large and extreme events and to consider questions about regional patterning of disturbance and ecosystem responses across a range of ecosystem types. Current sites provide opportunities for observation in ecosystems, including tropical and temperate forests, coastal barrier islands and wetlands, and an urban site.

Some scientists believe the U.S. East Coast is entering a period of several decades when the frequency of hurricanes making landfall will increase (Goldenberg et al. 2001). LTER scientists at potentially affected sites should make contingency plans to study new storms and their ecological impact by standardizing some of the methodologies that have been used in earlier studies. Preexisting conditions and common impact indicators should be carefully specified. Hurricanes are not the only possible focus of study. Other individual LTER sites or groups of sites may sample other climate disturbance processes. Diagrams such as figure 1.2 may be used to identify such groups of sites. As the LTER program extends into the future, it may also be possible to use available climate forecasts, such as those for the state of the ENSO, to design experiments that use the natural climate extremes of this quasi-quintennial phenomena. A visionary might even conceive of the ability to forecast the state of the interdecadal-scale PDO and its resulting ecosystem responses.

Another question related to cross-site comparison concerns the possibility that some sites might be more susceptible to climate variability than others. At first sight, the Sevilleta LTER site in New Mexico might be said to be more susceptible to an El Niño climate signal of a similar size than the Andrews site in the PNW. Could it be, for example, that the sites on the extreme outside edge of the cluster of LTER sites shown in figure 1.2 are likely to show a more marked response to climate variability than those sites near the center of the cluster? A related matter is the question of "redundance" of climate variability. There are many situations where the variability of a climate variable is of little importance to the ecosystem. For example, at the Andrews LTER, as long as no flooding occurs, greater than average January precipitation does little except run off from a system that is already fully charged with water physically and biologically. Where it is not already obvious, the identification of climate variability redundance would permit investigators to focus their resources on other parts of the ecosystem.

3. We must use our increasing knowledge at LTER sites and our cross-site comparisons to identify important generalities, often related to process, that are more specific in nature than our comments about the importance of scale. For example, we should pay more attention to critical thresholds such as those related to the ice/liquid boundary or to plant rooting depth. Other thresholds might include the precipitation duration-intensity necessary to trigger landslides or thresholds related to the phenology associated with achieving good seed crops. Another general concept has to do with the residence time of communities and individuals at a site and the sensitivity of a site to climate disturbance. One way to achieve the identification of these generalities is to hold workshops on them individually. Experience from some of the workshops that led to this volume has shown us that important concepts will emerge from such workshops.

4. We must start to develop multidimensional approaches to the issues of climate variability and ecosystem response. It is a rare case that there is only one aspect of climate variability occurring at any given time. At Coweeta, for example, both droughts and windstorms occur from time to time, but they each favor the development of different types of microhabitats. Drought effects, such as increased standing necromass, favor the development of some microhabitats, whereas windthrows favor others. The two situations exist simultaneously in the forest, and investigators must find ways to treat the parallel ecosystem responses and their possible interactions.

5. We must begin to confront the climate signal detection problem. The detection

of climate signals embedded in the ecosystem and realized as cascades of time-varying ecosystem properties is exceptionally complex. The problem requires the application of the full spectrum of analytical tools available to scientists and an ever-growing resource of long-term data, especially on ecosystem dynamics. Endeavors in this area will represent another fortunate congruence between climatologists and ecologists. This is especially so for paleoclimatologists, who have long used a wide variety of proxy ecological data such as tree rings. However, as difficult and sophisticated as the interpretation of tree rings is, we envision the problem of climate signal detection as being much more complex because it involves multiple levels in the ecosystem cascades. There is also the problem of the overprinting of a variety of climate impacts as one moves from shorter to longer timescales. Part of the task in climate signal detection is to gain a clearer picture of how phenomena at focal scales are affected by phenomena at adjacent or other scales. Hierarchy theory (Ahl and Allen 1996) will be one of the analytical tools for this task.

6. We must seriously consider how ecosystems may respond to global trends. In particular, we need to understand how ecosystem response may have either positive or negative feedback on a climate change. For example, shifts in polar ecosystems (melting of sea ice and permafrost, changes in snow cover, etc.) will, in turn, have an impact on climate. An understanding of such feedback mechanisms is of enormous ecological and social importance.

7. We must continue to refine the principles that emerge from the studies in this book. Quantitative modeling studies backed by carefully collected field data will help achieve this goal. More quantification will also help address some of the framework questions of this study that have been largely neglected.

8. At least as far as this list is concerned, one of the exciting realizations emerging from this volume is the possibility of ecosystems moving into resonance with climate variability at the quasi-quintennial and longer timescales. This seems to be a very fruitful idea worthy of further development and investigation. The growing maturity of LTER sites places researchers in a good position to examine the existence of such resonance in ecosystems other than those we have identified. In the cases we have already identified, the subtleties of the resonance may be examined more thoroughly. Such investigations exemplify the central core of the character of LTER research. This is true both in the sense of capitalizing on long-term research already completed and in the sense of opening up exciting new areas of investigation not envisioned when the LTER program began.

References

Ahl, V., and Allen, T. F. H. 1996. *Hierarchy Theory: A Vision, Vocabulary and Epistemology.* Columbia University Press, New York.

Bryson, R. A. 1997. The paradigm of climatology: An essay. *Bulletin of the American Meteorological Society* 78:449–455.

Casey, R. E., A. L. Weinheimer, and C. O. Nelson. 1989. California El Niños and related changes of the California current system from recent and fossil radiolarian records. Pages 85–92 in D. H. Peterson, editor. *Aspects of Climate Variability in the Pacific and the Western Americas.* American Geophysical Union.

Clark, W. C. 1985. Scales of climate impacts. *Climatic Change* 7:5–27.

Goldenberg, S.B., C. W. Landsea, A. M. Mestas-Nuñez, and W. M. Gray. 2001. The recent increase in Atlantic hurricane activity: Causes and implications. *Science* 293:474–479.

Lorenz, E. N. 1963. Deterministic non-periodic flow. *Journal of Atmospheric Sciences* 20:130–141.

Mann, M. E. 2001. Climate during the past millennium. *Weather* 56:91–102.

Nakamura, F., F. J. Swanson, and S. M. Wondzell. 2000. Disturbance regimes of streams and riparian systems—A disturbance-cascade regime. *Hydrological Processes* 14:2849–2860.

Parmesan, C. T., L. Root, and M. R. Willig. 2000. Impacts of extreme weather and climate on terrestrial biota. *Bulletin of the American Meteorological Society* 81(3):443–450.

Perry, D. A., and J. G. Borchers. 1990. Climate change and ecosystem response. *Northwest Environmental Journal* 6(2):293–313.

Phillips, J. D. 1999. *Earth Surface Systems*. Blackwell, Malden, Massachusetts.

Robertson, D. M., W. Anderson, and J. J. Magnuson. 1994. Relations between El Niño/Southern Oscillation events and the climate and ice cover of lakes in Wisconsin. Pages 48–57 in D. Greenland, editor. *El Niño and Long-Term Ecological Research (LTER) Sites*. LTER Publication No. 18. LTER Network Office. University of Washington. College of Forest Resources. AR-10. Seattle, Washington.

Scatena, F. N. 1995. Relative scales of time and effectiveness of watershed processes in a tropical montane rain forest of Puerto Rico. Natural and anthropogenic influences in fluvial geomorphology. *Geophysical Monograph* 89:103–111, American Geophysical Union, Washington, D. C.

Schlesinger, W. H., J. F. Reynolds, G. L. Cunningham, L. F. Huenneke, W. M. Jarrell, R. A. Virginia, and W. G. Whitford. 1990. Biological feedbacks in global desertification. *Science* 247(1990):1043–1048.

Schmidt, K. M., and R. H. Webb. 2001. Researchers consider U.S. Southwest's response to warmer, drier conditions. *EOS Transactions of the American Geophysical Union* 82(41):475 and 478.

Shaver G. R, J. Canadell, F. S. Chapin, J. Gurevitch, J. Harte, G. Henry, P. Imeson, S. Jonasson, J. Melillo, L. Pitelka, and L. Rustad. 2000. Global warming and terrestrial ecosystems: A conceptual framework for analysis. *BioScience* 50:871–882.

Sinton D. S., and J. A. Jones. 2002. Extreme winds and windthrow in the western Columbia River Gorge. *Northwest Science* 76(2):173–181.

Swanson, F. J., S. L. Johnson, S. V. Gregory, and S. A. Acker. 1998. Flood disturbance in a forested mountain landscape. *BiosScience* 48(9):681–689.

Swetnam, T. W., and J. L. Betancourt. 1990. Fire-Southern Oscillation relationships in the southwestern United States. *Science* 249:1017–1020.

Swetnam, T. W., and J. L. Betancourt. 1998. Mesoscale disturbance and ecological response to decadal climate variability in the American Southwest. *Journal of Climate* 11:3128–3147.

Tilman, D., and J. A. Downing. 1994. Biodiversity and stability in grasslands. *Nature* 367:363–365.

Wemple, B. C., F. J. Swanson, and J. A. Jones. 2001. Forest roads and geomorphic process interactions, Cascade Range, Oregon. *Earth Surface Processes and Landforms* 26:191–204.

Woodward, F. I. 1987. *Climate and Plant Distribution*. Cambridge University Press, Cambridge.

Index